Frontiers of Biogeography

New Directions in the Geography of Nature

Frontiers of Biogeography

New Directions in the Geography of Nature

Edited by

Mark V. Lomolino and Lawrence R. Heaney

Sinauer Associates, Inc. Publishers
Sunderland, Massachusetts

Published in association with the *International Biogeography Society*

About the Cover

The icon seen on the cover and throughout the book is an artist's rendering of Bernard Cahill's 1909 "butterfly map" of the world. The butterfly layout is one of many projections used to transform Earth's spherical three-dimensionality onto a planar surface.

For orders and information address Sinauer Associates, Inc.,
23 Plumtree Road, Sunderland, MA 01375 U.S.A.
FAX: 413-549-1118
Email: orders@sinauer.com, publish@sinauer.com

International Biogeography Society: www.biogeography.org

Library of Congress Cataloging-in-Publication Data
Frontiers of biogeography: new directions in the geography of nature / edited by Mark V. Lomolino and Lawrence R. Heaney.
p. cm.
Includes bibliographical references (p.   ) and index.
ISBN 0-87893-479-0 (casebound) — ISBN 0-87893-478-2 (paperbound)
1. Biogeography. I. Lomolino, Mark V. II. Heaney, Lawrence R.

QH84.F76 2004
578′.09—dc22   2004021324

Printed in U.S.A.
5 4 3 2 1

Contents

Contributors *viii*

Preface *x*

Introduction Reticulations and Reintegration of Modern Biogeography 1
Mark V. Lomolino and Lawrence R. Heaney

PART I: Paleobiogeography 5

Introduction by Stephen T. Jackson

1. Cenozoic and Mesozoic Paleogeography: Changing Terrestrial Biogeographic Pathways 9
   Christopher R. Scotese
2. Arid Lands Paleobiogeography: The Rodent Midden Record in the Americas 27
   Julio L. Betancourt
3. Quaternary Biogeography: Linking Biotic Responses to Environmental Variability across Timescales 47
   Stephen T. Jackson
4. Biogeography on a Dynamic Earth 67
   Christopher J. Humphries and Malte C. Ebach

PART II: Phylogeography and Diversification 87

Introduction by Brett R. Riddle and Vicki Funk

5. The Past and Future Roles of Phylogeography in Historical Biogeography 93
   Brett R. Riddle and David J. Hafner
6. Range Expansion, Extinction, and Biogeographic Congruence: A Deep Time Perspective 111
   Bruce S. Lieberman
7. Reticulations in Historical Biogeography: The Triumph of Time over Space in Evolution 125
   Daniel R. Brooks

PART III: Diversity Gradients 145

Introduction by Dov F. Sax and Robert J. Whittaker

8. Beyond Species Richness: Biogeographic Patterns and Biodiversity Dynamics Using Other Metrics of Diversity 151
   Kaustuv Roy, David Jablonski, and James W. Valentine
9. The Global Diversity Gradient 171
   John R. G. Turner and Bradford A. Hawkins
10. Diversity Emerging: Toward a Deconstruction of Biodiversity Patterns 191
    Pablo A. Marquet, Miriam Fernández, Sergio A. Navarrete, and Claudio Valdovinos
11. Dynamic Hypotheses of Richness on Islands and Continents 211
    Robert J. Whittaker

PART IV: Marine Biogeography 233

Introduction by John C. Briggs, Brian W. Bowen, and Michael A. Rex

12. Island Life: A View from the Sea 239
    Geerat J. Vermeij
13. A Marine Center of Origin: Reality and Conservation 255
    John C. Briggs
14. Pattern and Process in Marine Biogeography: A View from the Poles 271
    J. Alistair Crame

PART V: Conservation Biogeography 293

Introduction by Mark V. Lomolino

15. How Do Biological Invasions Alter Diversity Patterns? 297
    Julie L. Lockwood
16. GIS-Based Predictive Biogeography in the Context of Conservation 311
    Víctor Sánchez-Cordero, Mariana Munguía, and A. Townsend Peterson
17. Applying Species-Area Relationships to the Conservation of Species Diversity 325
    Michael L. Rosenzweig
18. Conservation Biogeography in Oceanic Archipelagoes 345
    Lawrence R. Heaney

Concluding Remarks 361
James H. Brown

Literature Cited 369

Index 421

Contributors

Julio L. Betancourt U.S. Geological Survey and University of Arizona Desert Laboratory, 1675 W. Anklam Rd., Tucson, AZ 85745 USA

Brian W. Bowen Hawaii Institute of Marine Biology, University of Hawaii, Kaneohe, HI 96744 USA

John C. Briggs Georgia Museum of Natural History (Emeritus), University of Georgia, Athens, GA 30602 USA

Daniel R. Brooks Department of Zoology, University of Toronto, Toronto, ON M5S 3G5 CANADA

James H. Brown Department of Biology, University of New Mexico, Albuquerque, NM 87131 USA

J. Alistair Crame British Antarctic Survey, High Cross, Madingley Road, Cambridge CB3 0ET UNITED KINGDOM

Malte C. Ebach Department of Botany, The Natural History Museum, Cromwell Road, London SW7 5BD UNITED KINGDOM

Miriam Fernández Center for Advanced Studies in Ecology and Biodiversity (CASEB) and Departamento de Ecología, Facultad de Ciencias Biológicas, Pontificia Universidad Católica de Chile, Casilla 114-D, Santiago CHILE

Vicki Funk U.S. National Herbarium, National Museum of Natural History, Smithsonian Institution, Washington, DC 20013-7012 USA

David J. Hafner New Mexico Museum of Natural History, 1801 Mountain Road NW, Albuquerque, NM 87104 USA

Bradford A. Hawkins Department of Ecology and Evolutionary Biology, University of California, Irvine, CA 92697 USA

Lawrence R. Heaney Field Museum of Natural History, 1400 So. Lake Shore Drive, Chicago, IL 60605 USA

Christopher J. Humphries Department of Botany, The Natural History Museum, Cromwell Road, London SW7 5BD UNITED KINGDOM

David Jablonski Department of Geophysical Sciences, University of Chicago, 5734 S. Ellis Avenue, Chicago, IL 60637 USA

Stephen T. Jackson Department of Botany, Aven Nelson Building, University of Wyoming, Laramie, WY 82071 USA

Bruce S. Lieberman Department of Geology and Department of Ecology and Evolutionary Biology, 1475 Jayhawk Blvd., 120 Lindley Hall, University of Kansas, Lawrence, KS 66045 USA

Julie L. Lockwood Department of Ecology, Evolution, and Natural Resources, Rutgers University, New Brunswick, NJ 08901 USA

Mark V. Lomolino SUNY College of Environmental Science and Forestry, Syracuse, NY 13210 USA

Pablo A. Marquet Center for Advanced Studies in Ecology and Biodiversity (CASEB) and Departamento de Ecología, Facultad de Ciencias Biológicas, Pontificia Universidad Católica de Chile, Casilla 114-D, Santiago CHILE

Mariana Munguía Departamento de Zoología, Instituto de Biología, Universidad Nacional Autónoma de México, Aptdo. Postal 70-153, México, D.F. 04510 MÉXICO

Sergio A. Navarrete Center for Advanced Studies in Ecology and Biodiversity (CASEB) and Departamento de Ecología, Facultad de Ciencias Biológicas, Pontificia Universidad Católica de Chile, Casilla 114-D, Santiago CHILE

A. Townsend Peterson Natural History and Biodiversity Research Center, The University of Kansas, Lawrence, KS 66045 USA

Michael A. Rex Department of Biology, University of Massachusetts, Boston, MA 02125 USA

Brett R. Riddle Department of Biological Sciences, University of Nevada, Las Vegas, NV 89154-4004 USA

Michael L. Rosenzweig Department Ecology and Evolutionary Biology, University of Arizona, Tucson AZ 85721 USA

Kaustuv Roy Section of Ecology, Behavior, and Evolution, University of California, San Diego, La Jolla CA 92093-0116 USA

Víctor Sánchez-Cordero Departamento de Zoología, Instituto de Biología, Universidad Nacional Autónoma de México. Aptdo. Postal 70-153, México, D.F. 04510 MÉXICO

Dov F. Sax Department of Ecology, Evolution and Marine Biology, University of California, Santa Barbara, CA 93106 USA

Christopher R. Scotese PALEOMAP Project, Department of Geology, University of Texas at Arlington, Arlington, TX 76019 USA

John R. G. Turner School of Biology, University of Leeds, Leeds LS2 9JT UNITED KINGDOM

Claudio Valdovinos Unidad de Sistemas Acuáticos, Centro de Ciencias Ambientales EULA-Chile, Universidad de Concepción, Casilla 160-C, Concepción CHILE

James W. Valentine Department of Integrative Biology, University of California, Berkeley CA 94720 USA

Geerat J. Vermeij Department of Geology, University of California at Davis, Davis, CA 95616 USA

Robert J. Whittaker Biodiversity Research Group, School of Geography and the Environment, University of Oxford, Mansfield Road, Oxford OX1 3TB UNITED KINGDOM

Preface

Biogeography is, in essence, a fundamentally broad-based and integrative field of endeavor, drawing not only from the full spectrum of biology but also from such disparate fields as geology, socio-economics, climatology, and philosophy. The challenge of integrating these fields to develop a comprehensive understanding of the origins and maintenance of global patterns of biodiversity is, for some scientists, one of biogeography's attractions; but the isolation of people with such interests into separate disciplines has represented a major hurdle to progress.

In spite of the limitations to communication and synthesis, in the past 25 years biogeography has become an increasingly prominent cornerstone of modern ecology, evolution, and conservation biology. The discipline explores the enormously diverse patterns in the geographic variation of nature—from physiological, morphological, and genetic variation among individuals and populations, to differences in the diversity and composition of biotas along geographic gradients, to the historical and evolutionary development of those patterns. The *International Biogeography Society* (IBS) was established to address the need for greater communication and collaboration between biogeographers in disparate academic fields and to promote the training and education of biogeographers so that they can develop and implement effective strategies for studying and conserving the world's biotas. *Frontiers of Biogeography: New Directions in the Geography of Nature* resulted from plenary papers presented at the inaugural meeting of the *International Biogeography Society* in 2003 (www.biogeography.org). Royalties from this volume will be donated to the general fund of IBS.

It is difficult not to be enthusiastic about recent advances in biogeography and the field's promise for answering some of science's greatest challenges, including explaining the origins, spread, distribution, and diversification of life. A tall order, granted, but our colleagues and our science have made great strides in recent decades and we are entirely confident that the next generation of biogeographers will not disappoint us. The contributors to this volume have not shied away from our challenge of being bold, identifying the most promising albeit often not yet mainstream or generally accepted lines of research. As members of a team organizing this effort, we are gratified and admittedly humbled by the compilation of stimulating and insightful contributions. If there is one overriding theme common to these works, it is that the frontiers are being defined and driven by the push to develop more general, more complex, and more integrative theories of the distributions and diversity of life—theories based on the dynamics of the earth, its environments, and its species.

Each of the contributions to this volume, and indeed nearly all of the most important advances over the past few decades, have been based on the knowledge that the Earth is dynamic, that environmental conditions vary from place to place in a highly non-random manner, and that this dynamism creates a changing geographic template to which all species must respond. Little wonder that Chris Scotese's reconstructions and visualizations of Earth's dynamic history captivated the audience at the meeting. His presentation here becomes Chapter 1; the chapter, along with his Web-based resources (www.scotese.com), will no doubt be invaluable to anyone interested in the dynamic Earth and the evolution of its life forms.

We have divided *Frontiers of Biogeography* into five units, corresponding to five of the field's most active and important subdisciplines: Plate Tectonics and Paleobiogeography; Phylogeography and Diversification; Diversity Gradients; Marine Biogeography; and Conservation Biogeography. Chapters in each of these units are preceded by an introduction to the subdiscipline, describing its foundations, current status, and frontiers (including, where appropriate, emerging areas of research that have not been addressed in the chapters). A number of the themes are woven across the units and chapters of this volume; indeed, many of these chapters could have been seamlessly placed in another unit of the book. We think this is a very encouraging reflection on the new emphasis on reintegration and syntheses across the subdisciplines of biogeography, an emphasis that is no doubt at least partially responsible for the re-emergence of biogeography as a science providing fundamental insights to those attempting to understand the distributions and development of biological diversity.

Finally, we cannot help but observe that an overwhelming majority of the contributors to this volume, regardless of their particular theme or focus, conclude with clarion calls to apply these lessons: to better predict the biogeography of the future, and to contribute to conserving the world's extant biological diversity. While we did ask that each contributor provide their views of the frontiers of their subject, their calls to conserve biological diversity were entirely unsolicited. We found this commendable and heartening.

MARK V. LOMOLINO AND LAWRENCE R. HEANEY

INTRODUCTION

Reticulations and Reintegration of Modern Biogeography

Mark V. Lomolino and Lawrence R. Heaney

For centuries, the geography of nature has in many ways served as a Rosetta Stone, providing key insights into the many mysteries of what we now call biological diversity. That the natural world varied over space and time was ancient knowledge, information essential to the survival of hunting and gathering societies. But these ancient observations—that the types and numbers of organisms encountered varies with area searched and with distance traveled, with elevation of land and depth of waters—did not coalesce into a true science until the Age of European Exploration. With the procession of natural scientists from Linnaeus, Forster, and Buffon during the eighteenth century, to Darwin, Hooker, Sclater, Wallace, and their colleagues in the nineteenth and twentieth centuries, our knowledge of the patterns and causal explanations for the geography of nature accumulated and matured into the science of biogeography.

This period of discovery marked the origins of the natural sciences in terrestrial and marine environments—geology, climatology, systematics, paleontology, evolution, and ecology—all practiced by biogeographers (i.e., scientists studying the spatial variation in the Earth's past and present life forms). From these distinguished origins, historical and ecological biogeography and their many subdivisions evolved, diverged, and eventually flourished (or languished) as increasingly more distinct disciplines. With this diversification and growth of distinctive scientific disciplines came a presumed "need" to specialize that resulted in more and more splintering. The grand view, the ultimate synthesis across space and time, became murky and more elusive.

The greatest strides we can make in unlocking the mysteries and complexities of nature in this fundamentally interdisciplinary science are those from new syntheses and bold collaborations among scientists across the many descendant disciplines, long divergent but now reticulating within a strong spatial context—the *new biogeography*. From the vicariance hypotheses of Joseph Dalton Hooker in the mid 1800s to the most recent methods of analyzing reticulating phylogenies and phylogeographies, geographic variation over time and space is the key. How life forms vary across kingdoms, from the unicellular organisms to the greatest beasts, and from ancient to current (and to the future), how all this varies across geographic gradients—this is the realm of the new biogeography.

To better assess and guide the frontiers of biogeography, we can learn much from its historical development. For some five decades in the first half of the twentieth century, Wegener's theory of continental drift was rejected and at times ridiculed in favor of more simplistic and seemingly more parsimonious explanations based largely on long-distance dispersal. The lesson learned from this experience applies not just to plate tectonics and historical biogeography, but to biogeography in general: perhaps too often we see beauty in simplicity, seeking the equivalent of a unified field theory—a relatively simple model that explains all, but masks underlying complexity, deterministic processes, and highly non-random patterns. It is ironic that we, the very scientists who so value diversity and spend our careers studying it, often still favor simple, and arguably simplistic, explanations for the essential complexity of nature.

We believe that the best means for advancing the frontiers of our science is to foster reintegration and reticulations among complementary research programs. The new series of syntheses—more complex, scale-variant, and multi-factorial views of how the natural world develops and diversifies—may be less appealing to some researchers, but it is likely to result in a much more realistic and more illuminating view of the complexity of nature.

Emerging Frontiers for the Next Generation

The insights presented in subsequent chapters make it clear that biogeography is experiencing a revitalization, and quite possibly a scientific revolution. The revitalization will continue in earnest, largely through the efforts of broad-thinking scientists who no longer shy away from but embrace the complexity of nature, and who foster collaborations and conceptual reticulations in modern biogeography. In addition to new syntheses and reintegration of its divergent but descendant disciplines, new disciplines and new challenges are emerging. We offer here a list of what we believe are among the most important challenges and opportunities for advancing biogeography.

1. Increase attention to scale dependence and to developing process-based linkage across scales of time, space, and biological complexity.
2. Encourage creative development and applications of the comparative approach, deconstructing and reassembling more comprehensive explanations for the diversity and distribution of biotas (guided largely by relevant theory and multi-scale linkages).

3. Avoid the sirens' call to specialize and splinter into ever more isolated and independent subdisciplines.
4. Address the "Wallacean Shortfall" (the paucity of information on the geographic variation of nature) and intensify our efforts to understand the biogeography of imperiled species as well as the less accessible biotas and ecosystems, including such elusive or novel groups as:
   - parasites, parasitoids, microbes and bryophytes
   - biotas of remote recesses of the biosphere (e.g., those of the abyssal zones and deep soils)
   - genetically engineered and other anthropogenic species
   - invasive and now "naturalized" species
   - species inhabiting, and in many ways surviving only in anthropogenic ecosystems (especially zoos, fragmented landscapes and the matrix of anthropogenic habitats, and nature reserves)
5. Fully utilize the rapidly expanding databases (such as those in marine paleontology, palynology, and integrated museum collections data) and increasingly powerful technology (e.g., Geographic Information Systems, gene sequencing techniques, and paleomapping) in the service of integrated, conceptually based biogeography.
6. Develop an integrated theory of the geography of differentiation and diversification—that is, a comprehensive view of how the production of living diversity is influenced by geographic circumstances.
7. Develop a theory of immigration, which considers the selective nature of immigration and its potential influence on the development and distributions of biotas.
8. Develop a more comprehensive understanding of the geography of extinction and how population persistence varies across the major geographic gradients (area, isolation, latitude, elevation, and depth) or with position (e.g., center versus periphery) within geographic ranges.
9. Advance our understanding of the dynamic geography of our own species: how we, our commensals, associated diseases, fragmentation and other anthropogenic environmental disturbances have and will continue to expand across and transform native landscapes.
10. Apply these approaches and increased understanding to developing better strategies for conserving not just isolated samples of imperiled life forms, but their geographic, evolutionary, and ecological context as well.

These are challenges indeed, but we are greatly encouraged by the modern renaissance and recent advancements in biogeography, richly evidenced by the distinguished contributions to this volume. We are equally confident that the current and future generations of biogeographers will continue to advance its frontiers, developing a more comprehensive understanding of, and more successful strategies for conserving, the geography of nature.

PART I

PALEOBIOGEOGRAPHY

Stephen T. Jackson

Paleobiogeography, like biogeography in general, covers a vast, heterogeneous territory of concepts, methods, data, and disciplines. In joining the prefixes *paleo* and *bio*, paleobiogeography encompasses the entire duration of life on Earth (i.e., the past 3.5 billion years), although it is more specifically applied to the past 550 million years of the Phanerozoic (the period of well-fossilized multicellular life). Processes relevant to biogeography occur across a wide range of timescales, from the seasonal, annual, and decadal timescales of ecological and microevolutionary processes, to multimillennial glacial/interglacial cycles, to the geological timescales of macroevolutionary patterns. Scaling up of patterns and processes is not straightforward because life and the Earth system itself are in continual flux at all timescales. Thus, the past century is a poor model for decadal-scale dynamics of 10,000 years ago, orbital-scale glacial/interglacial cycles of the late Quaternary are unrepresentative of orbital-scale climate variations of the Miocene, and biotic recovery following the end-Cretaceous mass extinction occurred in a fundamentally different world than that following the end-Permian mass extinction of 178 million years earlier. Paleobiogeography can be approached from multiple directions, most commonly using either the fossil record or the historical signatures evident in modern distribution and phylogenetic patterns.

Paleobiogeography represents a complex intersection of disciplines, and may appear to some to be a traffic jam in need of a policeman. But calling in the cops, whether to impose a one-size-fits-all research strategy or to outlaw certain processes or approaches, is likely to be unproductive. The apparent anarchy in the field is a necessary consequence of the diversity and complexity of patterns, processes, and phenomena in space and time, and of the limited scope of any single data type, analytical approach, or inferential tool. Certainly, better order can emerge from better communication among scientists using different tools, working at different timescales, and focusing on different phenomena. One of the goals of this book is to foster that exchange, and to encourage biogeographers to integrate diverse viewpoints. To draw an analogy, the science of geology is concerned with explaining the current configuration of the Earth's crust—its three-dimensional patterns of rock types and surface cover. These patterns, like those studied by biogeographers, require historical explanation. And geological patterns, like biogeographic patterns, are vestiges of diverse processes and events that occur over a broad range of spatial and temporal scales. Geology, generally regarded as a mature and unified science, lacks a single methodology or approach, instead comprising a diverse and often chaotic constellation of methods and approaches. If biogeography is to mature and unify, geology may be a better model discipline than physics, with its search for single unified theories.

Historical explanation, invoking both the history of place and the history of taxa, has been a central part of biogeography for more than two centuries (Browne 1983; Brown and Lomolino 1998). For example, the great seventeenth-century naturalist John Ray discussed former land connections in explaining the occurrence of wolves, foxes, and bears in Britain and on other islands (Ray 1721). More recently, the most fundamental and illuminating insights in paleobiogeography have come mainly from three sources: the fossil record of past biotic distributions; the geologic record of past continental configurations and environmental changes; and the modern spatial distribution of haplotypes, species, clades, and other biotic entities.

The four chapters that follow make use of all three of these sources, but an entire book would be required to cover them comprehensively. In recent decades, paleobiogeographic patterns have come into their own right as an object of study, with the recognition that macroevolutionary and macroecological patterns may not be predictable from extrapolation of modern patterns and processes. Although these chapters do not cover this important area, chapters in the unit on marine biogeography by Alistair Crame, John Briggs, and Kaustuv Roy and his colleagues discuss some of the linkages among paleobiogeography, macroecology, and macroevolution in the marine realm. Paleobiogeography in "deep time" (i.e., timescales of 10^6–10^8 years) provides views of global and regional diversity patterns and processes, and is an important and lively frontier of biogeography (e.g., Miller 1997a, 1998; Jablonski and Roy 2003).

Paleobiogeography, whether based on the fossil record or modern biogeographic patterns, requires accurate maps of the past configurations of continents, ocean basins, ice sheets, and climatic conditions. In his chapter, Christopher Scotese presents paleomaps for various time-slices in the Cenozoic and Mesozoic, together with a clear exposition of their foundations and uncertainties. These maps are an invaluable resource for both paleontologists and biogeographers. A major frontier lies in integrating pre-Quaternary paleogeography and environmental history with the fossil record to understand the origin, development, and history of biogeographic entities ranging from individual species to high-level clades and biogeographic "provinces." Recent reviews of the Cenozoic history of the North American flora (Burnham and Graham 1999; Manchester 1999; Dilcher 2000) provide a glimpse of the activity in just one small area of paleobiogeography; other regions, time periods, and taxonomic groups have substantial literature and activity.

Another important frontier of biogeography lies in the articulation and potential integration of phylogenetic biogeography, paleobiogeography, and earth history. Bruce Lieberman's chapter reviews recent and ongoing advances in this area (see also Lieberman 2002, 2003a) and makes an important distinction between dispersal and "geo-dispersal." It is essential to identify such distinctions as biogeographers working at different temporal and spatial scales try to communicate.

Integration of spatial patterns of genetic variation (e.g., phylogeography) with fossil records and environmental history represents another fast-moving frontier area. The chapters by Stephen Jackson and Julio Betancourt discuss this for the Quaternary, but many phylogeographic patterns predate this period (Avise 2000; Riddle and Hafner, this volume). More communication and collaboration among phylogeographers, historical geologists, and paleontologists will advance our understanding of how these patterns have arisen. Phylogeographic and paleontological data sets are, like all historical data sets, inherently incomplete: phylogeographic data are the vestiges of long, complex histories of range and population dynamics, and paleontological data have limitations of spatial and temporal coverage and usually do not reveal fine-scale genetic structure. Together these approaches can complement each other (e.g., Petit et al. 2002a,b) and provide some truly fundamental insights.

The chapters by Jackson and Betancourt also use the paleontological and paleoclimatic records of the late Quaternary to provide historical context for modern biogeographic distributions in temperate latitudes, with Jackson focusing on humid regions and Betancourt on arid regions. These chapters mine the extraordinary spatial, temporal, and biological detail available from the past 40,000 years of Earth history to draw some generalizations about the influence of climate variation on biogeographic patterns. Both chapters identify several frontiers in Quaternary paleobiogeography, and note some diverse areas that are ripe for advance. But these relatively brief chapters capture only a fraction of the frontiers of Quaternary biogeography. Other important ones include the Quaternary history of the tropics and subtropics (e.g., see Hooghiemstra and van der Hammen 1998; van der Hammen and Hooghiemstra 2000; Bush 2002), the dynamics of terrestrial faunas (FAUNMAP Working Group 1996; Stafford et al. 1999), the interplay of ecological and evolutionary responses (Davis and Shaw 2001; Ackerly 2003), and human impacts (Steadman 1995; Burney et al. 2001; J. B. C. Jackson et al. 2001).

One of the major insights that has emerged from the past three decades of research in Quaternary paleobiogeography is the recognition that species ranges are extraordinarily dynamic at timescales of 10^3–10^4 years (discussed in the chapters by Jackson and by Betancourt; see also Huntley and Webb 1988; Hengeveld

1989). These dynamics, which represent responses to changes in climate and other environmental factors (soils, atmospheric [CO_2]), are well documented for the last postglacial period in much of the world. However, climate variation at these scales, driven by variation in Earth's orbital parameters (Milankovitch cycles) and by internal feedbacks in the Earth system, extends far beyond the last glacial maximum to the entire Quaternary, and is also documented for much of the pre-Quaternary history of the Earth. It is now apparent that the biogeographic dynamics and community flux of the past 20,000 years are representative of the entire Quaternary, and probably the Neogene as well (e.g., Hooghiemstra and Van Cleef 1995; Allen et al. 1999; Willis et al. 1999; Whitlock et al. 2000). Unfortunately, at present we have relatively few paleobiogeographic records spanning periods of 10^5 years or more with sufficient temporal resolution to examine these (and other) high-frequency patterns. This leaves open a question that is critical to evolution and ecology: Are the high-frequency biogeographic dynamics of the Quaternary characteristic of earlier periods of Earth history?

An additional, related frontier is the linking of Milankovitch-scale dynamics to lower-frequency dynamics driven by tectonic-scale climate trends and continental movements. The remarkable pollen sequences from the Bogotá Basin show the superimposition of Milankovitch-scale biogeographic dynamics on lower-frequency changes driven by tectonic uplift and intercontinental biotic invasions (Hooghiemstra 1994; Hooghiemstra and Van Cleef 1995). Opportunities to bridge the gap between Milankovitch timescales and tectonic timescales in Earth history may be rare, but should be actively sought in different regions and time periods. Broadly important and previously unsuspected patterns may emerge.

In the final chapter of this unit, Christopher Humphries and Malte Ebach continue a long tradition in historical biogeography by advocating application of area cladistics to biogeography. They do so in order to develop an explicit history of geographic areas and their permanently resident biotas, in parallel with efforts by historical geologists such as Christopher Scotese (this volume), by examining the phylogenetic relationships of the taxa that occupy those areas. In this context they offer the general suggestion that biogeographers should strive to "accomplish an agreed methodology," a goal that, however admirable, may be elusive in view of the vast array of spatial and temporal scales at which biogeographic patterns occur and processes play out.

As the contributions to this and other units indicate, integrations among history of place, history of environment, and history of species are within reach. Since John Ray's time, natural history has diverged into various disciplines and subdisciplines, each with different interests, approaches, and cultures. However, we are now in a period where natural scientists are increasingly willing to peer over and even scale the walls that separate disciplines and subdisciplines, engaging their neighbors in constructive dialogue and collaboration. The coming decade should see the fruits of these efforts in the form of new syntheses that span the full array of temporal, spatial, and taxonomic scales relevant to biogeography.

CHAPTER 1

Cenozoic and Mesozoic Paleogeography: Changing Terrestrial Biogeographic Pathways

Christopher R. Scotese

Introduction

The geographic distribution of terrestrial plants and animals depends on many factors. The movement of the continents, together with the rise and fall of sea level, radically alters the configuration of land and sea, perpetually giving rise to new corridors for migration and, conversely, erecting new barriers that prevent free movement of terrestrial plants and animals. What makes the Earth an especially interesting place, from a biogeographic point of view, is that as a consequence of plate tectonics, these corridors and barriers continually change through time. It is not unlike the situation at Hogwarts School where the staircases leading to the student dormitories constantly change position (Rowling 1997), often diverting students to unexpected destinations. On our planet, it is the continents and ocean basins that are in constant motion. The movement of the continents, together with the rise and fall of sea level, radically alters the configuration of land and sea, perpetually giving rise to new corridors for migration and, conversely, erecting new barriers that prevent free movement of terrestrial plants and animals.

In this chapter, 18 plate tectonic reconstructions are presented that illustrate the movement of the continents since the breakup of Pangea 240 million years ago. These maps also illustrate the changing configuration of land, uplands, shallow seas, and deep ocean basins during the Mesozoic and Cenozoic eras. Figure 1.1 summarizes major plate tectonic events. Across the top, arrayed horizontally, are

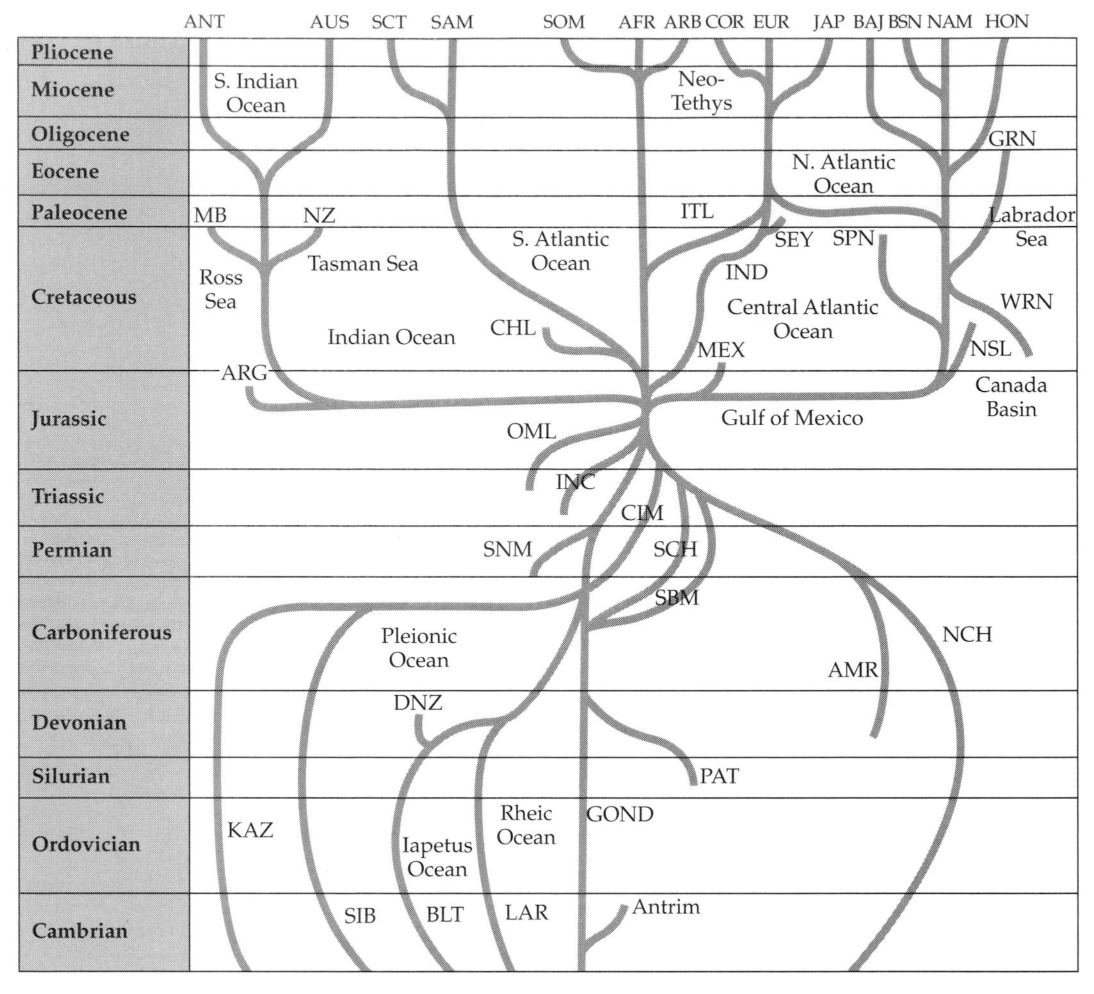

AFR = Africa
AMR = Amuria
ANT = Antarctica
ARB = Arabia
ARG = Argoland
AUS = Australia
BAJ = Baja California
BLT = Baltica
BSN = Basin and Range
CHL = Chile
CIM = Cimmeria
COR = Corsica and Sandinia
DNZ = Donetz Basin
EUR = Europe
GOND = Gondwana
GRN = Greenland
HON = Honduras
INC = Indochina
IND = India
ITL = Italy
JAP = Japan
KAZ = Kazakhstan
LAR = Laurentia
MEX = Mexico
MB = Marie Byrdland
NAM = North America
NCH = North China
NSL = North Slope of Alaska
NZ = New Zealand
OML = Omolon
PAT = Patagonia
SAM = South America
SBM = Sibumasu
SCH = South China
SCT = Scotia Arc
SEY = Seychelles
SIB = Siberia
SNM = Sonomia
SOM = Somalia
SPN = Spain
WRN = Wrangellia

◀ **FIGURE 1.1** Plate tectonic tree diagram illustrating the episodic formation and breakup of supercontinents, called the Wegener Cycle (after the German scientist Alfred Wegener; see Wegener 1912). Branching events represent the breakup of continents and the formation of new ocean basins. "Roots" are continental collisions. A branch that terminates represents an ocean basin that has stopped opening (for example, GRN = Labrador Sea between Greenland and North America) or the cessation of motion along a major strike-slip fault.

abbreviations signifying the modern continents. The "branches" that represent each modern continent can be traced downwards across the diagram and backwards through time. For instance, though South America (SAM) and Africa (AFR) today are separate continents, if we follow their branches backward through time we see that they come together in the early Cretaceous. The opening of the South Atlantic Ocean coincides with the split of the South American and African continental branches.

The most striking feature of the plate tectonic tree diagram is the constricted waistline that spans the Triassic and early Jurassic time interval. The convergence of all the branches back to a single "trunk" signifies that all land areas were joined together in one super-sized landmass, the supercontinent known as Pangea (Wegener 1912). Prior to the time of Pangea, the world's continents were dispersed, much like today's world. However, there is a hint that the roots of the plate tectonic tree converge yet again, in the late Precambrian. This late Precambrian supercontinent is poorly known and goes by the names Greater Gondwana, Pannotia, or Rodinia (Dalziel 1991, 1997; Moores 1991; Powell et al. 1993; Hoffman Van der Voo 1994).

The PALEOMAP Project

The maps presented following page 19 illustrate the plate tectonic events depicted in the top half of Figure 1.1. This history encompasses the formation and breakup of Pangea. I also discuss some of the biogeographic consequences of these plate tectonic and paleogeographic events.

These maps are the result of more than 25 years of plate tectonic model-building and data synthesis (Scotese and Baker 1975; Scotese et al. 1979; Scotese et al. 1981; Ziegler et al. 1983; Scotese and Sager 1988; Scotese et al. 1988; Scotese and Golonka 1992; Jurdy et al. 1995). These plate tectonic reconstructions have been used by numerous authors to illustrate paleogeographic (Ronov et al. 1984, 1989; Ziegler 1989; Cook 1990; Kazmin and Natopov 1998; Golonka 2000; Blakey 2003), paleoclimatic (Otto-Bliesner et al. 1994; Golonka et al. 1996; Boucot et al. 2004), and biogeographic change (McKerrow and Scotese 1990; Cocks and Scotese 1991; Chatterjee and Scotese 1999). The plate tectonic reconstructions of the PALEOMAP Project are similar to other compilations (Zonenshain et al. 1990; Dercourt et al. 1993; Smith et al. 1994; Hay and Bouyesse 1999; Vrielynck et al. 2002) because similar geophysical data sets were used by these research groups to construct the maps.

Plates 1–18 illustrate the plate tectonic development of the ocean basins and continents as well as the changing distribution of land and sea during the past 240 million years. The 18 time periods in this compilation cover the Mesozoic and Cenozoic eras. There are two maps for the Triassic, three for the Jurassic, six for the Cretaceous, and seven for the Cenozoic.

The Paleogeographic Method

The study of paleogeography has two principal goals. The first is to map the past positions of the continents and smaller terranes (e.g., the Philippines). This is done by reconstructing the tectonic history of the continents and ocean basins. The second goal is to illustrate the changing distribution of mountains, lowlands, shallow seas, and deep ocean basins through time by mapping the distribution of paleoenvironments as revealed through the geological record.

Mapping the Past Positions of the Continents

The plate tectonic history of the continents and ocean basins during the last 240 ma is reconstructed using the following six lines of evidence:

1. Linear magnetic anomalies produced by seafloor spreading
2. Paleomagnetism
3. Hot spot tracks and large igneous provinces (LIPS)
4. Tectonic fabric of the ocean floor as mapped by satellite altimetry
5. Lithologic indicators of climate (e.g., coals, salt deposits, tillites)
6. The geologic record of plate tectonic history

LINEAR MAGNETIC ANOMALIES PRODUCED BY SEAFLOOR SPREADING. Like the sun's magnetic field, the Earth's magnetic field "flips," or reverses polarity. As new ocean floor cools at mid-ocean ridges, it is magnetized in the direction of the prevailing magnetic polarity (normal or reverse). Fluctuations, or "anomalies," in the intensity of the magnetic field occur at the boundaries between normally magnetized sea floor and portions of the sea floor that were magnetized in the "reverse" direction. The age and duration of these linear magnetic anomalies can be determined using fossil evidence from the sea floor and radiometric techniques. Because these magnetic anomalies form at the mid-ocean ridges, they tend to be long, linear features (hence the name "linear magnetic anomalies") that are symmetrically arranged about the mid-oceanic ridge axes. The past positions of the continents during the last 150 ma can be directly reconstructed by superimposing linear magnetic anomalies of the same age from opposite sides of an ocean basin.

During the decades between 1965 and 1995, the pattern of linear magnetic anomalies in 80% of the world's ocean basins was mapped by marine geophysical surveys. This data was initially synthesized and summarized by the PALEOMAP Project, and was ultimately used to produce a detailed map of the age of the ocean basins (Mueller et al. 1996). For an ocean-by-ocean summary of the plate tectonic development of the world's ocean basin, see Scotese and Sager 1988.

PALEOMAGNETISM. By measuring the remanent magnetic field often preserved in iron-bearing minerals in rock formations, paleomagnetic analysis can determine whether a rock was magnetized near the North or South Pole or near the Equator. Thus, paleomagnetism provides direct evidence of a continent's north-south (latitudinal) position, but does not constrain its east-west (longitudinal) position. The paleomagnetic data used in this chapter to reposition the continents was compiled by Van der Voo (1993), and re-evaluated in light of the PALEOMAP plate tectonic model by Bocharova and Scotese (1993). For recent summaries of paleomagnetic data and procedures, see McElhinny and McFadden (1998) and Schettino and Scotese (2004).

HOT SPOT TRACKS AND LIPS. Plate tectonics is a description of the movement of the lithosphere—the rigid, outermost layer of the Earth. Most volcanic eruptions are derived from magmas that have their source within the lithosphere; however, a significant subset, "hot spot" volcanics, are derived from much deeper magma sources. Some of these deep sources of magma may originate near the boundary between the Earth's liquid core and lower mantle and ascend to the surface in mantle plumes ("hot spots") (Duncan 1981; Morgan 1981; Duncan and Richards 1991).

Hot spots literally burn through the lithosphere and erupt at the surface forming large, but rarely explosive, volcanoes (e.g., Mauna Loa). When a hot spot plume first breaks through to the surface, the large volume of magma in the plume head can spread across a continent as a flood basalt, or large igneous province (LIP) (Mahoney and Coffin 1997; Eldholm and Coffin 2000). As a plate moves over the more-or-less stationary hot spot, a sequential chain of volcanic islands, or "hot spot track," is formed. The Hawaiian Islands and Emperor seamount chain were formed in this manner (Wilson 1963).

Hot spots are useful in reconstructing past plate motions for two reasons. First, they date the start of continental rifting and sea floor spreading; and second, hot spot tracks provide an independent estimate of both north-south and east-west plate motion (Mueller et al. 1993). Hot spots appear to be relatively stationary for long periods of time, and in only a few cases has significant motion of hot spots been documented (Norton 1995, 2000).

TECTONIC FABRIC OF THE OCEAN FLOOR. Satellites equipped with radar altimeters can precisely measure the distance from an orbiting satellite to the Earth's surface. These distance measurements can be used to create detailed digital topographic and bathymetric maps. The pioneering work of Haxby (1987), together with the more recent digital maps of Smith and Sandwell (1997), have mapped out the location of mid-ocean ridges, oceanic fracture zones, volcanic seamounts, extinct spreading ridges, deep sea trenches, and submarine plateaus.

Information from satellite altimetry has played a critical role in the plate tectonic reconstruction of ocean basins (Gahagan et al. 1988). The long, linear fracture zones defined by this method can be used to progressively guide the continents back together (Reeves and deWit 2000). In this way, the past spreading history of the ocean basins can be mapped, and extinct spreading centers located. Moreover, from the pattern and distribution of seamounts and hot spot tracks, the relative motion between the plates and the Earth's interior can be deduced.

LITHOLOGIC INDICATORS OF CLIMATE. The Earth's climate is primarily a result of the redistribution of the sun's energy across the surface of the globe: it is warm near the Equator and cool near the Poles. Wetness, or rainfall, also varies systematically from the equator to the poles. Usually, it is wet near the equator, dry in the subtropics, wet in the temperate belts, and dry near the poles. Certain kinds of rocks form under specific climatic conditions. For example, coals occur where it is wet, bauxite occurs where it is warm and wet, evaporites and calcretes occur where it is warm and dry, and tillites occur where it is wet and cool (Parrish et al. 1982; Scotese and Barrett 1990; Parrish 1994; Boucot et al. 2004). The ancient distribution of these rock types can tell us how global climate has changed through time (Ice House versus Hot House), and provide an independent estimate of the north-south position of a continent (paleolatitude).

GEOLOGIC RECORD OF PLATE TECTONIC HISTORY. Finally, the best source of information—but often the most difficult to interpret—is the geologic record left by plate tectonic activity. The rock record has information about environments of deposition (deep sea, shallow sea, lowlands, and uplands) as well as the history of plate tectonic activity (faults, folding, thrusting, and rifting). During the past 250 years geologists have mapped out the age, environment of deposition, and tectonic history of the rocks at the Earth's surface. This knowledge of regional tectonics and geology allows us to map ancient sites of rifting, subduction, continental collision, and other important plate tectonic events.

Estimating Uncertainties in the Past Positions of the Continents

The plate tectonic maps presented here are visual hypotheses that seek to explain incomplete and sometimes contradictory observations. Therefore, it is reasonable to ask the question, "How accurate are the positions of the continents shown on these maps?" The techniques described in the previous section often have statistical methods for measuring error and estimating uncer-

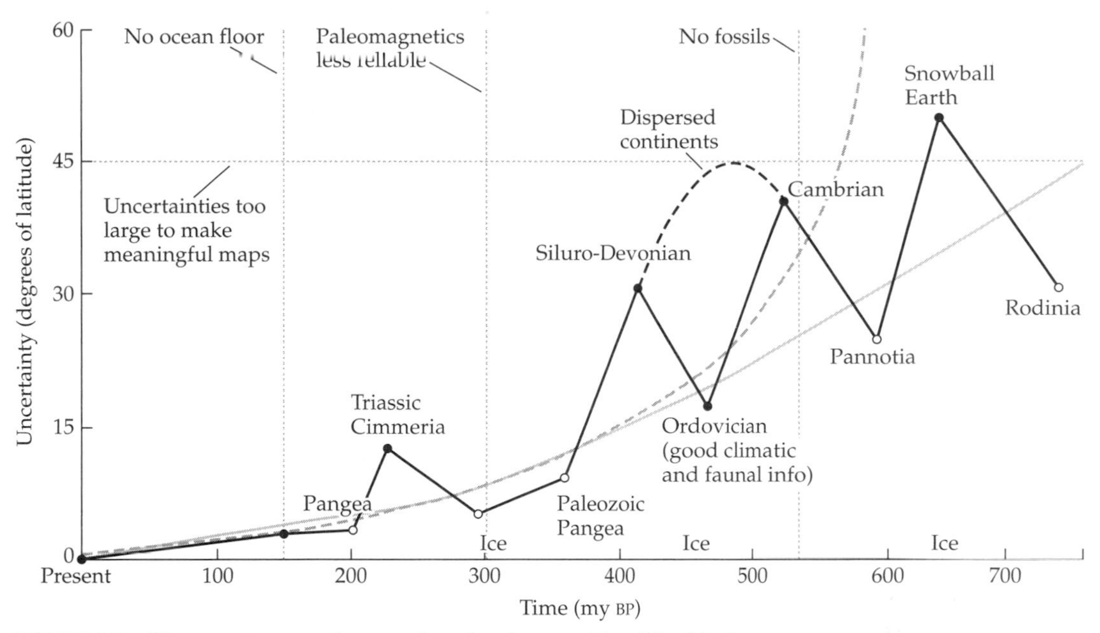

FIGURE 1.2 Three estimates of uncertainty in plate position. The black trace represents specific uncertainty estimates for given time intervals; uncertainty decreases when continents are gathered together into supercontinents and increases when continents are dispersed. The dashed gray curve is an averaged estimate based on the assumption that there were no supercontinents during the late Precambrian. The solid gray curve is also an averaged estimate but assumes the existence of the supercontinents Pannotia and Rodinia during the late Precambrian. See text for detailed explanation.

tainty. For example, in paleomagnetic studies, individual paleomagnetic poles are often combined into mean, or averaged, paleopole positions. These averages include estimates of precision (alpha 95) and dispersion (*k*) (Van der Voo 1993; McElhinny and McFadden 1998). Similar statistics can be used to evaluate how well lithologic indicators of climate match the paleolatitudes predicted by a given plate reconstruction (Scotese and Barrett 1990). Likewise, the goodness of fit between matching magnetic anomalies, or the trend of a hot spot track with the trajectory of plate motion, can also be calculated. However, there is no simple or direct way to summarize all these estimates of uncertainty into a single meaningful, quantitative estimate of error.

Despite this problem, it is still worthwhile to discuss the probable uncertainties associated with plate tectonic reconstructions. Figure 1.2 provides an estimate of the uncertainty in the positions of the continents back through time and indicates some of the criteria used to estimate these uncertainties.

In Figure 1.2, geological time lies along the horizontal axis. The vertical axis expresses uncertainty in the position of the continents in terms of degrees of latitude. The left-hand side of the graph starts at the present, with zero degrees of uncertainty—we know exactly where the continents are today. As we scan to the right, going backward in time, the uncertainty increases, so that at 200 my BP the uncertainty in plate positions is ±5° of latitude, and at 300 my BP the uncertainty in plate position is ±10° of latitude. Simply stated, the location of any continent on the plate tectonic reconstruction for 200 my BP might be 5° (550 km) too far north or too far south. At 300 my BP, the location of any continent might be 10° (1100 km) too far north or too far south.

It is clear from Figure 1.2 that uncertainty in plate position increases exponentially backward through time. In fact, three different estimates of uncertainty are shown. The two gray curves are "average uncertainties"; the dashed gray curve is an estimate of uncertainty based on the assumption that there were no late Precambrian supercontinents. In this case, no reliable reconstructions can be made for the late Precambrian because uncertainties in plate position exceed ±45° of latitude. The solid gray curve is a more optimistic estimate that supposes there were two late Precambrian supercontinents, Pannotia and Rodinia. In that case, the overall uncertainty in plate position is lower because it is easier to know the orientation of one large continent than it is to know the location of several widely dispersed continents (the Pangea Principle).

The solid black trace is a more detailed estimate of the uncertainties for specific time intervals. For the last 200 million years, the black curve falls along the gray curves. The uncertainties in plate position for the time interval covered are quite acceptable, ranging from ±5° (550 km) to a maximum of ±12° (1330 km) in the Triassic, when the continents in the Eastern Hemisphere (Cimmeria and Cathaysia) were more widely dispersed.

Mapping the Changing Distribution of Mountains, Lowlands, Shallow Seas, and Deep Ocean Basins

In addition to the changing position of the continents and ocean basins deduced from plate tectonics, another important feature of the maps is the changing paleogeography. Again, paleogeographers study the constantly changing distribution of deep ocean basins (>200 m), shallow seas (<200 m), lowlands (<800 m), and uplands (>1500 m).

Some paleogeographic features change very slowly and are easy to map. Other paleogeographic features change very rapidly, and therefore any map is at best an approximation. In this regard, the Earth, since the early Precambrian, has been divided into deep ocean basins (average depth 3.5 km) and high-standing continents (average elevation about 800 m). Continental lithosphere, because it is less dense, is more buoyant and is not easily subducted back into the Earth's interior. As a result, continents are made up of very old rocks—some dating back 3.8 billion years. The amount of continental lithosphere has probably changed very little during the last 2.6 billion years (possibly increasing 10%–15%). What has changed is the shape as well as the distribution of continents across the globe. The ocean basins, on the other hand, are all less than 150 million years old. And because it is denser, oceanic lithosphere is continually recycled back into the interior of the Earth by subduction. Except for a few remnants of oceanic lithosphere caught up in continental collisions (ophiolites), all oceanic lithosphere older than 150 million years has been subducted.

In contrast to the continents and ocean basins, which are long-term geographic features, the height and location of mountain belts and the shape of the Earth's shorelines change constantly. Mountain belts form either where oceanic lithosphere is subducted beneath the margin of a continent, giving rise to a linear range of mountains like the Andes of western South America; or where continents collide, forming high mountains and broad plateaus like the Himalayas and the Tibetan Plateau of central Asia. Less extensive mountains can also form when continents rift apart (e.g., the East African Rift), or where hot spots form volcanic uplifts.

In most cases, mountain ranges take tens of millions of years to form and, depending on the climate, may last for hundreds of millions of years. For example, the Appalachian Mountains of the eastern United States were formed over 300 million years ago when North America collided with northwestern Africa. A few peaks in this old mountain belt are still more than 2000 meters high. The Himalayas—the world's tallest mountain range—began to rise from the sea nearly 50 million years ago when northern India collided with Eurasia. On paleogeographic maps the extent of the mountain ranges increases during the collisional phase, but is slowly reduced (by erosion) on subsequent maps.

It is important to note that the shoreline, though the edge of land, is not the edge of the continent. In most cases, the continent extends seaward hundreds of kilometers beyond the shoreline. The actual edge of the continent is marked by the transition from the continental slope to the continental rise. This steep bathymetric gradient marks the boundary between continental lithosphere and oceanic lithosphere and is indicated on the maps by the transition from light blue shading (shallow shelf) to areas shaded in darker blue (deep ocean).

The position of the shoreline is a function of both continental topography and sea level. Though topography changes slowly (tens of millions of years), global sea level can change rapidly (tens of thousand of years). In comparison to topographic features such as mountain ranges, Earth's shorelines are ephemeral. The familiar shapes that characterize today's shorelines, such as Hudson Bay, the Florida peninsula, or the numerous fjords of Norway, are all less than 12,000 years old. The shape of the modern coastlines is the result of a 120 meter rise in sea level that took place during the last 18,000 years, after the ice sheet that covered much of North America and Europe melted. The modern coastlines (bold black lines) are shown on these paleogeographic maps only for reference purposes.

Several factors can affect sea-level change, including the amount of ice covering the continents. At times when the continents were covered by great ice sheets, sea level was low and the continents were exposed. For the last

20 million years, the continents have been largely high and dry due to extensive mountain building in Asia and the accumulation of extensive ice fields on Antarctica. Other important global episodes of glaciation occurred 300, 450, and 650 million years ago. The oldest known glacial episode occurred in the Precambrian, approximately 2.2 billion years ago.

Sea level also changes more slowly (tens of millions of years) due to changes in the volume of the ocean basins. More than 3 billion years ago, water from the Earth's interior erupted as volcanic gas and condensed on the cooling surface of the planet to form the oceans. However, we assume that there has been no significant addition to the volume of water on Earth since early Precambrian times. Therefore, long-term changes in sea level are not due to changes in the amount of water on Earth, but rather are due to changes in the shape and size of the ocean basins. If the volume of the ocean basins increases, then sea level will fall. Conversely, if the volume of the ocean basin decreases, global sea level will rise.

Sources of Information for the Maps

The interpretation of the past location of deep sea, shallow sea, lowlands, and uplands is based primarily on a synthesis of published works. Though there are few global compilations to rely on, numerous regional studies have mapped the changing paleogeography for continent-sized areas. The principal sources of paleogeographic information are shown in Table 1.1.

Chronological Review of the Atlas

Triassic Plate Tectonics and Paleogeography (Plates 1 and 2)

The oldest map in this series (Plate 1) illustrates the early Triassic supercontinent, Pangea. *Pangea* literally means "all land." Though certainly a supercontinent by any definition, the early Triassic Pangea probably did not include all the landmasses that existed at that time. In the Eastern Hemisphere, along the Equator, there were "continents" that remained isolated from western Pangea. These continents comprised a long "windshield-wiper" shaped terrane known as Cimmeria (Sengor and Natalin 1996). Cimmeria consisted of parts of Turkey, Iran, Afghanistan, Tibet, Indochina, and Malaya. It appears to have rifted away from the Indo-Australian margin of Gondwana during the late Paleozoic. Cimmeria moved northward towards Eurasia (Plate 2), ultimately colliding along the southern margin of central Asia during the late Triassic–earliest Jurassic (Yin and Nie 1996) (Plate 3). It was only after the collision of these Asian fragments during the early Jurassic that all the world's landmasses were joined together in a supercontinent truly deserving of the name Pangea.

TABLE 1.1 *Some sources of paleogeographic information*

Location (timeframe)	Reference
Global (Phanerozoic)	Ulmishek and Klemme 1990
	Scotese and Golonka 1992
	Golonka, Ross, and Scotese 1996
Global (Mesozoic and Cenozoic)	Ziegler et al. 1983
	Ronov et al. 1989
	Smith et al. 1994
Gondwana (Permo-Triassic)	Veevers and Powell 1994
Africa (Cretaceous)	Hulver 1985
Northeast Africa	Schandelmeier and Reynolds 1997
African basins	Selley 1997
Europe	Ziegler 1989, 1990
British Isles	Cope et al. 1992
Tethys region	Dercourt, Ricou, and Vrielynck 1993
Former Soviet Union	Vinogradov et al. 1968a,b, 1969
Northern Eurasia	Kazmin et al. 1998
Asia	Wang 1985
Australia	Veevers 1984
	Cook 1990
	Paleogeographic Atlas of Australia (10 volumes, various authors)
North America	Cook and Bally 1975
Rocky Mountains	Mallory (ed.) 1972
Western Canada	Mossop and Shetsen 1994
South America	Tankard et al. (eds.) 1995
	Walsh 1996
	Cordani et al. 2000
Northern South America	Pindell et al. 1998

From a biogeographic point of view, Pangea was not only a supercontinent but also a "superhighway." Sea level was low during most of the Triassic, and the emergent land areas provided access to every corner of Pangea. As has long been noted, a variety of reptile groups were able to spread across the supercontinent during the Triassic. The best example is *Lystrosaurus*, which has been found in Antarctica, South Africa, India, central Asia, and China.

Though a large, contiguous landmass facilitated dispersal of many plants and animals, there were important barriers to migration of terrestrial organisms during the Triassic. Most notable were the equatorial Central Pangean mountain range and the extensive subtropical Pangean deserts. Rivaling the modern Sahara in extent, these deserts occupied much of the northern and southern subtropics, covering central North America as well as the Amazon basin and west-central Africa. The Cimmerian terrane, which crossed the

Equator during the Triassic, was covered by marine limestones whose cosmopolitan fauna is typical of the Tethyan Realm.

Jurassic Plate Tectonics and Paleogeography (Plates 3–5)

The plate tectonic and paleogeographic situation in the early Jurassic was very much like the late Triassic. Sea level was low and Pangea was intact. Cimmeria, which had crossed Tethys during the Triassic, was in collision with south-central Asia.

Though no ocean floor had yet formed, the breakup of Pangea began in the middle Jurassic (Plate 4). A global rise in sea level accompanied the earliest phases of rifting. This rise in sea level flooded much of northern Europe and western Siberia, creating low-lying, isolated groups of islands that would have been a barrier to the migration of terrestrial organisms among continents in the Northern Hemisphere and between northern Africa and Eurasia.

After an episode of igneous activity along the east coast of North America and the northwest coast of Africa, the central Atlantic Ocean opened as North America moved to the northwest, away from Africa (175 my BP; Plate 5). This movement also gave rise to the Gulf of Mexico as North America moved away from South America. At the same time, on the eastern side of Africa, extensive volcanic eruptions in east South Africa, Antarctica, and Madagascar heralded the formation of the western Indian Ocean. During the middle and late Jurassic (180–160 my BP), the eastern half of Gondwana (India, Madagascar, Antarctica, and Australia) slid southward relative to western Gondwana (Africa and South America).

Throughout the Mesozoic, North America and Eurasia were one landmass, sometimes called Laurasia. As the central Atlantic Ocean opened (180–150 my BP), Laurasia rotated clockwise sending North America northward and Eurasia southward. Coals, which were abundant in eastern Asia during the early Jurassic, were replaced by deserts and salt deposits during the late Jurassic as Asia moved from the wet temperate belt to the dry subtropics. This clockwise, see-saw motion of Laurasia also led to the narrowing of the wide, V-shaped Tethys Ocean that separated Africa and India from Eurasian continents and opened a circumequatorial current.

PLATES 1–18
Reconstructions showing the movement of continents, starting with the breakup of Pangea. Tan areas represent highlands (>1500 m); other land masses are shown in green. The shallow oceans (< 200 m) are in light blue, and the ocean basins are darker blue. Black lines within the land masses represent faults and important tectonic plate boundaries. The text discusses each map in detail. (Copyright © 2004 by C. R. Scotese.) ▶

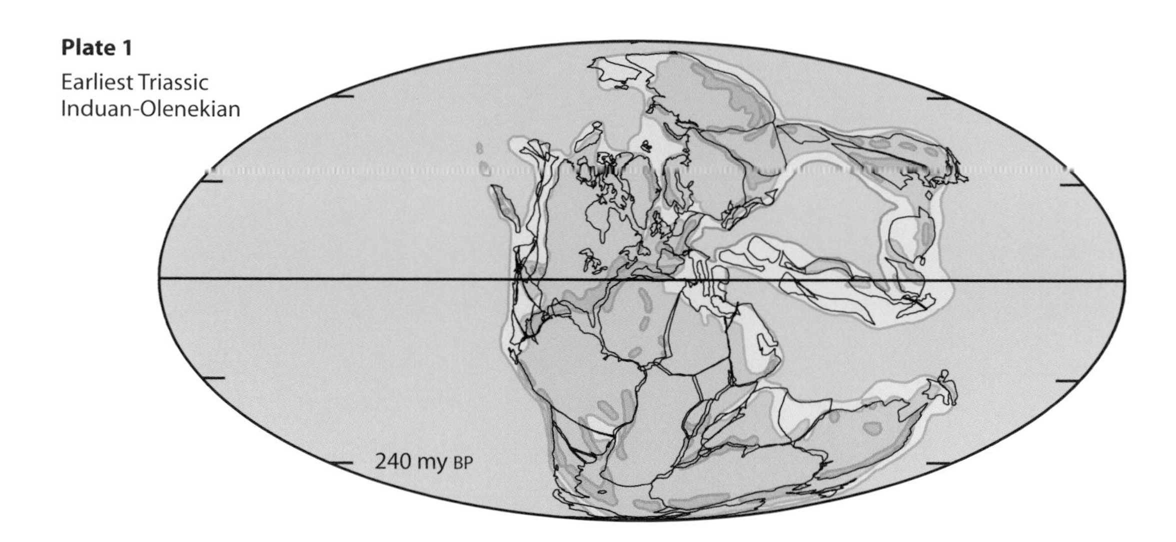

Plate 1
Earliest Triassic
Induan-Olenekian

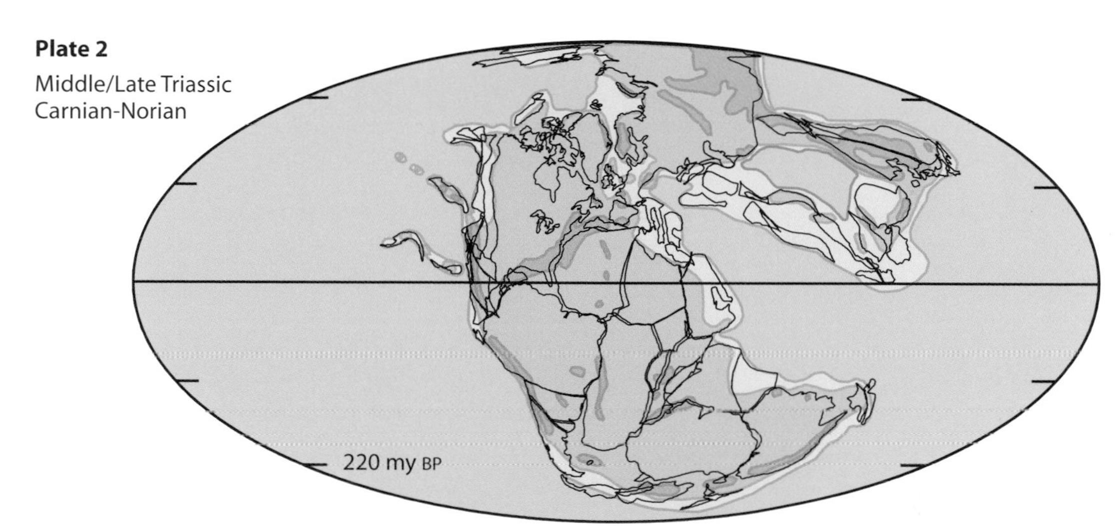

Plate 2
Middle/Late Triassic
Carnian-Norian

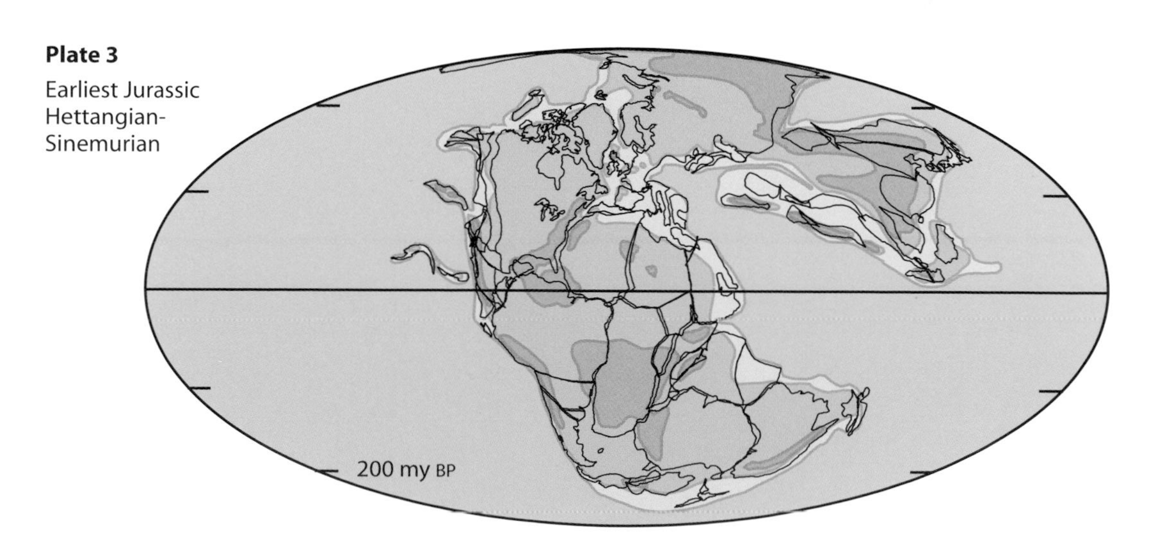

Plate 3
Earliest Jurassic
Hettangian-
Sinemurian

Plate 4

Early/Middle Jurassic
Toarcian-Aalenian

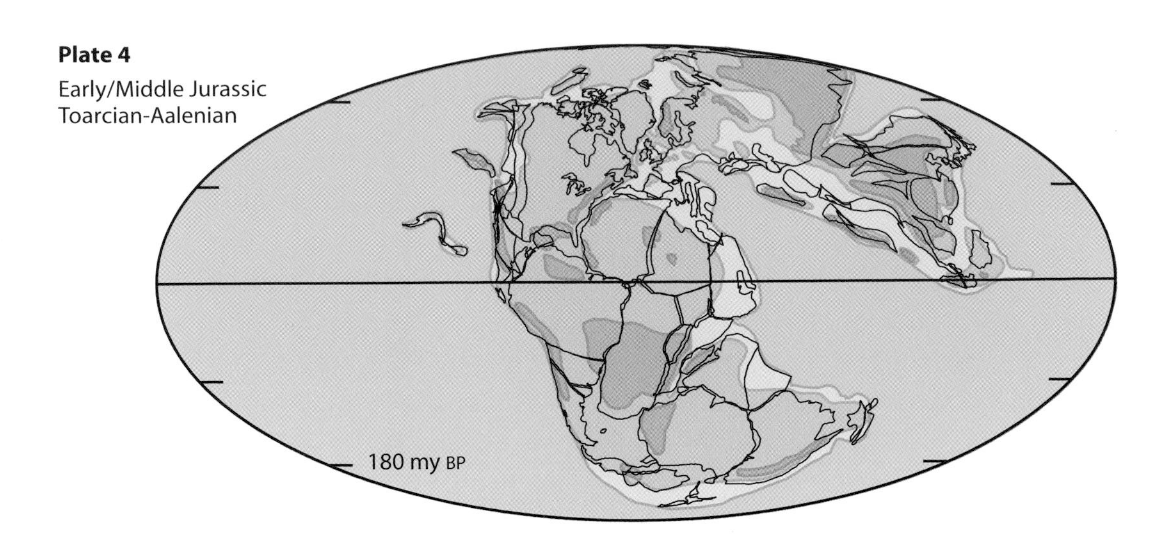

Plate 5

Middle/Late Jurassic
Callovian-Oxfordian

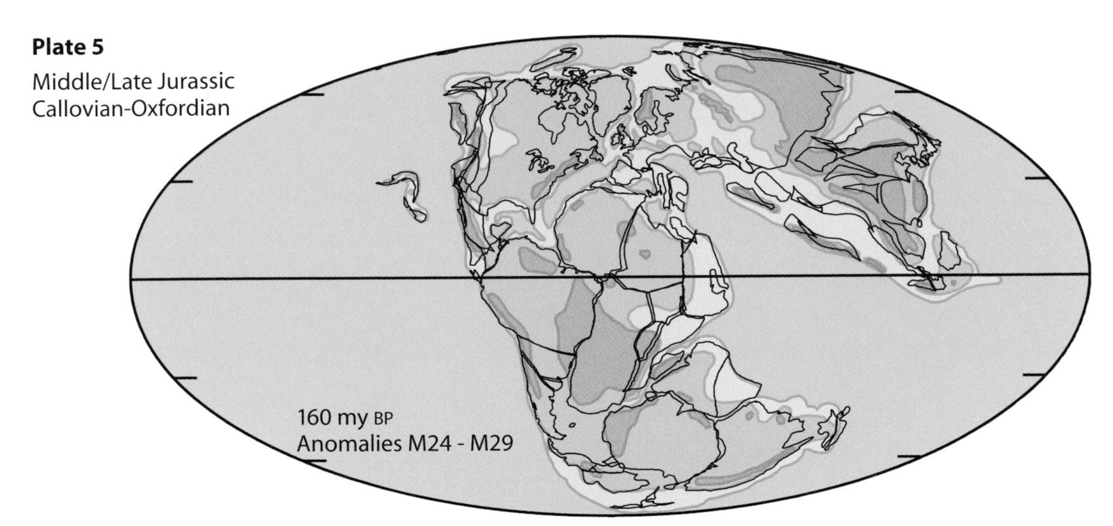

Plate 6

Earliest Cretaceous
Berriasian

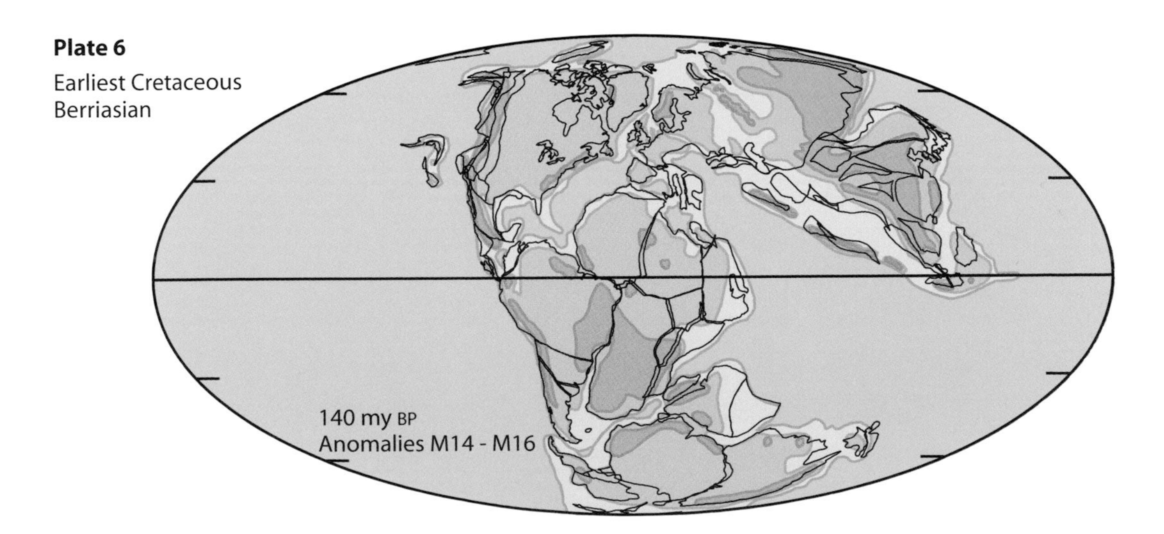

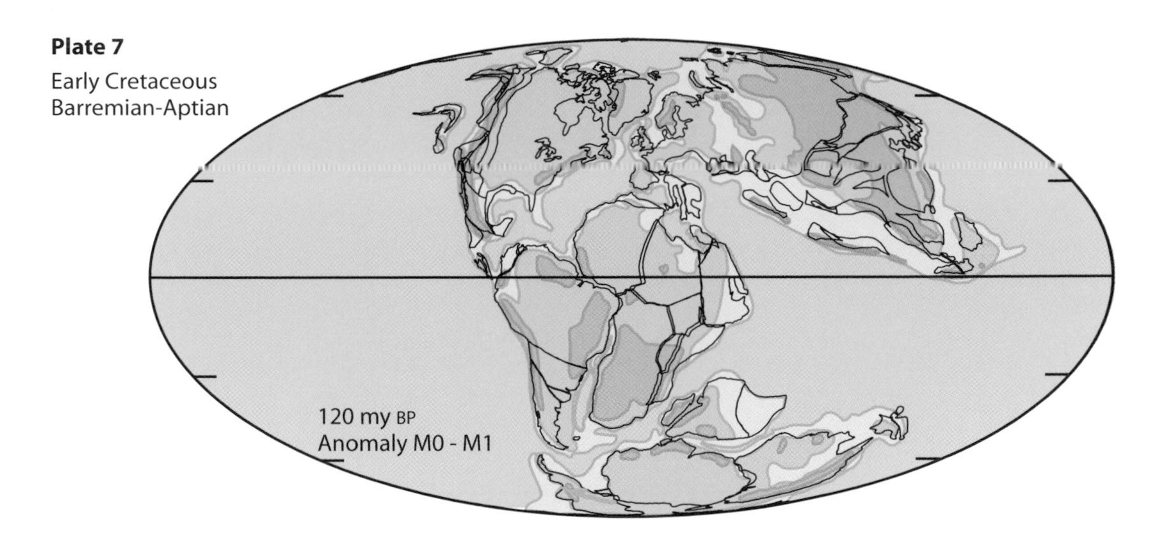

Plate 7
Early Cretaceous
Barremian-Aptian

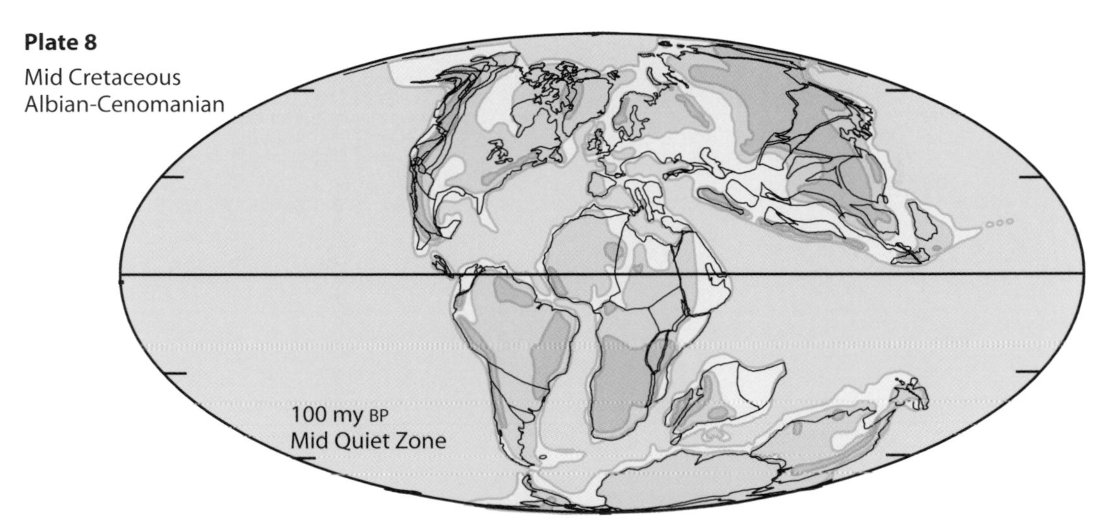

Plate 8
Mid Cretaceous
Albian-Cenomanian

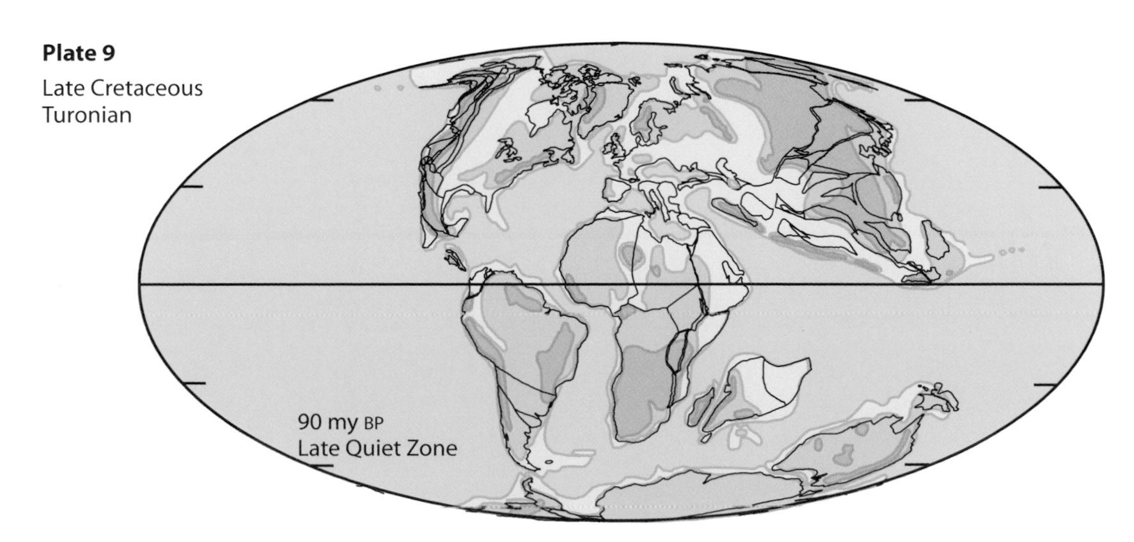

Plate 9
Late Cretaceous
Turonian

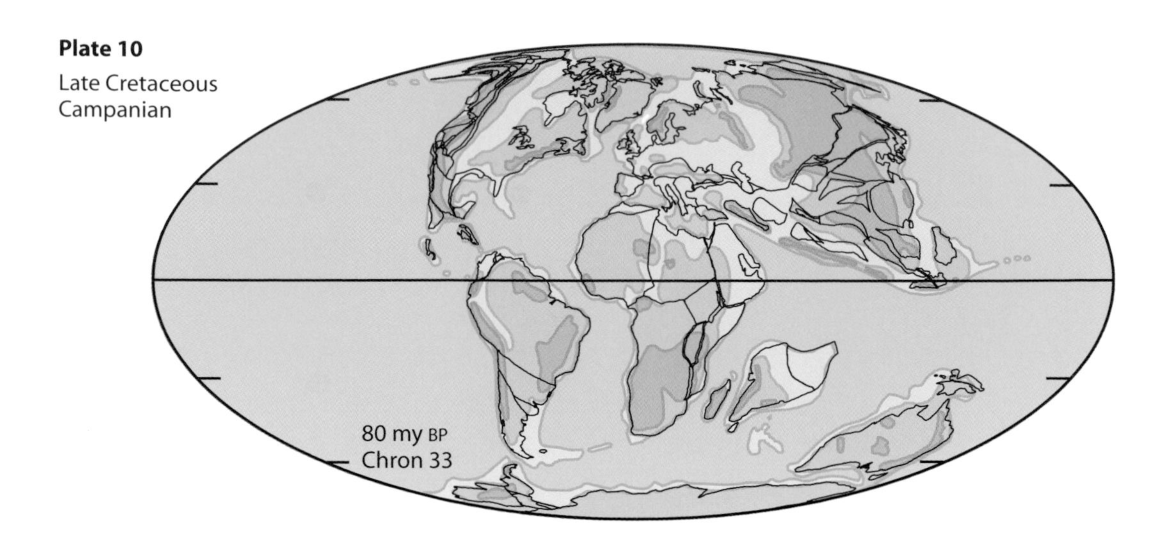
Plate 10
Late Cretaceous
Campanian
80 my BP
Chron 33

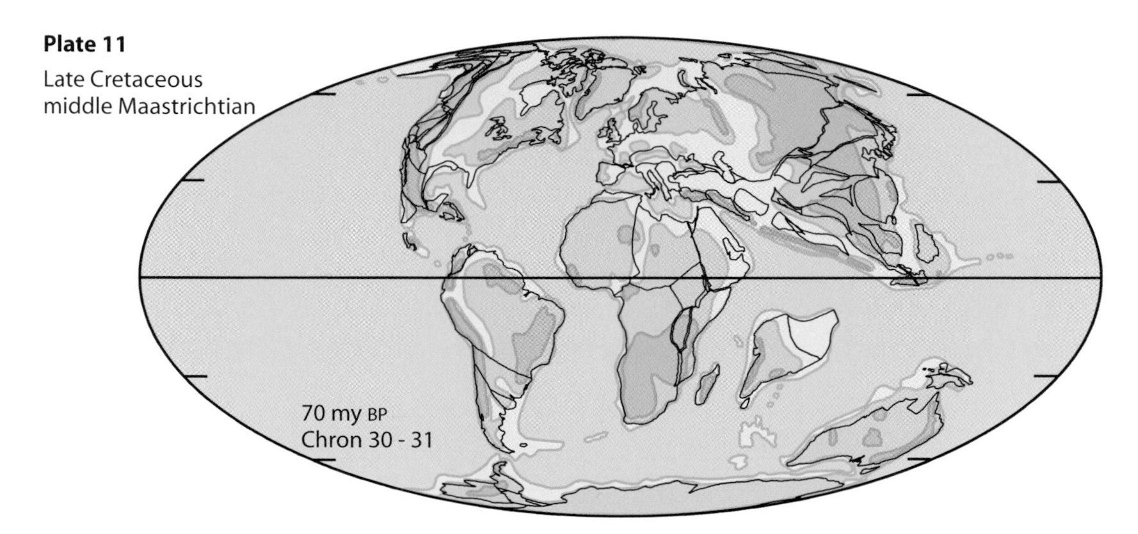
Plate 11
Late Cretaceous
middle Maastrichtian
70 my BP
Chron 30 - 31

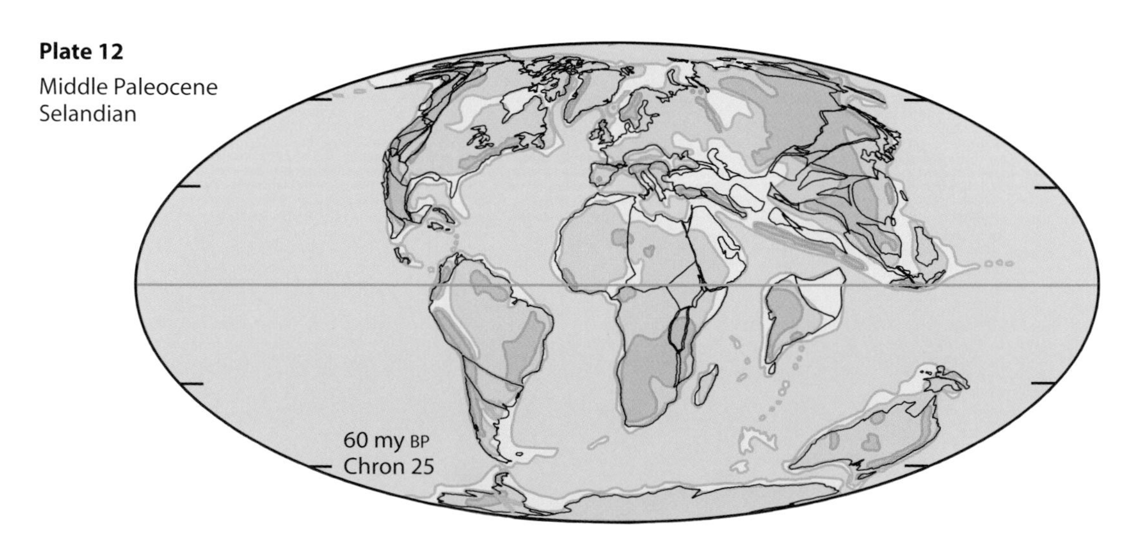
Plate 12
Middle Paleocene
Selandian
60 my BP
Chron 25

Plate 13

Early Eocene
Ypresian

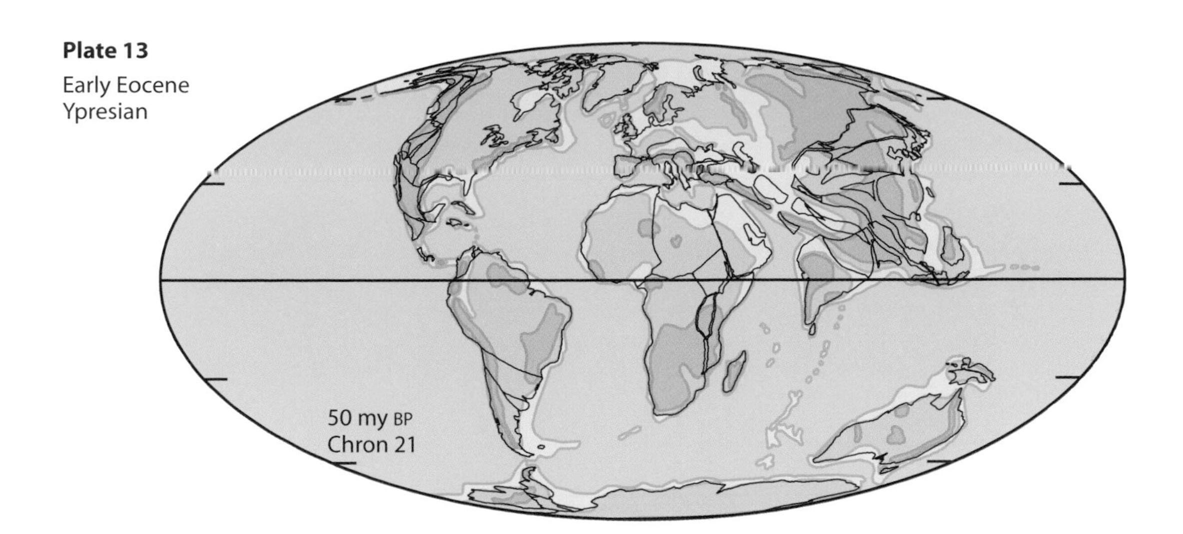

Plate 14

Late Eocene
Bartonian

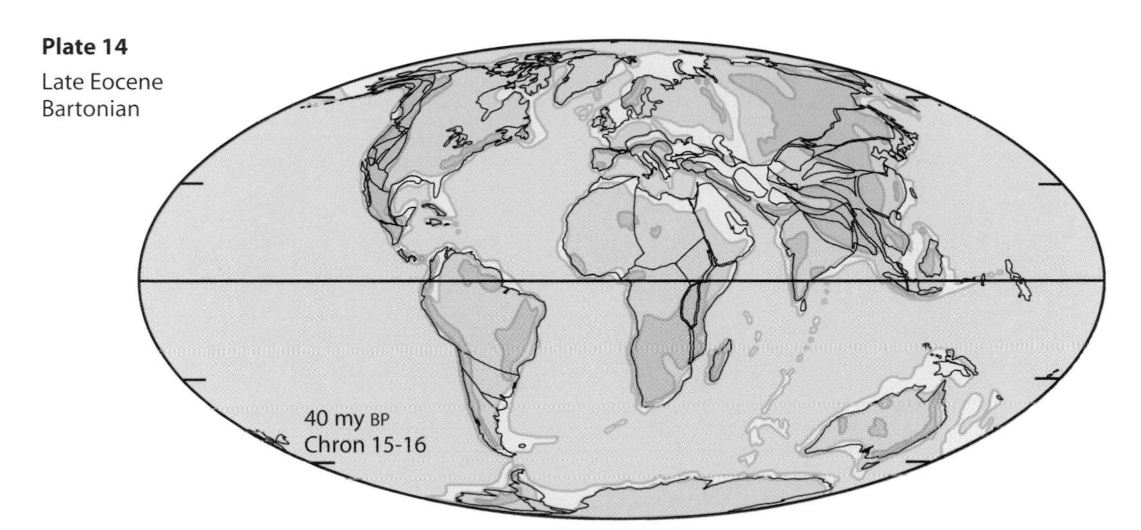

Plate 15

Early Oligocene
Rupelian

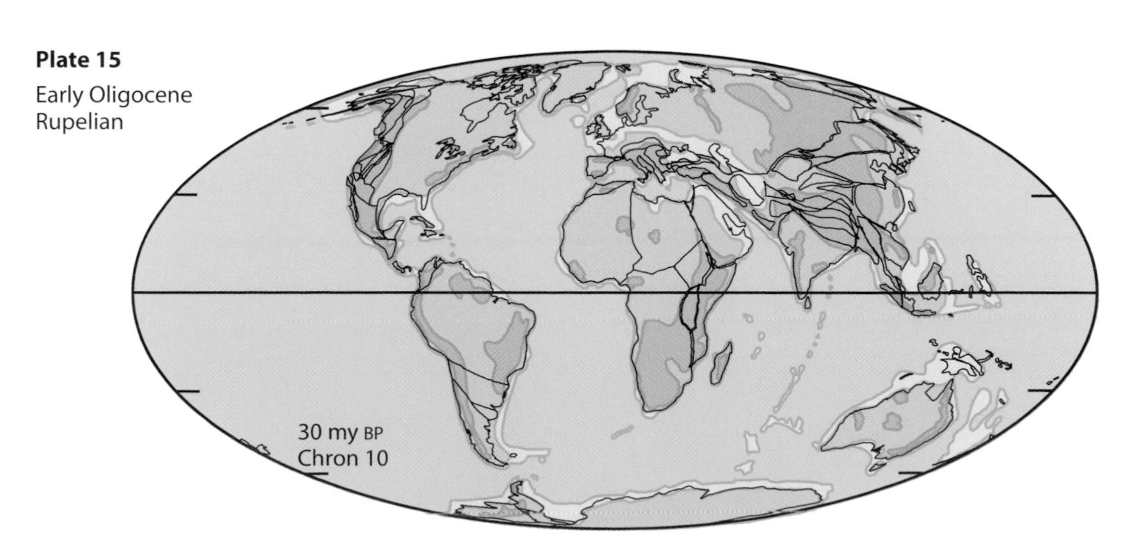

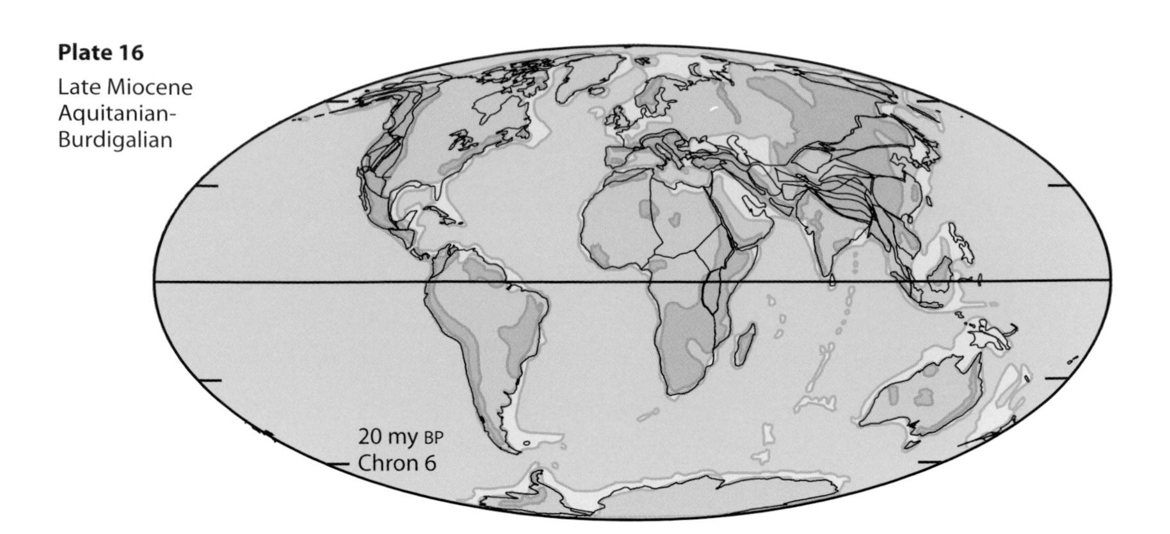

Plate 16
Late Miocene
Aquitanian-
Burdigalian

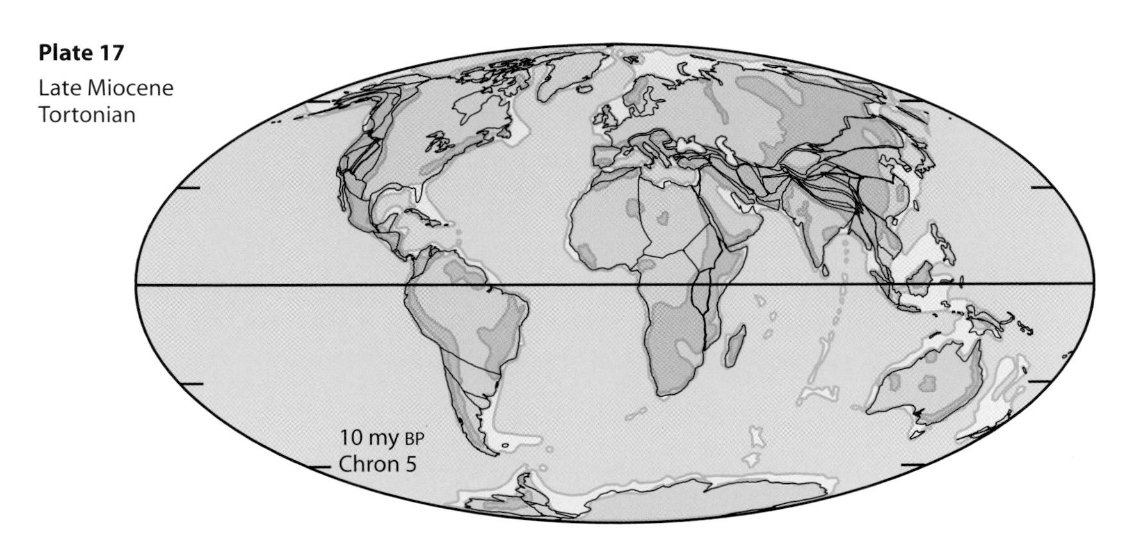

Plate 17
Late Miocene
Tortonian

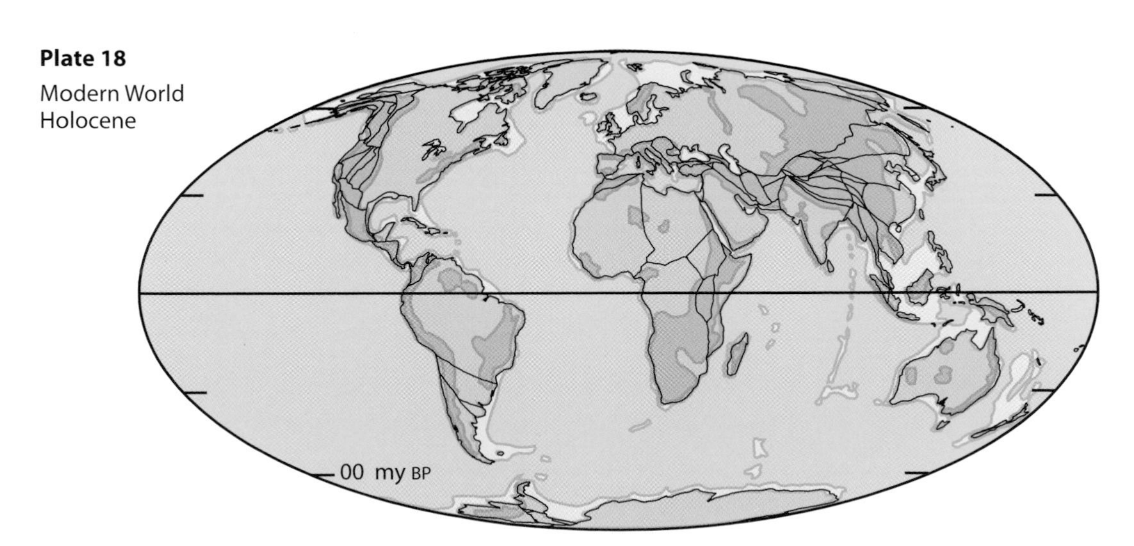

Plate 18
Modern World
Holocene

Though the new ocean basins of the late Jurassic were relatively narrow, they created a formidable barrier to overland migration. Coupled with a rise in sea level, the late Jurassic saw the end of the "Pangean Superhighway." For the first time since the formation of Pangea, the Northern and Southern Hemispheres were nearly isolated. A narrow strip of land connected northern South America with southern Mexico, and intermittent exchange may have occurred during times of low sea level across the broad carbonate shelves between North Africa and southwestern Europe (Plate 5).

It is interesting to note that throughout the Jurassic, Asia was divided into two large landmasses. North-central Asia (Siberia) was separated from south-central Asia (China and Southeast Asia) by the V-shaped Amurian seaway (Plates 3–5). These two landmasses were joined by a relatively narrow strip of mountainous terrain in western China (Tarim). The Amurian seaway closed in the earliest Cretaceous (130 my BP; see Plate 7).

Cretaceous Plate Tectonics and Paleogeography (Plates 6–11)

Six paleogeographic maps illustrate the plate tectonic and paleogeographic events that took place during the 80 million-year-long Cretaceous. Pangea, which had divided into three large continents during the Jurassic (Laurasia, West Gondwana, and East Gondwana), continued to subdivide as each of these continents split apart. During the Early Cretaceous, West Gondwana broke apart as South America separated from Africa, opening the South Atlantic. The South Atlantic did not open all at once, but rather progressively "unzipped" from south to north. In East Gondwana, India together with Madagascar rifted away from Antarctica and Australia, opening the Eastern Indian Ocean (Plates 6–8).

During the mid-Cretaceous (~95 my BP), while sea floor spreading continued in the South Atlantic and Eastern Indian Ocean, a new rift formed between India and Madagascar (Plate 9). During the remainder of the Cretaceous, India was pulled rapidly northward by the subduction of oceanic lithosphere beneath the southern margin of Asia. For much of the late Cretaceous, India appears to have been an island continent, separated from all other landmasses. The apparent plate tectonic isolation of India during the late Cretaceous is a biogeographic enigma (Chatterjee and Scotese 1999; Briggs 2003a). Though isolated from the other Gondwanan continents, the Campanian-Maastrichtian dinosaur faunas of India are similar to those of Africa, South America, and Europe, suggesting a migration route between these continents. Two different migration routes have been suggested. Sampson et al. (1998) proposed a southerly connection into India from Antarctica-South America via the Kerguelen plateau in the late Cretaceous. Alternately, the "Greater Somalia" hypothesis of Chatterjee and Scotese (1999) describes a connection to Africa via a narrow landbridge that extended eastward from Somalia. India collided with the Greater Somalia landbridge in the late Cretaceous (~65 my BP), allowing an influx of dinosaurs and other tetrapods.

In western North America, exotic oceanic terranes and displaced continental terranes collided to form the Canadian Cordillera. These terrane collisions began in the late Jurassic (Stikinia; 160 my BP) and continued through the middle of the Cretaceous (Wrangellia; 100 my BP) (Nokleberg et al. 2001). Unfortunately, nothing is known of the terrestrial fauna that may have arrived on these terranes.

The final phase in the breakup of Pangea took place during the late Cretaceous (Plates 9–11). Rifting between Australia and Antarctica began at about the same time that North America and Europe parted company (~100 my BP). At the end of the Cretaceous, the North Atlantic and Southeast Indian Ocean were narrow oceans, only about as wide as the modern Red Sea (Plate 11).

Other important plate tectonic and paleogeographic events took place during the latest Cretaceous. The Greater Antilles island arc, which earlier in the Cretaceous occupied a position between North and South America—much like modern Panama—began to move into the Caribbean, eventually colliding with Bahamas platform by about 55 million years ago. In the Mediterranean region, the Adriatic promontory, which is the northernmost extension of Africa, began to collide along the southern margin of Europe, forming the Alps.

Globally, the climate during the Cretaceous Period, like the Jurassic and Triassic, was much warmer than today. Dinosaurs and palm trees were present north of the Arctic Circle and in Antarctica and southern Australia. These mild climatic conditions were due in part to the fact that shallow seaways covered the continents during the Cretaceous. Warm water from the equatorial regions was also transported northward, warming the polar regions. These seaways also tended to make local climates milder, much like the modern Mediterranean Sea, which has an ameliorating effect on the climate of Europe.

Though winter snow and ice may have covered the polar regions during the late Jurassic and early Cretaceous, there is no direct evidence for large, permanent ice sheets at either pole at any time during the Mesozoic era.

Warmer polar climates would have permitted free migration across high-latitude continental corridors. The most important of these corridors were the western North America-northeast Asia connection across the Bering Sea landbridge (Plates 9 and 10), and the south polar migration route from southern South America across Antarctica and into India and Australia (open only during early Cretaceous, ~90 my BP; Plates 7–9).

With no polar ice caps to steal away water, the world's oceans were filled to the brim. Throughout the mid- and late Cretaceous, sea level was 100–200 meters higher than today (Haq et al. 1987). Shallow seaways covered most of the continents (Plates 9–11) with the exception of north-central Asia, which remained above sea level throughout the Jurassic and Cretaceous. The emergence of Asia during the late Mesozoic was due in part to mountainous areas in its interior that were the remnants of early Mesozoic collisions (Cimmeria and Cathaysia), and to active subduction along its eastern margin throughout the Jurassic and Cretaceous.

Cenozoic Paleogeography (Plates 12–18)

The third and final phase in the breakup of Pangea took place during the early Cenozoic (65–55 my BP). The North Atlantic Ocean opened as rifts propagated on either side of Greenland (Plates 11 and 12). Rifting on the west side of Greenland stopped in the late Paleocene, while sea floor spreading continued between East Greenland and Scandinavia. A mid-latitude connection between Europe and North America through northern England, the Faeroe Islands, and Greenland was maintained through the Paleocene. An intermittent, high-latitude migration route from Svalbard to the Canadian Arctic islands (Plates 10–12) may have linked Europe to North America through much of the early and mid-Tertiary (Paleocene to Oligocene).

In the Southern Hemisphere, rifting between Antarctica and Australia had isolated Australia from the other continents by the earliest Paleocene. The last land connection between these two continents was through Tasmania (Plate 11). During the late Cretaceous and Paleocene, sea floor spreading in the Southeast Indian Ocean was slow. Only after the collision of India with Asia about 50 my BP did Australia begin to move rapidly northward (Plates 14 and 15).

A land connection between southern South America and western Antarctica was maintained through the late Cretaceous and into the Paleocene. By the early Eocene (45 my BP) the Drake Passage had opened, completely isolating Antarctica from all other continents. In northern South America, only a tenuous connection with Central and North America existed through the proto-Panama archipelago. The consolidation of the Panamanian Isthmus in the Pliocene (3.5–5 my BP) resulted in the Great American Interchange (see summary and references in Brown and Lomolino 1998).

The recent rifting events, all taking place within the last 30 million years include: the rifting of Arabia away from Africa, opening the Red Sea (Plates 17 and 18); the creation of the East African Rift system; the opening of the Sea of Japan as Japan moved eastward into the Pacific; and the northward motion of California and northern Mexico, opening the Gulf of California.

Though several new oceans have opened during the Cenozoic, the last 66 million years of Earth's history are better characterized as a time of intense continental collision. The most significant of these collisions has been that between India and Eurasia (50 my BP; Plate 13). During the late Cretaceous, India approached Eurasia at rates of 15–20 cm per year—a plate tectonic speed record. After colliding with marginal island arcs in the late Cretaceous, the northern part of India—Greater India—began to be subducted beneath Eurasia, raising the Tibetan Plateau (Plate 14). Interestingly, Asia, rather than

India, has sustained most of the deformation associated with this collision. This is because India is a solid piece of continental lithosphere riding on a plate that is primarily made up of stronger oceanic lithosphere. Asia, on the other hand, is a loosely knit collage of continental fragments. The collision zones, or sutures, between these fragments are still warm, and hence can be easily reactivated. As India collided with Asia (Plates 13–16), these fragments were squeezed northward and eastward out of the way, along strike-slip faults that followed older sutures. Earthquakes along these faults continue to the present day.

The collision of India with Asia is just one of a series of continental collisions that has all but closed the great Tethys Ocean. From east to west, these continent-continent collisions were: Spain with France, forming the Pyrenees; Italy with France and Switzerland, forming the Alps; Greece and Turkey with the Balkan states, forming the Hellenide and Dinaride Mountains; Arabia with Iran, forming the Zagros Mountains; India with Asia, forming the Himalayas; and finally the most recent collision, Australia with Indonesia.

This phase of continental collision has raised high mountains by horizontally compressing the continental lithosphere. Though the continents occupy the same volume, their area has decreased slightly. Consequently, on a global scale the area of the ocean basins has increased slightly during the Cenozoic, at the expense of the continents. Because the ocean basins are larger, they can hold more water. As a result, sea level has fallen during the last 66 million years. In general, sea level is lower during times of continental collision (early Devonian, late Carboniferous, Permian, and Triassic).

During times of low sea level, the continents are emergent, land faunas flourish, migration routes between continents open up, the climate becomes more seasonal, and, probably most importantly, the global climate tends to cool off. This is largely because land tends to reflect the sun's energy back into space, while the darker oceans absorb the sun's energy. Also, landmasses permit the growth of permanent ice sheets, which have a high albedo and reflect more energy back to space. The formation of ice on the continents, of course, lowers sea level even further, which exposes more land, which cools the Earth, forming more ice, and so on.

During the last half of the Cenozoic, the Earth began to cool off. Ice sheets formed first on Antarctica (~45 my BP) and then spread to the Northern Hemisphere (~20 my BP). For the last 5 million years the Earth's climate has been in an "Ice House" phase, punctuated by alternating glacial and interglacial periods. Unlike the Mesozoic and early Cenozoic when cross-polar migration was common, the polar routes of terrestrial migration have been largely blocked by ice and snow during the last 20 million years. An exception has been the movement of cold-temperate mammal groups across the Bering Straits during the glacial maxima of the Pleistocene.

The continental rifting and collisions that began in the late Cenozoic are continuing today. Most notable are the opening of the Red Sea and Gulf of

Aden, the rifting of East Africa, the opening of the Gulf of California and the northward translation of California west of the San Andreas Fault, and the incipient collision of Australia with Indonesia giving rise to the mountain ranges of New Guinea.

Concluding Remarks

This chapter paints a broad-brush portrait of the plate tectonic and paleogeographic changes that have taken place during the last 240 million years, briefly discussing the consequences these changes may have had on terrestrial biogeographic pathways. Though the tectonic and paleogeographic events depicted in the accompanying maps have controlled the pattern and mode of faunal and floral dispersal, we have just begun to understand the interaction between plate tectonics, paleogeography, and biogeographic change. The complex pattern of biogeographic change has barely been described. Hopefully these maps, in conjunction with the traditional tools of biogeography and the new tools of phylogeography, will be used to explain biogeographic patterns that are just beginning to emerge as well as patterns that have yet to be recognized.

Though the maps shown here are an attempt at a comprehensive review of plate motions and paleogeographic change during the past 200 million years, they are little more than a sketch of what could be done, and what I believe will be done, during the next decades. As on-line digital geological databases become increasingly available, and as Geographic Information Systems (GIS) technology becomes more useful and widely used, it is likely that we will see a renaissance in paleogeographic mapping. In the future, these high-resolution digital reconstructions of the Earth's changing features will provide a new framework for a greater understanding of Earth System History. This spatial/temporal framework is necessary if we are to understand the complex history of climate change, oceanographic change, biogeographic change, and the evolution of life on this amazing planet.

Over 500 years ago, Christopher Columbus began a voyage into the unknown. As a result, the true extent and geography of the world was discovered. We are now preparing to take a similar voyage. But this voyage of discovery will take us on a journey through time, rather than space.

This voyage has already begun. Interested readers are directed to the following websites where interactive plate tectonic mapping and animations are available:

http://www.odsn.de/odsn/services/paleomap/paleomap.html

http://www.itis-molinari.mi.it/Intro-Reconstr.html

http://www.scotese.com.

Acknowledgments

The author would like to thank the organizers of the First International Biogeography Symposium, in particular Mark Lomolino, Jim Brown, and Julio Betancourt for the invitation to present and submit this work for publication. Thanks are also extended to Lawrence R. Heaney, Mark Lomolino, and Bruce Lieberman for their editorial help and constructive comments on this manuscript, as well as the industrial sponsors of the PALEOMAP Project.

CHAPTER 2

Arid Lands Paleobiogeography: The Rodent Midden Record in the Americas

Julio L. Betancourt

Introduction

Much of what we know about the late Quaternary vegetation history of the American deserts originated with the advent of rodent midden analysis in the 1960s for North America and the late 1990s for South America. This late development in desert paleobotany has several explanations. Unlike humid regions, deserts harbor few permanent bodies of water suitable for continuous pollen deposition, and the typical alkalinity of desert sediments is bad for pollen preservation. Just as critical, warm desert floras have large proportions of insect- or animal-pollinated plants, including the most prevalent species. Entomophilous and zoophilous species tend to produce less pollen than wind-pollinated ones, and what little is produced usually is consumed. Imagine the broad expanses of creosote bush (*Larrea*, Zygophyllaceae) that now characterize the Chihuahuan, Sonoran, and Mojave Deserts of North America and the Monte Desert of South America. Even if its pollen were well preserved in desert sediment, there would be very little way of inferring *Larrea* dominance from the pollen stratigraphy of monotonous stands in any of these deserts.

The paleoecological worth of woodrat or packrat (*Neotoma*) middens was first discovered in 1961 at the Nevada Test Site near Las Vegas. The initial find was made by Phil Wells and Clive Jorgensen, two young biologists on a team contracted by the Atomic Energy Commission to study the biological effects of fission-type nuclear detonations. The revelation occurred happenstance on a hike to determine if the summit of a small, desert mountain was wet enough to support oaks and junipers. There were no trees on the summit, but underneath a limestone overhang they found a dark, urine-cemented mass full of juniper twigs and woodrat fecal pellets. A few months later, W. F. Libby's Radiocarbon Laboratory at UCLA produced a date of 10,590 ±400 years BP. In their seminal paper, which reported on ten other nearby middens dated between 40 and 8.7 ky BP, Wells and Jorgensen (1964, p. 1172) prophesied that woodrat middens "may have unique value as a check of the palynological approach to Pleistocene ecology in the arid Southwest."

With the advent of woodrat midden analysis in the 1960s, there has been no shortage of late Quaternary fossil plant and animal records in the North American deserts. Quite the contrary, in the last 40 years fossil rodent middens in the Americas have developed into one of the more versatile and informative paleorecords worldwide. The body of knowledge generated from fossil rodent middens has been summarized elsewhere (Betancourt et al. 1990; Thompson et al. 1993; Thompson and Anderson 2000; Rhode 2001; Pearson and Betancourt 2002; Betancourt and Saavedra 2002). I will use this chapter to update and reflect on the significance of this knowledge for contemporary issues in biogeography, and to feature extension of the method to the South American continent. I will conclude by pointing out the more promising frontiers for rodent midden analysis and late Quaternary biogeography in deserts of the world.

Fossil Rodent Middens: Strengths and Weaknesses

Fossil rodent middens represent rich amalgamations of plant fragments, pollen, fecal pellets, bones, and insects imbedded in crystallized rat urine and preserved in rock shelters, crevices, and caves for tens of millennia. Since 1961, about 2500 woodrat middens have been dated and analyzed in arid and semi-arid regions of North America, from northern Mexico to southern Canada (Betancourt et al. 1990; Pearson and Betancourt 2002). More recently, a parallel effort in arid and semi-arid South America has relied on analogous deposits made by leaf-eared mice (*Phyllotis*, Muridae), vizcachas (*Lagidium*, Chinchillidae), chinchilla rats (*Abrocoma*, Abrocomidae), and chozchori or brushy-tailed rats (*Octodontomys*, Octodontidae). About 500 middens have been dated and analyzed from Argentina, Bolivia, Chile, and Peru, mostly in the last decade (Markgraf et al. 1997; Betancourt et al. 2000; Holmgren et al. 2001; Betancourt and Saavedra 2002; Latorre et al. 2002; Betancourt et al. 2003).

Advantages of rodent midden analyses include the high taxonomic resolution (often to species and potentially to genotype); the restricted origin of plant and animal material (generally within 100 m of the midden); the abundance of plant and fecal material for radiocarbon and other morphological, geochemical, and genetic analyses; and the easy replicability within and across areas. Although best known for their rich plant macrofossil assemblages, rodent middens have also contributed useful information about the past distributions of vertebrates and arthropods (e.g., Van Devender et al. 1987; Van Devender and Hall 1994; Elias et al. 1995).

The taxonomic and spatial accuracy associated with rodent middens can provide unprecedented resolution. For example, a late glacial hybrid zone between two species of pinyon pine can be pinpointed to particular hillslopes (Lanner and Van Devender 1998); a specific founder event and its genetic consequences can be identified and tested for a pinyon pine isolate (Betancourt et al. 1991); and the age of a clonal stand of starleaf Mexican-orange (*Choisya dumosa*, Rutaceae) can be inferred from its continuous presence throughout a 20,000-year midden series, pending confirmation by ancient DNA analysis that it is the same individual (Betancourt et al. 2000).

Disadvantages of rodent middens include dietary selectivity of different rodents, a clear bias for vegetation in bedrock escarpments and against more open terrain, occasional temporal mixing of midden assemblages, and the episodic and discontinuous nature of midden deposition. Although woodrats are considered generalists, dietary bias has been identified among some of the species that produce middens (Dial and Czaplewski 1990). This is apt to be more problematic in South America, where dietary preferences for midden agents cut across four families of rodents (Betancourt and Saavedra 2002).

In most cases, individual middens represent discrete and isolated depositional episodes rather than continuous stratigraphic accumulation. Temporal averaging can be caused by contamination from older and younger material, though at least some mixed assemblages can be detected by Accelerator Mass Spectrometer (AMS) dating of individual plant fragments (Van Devender et al. 1985). The duration of the depositional episode is presumed to be bracketed by the standard deviation of a ^{14}C date (say, ±100 years), but there are currently few ways to distinguish between episodes that lasted a few months, a few decades, and in some cases a few centuries. Among other pitfalls, uncertainty about the duration of the depositional episode is why midden inferences rely mostly on presence-absence, and put less stock on taxonomic abundance or diversity within single middens. This should serve as a note of caution for unscrupulous mining of the midden database (http://climchange.cr.usgs.gov/data/midden/) in support of macroecological studies.

In effect, rodent middens represent individual snapshots of vegetation that have to be collated into series or chronologies for any particular site or area (Betancourt et al. 1990). Although the success of these midden surveys still relies on personal skill and field experience, GIS-based approaches such as Weights of Evidence are now being used as a first approximation of suitable areas for midden preservation (Mensing et al. 2000). We now deploy larger field crews to comb extensive outcrops and to generate series of 50 to 100 middens from a single area (Latorre et al. 2002), or multiple chronologies from many sites along a presumed pathway of plant migration (Lyford et al. 2003b). Ready access to vacuum gas lines for pretreatment of plant macrofossils to gas-CO_2 and graphite targets, and the associated discounts for AMS dating, now make these synoptic approaches economically feasible.

Although there was some early debate on the methods and assumptions of midden analysis (Wells 1976 vs. Van Devender 1977), most authors tend to follow accepted methods for processing and analysis (Spaulding et al. 1990) with hardly a wrinkle. There have been few efforts to calibrate plant macrofossil assemblages in rodent middens with modern vegetation on par with the much more extensive and necessary empirical validation of pollen percentages in lake sediments (see discussion by Jackson, this volume). Two notable exceptions are Nowak et al.'s (2000) computation of probabilities that a taxon absent from a woodrat midden is necessarily absent from the paleolandscape (within a 100-m radius of the midden), and a comparable analysis by Lyford et al. (2003b) to evaluate the significance of Utah juniper (*Juniperus osteosperma*) presence and absence in midden series to support its migrational history. Both of these studies had comforting outcomes, but they still need to be repeated for South American middens.

Late Quaternary Biogeography of the North American Deserts: Highlights

In North America, the midden record spans a large area from the southernmost sites in Durango, Mexico to British Columbia in Canada. The majority of the work, though, has been done in the core deserts: the warm deserts (Chihuahuan, Sonoran, and Mojave Deserts) and the floristic Great Basin (the Colorado Plateau, the physiographic Great Basin, and the Wyoming Basins extending into southern Montana). Winter precipitation in the warm deserts is highly variable from year to year, but well synchronized across the three deserts and closely teleconnected to the tropical Pacific (wet during El Niño and positive Pacific Decadal Oscillation [PDO]; dry during La Niña and negative PDO) (Sheppard et al. 2002). The warm deserts are all characterized by a very arid foresummer (April–June). Summer monsoonal precipitation (July–September) starts abruptly, is less variable from year to year but patchier than winter, and decreases in importance from the Chihuahuan to the Mojave Desert. Most of the summer air masses in July originate in the Gulf of Mexico, while those in August tend to come from the Pacific and Gulf of California. The floristic Great Basin is in the transition between Southwest-like (more precipitation) and Pacific Northwest-like (less precipitation) responses to El Niño.

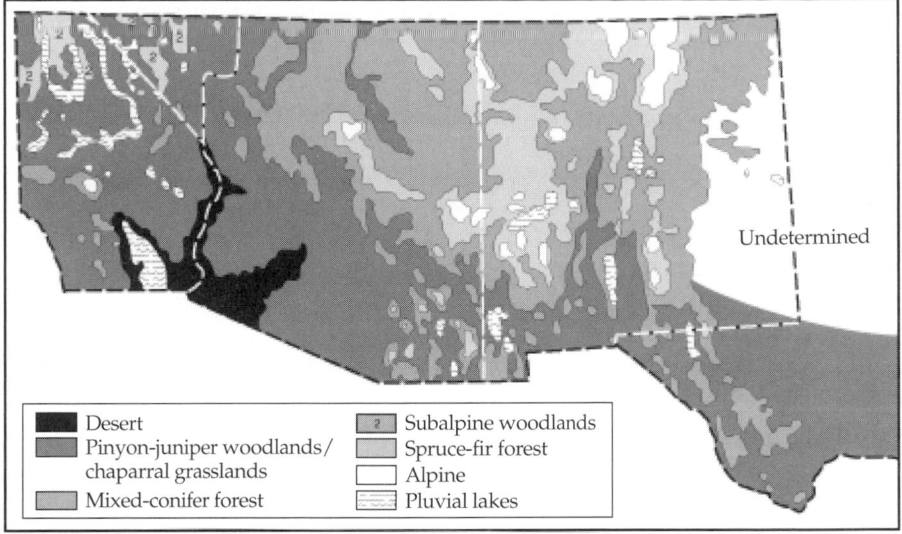

(B) Modern

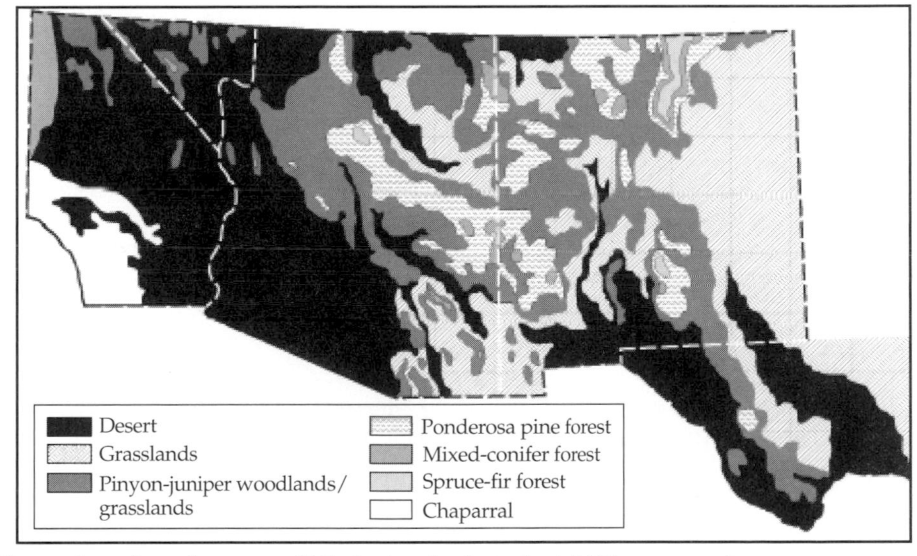

FIGURE 2.1 Vegetation of southwestern U.S. during the last glacial (A) compared to modern (B). (From Swetnam et al. 1999.)

The North American midden record has matured to the point of requiring constant updates. During the last glacial, desert vegetation of North America was restricted to a few areas below 300 m in Death Valley and around the Gulf of California (Figure 2.1). Pinyon-juniper-oak woodlands covered large expanses of what is now classified as desertscrub between 300 and 1700 m. Mixed-conifer woodlands and forests in turn dominated rocky terrain above

1400 m in the Colorado Plateau and Great Basin, which now harbor ~20 million ha of pinyon-juniper woodlands. Now restricted, spruce-fir forests once covered extensive areas above 2000 m in the central and southern Rockies, the Mogollon Rim, the High Plateaus of Utah, and the Sierra Nevada and adjoining mountain ranges of the Great Basin. Forest and woodland expansions, glacial advances, and pluvial lake highstands have all been attributed to a 5°C–10°C decrease in summer temperatures and significant increases in winter precipitation (and probably reduced variance) with a 10° southward displacement in the average winter position of the upper-air westerlies and the winter storm track (Bartlein et al. 1998). Summer precipitation was much reduced over the western U.S., but remained a significant portion (~45%) of the annual precipitation in the northern Chihuahuan Desert (Connin et al. 1998; Pendall et al. 1999; Holmgren et al. 2003). The July–September greenup, when areas to the north were experiencing seasonal drought and fire, may explain the rich assortment of megafaunal and Paleoindian sites in the northern Chihuahuan Desert.

Pluvial lake and groundwater levels dropped dramatically at 14 ky BP, but vegetation held steady until ~13 ky BP, around the time of the major megafaunal extinction. Holocene deserts replaced Pleistocene woodlands, and in return Holocene woodlands replaced Pleistocene forests. If the Pleistocene ended with a "Clovis" megadrought ~13 ky BP, vegetation change at the Pleistocene–Holocene transition must have entailed massive tree die-offs, widespread insect outbreaks, and catastrophic wildfires in forests above 2000 m. These disturbances cleared the way for forests of ponderosa pine, probably produced enigmatic and temporary fluctuations in groundwater levels, and must have translated large pulses of sediment from the mountains to the lowlands. Ponderosa pine, currently the dominant forest type across the southern and central Rockies, was surprisingly restricted to south of 35° N (i.e., south of Albuquerque, New Mexico and Flagstaff, Arizona) during the last glacial. Paradoxically, its postglacial expansion produced the considerable genetic diversity now evident across much of its northern range (Rehfeldt 1999). Interestingly, late Pleistocene extinctions of several key grazers (e.g., horses, camels, and mammoths), and the resulting increase in "grassiness," probably enabled frequent low-intensity fires that favor thick-barked and fire-adapted ponderosa pine at the expense of other conifers.

The late Quaternary history of North American pinyon-juniper woodlands is better known than for any other vegetation type worldwide (see summary in Lanner and Van Devender 1998). Most of the pinyon and juniper species are indistinguishable in pollen, but can be identified to species and even subspecies in macrofossils. The high taxonomic resolution afforded in middens has permitted a species-specific history of past distributions for pinyon and juniper species, including zones of past and present hybridization. Pinyon pines are a good example. During the last glacial, what are now desert lowlands (below ~1500 m) supported *Pinus remota* and

P. edulis in the Chihuahuan Desert, *P. edulis* var. *fallax* in the Sonoran Desert, *P. monophylla* in the Mojave Desert, and *P. juarensis* in central Baja California. Hybrids of *P. juarensis* × *californiarum* var. *fallax* occurred in Joshua Tree National Monument, southeastern California. The northern limits of *P. edulis*, the dominant species today in the southern Rockies and Colorado Plateau, were along the Little Colorado River Valley. Possible hybrids of *P. edulis* × *monophylla* occurred in the Lower Grand Canyon, and of *P. edulis* × *P. edulis* var. *fallax* on the south slope of the Santa Catalina Mountains near Tucson. At desert elevations, pinyons were extirpated from desert elevations by ~12.5 ky BP, *Pinus remota* retreated upslope to a few mountain refugia in west Texas and Coahuila, and *P. edulis* var. *fallax* wedged itself between 1250–1800 m and below *P. edulis* at the base of the Mogollon Rim.

The Holocene migrational histories of pinyon and juniper species are dynamical, modulated by dispersal, climate, and physiography. The climate space for these pinyon and junipers species narrows considerably to the north, so that secondary climatic changes during the Holocene have either retarded or accelerated northward migration. At their northernmost sites, pinyons and junipers have been spreading into progressively narrower elevational wedges, constrained at the bottom by relatively high base levels, and at the top by cooler summers and competitive exclusion by higher elevation conifers. These wedges become available for invasion only during brief warm and dry episodes. For example, *Juniperus osteosperma* abutted against the Rockies in northeastern Utah/southwestern Wyoming by 10 ky BP, jumped over them to reach its northernmost sites in southern Montana by ~6 ky BP, occupied only a fraction of its present range in Wyoming/southern Montana during the wet Neoglacial (~4–2.5 ky BP), and then backfilled explosively across the intervening Wyoming Basins in the last 2000 years (Lyford et al. 2003a). Founder events in the last 1000 years explain northern disjuncts of *P. monophylla* in northwestern Nevada (Nowak et al. 1994), *P. edulis* in northeastern Utah (Jackson et al., in review), and northern Colorado (Betancourt et al. 1991). In the northern peripheries of pinyon and juniper distributions, population expansion and infilling associated with ongoing, natural invasion could account for some of the increasing stem density and encroachment of grassland attributed to post-settlement grazing and fire suppression. Large-scale changes in climate are responsible for this suite of late Holocene northward migrations, and they should also be evident in the migrational histories for other woodland and desert species in North America.

Holocene migration of desert species has also been time-transgressive. For example, much of the saguaro's (*Carnegiea gigantea*) range was established quickly through migration northward from unidentified late Pleistocene refugia in coastal Sonora, northwestern Mexico. Saguaro established its northernmost limits by 12.3 ky BP (McAuliffe and Van Devender 1998). Such rapid migration may be due to long-distance dispersal of its small seeds by frugivorous birds, such as white-winged doves (*Zenaida asi-*

atica). By contrast, creosote bush (*Larrea divaricata* var. *tridentata*) was relatively slow in expanding to its northern limits in the Chihuahuan, Sonoran, and Mojave Deserts (Hunter et al. 2001).

Identification of teeth from middens also indicates dramatic shifts in the distribution of some *Neotoma* species during the last glacial-interglacial cycle. For example, with postglacial warming, *N. cinerea* contracted much of its southern glacial range in what is now the Chihuahuan Desert, while other species (*N. albigula, N. lepida,* and *N. mexicana*) expanded to the north (Harris 1984). In northern Utah, *N. cinerea* abandoned its lowest elevations to *N. lepida* at ~9.2 ky BP only to return just prior to 1 ky BP (Grayson 2000; Grayson and Madsen 2000). Just to the north, *N. lepida* overshot its northern limits, becoming sympatric with *N. cinerea* in southern Idaho during the latter part of the middle Holocene (Smith and Betancourt 2003). *N. mexicana* may still be expanding north in Colorado.

Late Quaternary Biogeography of the South American Deserts: Update on First Results

The Atacama Desert

The hyperarid Atacama Desert extends along the Pacific Andean slope from the southern border of Peru (18°S) to Copiapó, Chile (27°S). The region's hyperaridity is due to the extreme rainshadow of the high Andes, which blocks the advection of tropical/subtropical moisture from the east; the blocking influence of the semi-permanent South Pacific Anticyclone, which limits the influence of winter storm tracks from the south; and the generation of a temperature inversion at ~1000 m by the cold and north-flowing Humboldt Current, which constrains inland (upslope) penetration of Pacific moisture. Seasonal and annual precipitation totals in the region are determined by the number of precipitation days, and associated circulation anomalies during those days. Precipitation variability in both summer and winter is modulated primarily by Pacific sea surface temperature (SST) gradients and associated upper-air circulation anomalies. These anomalies promote either greater spillover of summer moisture from the Amazon Basin and the Gran Chaco to the east, or conversely, greater penetration of winter storm track from the southwest. Throughout the Quaternary, the most pervasive climatic influence has been millennial-scale changes in the frequency and seasonality of the scant rainfall, and associated shifts in plant and animal distributions with elevation.

Midden series are being developed along a 1100-km transect (16–26°S) in the Atacama Desert (Betancourt et al. 2000; Holmgren et al. 2001; Latorre et al. 2002, 2003). The seasonality of precipitation along the transect changes from 90% winter at 22° S to 90% summer at 26° S—a distance of only 450 km. The midden focus has been on this latitudinal transition, specifically in the

FIGURE 2.2 Midden sites near edge of Absolute Desert, a waterless and plant-less terrain in the central Atacama Desert (23.5°S). Middens more than 40,000 years old were retrieved from small overhang shelters as well as from underneath desk-size boulders on hillslopes.

boundary from sparse, prepuna vegetation into absolute desert, an expansive waterless terrain that extends from just above the coastal fog zone (800 m) to more than 3000 m in the most arid sectors (Figure 2.2). The most compelling midden evidence for past changes involves elevational displacements of indicator plants and large-scale invasions of vegetation across what is now the edge of the absolute desert (Figure 2.3).

Midden results for the central Atacama (~22–24° S) indicate maximum summer precipitation and 500–900 m lowering of plant distributions between 14 and 10.5 ky BP (Figure 2.4) (Latorre et al. 2002), contrasting with pluvial lake highstands on the adjacent Bolivian Altiplano from 26 and 15 ky BP according to one group (Baker et al. 2001a) or 19–14 ky BP according to another (Fornari et al. 2001). The disparity may have to do in part with different sources and mechanisms of summer precipitation across the central Andes. At stake in the timing is the extent to which pluvial events are driven by increasing moisture in source areas (the Amazon or the Gran Chaco), summer insolation maxima over the Bolivian Altiplano intensifying the Bolivian High and increasing convection, or by changing tropical Pacific SST gradients that intensify the easterlies and enhance large-scale advection of

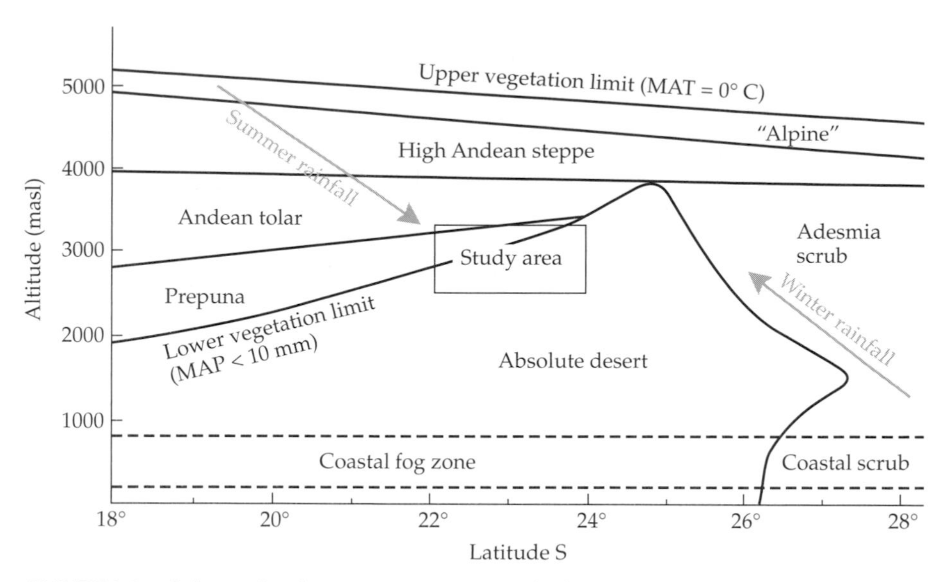

FIGURE 2.3 Schematic of vegetation zones with elevation and latitude in the Atacama Desert of northern Chile, showing its relationship to Absolute Desert lacking rain and vascular plants. (From Latorre et al. 2002.)

moist air masses. For those of you not following the most recent development in paleoclimatology, there is now compelling evidence that SSTs in the so-called warm pool of the western Pacific have fluctuated by ~4°C, in synchrony with global atmospheric CO_2 variations, leading Northern Hemisphere deglaciation by 2000–3000 years (Visser et al. 2003). If insolation variations drive the ice ages and other climatic variations, they would surely have to involve the tropical Pacific and its complex interaction with the extratropics.

In the driest sector of the southern Atacama (25.5°S), where the scant precipitation occurs mostly in winter, midden series were developed from two elevational belts (2650–2850 m and 3450–3500 m) (Maldonado 2003; Betancourt et al., in preparation). The lower site registers only the most extreme wet events, evident in full-glacial (23–18 ky BP) expansion of steppe grasses and other high-elevation shrubs >800 m downslope into absolute desert. The higher site registers another, perhaps lesser, pluvial event between 17–10 ky BP evident in an elevational displacement of steppe grasses by at least 100 m. The full glacial pluvial (perhaps due to a northward expansion of the westerlies and increased winter precipitation) recorded at the lower site did not produce wholesale expansion of steppe grasses in the central Atacama. Conversely, the late glacial/early Holocene summer pluvial, which yielded up to a 1000 m expansion of steppe grasses in Salar de Atacama/Calama Basins, produced a lesser expansion recorded only at the higher site in the southern Atacama.

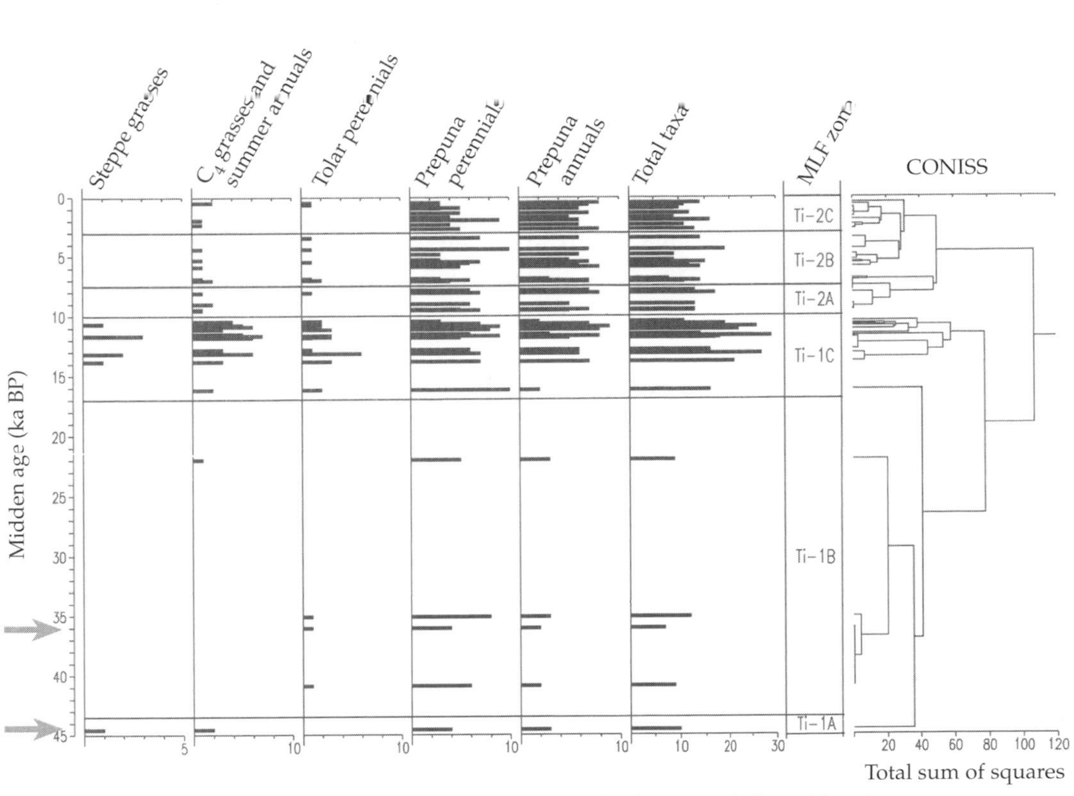

FIGURE 2.4 Summary diagram for number of species, by functional class, identified from series of middens between 2400 m and 3000 m in elevation in the central Atacama Desert. (From Latorre et al. 2002.)

The Monte Desert

Rodent middens are also being studied east of the Andes in the Monte Desert of Argentina. The Monte Desert covers ~40 million ha from 24–43°S, and is surprisingly uniform in physiognomy and floristic composition. It is rich in endemic genera and species of plants and insects. The Monte is a xerophytic scrubland dominated by three species of creosote bush or "jarilla" (*Larrea divaricata, L. cuneifolia,* and *L. nitida*) in association with a wide variety of mostly microphyllous, dry-deciduous, and aphyllous shrubs. There are relatively few herbaceous elements in the Monte flora, and annuals consist mostly of summer-flowering grasses. The Monte is a warm desert and more than 60% of the precipitation falls in the summer months. There are conspicuous floristic similarities between the Monte and Sonoran Deserts, including 10% overlap in plant genera, most of which comprise the dominant vegetation in both deserts (e.g., *Larrea, Prosopis, Acacia, Celtis, Cercidium, Condalia, Ziziphus, Cereus,* and *Opuntia*). During the International Biological Program's Structure of Ecosystems Subprogram (IBP-SES) from 1968 to 1974, the likeness between the Sonoran and Monte Deserts inspired studies of how communities in similar but geographically distant habitats converge in composition and function given the same environmental pressures (Orians and Solbrig 1977). One of the glaring differences, the dominance of seed-hoarding heteromyids in the Sonoran and the lack of granivorous rodents in the Monte (Mares and Rosenzweig 1978), has fueled a persistent dialog about worldwide differences in desert granivory (Lopez de Casenave et al. 1998; Marone et al. 2000; Saba and Toyos 2003). The midden record from the Monte extends this dialog to further consider the hand of history.

Unlike the Atacama and North American deserts, midden chronologies from the Monte exhibit surprising stability. In the Monte, evidence is lacking for the kinds of latitudinal and elevational displacements documented for the other deserts. Despite evidence for significant glacial advances down major Andean valleys at 33–35°S (Espizua 1999), Andean floras were not displaced hundreds of meters downslope. Also, there apparently was no wholesale, northward expansion of forest elements (*Araucaria*, *Austrocedrus*, and *Nothofagus*) that today reach their northern limits at ~38–40°S. Furthermore, dung deposits in Monte caves show that, prior to 12 ky BP, ground sloths browsed many of the same shrubs and herbs now growing at the site (Garcia and Lagiglia 1999; Hofreiter et al. 2003). Holocene midden series from the Patagonian-Monte elevational transition, the southern equivalent to the desert-woodland ecotone, display none of the sequential departures and arrivals typical of North American midden series (Betancourt et al. 2003).

While the modern Sonoran and other North American deserts were mostly wooded during Quaternary glaciations, the Monte stayed desert throughout the last glacial-interglacial cycle. Although the southern westerlies and winter storm tracks may have shifted far enough north to drive vegetation changes in the southern Atacama during the Last Glacial Maximum (LGM), the imposing rainshadow muted the impact on the leeward side of the Andes. The seasonality of precipitation in the Monte remained unchanged. During the glacial period, summer precipitation in the Monte Desert must have decreased with continental cooling, but the decrease was probably offset by cooler summers. Effective moisture thus remained more or less the same during glacial and interglacial stages.

So what does this have to do with desert granivory? First, I concur with Lopez de Casenave et al. (1998), who suggest that granivory in the Monte is not abnormally depressed compared to most deserts, but that it is exceptionally high in the Sonoran Desert due to some quirk of history. That quirk probably involved the evolutionary consequences of cyclical migration of the northern westerlies and winter storm tracks to lower latitudes. Today, annual plant species comprise 50% of local floras in the Sonoran Desert, and of these, about 60% to 80% are winter annuals (Venable and Pake 1999). This is in spite of the fact that summer rainfall generally exceeds 50% of the annual average, and is much more predictable from year to year than winter precipitation. After abundant winter rains, mostly in the occasional El Niño year, spring-blooming annuals add millions of seeds to the soil seed bank, renewing the major food source for granivores. The blooms clearly stimulate spring reproduction in rodents and the highly synchronized and broad-scale masting occurs in the arid foresummer, when there is little else to eat. It is easy to imagine how climatic fluctuations during the Quaternary could have driven an impressive radiation of winter annuals and seed-hoarding heteromyids, which flourished in pinyon-juniper woodlands during the winter-wet glacial periods and persisted in desertscrub during dry interglacials. The only comparable climatic histories include smaller areas in the southern Atacama, southern Namibian, and the Israeli deserts, where granivory is also high. The other deserts of the world, like the Monte, were presumably stable throughout the Quaternary and exhibit average levels of granivory (Betancourt et al. 2003).

Bioclimatic Modeling with Midden Data

Various types of models are currently being used in the western U.S. to understand and predict large-scale vegetation responses to past and future climate change, including shifts in the distribution of biomes and individual species. At regional scales, deterministic biogeographic models can simulate the distribution of biomes, including deserts. Models such as BIOMES (Haxeltine et al. 1996), DOLY (Woodward et al. 1995), and MAPPS (Neilson 1995) are deterministic renditions of the climate-life zone correlations implied in the Holdridge (1967) classification. These models simulate the distribution of growth forms across spatial grids of 0.5° resolution for global vegetation and 10-km resolution for subcontinental scales. Because they are deterministic, these models can be verified by comparing simulated versus actual biome distributions. Such physiognomic or biome models are difficult to use for predicting finer-scale changes within deserts, which usually encompass many functional types and dominant species.

In the absence of complete knowledge about physiological tolerances, species-based biogeographic models tend to be statistical, and entail the construction of climatic envelopes around species distributions and abundances (see summaries in Franklin 1995 and Scott et al. 2002). These statistical models are intended to describe "climate spaces" occupied by individual species. They are arguably circular, but nevertheless useful for exploring the bioclimatology of modern species, and for generating climatic reconstructions from distributional shifts through time.

The most commonly used statistical biogeographic model in the western U.S. (referred to herein as the USGS-OSU model) envisions plant distributions as stochastic, spatial realizations of response surfaces—functions that describe the way in which each species' expected distribution and abundance depends on the combined effects of several environmental variables

(Bartlein et al. 1998; Thompson et al. 1998, 2000a,b,c). Solutions of the climate space occupied by a particular species vary widely, but generally start with development of a climate database for the region of interest, interpolated into grid cells at the desired resolution. The USGS-OSU model uses a least-squares estimation technique to develop regression equations for estimating the monthly and annual values for temperature and precipitation at each point in an equal-area 25-km grid as functions of location and elevation. A digital elevation model, or DEM, is used to calculate the mean elevation across each grid cell, and climatic variables for that elevation (grid point) are estimated from local relationships between elevation and climate. This should only be considered a first approximation. Mean climate at the mean elevation is hardly representative of the whole suite of elevations that can occur in a 25-km grid cell, roughly the same area as some desert mountain ranges. In such cases, the mean elevation may be 1500 m, but the grid cell can encompass desert (0–1400 m) to subalpine (>3000 m) elevations. Using digitized range maps, the USGS-OSU model determines at which of the points on the 25-km grid a species is present to create a presence-absence matrix of plant distributions associated with modern climate data. Because the distributional data are gleaned from a flat range map and thus are not point and elevation-specific, the same climatic value (the mean climate of the mean elevation of one grid cell) could be assigned to a saguaro (*Carnegiea gigantea*) as well as to a corkbark fir (*Abies lasiocarpa* var. *arizonica*) growing at the base and summit of the Santa Catalina Mountains in southern Arizona. I say this not to be disparaging—the USGS-OSU model is a good first approximation and lays out some of the necessary groundwork—but rather to point out critical data and modelling needs in the future.

Response surfaces from the USGS-OSU model can be mapped for times in the past using the output from general circulation models (GCMs) forced by changing boundary conditions such as insolation and CO_2 levels. This top-down modeling approach can be used to determine the probability of species occurrence for simulated modern climate. For most western U.S. species, the modern range is aligned with the 0.4 probability of occurrence in the simulated distribution for species in the western U.S. (Thompson et al. 2000a,b,c). A related approach involves biomization, or the grouping of individual plant taxa in plant functional types (PFTs), specification of the set of PFTs that can occur in a biome, and then construction of climatic envelopes for each of the biomes (Thompson and Anderson 2000).

The USGS-OSU model has been used to extract climatic information from individual species, as well as biome distributions (gleaned from both midden and lake pollen data), in western North America since the last glacial maximum (Bartlein et al. 1998; Thompson and Anderson 2000; Sharpe 2002). Interestingly, plant macrofossils from packrat middens and pollen spectra from lakes, when available from the same region, yielded identical biome reconstructions (Thompson and Anderson 2000). This explains the long-recognized agreement between late glacial versus modern vegetation maps of the southwestern U.S. based on pollen in lakes (Martin and Mehringer 1965) and woodrat middens (Swetnam et al. 1999). Finally, other geostatistical approaches using point-specific plant distributional data have been used to develop climatic envelopes and reconstruct paleoclimates from Sonoran Desert middens (Arundel 2002; Craig et al. 2003).

Rewards of a Rich Archive: Morphological, Isotopic, and Genetic Studies

A side benefit of the midden work has been the accumulation of a rich archive of plant and animal remains identified to species and spanning one of the most dramatic transitions in the geologic record. Between 15 and 12 ky BP, atmospheric CO_2 concentrations increased by 30%, land temperatures increased by 5°–10°C, seasonal precipitation regimes shifted, and there were broadscale changes in biotic distributions, at least in some deserts. Midden researchers are taking advantage of this rich archive by conducting a variety of morphological, geochemical, and molecular analyses to develop unique insights about evolutionary and ecophysiological responses to global change.

For example, two characteristics of plant gas exchange and water relations that are quantifiable by measurement in modern and ancient plant leaves are stomatal density and stable isotope variation. These parameters are sensitive to changes in climate as well as atmospheric CO_2 levels, such as have occurred since industrialization (280–360 ppmv), and during the last deglaciation (200–280 ppmv). To evaluate ecophysiological responses to environmental variability, stomatal density and stable isotopes (i.e., δD, $\delta^{18}O$, and $\delta^{13}C$) are being measured across individuals, across environmental gradients, through time in modern populations, in time series of herbarium specimens, and via fossil leaves of species with different life histories and photosynthetic pathways (i.e., C_3, C_4 and CAM) (Van de Water et al. 1994; Pendall et al. 1999; Pedicino et al. 2002; Terwilliger et al. 2002; Van de Water et al. 2002). There are many take-home messages from this body of work. For example, plants responded to CO_2 enrichment during deglaciation by reducing stomatal density and increasing water use efficiency (WUE) (Van de Water et al. 1994). These savings in WUE, however, were not enough to offset increasing aridity, so Holocene deserts quickly replaced Pleistocene woodlands. At some threshold in effective moisture, climate trumps direct CO_2 effects on vegetation dynamics.

Stomatal characters in fossil leaves from middens also have allowed detailed mapping of polyploidy races of creosote bush since the Last Glacial Maximum (Hunter et al. 2001). North American creosote bush (*Larrea divaricata* var. *tridentata*) generally exhibits a classic and much discussed gradient of diploidy in the Chihuahuan Desert, tetraploidy in the Sonoran Desert, and hexaploidy in the Mojave Desert. Stomatal guard cells increase in size

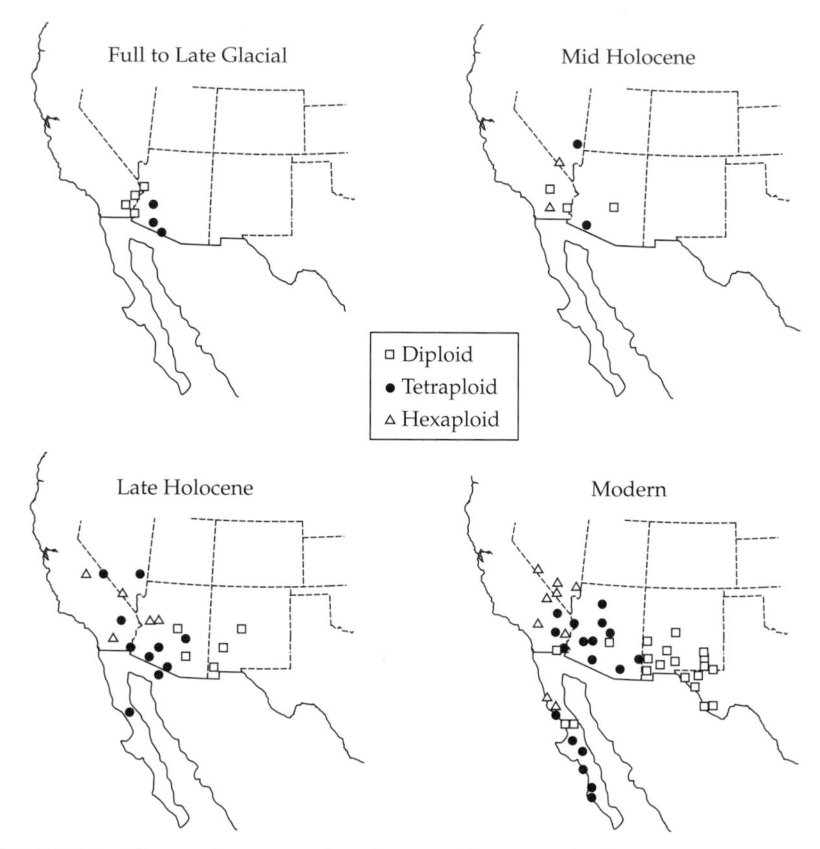

FIGURE 2.5 Past and present distribution of creosote bush (*Larrea divaricata* var. *tridentata*) ploidy races inferred from stomatal guard cell size. (From Hunter et al. 2001.)

with increasing amounts of DNA and ploidy level, allowing ploidy levels to be inferred from fossil leaves. Using the extensive midden record for creosote bush, Hunter et al. (2001) showed that the higher ploidy races had not evolved with Holocene migration from Mexico into the southwestern U.S., as previously thought, but rather were already present during the last glacial and simply reorganized their distributions in the Holocene (Figure 2.5). A moral to this story is that populations, and not whole species, migrate. Modeling the response of creosote bush to climate change would produce quite different outcomes by simply changing the unit of analysis from the whole species to the individual ploidy race.

The midden record also provided a novel opportunity to gauge the influence of climate on a key animal, the woodrat itself. Woodrats are particularly sensitive to environmental temperature and, as predicted by Bergmann's Rule, a strong correlation exists between mean adult body mass and ambient temperature across populations in the western U.S. It turns out that the diameter of woodrat fecal pellets, which are abundant in individual middens, are highly correlated with body size. How then did body size change in particular species of woodrats with late Quaternary fluctuations in ambient temperature? At hand was the opportunity to secure quantitative measures of the magnitude and rates of genetic change (versus phenotypic plasticity) associated with the evolution of a basic determinant of physiology, ecology, and behavior—that is, body size. In the first midden study, a 6°C warming was inferred from body size changes in *Neotoma cinerea* during the last deglaciation, a temporal test of Bergmann's Rule (Smith et al. 1995). There have been follow-up efforts to hone this apparent paleothermometer, apply it throughout the Holocene, and use fecal pellet diameter to delineate areas of past sympatry between different species of woodrats (Smith and Betancourt 1998, 2003). Advances in sequencing ancient DNA could add a genetic dimension to these body size studies.

The advent of polymerase chain reaction (PCR) has not only enhanced molecular phylogeography, it has also made possible identification of ancient DNA sequences from bones and from plant and animal tissues. There have been a few technological hurdles to overcome, however. PCR amplification of the DNA is apparently blocked by cross-links between reducing sugars and amino groups. These cross-links are generally the product of Maillard reactions, a series of condensation reactions that bind proteins and carbohydrates with long-term preservation of plant and animal tissue. Recently, it was discovered that such cross-links can be cleaved by the chemical reagent, *N*-phenacylthiazolium bromide (PTB) (Vasan et al. 1996), permitting routine amplification of both mitochondrial DNA (mtDNA) and chloroplast DNA from a variety of depositional settings, including dry caves in the Americas (Poinar et al. 1998; Hofreiter et al. 2000). Such dry caves had not yielded nuclear DNA, which until recently was thought to preserve only in permafrost. Nuclear DNA is typically longer and contains more information than mtDNA, which could allow resolution of phylogenies, determination of sex, and characterization of genes in phenotypic traits. A major breakthrough was made recently when Poinar et al. (2003) recovered single-copy nuclear DNA from the dung of Shasta ground sloth (*Nothrotheriops shastensis*) at Gypsum Cave, Nevada. In this particular instance, nuclear DNA resolved a previously intractable phylogenetic problem, mainly that the extinct Shasta ground sloth and extant three-toed sloth (*Bradypus*) share a more recent ancestor than do the Shasta and extant two-toed sloth (*Choeloepus*).

It is only a matter of time before rodent middens, surely the richest sources of well-preserved and dated plant and animal tissue in the American deserts, become one of the more popular targets for such genetic studies. The first successful studies involved two South American middens. The Monte midden was made by a vizcacha, but also incorporated numerous fragments of larger herbivore dung dated at 17.3 and 16.2 ky BP. Phylogenetic analyses of the mitochondrial 12S rDNA show that the dung originated from a small ground sloth species not yet represented by skeletal material, and not closely related to any of the four previously sequenced extinct (*Nothrotheriops* and *Mylodon*) and extant (*Choelopus* and *Bradypus*) sloth genera (Hofreiter et al. 2003).

The second genetic study showed that *Phyllotis limatus*, a leaf-eared mouse, was the principal agent in accumulation of an 11.7 ky BP midden from the Atacama Desert (Kuch et al. 2002). The modern and ancient sequences reported in this paper reinforce the proposal that *P. limatus* was derived very recently from a western lineage of *P. xanthopygus rupestris* (Steppan 1998). The midden agent apparently was part of the ancestral *P. limatus* populations that extended at least 100 km further south than today. This expansion happened during the summer pluvial event that affected the northern and central Atacama between 14 and 10.5 ky BP.

Frontiers of Rodent Midden Analysis and Desert Paleobiogeography

The growing midden database for North and South America has great potential for populating and testing a wide variety of biogeographic models to address key questions such as the role of history in community assembly, species coexistence, and biodiversity. For now, applications of biogeographic models have been limited to estimating paleoclimates from past distributions and predicting future vegetation from global change. The bioclimatic models that already have been applied in the western U.S. are elegant, but leave much room for improvement. Probably the greatest limitation in deserts is the lack of point-specific, georeferenced data for plant distributions, with reliance placed instead on digitized flat range maps. This will be less of a problem to remedy in the Atacama and Sonoran Deserts, where there are at least some digitized atlases of plant distribution (e.g., Turner et al. 1995; Marticorena et al. 1998).

In the South American deserts, appropriately gridded climatological data are lacking that can be extrapolated to all points of species occurrence across complex topographies. The western U.S., however, has a well-honed and accessible analytical tool (PRISM: Parameter-Elevation Regressions on Independent Slopes Model) that uses point data, a digital elevation model, and other spatial datasets to generate gridded estimates of monthly, yearly, and event-based climatic parameters, such as precipitation, temperature, snowfall, degree days, and dew point (Daly et al. 1994). The low density of weather stations may thwart extension of PRISM to South America.

At present, few of the biogeographic models integrate physical parameters other than climate, even though soils and substrate, for example, can explain a large portion of both the local and regional variance in plant distributions across complex landscapes typical of some deserts (McAuliffe 1994; Lyford et al. 2003b). Biogeographic models also presently ignore that the distribution of a given species to some degree reflects interactions with other plant and animal species, such as competition and plant-herbivore interactions (Davis et al. 1998).

Existing biogeographic models are only static comparisons of past, present, and future distributions, with little regard to how a species might actually shift from one distribution to the other. The real breakthroughs will come with the marriage of bioclimatic models to cell automaton and other spread models (e.g., Aassine and El Jaï 2002) that incorporate both short- and long-distance dispersal capabilities, population dynamics in the context of varying and not just changing climate, and the permeability of heterogeneous landscapes to invasion. Now that we know the migrational histories of many species in western North America, perhaps it is time to bring back, and apply, the chronosequence approach in ecological field studies. To what extent are population, community, and ecosystem properties (e.g., genetic distance from source populations, biodiversity, heterogeneity of soil resources, erosion potential, etc.) a function of time since arrival of the dominant plant species? Could integrated biogeographic models be developed and tested as formal platforms for exploring the spatial-temporal context of modern phylogeographic structure in populations of species with well-known migrational histories such as woodrats, creosote bush, Utah juniper, and pinyon pine?

Phylogeographers increasingly rely on the historical record to evaluate the plausibility of their interpretation (e.g., Mitton et al. 2000); still, many biogeographic and evolutionary models based on modern phylogeography remain unverified. Rather than voicing general skepticism about phylogeographic methods, I think it is this lack of verification that prompted Brown (1995, p. 191) to lament that "it would be difficult or impossible to reconstruct the spatial pattern of the history of lineage." In the American deserts, however, ancient DNA breakthroughs could provide an unprecedented opportunity to conduct phylogenetic analyses of both modern and ancient rodent and plant populations at comparable spatial resolutions.

Recent analyses of ancient cytochrome *b* fragments from fecal pellets have proven reliable in identifying rodent agents of fossil middens (Kuch et al. 2002). This development could provide a springboard for what I term paleophylogeography, permitting phylogeographic structure to be defined directly in fossil populations. This time-lapse view of molecular diversification could provide both the dynamics of datable range shifts and the timing of origination of phylogeographic structure observed today. This comparative effort can also furnish a set of empirical tests for the analytical methodology of population genetics and coalescence theory.

The molecular phylogeographic structures of several western species of *Neotoma*, based on mtDNA cytochrome *b* gene sequences, have been studied well enough in modern populations to construct detailed phylogenetic trees and generate a battery of historical hypotheses (Edwards et al. 2001; Patton and Alvarez-Castañeda 2002). Though fossil teeth already provide several genetic hypotheses, they are sporadically available, and preclude identification of subspecies or geographic clades. In the majority of dated middens, there are sufficient fecal pellets remaining in the archived material for mtDNA and, with a bit of luck, nuclear DNA analyses. More middens could be collected from key areas to supplement the existing archive or to focus in on geographic clade boundaries of special interest.

Even with the many advances in arid lands paleobiogeography, there remain a number of physical, conceptual, and technological frontiers for rodent midden research. Despite 40 years of midden surveys, there is still plenty of unexplored terrain in North America. For example, there has been surprisingly little work in Baja California and the Mexican mainland, and the coverage for some states (Arizona, California, Colorado, Idaho, and Utah) is patchy at best. Also, much of the fieldwork was done before 1990, when resources were scarce and chronologies consisted of at most two dozen middens. Precise timing has become critical in our ability to sort out leading explanations of both regional and global climate change in the late Quaternary, justifying the resampling of key midden sites. There is even more unexplored territory in South America ranging from Patagonia to northern Peru.

As might be expected, mammals also produce middens in other parts of the world. Paleoecologists have long nibbled at the midden record from Africa and Australia with modest results (see discussion in Pearson and Betancourt 2002), but midden potential exists even in the arid lands of central Asia and the Middle East. In the meantime, the rich midden archive keeps increasing its holdings, and our ability to make morphological, geochemical, and genetic measurements continues to improve. As a final example that is particularly germane to contemporary biogeography, there are still unlimited opportunities to explore how scaling of plant anatomy might have behaved under dramatically different temperature and atmospheric CO_2 levels.

CHAPTER 3

Quaternary Biogeography: Linking Biotic Responses to Environmental Variability across Timescales

Stephen T. Jackson

Introduction

The geographic distribution of a species derives from both contemporary and historical processes. Contemporary processes include the ecophysiological responses of individual organisms to the prevailing environment and the aggregate population consequences of these responses in the context of landscape texture, interactions among species, disturbance regimes, and propagule dispersal. Historical processes influence geographic distributions at a variety of spatial and temporal scales, ranging from decadal- to centennial-scale legacies of disturbance or dominance in local patches to deeply rooted phylogenetic divergences at continental to global scales. Environmental changes of the Quaternary, encompassing the past 1.6 million years of Earth history, have had dramatic biogeographic consequences, and signatures of Quaternary history are apparent in the spatial patterns of species distributions and genetic variation across the globe today.

Our understanding of Quaternary environmental history is currently undergoing revolutionary advances, owing to the development of powerful tools for paleoenvironmental inference and chronological placement, as well as the acceleration of research activity motivated by global change concerns. Rapid advances are also underway in Quaternary ecology and biogeography, driven by improved inferential and chronological tools, accretion of dense spatial networks of paleoecological data, and increasing ability to link paleoecological records with independently derived records of environmental change. In this chapter, I will discuss the relationships between biogeography and Quaternary history, outline the nature of the Quaternary paleoecological record, and summarize recent work in selected areas of paleoecological research relevant to biogeography.

A Brief History of Quaternary History

Biogeography and Quaternary sciences share common roots in natural history and philosophy of the late eighteenth and early nineteenth centuries (see Brown and Lomolino 1998; Lomolino et al. 2004). The foundations of our understanding that Earth has experienced substantial environmental variation in its most recent history were laid by active students of biogeography (Edward Forbes, Louis Agassiz) and by geologists who relied heavily on biogeographic evidence (Charles Lyell). By the mid- to late nineteenth century, such naturalists as Asa Gray, Charles Darwin, Joseph Hooker, and Alfred Russel Wallace were routinely attributing biogeographic disjunctions and vicariant distributions to climatic changes during and since the last "Ice Age." Wallace was particularly fascinated by the glacial episodes of the Quaternary, devoting the longest chapter of his classic biogeographic work *Island Life* to "the causes of glacial epochs" (Wallace 1892).

Well into the twentieth century, biogeography held an important place in the study of Earth's recent climatic history. "Relict" populations of arctic, alpine, and boreal species in temperate regions were used to gauge the magnitude of temperature depression during the last glacial period, and disjunct populations of xerophytic or thermophilic species were cited as evidence of warmer and drier conditions during the mid-Holocene. These biogeographic inferences long predated most of the fossil evidence for late Quaternary climate change.

Fossil evidence was used to infer Quaternary climate change as early as the 1840s, and studies of macrobotanical, molluscan, and vertebrate fossil assemblages were widespread in Europe and North America by the turn of the twentieth century. Pollen analysis was developed relatively late and was not widely applied until the 1920s and 1930s. Many biogeographers were quick to grasp the potential power of pollen stratigraphy for inferring past vegetation and climate. However, the critical test for the nascent field was whether pollen sequences revealed temporal patterns consistent with past climatic and biogeographic changes already inferred from modern distribution patterns (e.g., Transeau 1935).

The fields of biogeography and Quaternary environmental history diverged in the mid-twentieth century, although they have maintained loose connections since then. Biogeography was eclipsed by paleontology as a primary source of historical inference. In North America, this divergence was initiated by Deevey's influential review (Deevey 1949), in which he critiqued a number of biogeographically based paleoclimatic and paleobiogeographic inferences, and amassed paleontological evidence that contradicted these inferences. The role of Deevey's paper in the demise of biogeographically based historical inference is ironic in view of his elegant use of biogeographic and fossil evidence in tandem to understand biogeographic history.

Most paleoecologists trained since the 1950s have tended to view modern biogeographic patterns as "soft" evidence of history. Inferences based on modern distributions have been subject to examination in the cold light of the fossil record, which has been regarded as the paramount form of historical evidence. This hierarchy is eroding, however, as paleoecologists have come to recognize both the limitations of their own methods and the power of recent biogeographic patterns, particularly genetic markers, for inferring biogeographic history. As noted later in this chapter, a new synthesis may be emerging in which paleoecological and geological evidence are used together with biogeographic and genetic evidence to provide mutual tests and complementary inferences (see also chapters by Lieberman and Brooks, this volume).

The Nature of the Historical Record

Quaternary paleoecology and paleoclimatology rely on archives preserved in the geological record and, occasionally, in tissues of living organisms (trees, corals) and in written documents. These archives are highly variable in the quality of information they contain, their spatial and temporal precision, and their continuity in time. Tree rings, certain types of corals, and varved (stratified) sediments of deep lakes or ocean basins contain continuous records of past environment at annual or even seasonal resolution. Precise age assignments can be made based on simple counting of rings or laminations from a benchmark. Sediments of peatlands, lakes, and ocean basins contain continuous records with temporal resolution ranging from decades to millennia, depending on sediment accumulation rates and the degree of sediment mixing. Sediment ages can be assigned using interpolation among benchmarks provided by radiometric dates and/or event horizons (e.g., volcanic ash layers, magnetic field reversals, and pollen markers).

Many Quaternary records are from discontinuous sediments, which range from communities that were buried alive by flooding, glaciation, or debris flows, to sediments accumulated locally over a few decades or centuries (e.g., buried peats, fluvial deposits, and rodent middens; see chapter by Betancourt, this volume). Independent dates (e.g., from radiocarbon) are required to estimate ages of these types of assemblages. In certain cases where temporal density of these "snapshot" records is sufficiently high, individually dated snapshots can be stacked chronologically to provide a time-series analogous to a record from continuous sediments (e.g., Baker et al. 1996).

Quaternary paleoclimates can be inferred from a wide variety of sources, including biotic-assemblage composition, organic tissue growth rates, geochemistry, geomorphic features, and various sedimentary indices of net primary production, hydrology, and water chemistry. Many of these paleoclimatic data sources are well reviewed by Bradley (1999), although additional types of paleoclimatic data continue to be developed and applied. This is a rapidly advancing field, and emerging data sources can be identified by scanning recent issues of *Quaternary Research, Quaternary Science Reviews, The Holocene, Science*, and *Nature*.

Radiocarbon dating, based on β-decay of ^{14}C isotopes, has been the stock-in-trade for dating of late Quaternary sediments since the 1950s. Its use requires organic material, however, and is restricted to the past 40–50 thousand years. In addition, because ^{14}C content of the atmosphere has varied during the Quaternary, ages estimated from ^{14}C dates are not the same as calendar year ages. Despite these limitations, radiocarbon dating is still the most widely used form of Quaternary dating, and the relationship between ^{14}C ages and calendar-year ages is well calibrated for the past 20 ka (Stuiver and Reimer 1993). Development of Accelerator Mass Spectrometry (AMS) dating, which requires minute quantities (< 2 mg) of organic carbon, has greatly advanced dating quality and precision, allowing, for example, direct dating of individual macrofossils to assess contemporaneity or to circumvent problems arising from contamination or bias of other sediment fractions. Alternative methods of dating Quaternary sediments include both radiometric (K/Ar, U/Th, ^{210}Pb) and non-radiometric approaches (e.g., optically stimulated luminescence, amino acid racemization, event stratigraphy) (Noller et al. 2000). These methods serve well in cases where ^{14}C is unsuited for dating and as complements to ^{14}C.

Pollen analysis is the primary source of information about history of Quaternary vegetation and plant geography in most parts of the world. This method has a well articulated and empirically validated theoretical foundation (Prentice 1988; Jackson 1994; Sugita 1994; Jackson and Lyford 1999; Davis 2000). Pollen assemblages, generally expressed as percentages of a designated pollen sum (although pollen accumulation rates are also used), represent the composition and pattern of vegetation within a 20–30 km radius of the depositional site. Pollen source area depends on pollen dispersal properties (small grains travel farther in the atmosphere than larger grains) and basin surface area (small basins have a higher proportion of pollen grains deriving from nearby vegetation than larger basins). The coarse spatial resolution of most pollen assemblages can be construed as an advantage in that pollen data provide an integrated record of vegetation across the surrounding landscape; so local biases arising from fine-scale patchiness in vegetation are dampened. Use of small basins can amplify the vegetational signal from local patches when patchiness or local dynamics are of particular interest (e.g., Davis et al. 1998).

Pollen analysis has three critical limitations of direct relevance to biogeography. First, because pollen is transported in the atmosphere—potentially for long distances—a pollen assemblage consists of a mixture of individual grains from sources ranging from the basin margin to plants growing over 100 km away. Because we cannot discriminate local from distant sources for individual grains, it is difficult to assess whether trace amounts of pollen of a particular type represent small, scattered local populations of a species, or larger, distant populations. This hampers application of pollen analysis to studies of natural invasions and extirpations. Second, pollen assemblages are overwhelmingly biased towards wind-pollinated plants, and hence provide no records (or at best, poor records) of insect-pollinated taxa (e.g., *Liriodendron, Prunus*) and taxa for which pollen preserves poorly in sediments (e.g., *Populus*). Finally, taxonomic differentiation based on pollen morphology is often limited to the genus level (e.g., *Quercus, Picea, Pinus, Betula*) or even family level (e.g., Poaceae, Cyperaceae, Cupressaceae). Thus, biogeographic dynamics of individual species are often masked in the pollen record. Nevertheless, pollen data have made important contributions to our understanding of changes in species distribution and abundance (Davis 1976, 1981; Webb 1988; Huntley and Webb 1989; Williams et al. 2003).

Plant macrofossils provide complementary information about past vegetation composition and species distributions. Macrofossils have the advantage of being identifiable to species level in many cases, and in most depositional environments they derive unambiguously from local sources. Macrofossil data have their own unique biases, particularly in favor of species that produce copious numbers of small, readily dispersed and easily preserved plant organs (e.g., conifer needles, small seeds) and that tend to grow near depositional environments (e.g., along lake shores or in wetlands). The spatial precision of macrofossil assemblages also constrains inferences concerning regional vegetation in the absence of pollen data. Macrofossils have been analyzed from hundreds of lakes, peatlands, buried soils, and other settings worldwide, and mapped macrofossil distributions at individual time slices have been used to infer past species ranges and migration patterns (Jackson et al. 1997, 2000).

Plant macrofossils are abundant in fossil middens produced by woodrats (*Neotoma*) and other nest-forming rodents (see Betancourt, this volume). These middens are well preserved in bedrock uplands in arid and semi-arid regions, and have been critical to understanding the biogeographic and climatic history of these regions.

Quaternary sediments also frequently contain the remains of fossil insects, molluscs, and vertebrates, which are useful in paleoenvironmental reconstructions and as sources of biogeographic information in their own right (Elias 1994; FAUNMAP 1996; Preece 1997). The FAUNMAP project (FAUNMAP 1994) has compiled Quaternary vertebrate faunal records from North America to produce maps of past and present species distributions.

Quaternary biotic assemblages, most notably pollen assemblages, were burdened until the 1950s by serving simultaneously as sources of chronological and paleoecological information. Determination of spatial and temporal patterns of species distributions and biotic communities was hampered because the biotic assemblages were used to provide chronological control for the biogeographic questions of interest. Paleoecology underwent its first great liberation in the 1950s with the advent of radiocarbon dating, which allowed time to be measured independently of the biotic record. One of the most important results of this new dating process was the ability to examine spatial patterns of biotic change through time, including the migrations of plant species in response to climatic change (Davis 1976, 1981; Webb 1987, 1988).

Quaternary fossils (particularly pollen assemblages) remain encumbered, however, because they constitute a primary source of paleoclimate inference (e.g., Bartlein et al. 1984; Webb et al. 1993, 1998; Gajewski et al. 2000). This has limited their utility in understanding biotic responses to climatic changes of the past because of the circularity inherent in simultaneous use as paleoenvironmental indicators and response variables. Paleoecologists have applied clever strategies to circumvent this problem (e.g., Webb et al. 1987; Prentice et al. 1991), and there is general consensus in the field that climatic change is the primary driver of large-scale changes in community composition and species ranges at millennial timescales (Webb 1986). But the circularity problem has continued to lurk in the background and is particularly troublesome at finer temporal and spatial scales. Happily, paleoecology is now undergoing its second great liberation with the development and application of numerous other sources of paleoclimatic inference. One of the most exciting research challenges in Quaternary biogeography is the use of independent records of climate change in tandem with paleoecological records to develop a rich and detailed understanding of how species ranges, populations, communities, and ecosystems have responded to environmental forcings of varying magnitude, rate, and duration.

In the following sections, I discuss some significant advances and insights that have come from Quaternary paleoecology and environmental history, and indicate what I perceive as some areas ripe for major advances. My selection is admittedly biased, reflecting my own interests and perspectives as a paleoecologist who has worked primarily in temperate, montane, and semi-arid forests, woodlands, and wetlands of North America.

There Is No There in the Quaternary

Gertrude Stein, writing of her childhood hometown of Oakland, California, observed that "There is no there there" (*Everybody's Autobiography*, 1937). Her statement is an apt description of the Quaternary. The past three decades of research on Quaternary environmental history have revealed that climate varies continually at all ecologically relevant timescales, from interannual to multimillennial. Put simply, there is no modal condition for the past century, the past millennium, the Holocene, or the Quaternary.

Figure 3.1 shows a reconstruction of past precipitation variation estimated from widths of annual growth rings of Douglas-firs in northwestern New Mexico (Grissino-Mayer 1996). The upper diagram shows unsmoothed interannual variation during the past two centuries, and the lower diagram shows smoothed decadal-scale variation during the past two millennia. No

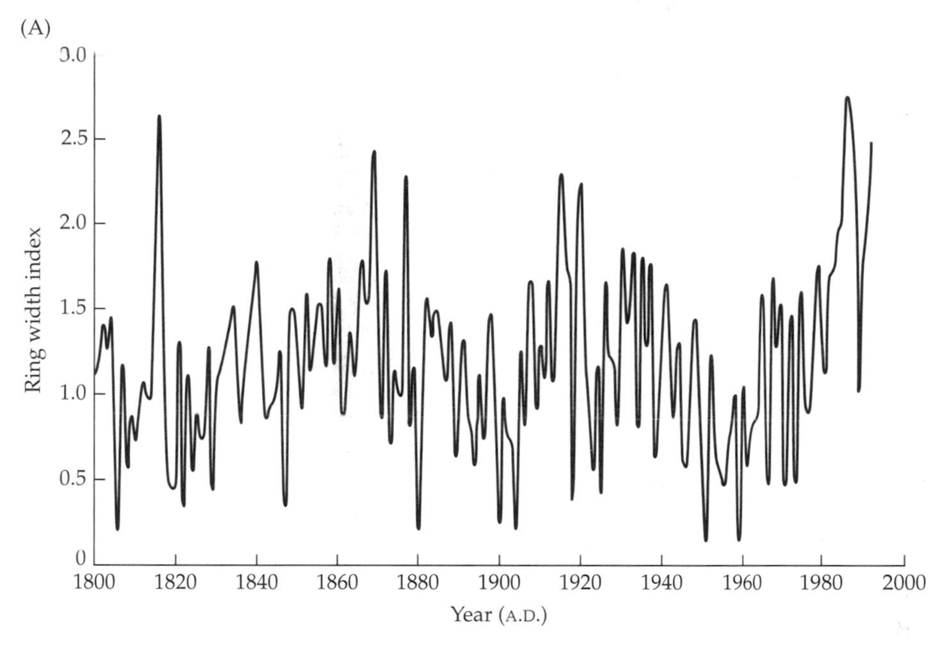

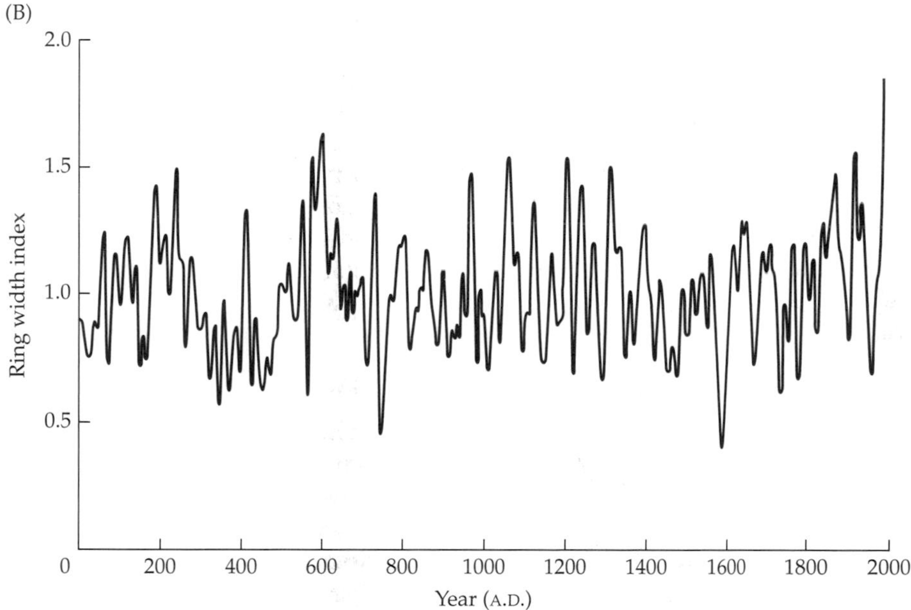

FIGURE 3.1 Composite tree-ring record of precipitation variation in northwestern New Mexico. In general, high ring-width index (RWI) corresponds to wet conditions, while low RWI relates to drought (e.g., note low RWIs during the historical drought of the 1950s and the well-documented mega-drought of the late sixteenth century). (A) Record for the nineteenth and twentieth centuries, plotted at annual resolution. (B) Precipitation record spanning the past 2 ka, smoothed using a 20-year cubic spline to show the decadal-scale variability. Note, for example, that the multi-year droughts of the 1900s and 1950s seen in (A) are also visible in the smoothed record of (B). (After Grissino-Mayer 1996.)

two individual years are exactly alike in any given decade, no two decades are alike in any given century, and no two centuries are alike in any given millennium. The concept of a modal or typical condition is elusive. Climate variability at annual to centennial timescales cannot be characterized as random fluctuation about a mean, but consist of continually evolving patterns, driven by complex interactions of atmospheric and oceanic circulation patterns occasionally punctuated by volcanic events and solar variations (Markgraf 2001). The patterns in Figure 3.1 are characteristic of high-resolution records of temperature and precipitation variation worldwide (e.g., Cook et al. 1999; Jones et al. 2001; Gray et al. 2003).

Substantial changes in temperature and precipitation at lower frequencies (millennial and beyond) are well documented for the Holocene (the past 10 ka) on all continents from a wide array of evidence (Wright et al. 1993; Markgraf 2001; Thompson et al. 2002). These changes, driven in part by Milankovitch variation in Earth's orbital parameters and (in the early Holocene) by position and size of remnant continental ice, had dramatic biogeographic consequences (Webb 1987, 1988; Huntley and Webb 1989; Overpeck et al. 1992; Jackson et al. 1997; Williams et al. 2003).

Extending our view to encompass the past 70 ka, new patterns of climate change emerge (Figure 3.2). The Earth was in the grip of the last glacial maximum 21.5 ka BP, when continental ice sheets were at their maximum extent and global sea levels were some 120 m below their current position. Paleoclimate records from Greenland ice cores and other Northern Hemisphere sources indicate that particularly rapid warming occurred 17–14 ka BP, followed by a dramatic plunge to cooler temperatures 12.8 ka BP. This cooling, the Younger Dryas interval, persisted for 1.2 ka, and ended with a rapid warming to the current interglacial (Alley and Clark 1999). The global extent of these events, particularly in the Southern Hemisphere, remains unclear. The rapid events of the Younger Dryas were not unprecedented; there were many such events in the 50 ka preceding the last glacial maximum (see Figure 3.2). Biogeographic changes associated with these rapid events are poorly understood, although they must have been large.

The last glacial maximum was only the most recent of five glacial maxima during the past 450 ka (Figure 3.3). Glacial maxima and interglacials have each typically lasted 10–15 ka, and the Earth has spent most (up to 80%) of the past half-million years in various states intermediate between glacial and interglacial (Petit et al. 1999). These intermediate states were not constant, but were characterized by substantial climate variation at millennial to multimillennial scales. Glacial/interglacial variation has occurred for at least the past 2 ma, with generally increasing duration and severity of glacial periods.

Regardless of the time span and timescale we adopt, there is no standard Quaternary environmental state. Glacial and interglacial maxima comprise modes that repeat in time, but the environment of each glacial period and interglacial period is unique, and these modes each represent a fraction (~10%) of late Quaternary time. Within the current interglacial, the climate

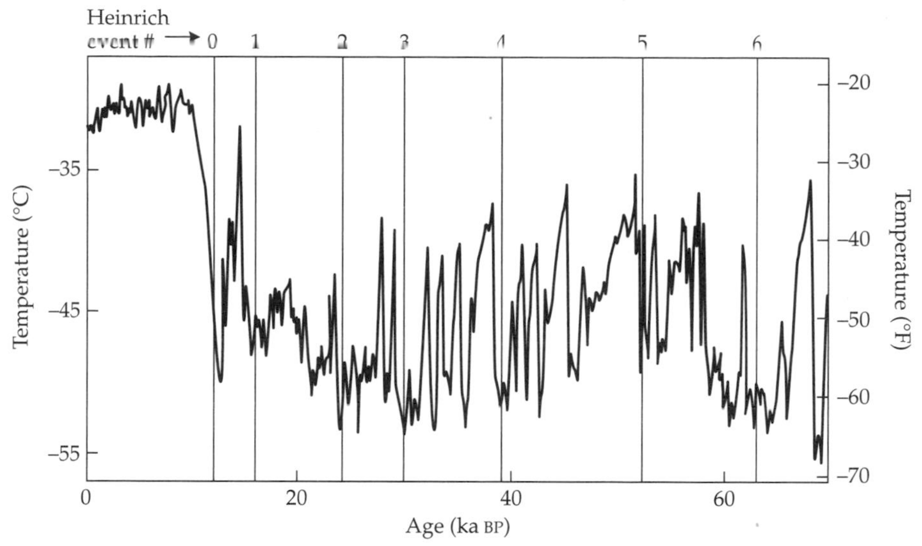

FIGURE 3.2 Temperature history of central Greenland for the past 70 ka, based on dO^{18} ice-core measurements. Note the abrupt changes before 10 ka BP. The Heinrich events represent massive discharges of icebergs into the North Atlantic, as recorded by layers of ice-rafted sediment; they are described more fully by Bond et al. (1992). Heinrich events and other abrupt changes had widespread climatic and ecological effects. Heinrich event #0 corresponds to the Younger Dryas interval, for which ecological and biogeographic effects are well documented (e.g., Shuman et al. 2002b). (Modified from Figure 12.3 of Alley 2000; original data from Cuffey and Clow 1997.)

has changed significantly at multimillennial and millennial timescales. The past millennium has witnessed major centennial and decadal variations in temperature and moisture regimes. A critical challenge for paleoclimatologists is to document the nature and underlying mechanisms of these climate variations across this range of timescales. A critical challenge for biogeographers is to document how these changes influenced geographic distributions of species and their genetic structure across this range of timescales.

The Environment: A Multidimensional Shape-Shifter

The environment experienced by individual organisms is multivariate. Certain pairs or groups of environmental variables may covary in space. For example, summer and winter temperatures decrease as latitude increases. Temperature decreases while precipitation increases along most elevational gradients. However, nature of the covariation—the slopes, *y*-intercepts, and variance of the relationships—can change through time. Changes in Earth's orbital parameters lead to changes in the seasonal and latitudinal distribution of insolation reaching Earth's atmosphere (see Figure 3.3), which in turn

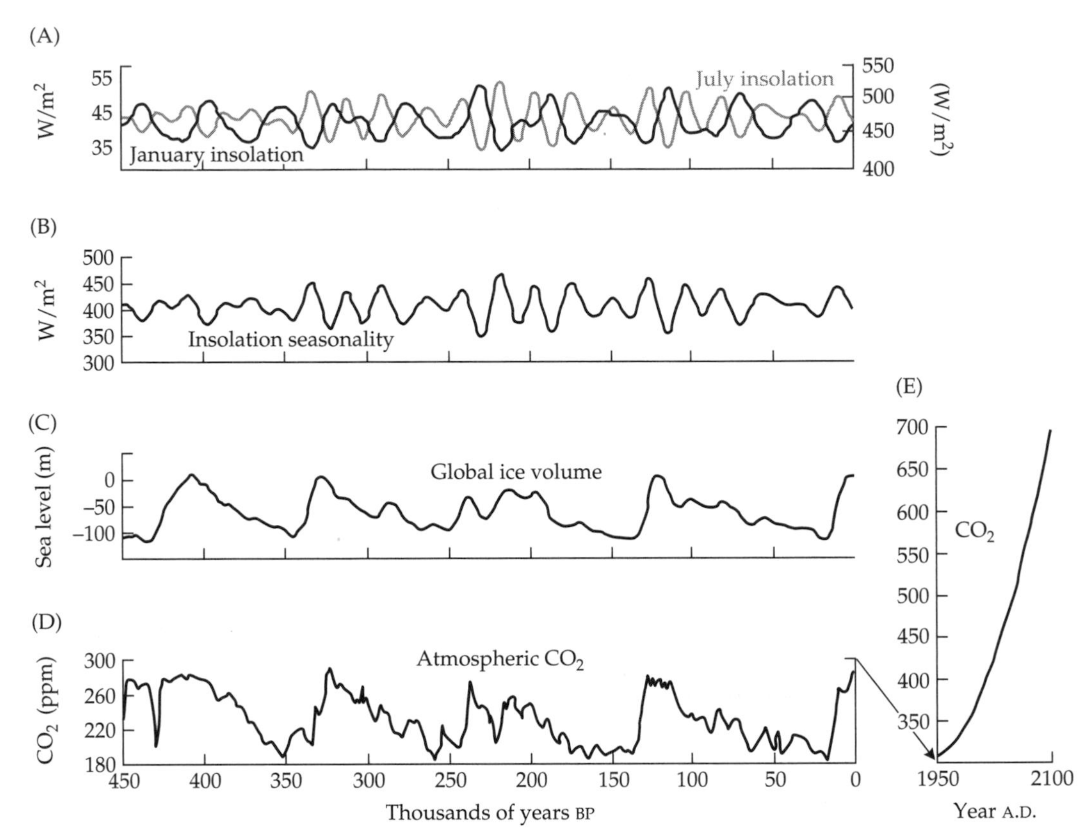

FIGURE 3.3 Time series (A–D) of some of the major climatic forcing variables for the past 450 ka (plotted at 1000-year intervals to 1950), and (E) for the period 1850 to 2100 (observed and projected at 25-year intervals). January and July insolation (A) are calculated for 60°N from Milankovitch forcing; global ice volume (C) is expressed in eustatic sea-level change based on marine dO^{18} records; and past CO_2 concentrations (D) are from ice-core data. Note the antiphased variation of January and July insolation (A), the repeated alternation of glacial and interglacial periods with extended periods of intermediate conditions (C), the pattern of relatively high CO_2 during interglacial periods and low CO_2 during glacial periods (D), and the unprecedented high CO_2 concentrations recorded during the past few decades and projected for the future (E). (See Jackson and Overpeck 2000 for details and sources.)

lead to changes in the patterns of covariation between summer and winter temperatures. For example, summer temperatures were warmer than today, and winter temperatures cooler, over much of the Northern Hemisphere during the early Holocene (10 ka BP) owing to increased summer and decreased winter insolation (Bartlein et al. 1998; Kutzbach et al. 1998; CAPE 2001). Furthermore, changes in atmospheric circulation patterns, including continental penetration of moist air masses, can lead to changes in the slope of temperature-precipitation relationships in interior mountain regions. Thus, the realized environmental space—the multivariate space representing the combinations of environmental variables that exist at a particular time—can change in shape and position (Jackson 2000; Jackson and Overpeck 2000).

Changes in the realized environmental space are well documented for terrestrial environments, particularly for biologically important climate variables (Jackson and Overpeck 2000; Williams et al. 2001; Jackson and Williams 2004). This environmental plasticity has important biogeographic consequences (discussed below). The extent to which different environments and habitats are characterized by environmental plasticity at various timescales is not well known, but it is important to understanding species and community responses to environmental change (Jackson 2000).

Habitat Tracking: Predictable Pathways and Some Surprises

Biotic consequences of environmental change can be predicted using ecological niche theory (Jackson and Overpeck 2000; Ackerly 2003). Depending upon the nature and magnitude of the environmental change relative to the fundamental niche, phenotypic plasticity, and genetic structure of a species population, the population can respond by simply staying put (toleration and/or evolutionary accommodation), or by moving somewhere more favorable (migration or habitat tracking). Toleration works well when the environmental change is small relative to the fundamental niche, provided the population has sufficient phenotypic plasticity and/or genetic variability to accommodate the changed environment. Evolutionary adjustments via natural selection may be part of the toleration response, depending on the degree of genetic variability for selection to act upon and the rate of environmental change (Davis and Shaw 2001). Environmental changes of the late Quaternary have frequently been sufficiently large and/or rapid to outstrip the capacity of many, perhaps most, species populations to remain in place via natural selection (Good 1931; Webb 1987; Huntley and Webb 1989; Huntley 1991, 1999). Unfavorable environmental changes at particular locations have led to extirpation of local populations, while conversion of unfavorable to favorable environments at other sites has led to colonization and establishment of new populations. These local processes scale up to shifts in species distributions that range from local habitat gradients (elevation, substrate) to continental-scale latitudinal and longitudinal gradients spanning hundreds of kilometers. Changes in population density and spatial dispersion may accompany all of these responses, from local toleration to continental migration.

All of these kinds of responses are observed in the late Quaternary fossil record (Jackson and Overpeck 2000). For example, populations of *Juglans nigra* and *Fagus grandifolia* have existed along the Louisiana–Mississippi border since the last glacial maximum (Givens and Givens 1987; Jackson and Givens 1994), though these populations have probably undergone changes in abundance and genetic composition during that time. Shifts of species along

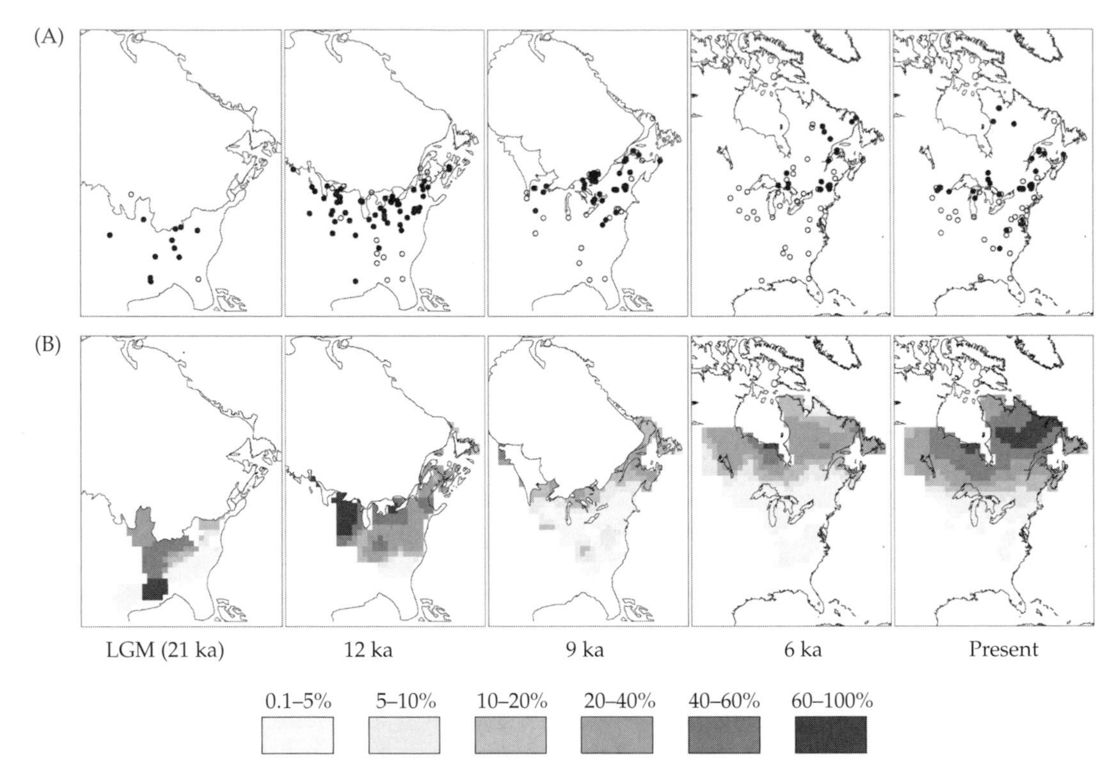

FIGURE 3.4 Changes in the geographic distribution and abundance of spruce (*Picea*) in eastern North America since the last glacial maximum (21.5 ka BP), as shown by maps of macrofossil presence/absence (A) and pollen percentages (B). Closed circles in (A) indicate sites where *Picea* macrofossils occurred during the specified time interval; open circles denote absence of *Picea* macrofossils. (Modified from Jackson et al. 1997, 2000.)

elevational and other environmental gradients within regions are also well documented; several tree species of northeastern North America have shifted elevational ranges by as much as 300 m in the past 6 ka (Jackson and Whitehead 1991; Spear et al. 1994). And continental-scale migrations of many tree and other species during the past 20 ka have occurred in North America and Europe (Figure 3.4) (Webb 1988; Thompson 1988; Huntley and Webb 1989; Betancourt et al. 1990; Jackson et al. 1997; Williams et al. 2003). Although most of the continental-scale migrations have occurred in a predictable fashion (i.e., most species have expanded northward since the last glacial maximum), a few surprises have been documented. For example, the geographic range of *Pinus remota* in west Texas and adjacent Mexico has contracted 300 km *southward* since the last glacial period (Lanner and Van Devender 1998). This and similar examples emphasize the subtlety of species responses to a shape-shifting, multivariate environment.

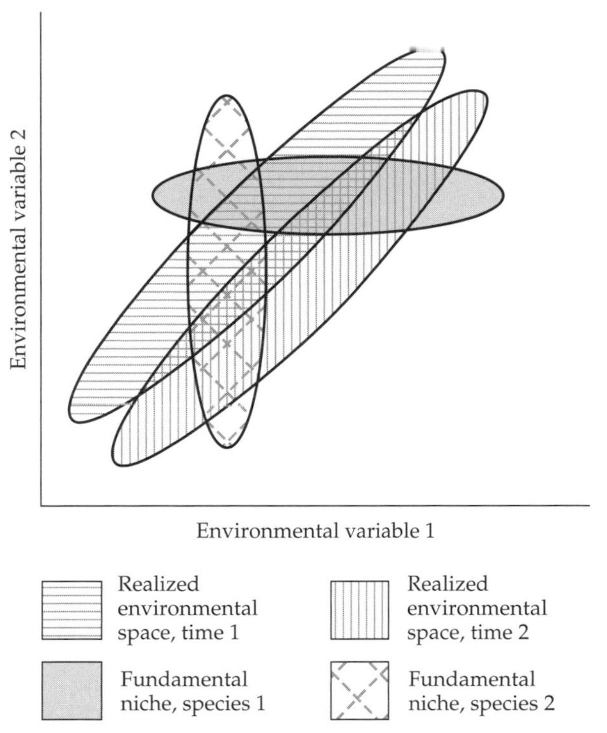

FIGURE 3.5 Conceptual diagram of the consequences of multivariate environmental change for species co-occurrence. At time 1, the intersection of the fundamental niches of species 1 and 2 overlaps with the realized environmental space. Thus, sites exist at time 1 where species 1 and 2 can co-occur. At time 2, the intersection of the fundamental niches of the two species does not overlap with any part of the realized environmental space. Thus, at time 2, the geographic ranges of species 1 and 2 will be disjunct. (From Jackson and Overpeck 2000, where fuller details and consequences of this conceptual model are discussed.)

Quaternary Terrestrial Communities: Ecological Disassembly Rules

Shape-shifting environments have important consequences for community composition (Figure 3.5). Species that differ in their environmental tolerances to multiple variables may occupy the same habitats at some times (when the area of overlapping tolerances between the species overlaps with the realized environment), and not others (when the realized environment does not overlap with the overlap in fundamental niches; see Figure 3.5). This can lead to ephemeral communities, where particular species assemblages emerge under particular environmental realizations, and then disassemble as those environmental realizations disappear (Jackson and Overpeck 2000). Just as most species are involved in a variety of species associations across their geographic ranges, they may also participate in a variety of associations through time, many of which may not exist today.

The ephemeral nature of particular plant communities in the Quaternary has long been noted by paleoecologists (West 1964; Davis 1976; Whitehead 1981; Webb 1987) and has received particular attention recently by the formal recognition of "no-analog" pollen assemblages (Huntley 1990; Overpeck et al. 1992; Williams et al. 2001; Jackson and Williams 2004). No-analog assemblages are judged, based on multivariate distance metrics, to have no counterpart among modern pollen assemblages, indicating that they represent vegetation unlike any modern vegetation in composition and/or structure (Overpeck et al. 1985, 1992). In North America, no-analog pollen assemblages were prevalent in a broad band from the northern Great Plains eastward to the eastern Great Lakes during the last glacial-to-interglacial transition (ca. 17–12 ka BP) (Williams et al. 2001). These assemblages, characterized by unique combinations of abundant *Picea* and herbaceous types, together with temperate deciduous trees, were associated with climatic realizations unlike those of today (Williams et al. 2001).

The continual assembly and disassembly of communities in response to changing environmental realizations appears to be characteristic of terrestrial plants, vertebrates, and at least some insects. It may not be characteristic of all other terrestrial communities, nor of certain kinds of habitats. Temporal turnover of species assemblages is probably contingent upon the extent of temporal environmental plasticity experienced by the species. A fuller discussion of this issue is provided in other papers (Jackson 2000; Jackson and Overpeck 2000).

The Consequences of Environmental Lumpiness

Quaternary paleoecology has a long history of attention to spatial heterogeneity and temporal variability in the environment. The ongoing explosion of information on past climatic variability at a range of timescales is leading to recognition that ecological and biogeographic patterns at relatively coarse spatial and temporal scales may be governed by dynamics occurring at finer scales (Jackson and Booth 2002; Shuman et al. 2002a; Lyford et al. 2003; Gray et al. 2003; Booth et al. 2004). Landscape texture, defined as the spatial pattern of suitable habitats for a species (With 2002), influences dispersal and gene flow and hence probably plays an important role in colonization by invading species (e.g., Lyford et al. 2003), as well as extirpation during range contraction. Successful seedling establishment and continued recruitment within a population often occurs under more restricted environmental conditions than adult survival, and hence colonization and migration patterns for species at millennial timescales may be strongly structured by climate variability at finer timescales (Jackson and Booth, in prep.). Species may fail to occupy suitable territory for decades, centuries, or even millennia simply because an insufficient number of recruitment opportunities—climatic episodes favorable for widespread establishment—have occurred.

Evolution and Extinction in the Quaternary

The profound environmental changes of the late Quaternary, and the biogeographic and ecological responses to them, inevitably led to a wide range of evolutionary changes. Alternating population expansion and contraction, isolation and coalescence, and colonization and extirpation, all influenced genetic structure and diversity. Populations of species were exposed to a continually changing array of selective regimes as physical environment, disturbance regimes, and assemblages of associated species changed. Isolation of populations created opportunities for speciation, and some species—and many populations—underwent bottlenecking or extinction as the environment changed.

Evolutionary consequences of Quaternary dynamics have been inadequately explored, and are poorly documented in the fossil record. Quaternary paleoecologists have tended to view populations and species as passive objects, pushed here and there by changes in the physical environment. This perspective seems accurate as a first approximation; migrations, habitat shifts, and population fluctuations in relation to environmental change are well documented worldwide (Bennett 1997; Jackson and Overpeck 2000). However, populations and species are active players in responding to environmental change, and adaptive changes must have accompanied past species migrations and population changes (Rehfeldt et al. 1999, 2001; Davis and Shaw 2001; Ackerly 2003). The extent to which the fundamental niches of individual species have changed in response to environmental changes of the Quaternary across a range of timescales (i.e., from 10^2–10^6 yr) is an important question deserving theoretical and empirical study (Huntley et al. 1989; Jackson and Overpeck 2000; Ackerly 2003).

Species extinctions are documented from the Quaternary for both vertebrates (Martin and Klein 1984; Stuart 1991) and for vascular plants (Godwin 1975; Watts 1988; Jordan 1997). The plant extinctions, which are not attributable to direct effects of human activities, are documented mostly from the early and middle Quaternary, although one North American species of *Picea* disappeared during the last glacial/interglacial transition (Jackson and Weng 1999). These extinctions may represent normal background extinctions, culling of Tertiary-adapted species unsuited for Quaternary environments, or bad-luck situations in which species were temporarily unable to track their habitats through dispersal. Davis (1983) and others have noted that species incapable of dispersing to and becoming established in new territory as the environment changes would not have survived the Quaternary.

It remains unclear whether such a form of species selection was responsible for Quaternary extinctions. The extent of Quaternary plant extinction may be underestimated owing to the dearth of detailed anatomical and morphological studies of plant macrofossils.

Biogeography and Paleoecology: Together Again?

The Quaternary biogeographic history of species and populations has left signatures in the form of spatial patterns of genetic variation. Genetics has a long and rich tradition of examining spatial variation in morphological, isozyme, and biochemical characters in the context of Quaternary history (e.g., Critchfield 1984). The recent development and application of molecular markers to natural populations in a biogeographic context is rejuvenating the application of modern biogeographic patterns to historical inference. Spatial patterns of genetic markers have indicated the locations of glacial-age populations and postglacial migration patterns of many plant and animal species, particularly in Europe (Hewitt 1999, 2000; Petit et al. 2002b) but also on other continents (McLachlan 2003; Rowe et al. 2004).

The advent of molecular markers in historical biogeography provides a potential basis for fusion of paleoecology and biogeography that will be beneficial to both fields. Patterns of genetic variation provide information that is absent or masked in the fossil record. With few exceptions, fossil evidence cannot provide information below the species level, is relatively ineffective at detecting small and scattered populations, and is sparse in many regions where glacial-age populations of important species were likely to have occurred (e.g., southeastern United States, southern Europe). Genetic markers can identify historical and biogeographic patterns below the species level; reveal the whereabouts of small, isolated populations in the past; and indicate where species of glaciated regions might have occurred during the last glacial maximum (Hewitt 2000). At the same time, paleoecological and paleoclimatic studies are required to validate assumptions and inferences based on genetic patterns, and such studies can indicate where specific genetic-marker studies are needed to resolve important questions. The potential for using genetic studies in tandem with paleoecological studies in a coordinated fashion (e.g., Petit et al. 2002a) is vast.

Much attention has been focused on attempts to extract DNA from ancient organic tissues and deposits. These efforts have met with mixed success; some of the most successful efforts have involved mummified material from arid regions (Poinar et al. 1998; Kuch et al. 2002) or frozen tissues from high latitudes (Leonard et al. 2000). Bulk samples of permafrost paleosols and cave sediments have recently yielded ancient DNA (Willerslev et al. 2003), and DNA has been extracted and sequenced from *Betula* seeds and *Fagus* leaves in late Holocene lake sediments (H. Poinar, S.T. Jackson, and R .K. Booth, unpublished data). Analyses of bulk sediments have the potential for detecting taxa that are poorly represented in pollen and macrofossil assemblages (e.g., Willerslev et al. 2003), although taphonomic studies in modern settings are needed to assess the nature of the information (time-averaging, spatial variation, sampling properties). Studies of molecular markers from ancient DNA can add a temporal dimension to genetic studies, and may revolutionize our understanding of the genetic consequences of dispersal, migration, and population fluctuations during the late Quaternary.

Plus Ultra! Selected Challenges for Future Research

As a beginning graduate student in ecology/paleoecology in the late 1970s, I thought a day might come in my lifetime when paleoecology would come to an end—when all the suitable sites in the world had been discovered, cored, and analyzed. There would be nothing left to do. This view seems incredibly naive in hindsight, some 25 years later. Paleoecology is more vibrant than ever, and not just because new sites are being discovered. New questions are arising as we learn more about the past, new techniques are being developed, new perspectives are emerging from other disciplines, new capabilities are issuing from bio/geoinformatics and modeling, and new demands are issuing from conservation biology and global-change sciences. I do not worry about running out of things to do in my lifetime, nor that my students will in their lifetimes.

Natural philosophers, the antecedents of modern scientists, adopted *plus ultra* ("more beyond") as a motto during the seventeenth century, implying that advancement of knowledge could continue indefinitely, without limit (Medawar 1984). Although the past decade has seen some fatuous statements about an imminent "end of science," there is certainly no end in sight for the Earth, environmental, and ecological sciences. These fields are flourishing, and the sheer complexity and interactions of the Earth and ecological systems, and the broad range of spatial and temporal scales at which they operate, ensures that scientific advances in these fields will continue indefinitely. It would be foolish to try to predict all of the questions paleoecology and biogeography will be pursuing in the coming decades. In this chapter, I have identified a few fertile areas that I think merit exploration and development. My choices are idiosyncratic, reflecting my own interests, and I know that I have omitted several areas that are ripe for advancement. I am sure I am completely blind to others that are emerging now. In closing, I draw attention to what I think is the most important and far-reaching challenge for paleoecologists, biogeographers, and other environmental scientists concerned with history: integration of patterns and processes across spatial and temporal scales.

Conceptual and empirical understanding of scaling issues, particularly in the temporal domain, remains vague. The most important gap is in linking between "real-time" ecology (daily, seasonal, annual, decadal processes) and "Q-time" paleoecology (the centennial- and millenial-scale patterns observed in Quaternary paleoecological records). Milankovitch-scale climatic change is clearly necessary to explain the dynamics of species ranges and abundance observed in the fossil record. Whether it is sufficient remains unclear. Does climate variability at intermediate timescales (decadal to millennial) play a vital role in governing biotic responses at longer timescales? How do organismal and population responses to the environment scale up to community and biome dynamics at millennial scales?

There is also a need for linking the well-documented Milankovitch-scale dynamics of the past 25 ka with records at longer timescales, extending into the Tertiary. Do the records of the past 25 ka encapsulate all of the important dynamics that might be observed in any comparable time slice within the past 500 ka? The past million? The past 10 million? What are the important dynamics that play out over longer timescales (e.g., among individual glacial/interglacial cycles, or between the early and late Quaternary, or during the Tertiary greenhouse-to-icehouse transition)? What are the relative contributions of climate and CO_2 variations to dynamics at glacial/interglacial and longer timescales?

These questions are important because the living biota of the planet are products of the ecological and evolutionary processes that have taken place at timescales ranging from 10^0–10^7 years during the Cenozoic. We need to understand those processes better in order to understand and conserve the remaining biota in a time of rapid environmental change. Climatic changes of the Quaternary at timescales of 10^3–10^5 years appear to have been more dramatic than in most of the Tertiary. Assuming for the moment that this statement is true, and given that the Quaternary is short (2 ma) relative to the rest of the Neogene (ca. 20 ma), to what extent are the living biota well adapted to ever-changing Quaternary environments? Are they the remnants of a Miocene/Pliocene biota, escaping Quaternary extinction by virtue of good fortune (blind luck, or exaptations such as ability to cope with low [CO_2] or rapid change)? How much adaptive change has occurred (e.g., selection towards broad niches, phenotypic plasticity, and dispersal ability)? Has the Quaternary been characterized by net gain or net loss of biodiversity?

We are entering a period of rapid environmental change, driven by greenhouse-gas accretion accompanied by widespread human expropriation of habitat and intercontinental exchanges of species. The Earth system has experienced rapid and dramatic climate changes in the Quaternary, as recently as the Younger Dryas interval that ended 11.7 kya. The past 10 ka have been characterized by local- to global-scale abrupt climate changes of lesser magnitude. We know very little about the biotic responses to these rapid changes, despite their clear relevance to ongoing and future environmental changes. How did extant species pass through the crucibles of rapid change in the past? What were the consequences of these rapid changes for genetic diversity and species diversity? By grappling with these questions, paleoecologists can contribute to an understanding of how we can maintain biodiversity in the face of global change.

Acknowledgments

Comments by Lawrence R. Heaney, Jack Williams, and an anonymous reviewer improved the manuscript. Steve Gray and Kathy Anderson helped with figure preparation.

CHAPTER 4

Biogeography on a Dynamic Earth

Christopher J. Humphries and Malte C. Ebach

> The number of biogeographers who confidently drew dispersal routes on fixed continent maps ten or more years ago and now just as confidently draw dispersals of the same organisms on continental maps must cause us to seriously question the procedures of biogeographers.
>
> G. F. Edmunds, *Annals of the Missouri Botanical Gardens* 62: 251 (1975)

Introduction

In discussing the current state of cladistic biogeography, this chapter describes the successes and theoretical steps that may lead to a deeper understanding of component analysis, and envisages what progress can be made in the near future with the adoption of recent modifications that are described herein as *area cladistics*. It is our belief that many technical arguments in the biogeographic literature have come about because different techniques all lay claim to the original principles enunciated by Rosa, Hennig, Brundin, and Croizat that continued as cladistic biogeography by Rosen and by Nelson and Platnick (Nelson and Platnick 1981). Brooks (this volume), for example, considers that there are two research programs, phylogenetic biogeography and cladistic biogeography, which have been classified by him and others as *a priori* and *a posteriori* techniques. We shall argue that the difference between the two revolves around whether there is a difference between cladograms and trees or whether they are the same.

The consequences of not making a distinction between cladograms and trees has many other implications, some of which are described in the justification of phylogeography by Riddle and Hafner (this volume), and also by Brooks in his method described as Brooks Parsimony Analysis (BPA) (this volume; see Wiley 1987). The old chestnuts of dispersal versus vicariance and the role of ancestors in determining centers of origin are good cases in point. Also, the postmodern versions of historical biogeography are creating a milieu of mixed messages that consider life to be separate from geophysical history. Old appositions constantly resurface in different guises and so we argue for the theory, methods, and implementation of historical biogeography from a pattern perspective that sees cladograms as different from phylogenetic trees; that ancestors and centers of origin cannot be ascertained; and that the distinction between dispersal and vicariance cannot be justified. We describe the crucial steps of theoretical and practical implementation of the methods of component analysis over the last 30 years or so.

We consider that the least understood aspect of historical biogeography is the determination of area relationships and what constitutes homology in area analysis. We believe that cladistic biogeography has its roots in the pattern approach advocated by de Candolle (1820) and that phylogenetic biogeography is a form of gradualism rooted in Darwinian evolutionary theory. We begin our chapter by asking whether the Earth should be viewed as static or dynamic, and whether it is biogeography that gives us theories about evolution, or vice versa. We believe that phylogenetic biogeography views the world as static, but cladistic biogeography (*sensu* Ebach and Humphries 2002) allows us to assess relationships of areas on a dynamic Earth.

Plate Tectonics

No one would disagree that Earth is a dynamic planet. Billions of years of volcanic activity, continental drift, erosion, sedimentation, mountain building, and ocean formation have provided Earth with an atmosphere, a climate, and a life-sustaining environment (see Scotese, this volume). Our planetary neighbors in the solar system all lack this vital life-support system. Importantly, the organic shifts caused by a geologically active Earth left patchy imprints of fossil distributions in a plethora of distribution mechanisms that created the present-day distribution patterns we now perceive. The processes for the distribution of species over time can be roughly divided into *dispersal* and *vicariance*, a dichotomy fuelling considerable debate in biogeography. Such processes may affect one species or might affect many different taxa simultaneously.

Catastrophists and Gradualists

For almost two centuries, biogeographers in the dispersalist tradition have considered life and Earth as separate entities. Dispersal from centers of origin and geophysical stasis, or at best, gradual change, was the enduring paradigm from the mid-nineteenth century until well into the twentieth century (Humphries 2004). Lyell (1830–1833), opposing biblical influences on geology, thought that catastrophe theories were wrong because they were based on interpretations of the *Book of Genesis*. Instead of rapid catastrophic changes in Earth's geometry over relatively short periods of time (anything between 6 days and 4000 years), he thought gradual changes through time accounted for fossil remains of extinct species. For him it was necessary to portray Earth's history on a long timescale. Oddly enough, catastrophe theories provided early clues to continental drift. Prior to the seventeenth century, maps of land areas on either side of the Atlantic Ocean were sufficiently clear and precise such that curious minds noted a certain parallelism in the layout of the coasts of Africa and South America (Romm 1994). Scientists of the time tried to find an explanation for the apparent "fit" of the two continents, and even four centuries ago Bacon (1620) considered that the west coast of Africa and the east coast of South America looked as if they would fit together like pieces of a jigsaw puzzle.

Early in the twentieth century, Taylor (1910) considered that the Atlantic was formed by the separation of two continents derived from a former contiguous landmass, but at a rate much slower than the early catastrophists implied. Like his predecessors, Wegener (1915) developed the idea for the whole globe, but biogeography stuck firmly to an Earth of Lyellian fixed geometry for another 50 years or more. Largely through the pioneering work of Holmes (1931) and Hess (1962), a whole range of geological observations finally coalesced to provide unequivocal evidence for continental drift. Today, there is still considerable discussion about some aspects of tectonics. For example, support for an expanding Earth to account especially for biogeographic patterns around the Pacific (see Nelson and Rosen 1981; Nur and Ben Avraham 1981; Shields 1979, 1983, 1991, 1996) is quite acceptable to some biogeographers but not to geophysicists.

We consider that these developments are the important former steps comprising area cladistics, and that they are based on a series of discoveries that span the last century, but mostly the last 30 years or so. Rosa (1918) and Croizat (1952, 1958, 1964) noted that spatial history and evolution shaped both past and present. When fused with cladistics, the background to vicariance biogeography was created (Nelson and Platnick 1981), and this in turn led to the development of methods to understand congruence and the relationships of areas at different times in geological history (Ebach and Humphries 2002). The crucial issue is that biogeography does not depend on geology; instead, detailed analysis of distribution patterns might actually tell us more about the interrelationships of areas.

Biogeographical Mechanisms and Patterns: Dispersal, Vicariance, Allopatry and Sympatry

In terms of biogeographical theory, it appears that dispersal has its roots as an explanation in a static Earth with gradual geological change, and vicariance has its beginnings in the dynamic Earth of plate tectonics and continental drift (Ebach 2003). Dispersal is the process of one taxon forming in a place of origin (a center of origin) and moving to another. A dispersal event is younger than its underlying rocks—thus, dispersal hypotheses are about shifting organisms across existing barriers. By contrast, vicariance is the splitting of one range into two or more ranges as a result of geological changes such as continental drift. Thus the creation of the geographical barrier divides the taxon into two or more populations.

Dispersal and vicariance cause geographical isolation (allopatry), which in time allows separated populations to evolve independently. A series of allopatric events form recognizable geographical patterns that are shared with one or more unrelated groups of organisms. Historical biogeographers rely on these common patterns to uncover biotic history. Without recognizable common patterns, historical biogeographers are faced with unique patterns of distribution that can neither be corroborated nor justified as being part of a general phenomenon. Repeating patterns in historical biogeography are the nearest thing biogeography has to repeatable experiments, but nevertheless they still represent retrodictive reconstructions. Uncovering shared common patterns makes historical biogeography an *empirical* science, albeit one open to questions of how one might go about applying it. On the other hand, explaining one-off (isolated) patterns is not empirical. Isolated patterns are possibly created by internal isolating mechanisms (behavioral, physiological, etc.), most of which are *sympatric*. Unlike allopatry, sympatry does not form shared common patterns but affects only members of a particular group. In other words, sympatry is invisible to, and not part of, historical biogeography. Vicariance is the visible, physical geographical isolation of gene flow between individuals of a formerly connected taxon. Even in methods that see the Earth as static (e.g., island biogeography), dispersal will still appear as vicariance, whatever caused the pattern.

In agreement with Nelson and Platnick (1980), vicariance is a causal explanation for biogeographical patterns, but it also functions as a general statement of taxic distribution. Dispersal, however, is an allopatric mechanism that may explain vicariance. The pattern is all that remains to be uncovered through rigorous analysis, whereas dispersal plays a part in explaining hypotheses of the various distributional scenarios. Cladistics can uncover *patterns,* although explanation of these patterns is a separate step and can involve various different scenarios. Taken from this perspective, the method of area cladistics uses component analysis to determine congruent patterns, remove geographic paralogy (i.e., geographical repeats and redundant information), and use classifications of relationships rather than phylogenies (i.e., cladograms rather than trees) to uncover patterns.

Area Cladistics: 30 Years in the Making

Cladistics (Hennig 1950, 1965, 1966) originally developed as a method for elucidating patterns of evolution by common descent, but it has since burgeoned into a plethora of different methods that space forbids us to explore here (see Humphries 2004). Since the late 1970s, different underlying philosophies have descended on cladistics and a wide range of implementations—especially the use of different computer programs for computing trees—has emerged (see, for example, Kitching et al. 1998; Schuh 2000). The subsequent critical developments in cladistics, such as component analysis, area analyses, and different means for solving incongruence, concomitantly affected the thinking of cladistic biogeography, and hence, area cladistics.

Hennig (1966) showed only marginal interest in biogeography and was concerned mostly with the way species evolved in space. His main concern was to locate the pattern of character change in space for particular groups of flies (Hennig 1966). That he was interested in determining the centers of origin of each particular genus meant that he accepted the prevailing view of the time that ancestors were considered paramount for determining centers of origin. The prevailing paradigm at the time of Hennig (1966) and Brundin (1966) was that phylogeny was unknowable without fossils (Patterson 1981a,b). Indeed, as Forey (1981) noted, centers of origin were implicitly tied to the notion of migration or dispersal from one place to another. Thus, just as characters could be optimized onto a cladogram, areas could also be optimized onto the internal nodes so as to find centers of origin at the root and migration routes along the branches (Figure 4.1).

It is important to note, however, that area cladograms can be viewed with a purely additive approach that implies vicariance rather than dispersal. In fact, Croizat's work (1952, 1958, 1964) on panbiogeography was critical in

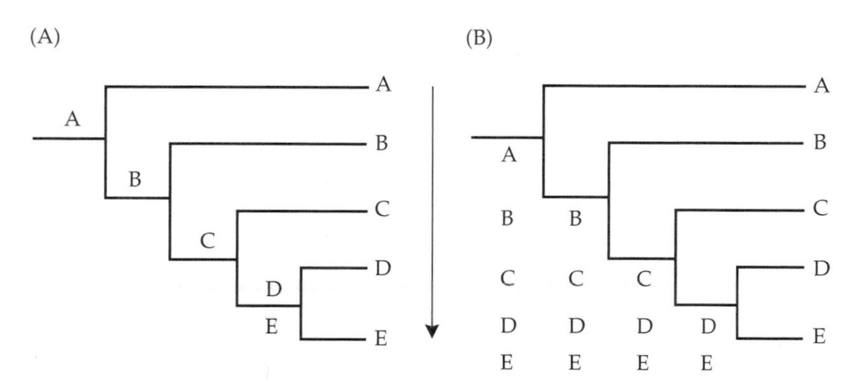

FIGURE 4.1 The cladogram on the left shows a dispersal from area A to areas D or E by optimizing the successive terminal nodes to the internal nodes of the cladogram. This is the technique used to determine centers of origin from the basal area. The cladogram on the right shows the cladistic (additive) interpretation of vicariance: an original area ABCDE that successively vicariates into five separate areas.

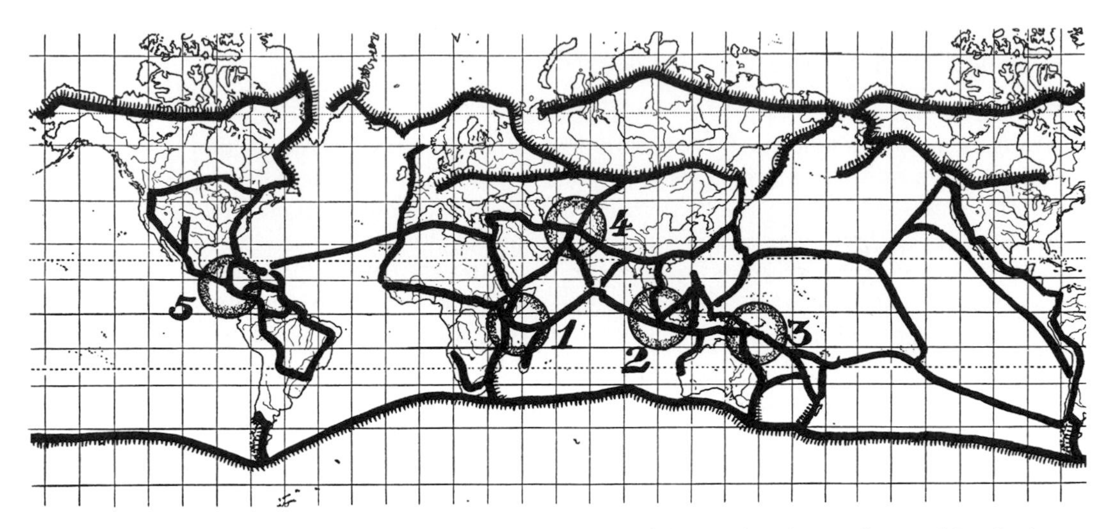

FIGURE 4.2 Paths of spatial distribution of the modern biota, depicted by Croizat as generalized "tracks." Hatched lines depict the "boreal" (Nearctic and Palearctic) and "austral" (Antarctic and Oceanic) tracks. The circles numbered 1–5 are main "nodes" on important intercontinental tracks. Tracks show connections between areas, but are non-directional: there are neither points of origin nor centers of origin. (From Croizat 1958, v. 2b, p. 1018, Figure 259.)

showing that it is impossible to determine migration routes from the spatial distribution of organisms (Figure 4.2; Craw and Page 1988).

Cladograms and Trees

Hennig's (1966) phylogenetic systematics had a dual aspect that became resolved in the late 1970s. Platnick (1979), Wiley (1981), and Patterson (1982) all described in various ways how Hennig's method contained circular reasoning. Hypotheses about characters (synapomorphy) and hypotheses about groups (monophyly) both appealed to ancestors for their justification, and were thus suspect to appealing to the mysterious and unknowable rather than being viewed as an empirical pursuit (Figure 4.3). "Transformed cladistics" (Platnick 1979) reformulated the equation to show that hypotheses about characters (homology) gave rise to hypotheses about groups (hierarchy)—akin to Hennig's *Semaphoronts* (see Figure 4.3; see Rieppel 2003). Thus, interpretations about ancestry could only be inferred from the cladogram using some criteria other than taxa and characters to interpret the cladogram as a phylogenetic tree. Consequently, internal nodes were only hypothetical constructs and could not be construed as real ancestors, morphotypes, or archetypes.

Such a discovery—the difference between cladograms and trees—is the main reason why there is a profound dichotomy between cladistic biogeography (Rosen 1978; Nelson and Platnick 1981; Nelson and Ladiges 1991a,b,c)

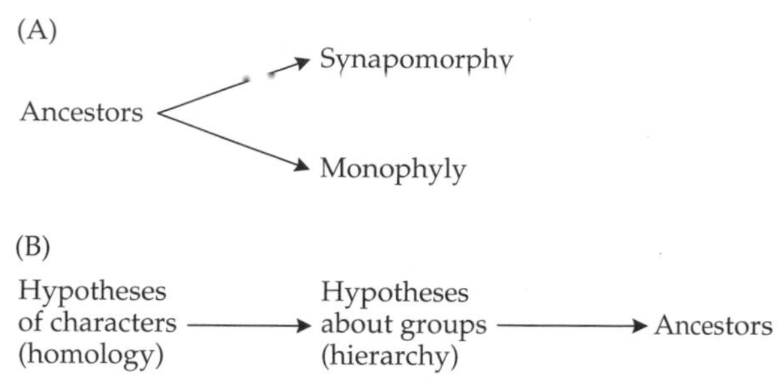

FIGURE 4.3 (A) Hennig's method appealed to ancestors to justify hypotheses about characters, as well as hypotheses about groups. (B) Transformed cladistics means that, to be empirical, hypotheses about characters gives us hypotheses about groups, thus relying on homology rather than common ancestors to determine hierarchy. Ancestors are unknowable and would have to be inferred from cladograms using auxiliary evidence not normally available to comparative biology.

and phylogenetic biogeography (Brundin 1966; Ebach et al. 2003a; Hennig 1966; van Veller et al. 2003 and in press). We believe the refusal by some to realize that ancestry must be separated from analysis in order for evolution to be studied has spawned the many different techniques described by Crisci et al. (2003). Although the main competing paradigms in historical biogeography during the 1970s and 1980s were cladistic biogeography and panbiogeography (Craw 1983), the departures from component analysis (e.g., Brooks Parsimony Analysis) are largely implementations that concentrate on evolutionary aspects. What is common to both panbiogeography and cladistic biogeography is the notion that life and Earth have evolved together: biology and the distribution of organisms are inextricably linked, and congruence amongst different generalized tracks (Croizat 1952), or area cladograms, is evidence of shared history (e.g., see Heads 1989, 1990).

Congruence

The first manifestation of vicariance biogeography came when Rosen (1979) wrote a pioneering series of papers between 1974 and 1978 on the fish genera, *Heterandria* and *Xiphophorus*. These papers, together with the earlier work of Nelson (1969, 1974, 1978, 1982, 1983, 1984, 1985, 1986), Nelson and Platnick (1980, 1981), Nelson and Rosen (1981) and Croizat et al. (1974) generated immense interest in the application and a standard data set that encouraged new implementations of vicariance biogeography. The aim of vicariance biogeography was to provide a method that classified areas in general area cladograms based on unrelated groups of taxa occupying the same or similar areas. To Nelson and Platnick, area relationships were based on taxic relationships, and the technique greatly clarified (for us at least) Croizat's dictum that life and Earth evolved together, especially as indicated by seeing repetitive patterns amongst disjunct (vicariant) distributions of

A(B(DE)) + B(C(DE)) + A(B(CE) = A(B(C(D(E)))

FIGURE 4.4 Components and consensus trees. In this simple example, three groups displaying seemingly different patterns can be combined by component analysis into one congruent solution. The components ABDE, BDE, DE, BCDE, CDE, DE, ABCE, BCE, and CE all combine to give one Nelson consensus tree, A(B(C(D(E))).

two or more groups of taxa. Thus, congruence amongst the patterns implied that there was a direct link between Earth history and the evolution of taxa. Incongruence was considered to be merely noise. Component analysis tried to maximize congruence, and any remaining incongruence was considered to be noise caused by stochastic events.

In other methods such as phylogeography (Riddle and Hafner, this volume) and BPA, dispersal and migration are used to explain departures from congruence. As with homoplasy in biological systematics, it is the signal versus noise problem that causes most doubt in the efficacy of methods for determining congruence amongst groups of taxa being compared for particular sets of areas.

Nodes on cladograms are potentially informative about the distribution history of organisms. Nelson and Platnick (1980) indicated that, by making a distinction between cladograms and phylogenetic trees, it was possible to see internal nodes as components specifying area relationships, thus providing a means of deriving general area cladograms from individual area cladograms through component analysis and consensus (Figure 4.4). Rosen's (1979) application to *Heterandria* and *Xiphophorus* exploited component analysis only to find that, because of slight differences between the distribution and species relationships of the two groups, it was not possible to obtain an unequivocal congruent general area cladogram.

Central to the thinking here was the idea that taxa and areas should be treated as similar units of comparison. Information regarding interrelationships of taxa should say something about the interrelationships of areas. The problems arose because, on comparison, species of each group of fishes were represented in some areas and not in others (i.e., some areas were missing, and several species were present in more than one area of endemism [the species were widespread]). Rosen's solution to the problem was to simply extract the pattern common to both groups of taxa (Figure 4.5). Naturally, this was incomplete and so the stages between converting the taxon cladograms into a general area cladogram (common pruned trees) lost information about the interrelationships of certain areas, which in this example were areas C and D.

Nelson and Platnick (1981) realized that congruence was more than just the common components on a cladogram. Taxa and areas are not equivalent, but are separate entities that are causally related, although with complications of dispersal, extinction, and paralogy. Platnick and Nelson (1978) made

A (B (C E)) + A (B (D E)) = A (B E)

FIGURE 4.5 Rosen's "pruned" tree combines only common components into the consensus tree.

a great intuitive leap when they realized that a cladogram showing the interrelationships of areas could be derived from the same cladogram showing the interrelationships of taxa, but at the same time the two were not necessarily saying the same thing with respect to relationships of areas. Taxa and areas could be treated separately.

The work of Rosen, Nelson, and Platnick was critical in the understanding of spatial relationships, but if anything was lacking it was the ability to deal with spatial proximity (Ebach and Humphries 2002). Biogeographers consider that the space–time processes that can most modify the distribution of organisms are speciation, extinction, dispersal, and vicariance. Because these processes can affect unrelated groups of organisms quite differently from each other, there is inevitable conflict between two or more groups of organisms occurring in similar areas. To overcome this problem, the importance of component analysis was enhanced by recognizing that the interrelationships of areas, even though derived from the taxon cladograms, might be quite different from that initially expected by mere inspection of individual cladograms (Figure 4.6).

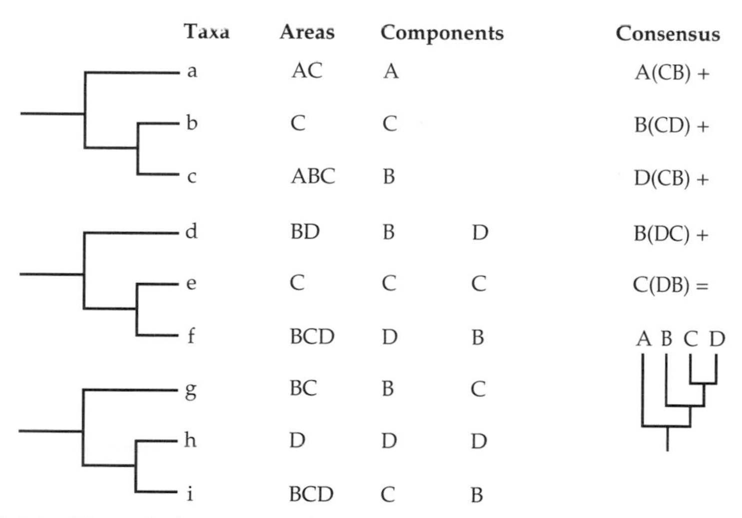

FIGURE 4.6 Three cladograms, each with one informative component: a(bc), d(ef), or g(hi). Each species occupies position 1, 2, or 3 among areas A, B, C, and D. The only informative components are shown on the right. When combined, these components suggest that, if there is a single history that explains the area patterns observed in the original, that history is A(B(CD)).

By developing the rather cryptic "assumptions 1 and 2," component analysis could resolve the conflicts caused by the different processes without invoking them *a priori*. Assumption 1 was merely a step in the development of a complete method (Humphries and Parenti 1999), but assumption 2 could deal with all causes of conflict using cladogram logic rather than invoking evolutionary interpretations—or *a priori* principles—that so bedevil methods such as BPA. The importance was that it represented a purely pattern-based cladistic approach—a brand-new protocol for systematics. Nelson (1984) argued the case for component analysis most eloquently when he offered his "intuitively opaque" example, wherein nothing but component analysis using assumption 2 and Nelson consensus produced a satisfying result (see Figure 4.6).

With the benefit of hindsight, we believe that the parting of the ways of phylogenetic biogeography and component analysis began here. Zandee and Roos (1987) developed a method called *component compatibility*. Being somewhat similar to BPA, this matrix method stated that actual distributions of organisms (assumption 0) should be taken as read, and assumptions 1 and 2, which separated areas and taxa, should not be applied. Wiley (1987) emphasized even further the dichotomy between the pattern approach of Nelson and Platnick by elaborating BPA to optimize incongruent taxa back onto area trees, saying the patterns were caused by dispersal. Treating cladograms as representative of species history meant that homoplasy was meaningful, and all kinds of dispersals, extinctions, and separate origins were employed to explain the distributions deviating from congruence. Geographical homoplasy led to a whole range of contortions and explanations for species in order to provide explanations for dispersal, extinction, and multiple origins (Page 1989a,b, 1990).

Relationships (3-Item Analysis), Paralogy, and Sub-Trees

Nelson and Platnick (1981) built upon the idea that one taxon cannot "give rise" to another. Rather than character-coding with recognition of a "transformation series" and the use of character optimization, they adopted Hennig's (1966) idea that three taxa is the minimum required to express a relationship. Thus, given the example AB(CD), there are two possible expressions of relationship, given that there is only one informative component: A(CD), B(CD). Their reasoning was that, given the idea that ancestral taxa have no place in analysis, the same must be true for ancestral characters. All cladists agree that cladistics is about grouping by synapomorphy and evidence of relationship. Three-item statement analysis codes data in the sense that "taxon" and "homology" represent the same relationship (Nelson 1994), but unlike all other methods, it focuses on the smallest possible unit of relationship—the three-item statement—and then finds the tree to accommodate the most statements (Nelson and Ladiges 1991d, 1994, 1995).

Three-item analysis in biogeography has its roots in component analysis and assumption 2, and it is about allowable and non-allowable relationships discovered by pure cladogram logic. Coupled with sub-tree analysis and the understanding that problems of paralogy in molecular biology have their equivalent in geography (Page 1993a,b), Nelson and Ladiges (1996) have come to realize that the problems of overlapping and repeated areas within and between groups and seemingly incongruent patterns, due presumably to dispersal, extinction, and speciation, can be resolved. By using informative sub-trees from a cladogram, it became possible to extract the cladistic signal from a sea of paralogous noise. Application of the method to the biogeography of *Nothofagus* (Nelson and Ladiges 2001), for example, showed that rather than the group being the key to southern hemisphere biogeography, it is in fact quite a boring taxon. As distinct from the elaborate scenario of dispersal and extinctions (Swenson et al. 2001), only two informative nodes amongst the 36 species show interrelationships among South America, Australia and New Zealand on one hand, and South America, New Guinea, and New Caledonia on the other (Figure 4.7).

Area Cladistics

As stated earlier, allopatry may be ascribed as the cause of similar geographical distribution patterns in different organisms. The interpretation in area cladistics is that shared patterns are due to the same historical events leading to geographical congruence (Ebach 1999; Ebach and Humphries 2002). The basic premise is that geographical congruence uncovers general patterns among unrelated groups of taxa. Thus, as a mode of discovery, it is quite the opposite to *ad hoc* explanations of single species or individual group distributions. The age of groups is crucial in determining whether ecological or historical biogeography is pertinent as the approach for understanding area patterns.

The way area cladistics works is simple. Endemic areas are treated like taxa that are made up of characters that share other endemic areas. These area characters are the taxic relationships of organisms that live within these areas. The endemic area, like a species, is chosen and described based on the theory of homology that all closely related taxa share a common ancestor. These hypotheses of relationship are tested using cladistics. For instance, reptiles were once considered to be monophyletic—a group of homologous taxa that shared a common ancestor. Cladistic analyses have since shown that particular groups of taxa (clades) share different sister groups. The Reptilia are presently considered to be non-monophyletic and therefore do not constitute a valid historical and evolutionary group within the Linnaean taxonomic classification scheme. Groups of areas (regions) can be classified in the same way using area cladistics.

Monophyletic cladograms found by cladistic analysis are mere representations of the proximal relationships of taxa. The cladogram A(B,C) states that B and C are more closely related to each other than they are to A. In area

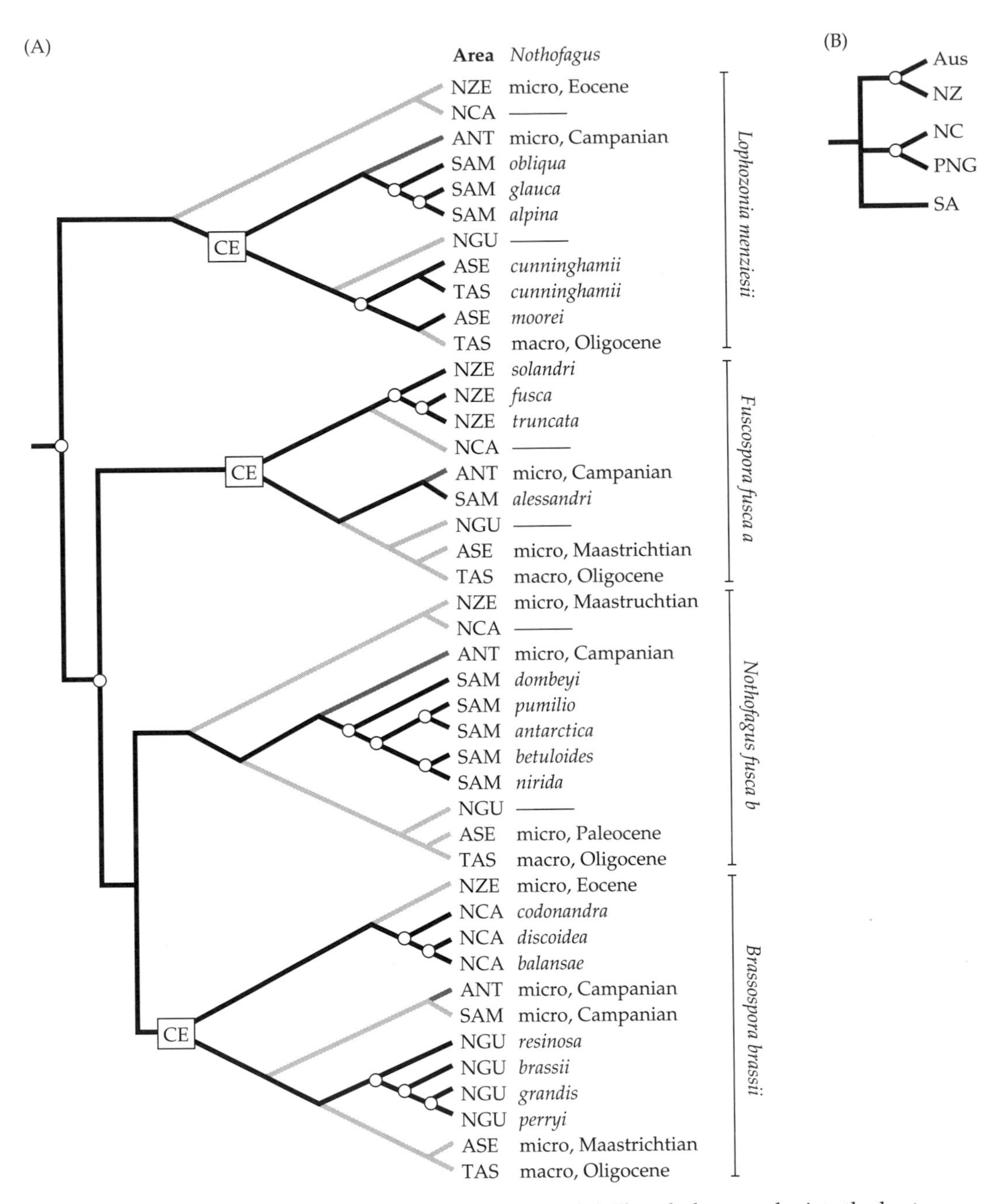

FIGURE 4.7 With and without paralogy. (A) This cladogram depicts the best reconstruction using a geological hypothesis that reflects the breakup of Gondwana in order to maximize the number of co-divergence events utilizing all that is known about modern and fossil relationships of *Nothofagus*. (After Swenson et al. 2001, Figure 5.) (B) This cladogram shows that by removing paralogy, two informative components remain: SA(AUS,NZ) and SA(NC,PNG). (After Humphries and Parenti 1999; Nelson and Ladiges 2001.)

cladistics, we replace the names of the taxa with the endemic areas in which they occur. The result is an *areagram*, a proximal representation of the areas. Note that the nodes of the areagram are not the same as the nodes on the cladogram. In any proximal representation, taxa and areas are the information. Nodes, components, or junctions show only proximal, rather than absolute relationships, which are unattainable. Proximal relationships are derived from cladograms and areagrams.

Geographical congruence reveals biotic relationships. To find geographical congruence, we look at areagrams of different organisms that occur in the same areas, preferably at the same time. By combining areagrams we may, or may not, find a common pattern. Areagrams are not patterns *per se*, but are merely *points of relationship*. A consensus of these points may produce a pattern, and that pattern is called a geographical congruence. Consider these three areagrams (set of points): A(B(B(B(B(C,D), A(B(C(C(C(C,D) and A(A(A(A(B,C). In each areagram, there is only one relationship: A(B(C,D). The repeating (paralogous) areas have no meaning other than that they state the same thing over again. Thus paralogy (Nelson and Ladiges 1996) is *not* information (Ebach and Williams 2004). The common pattern to all three points is A(B(C,D). Similarly, the three points A(B(C(D,E), A(F(G(D,E), and A(H(I(K(D,E) all contain one common pattern, A(D,E). The areas that do not overlap, or do not form patterns, are not expressing a common relationship; therefore, they are incongruent. In some cases, two or more patterns may emerge from a series of points through time. If areas are not temporally consistent (e.g., if area A consists of completely different biotas—50 kya and 100 kya), then they need to be treated as different areas. An 'area' in area cladistics is not the inorganic substrate of soil, rock, and sea, but rather the endemic biota. A case of two conflicting patterns may sometimes be resolved if the same area, over time, is treated as two different areas. This is known as *time-slicing*, an example of which is described below.

Once geographical congruence is found (the general areagram), interpretation of the biotic history may begin. Geographical congruence, as discussed above, is a result of the divergence of biota. Biotic divergence may occur in infinite different ways. Each individual instance of biotic divergence may have occurred differently—we simply cannot know. Biotic divergence, a physical event (allopatry), is herein called *vicariance*. Commonly used terms such as "vicariant event" or "dispersal" are explanatory devices to explain vicariance. A general areagram that shows geographical congruence is evidence of divergence (vicariance). In this sense, the congruent pattern A(B,C) suggests that B and C share biotic divergence. That means that faunas B and C are more closely related to each other and share far more characteristics with each other than they do with fauna A. We can interpret this to mean that faunas B and C at one time were closer together than they were to fauna A. An explanation for the actual event is impossible, simply because we have no evidence of any event. So we must be satisfied with discovering a historical association that can be translated to a palaeogeographical reconstruction (Figure 4.8). To assume anything more is merely "storytelling."

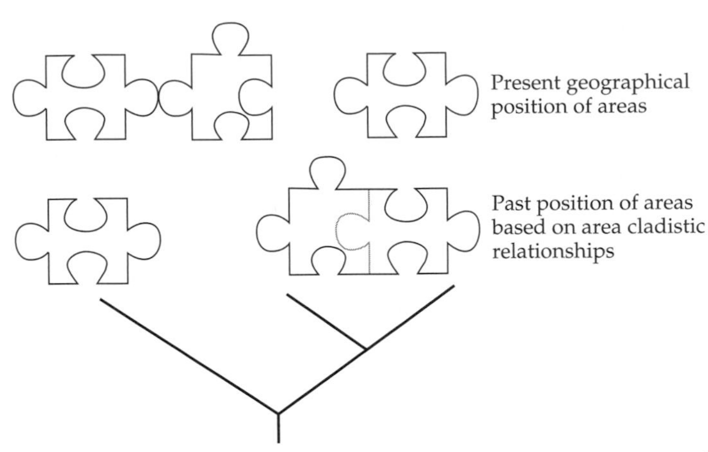

FIGURE 4.8 Area cladistics may find former geographical barriers responsible for taxonomic divergence.

An Applied Example: Area Cladistics of the Hawaiian Islands

Area cladistics combines the cladograms of monophyletic groups that occur in the same area. In the Hawaiian Island chain, for instance, we may combine the cladograms of 14 unrelated groups: Asteraceae (Funk and Wagner 1995, Figure 17.25), *Platydesma* (Funk and Wagner 1995, Figure 17.33), *Drosophila* (Funk and Wagner 1995, Figure 17.15), *Hesperomannia* (Funk and Wagner 1995, Figure 17.16B), *Kokia* (Funk and Wagner 1995, Figure 17.16C), *Neurophyllodes* (Funk and Wagner 1995, Figure 17.17), *Prognathogryllus* (Funk and Wagner 1995, Figure 17.20), *Laupa* (Funk and Wagner 1995, Figure 17.21), and *Tetragnatha* (Funk and Wagner 1995, Figure 17.23). Each area (Kaua'i, O'ahu, Moloka'i, Lana'i, East and West Maui, Kaho'olawe, and Hawai'i) is considered by the comparative biologist to be endemic, just as the taxa in each cladogram are considered to be homologous. If we combine all these cladograms and find geographical congruence, then we have found evidence for taxic divergence and area proximity (see Figure 4.8). In other words, two biotas have been geographically isolated by some means. We may use evidence of taxic divergence to test distributional theories in general biology after the analysis has taken place. The area cladistic analysis with TASS (Nelson and Ladiges 1995) uses sub-tree analysis (Nelson and Ladiges 1996) to find sub-trees using assumption 2 (Nelson and Platnick 1981). TASS combines the sub-trees in a data matrix and finds a minimal tree using NONA (Goloboff 1998). TASS is compared to the new computer software called *3item* (Ebach et al. 2003b), which implements sub-tree and three-item analysis (the *3item* "split trees" function implements assumption 2). *3item* differs from TASS in that it treats the three-item relationships of areas as "area characters" that may be selected for analysis. *3item* is the preferred method for area cladistics because it neither relies on the phenetic data matrix (area-

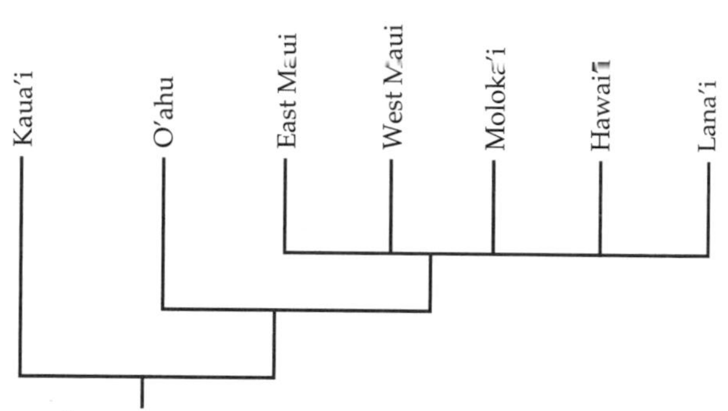

FIGURE 4.9 General areagram using TASS and NONA (trees = 9, ci 74, ri 65).

grams are manually input into the program) nor does it depend on optimization (NONA, etc.) or any other "black box" approach.

Area cladistics of the Hawaiian Islands shows us that there are non-endemic areas within the areas given above (Figure 4.9). The tree is resolved basally, but lacks any resolution for Moloka'i, Lana'i, East and West Maui (including Kaho'olawe), and Hawai'i. The analysis treats all islands as endemic except Maui, which is designated East Maui and West Maui. The distinction between East and West Maui, made by Desalle (1995), Funk and Wagner (1995), and Givnish et al. (1995) refers primarily to the current distributions of taxa. However, if the whole island of Maui acts as a potential distribution (with the ocean acting as a barrier), then confining the current distributions of taxa to both East and West Maui is unnecessary. In an area cladistic analysis treating Maui as one area, TASS finds one general areagram (Figure 4.10) in which Maui is more closely related to Lana'i than it is to Hawai'i, with all three being more closely related to each other than to Moloka'i. The result suggests that Lana'i shares a common biotic divergence with Maui rather than with Hawai'i. This result is consistent with the reconstruction provided by Carlquist (1995, Figure 2.2) showing the Maui Nui island complex giving rise to Lana'i and

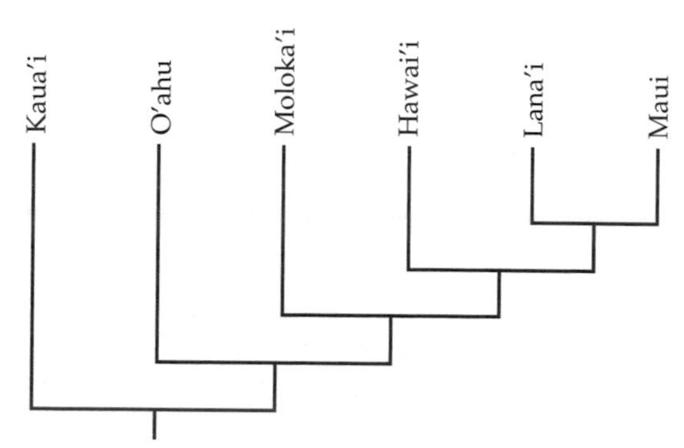

FIGURE 4.10 Single general areagram using TASS and NONA (ci 86, ri 85).

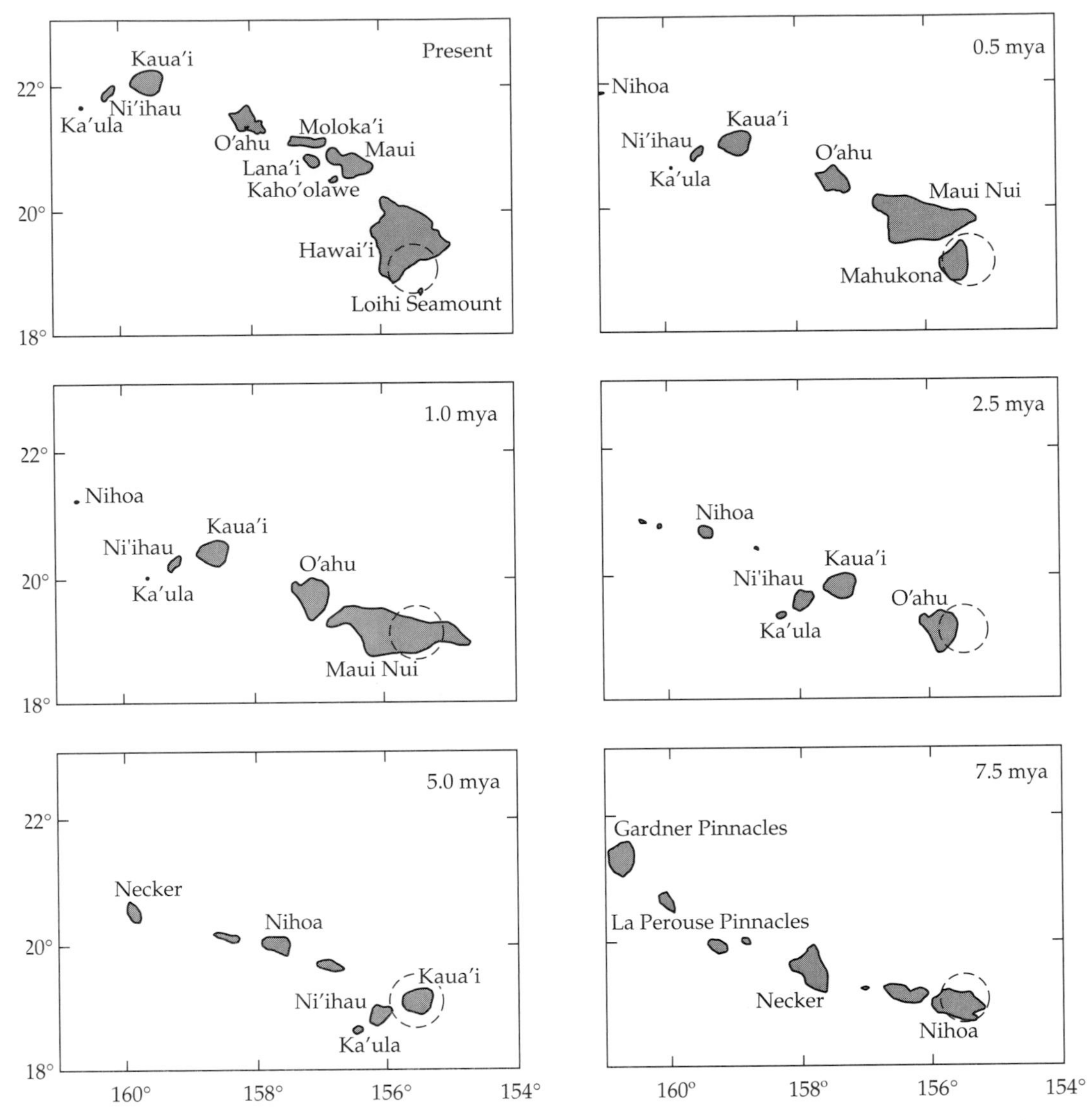

FIGURE 4.11 The positions of the Hawaiian Island chain over the last 7.5 million years. (From Carson and Clague 1995.)

Moloka'i during the time of the formation of Hawaii (Mahukona) (Figure 4.11). The taxic divergence of Maui and Lana'i suggests that the Hawai'i biota were already established on Maui Nui 1 mya and that the current (0.5 ma to present) Maui and Lana'i biota developed more recently.

Analysis using the *3item* software found 64 area characters, of which 27 are unique. The area character Kaua'i (Maui, O'ahu) was represented the most—9 times—followed by Kaua'i (Maui, Hawai'i), which was denoted 7 times. If all

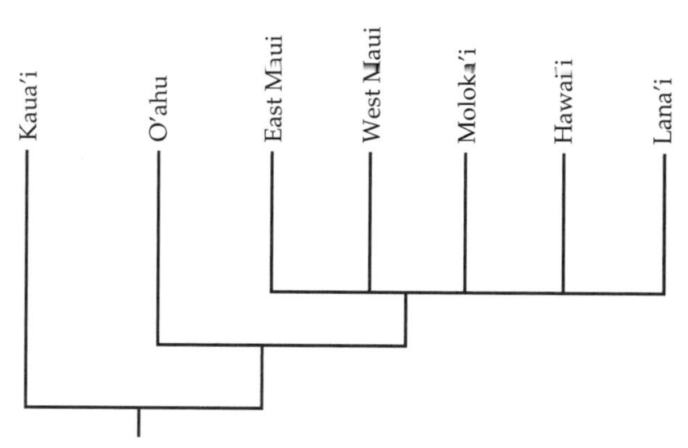

FIGURE 4.12 Analysis excluding any area characters with O'ahu as a basal area; general areagram using *3item* (score = 85%, node count = 9).

area characters are selected, *3item* found one unresolved general areagram (node count = 9) with a score (percentage of relationships represented) of 100%. In a separate analysis, 11 conflicting area characters (8 unique relationships), that include O'ahu as a basal area, were excluded. The analysis found 55 area characters and a general areagram that included 85% (55 out of 64) of relationships (Figure 4.12). The general areagram shows that Lana'i and Moloka'i are more closely related to each other than they are to O'ahu. Hawai'i, Maui, and Kaua'i form an unresolved basal node. The closer biotic relationship between Hawai'i and Kaua'i may suggest that present-day Hawai'i consists of an older, biotic community that vicariated 2.5 mya from the earlier Kaua'i biota. The biota on O'ahu may have vicariated 1.0 mya, or during the formation of Maui Nui that consisted of the future Hawai'i–Kaua'i biota.

The second analysis of the 11 area relationships with O'ahu as basal found a fully resolved general areagram (node count = 10) that includes 76% of all area relationships (Figure 4.13). The general areagram has a different

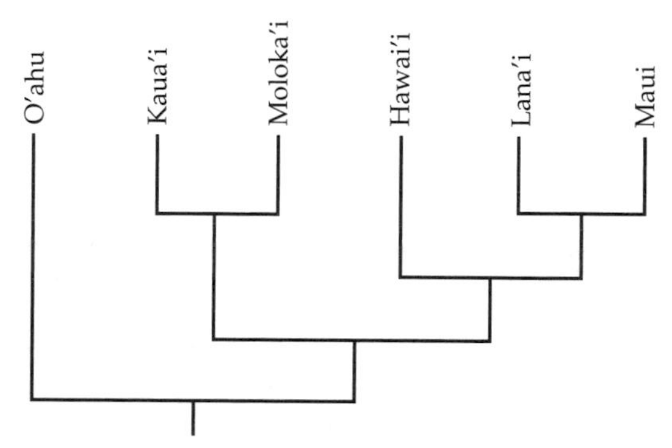

FIGURE 4.13 O'ahu analysis: general areagram using *3item* (score = 76%, node count = 10).

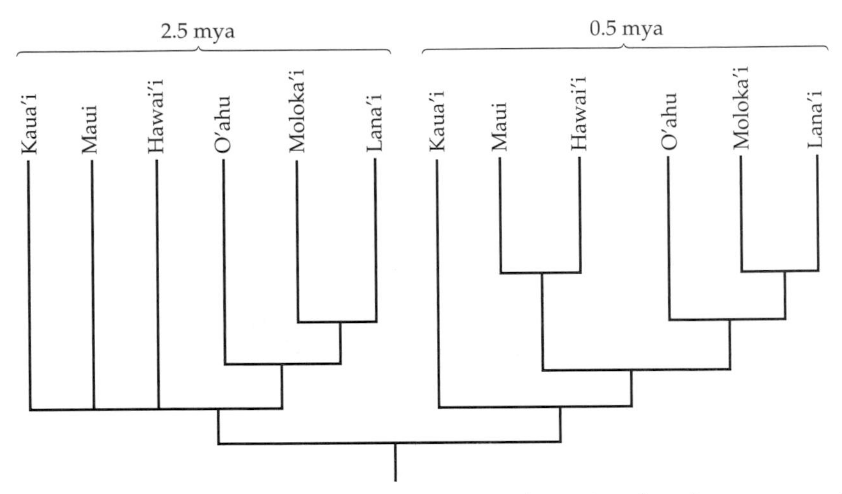

FIGURE 4.14 General areagram including a time slice. Note that there are now 12 areas.

area relationship: Maui is more closely related to Lana'i, forming a clade with Hawai'i that is sister to the Moloka'i and Kaua'i group. The difference between the two general areagrams suggests two different divergence times for the *Kokia*, *Drosophila*, and *Platydesma* biota. The second analysis suggests a more recent history (0.5 mya to present), because both clades match the present biotic proximity of the Kaua'i, O'ahu, and Moloka'i communities and the Hawai'i, Maui, and Lana'i biota. Ideally, a time slice should be implemented at the level of areas—namely, the areas in the 2.5 ma general areagram are *different* areas and should be labeled as such. In this sense, several area time slices can be shown within a single general areagram (Figure 4.14).

Area Cladistics and Time-Slicing

Time-slicing is essential in the Hawaiian Island example. The two general areagrams suggest that there were two distinctive divergences of biota at different times during the formation of the Hawaiian Island chain. Time-slicing (in cladistic biogeography), which was pioneered by Young (1984, 1995), delimits the ages of the *areas*. For instance, if one monophyletic group contains taxa that are part of a particular biota during different times, then the *areas which the taxa inhabit* need to be sliced. In the general areagram A(B(C(D(E,F,G), we find that E and F are equally related to G. Area G we discover, has different biotas that occur at different times, namely times 1 and 2. All areas occur in time 1 and only one area occurs in time 2. Therefore, the biota that existed in area G during time 1 will delimit endemic area G to that time. The same applies to area G during time 2. Hence areas G1 and G2 are two different endemic areas that inhabit the same inorganic geographical area (Ebach and Humphries 2002). The general areagram can then be adjusted accordingly to include the relationships of areas G1 and G2. Treating area G1 as an ancestral biota to area G2 (*sensu* DIVA of Ronquist 1988b) is erroneous. Biogeography is about spatial patterns. To assume that one biotic area evolves into another makes biogeography dependent on evolutionary and dispersalist assumptions and therefore *generates a result* as opposed to discovering a pattern (Ebach and Humphries 2002). Recent suggestions of a chronological paradigm in biogeography (Hunn and Upchurch 2001; Upchurch and Hunn 2002; Upchurch et al. 2002; Donoghue and Moore 2003) concentrate on time-slicing trees rather than areas. Spliced trees are mapped onto a stratigraphical column in much the same way as strato-phenetic and strato-cladistic methods (see Smith 1994). The time-slicing of the above hypothetical example will only leave one tree that includes area G1 and not area G2. Area G2 will not be included in the analysis; therefore, its spatial relationship with other areas through time will be ignored.

Contrary to other methods, area cladistics does not need a distributional mechanism to explain the diversification and divergence of biota. Such narratives are not necessary to explain biotic history, as one mass dispersal of a large number of species will not appear any differently on a general areagram, as a vicariance event, or any other known mechanism. Dispersal and other unobservable narratives are confined to explaining species histories, not for finding biotic histories and biogeography. Biogeography is about pulling the evolutionary and tectonic cart. Through biogeography, natural historians (i.e., Buffon 1776; Darwin 1859; Wallace 1876; Wegener 1915) have discovered natural selection and continental drift. By assuming that biogeography is part of evolutionary biology or geology is putting the evolutionary and tectonic cart in front of the biogeographic horse.

Areas and Congruence

Geographical congruence is a goal that can be reached only if the theory, methods, and data are applied rigorously. In most methods, for example, scant attention is given to the circumscription of endemic areas. Biogeography needs a clear revision of endemic area analyses attempted over the last 30 years (Rosen 1988; Hausdorf 2002; Linder 2001). The means of recognizing endemic areas are analogous to species concepts. Wilkins (2003) noted that there are as many as 22 species concepts, of which most biologists use only 5. Similarly, biogeographers use numerous area definitions, including geopolitical areas, dot maps, range maps, grid cells, polygons, and other arbitrary units such as vegetation zones and biomes. Endemic areas can be based upon total or partial overlapping range maps, clusters in phenograms, and host and parasite ranges. The lack of theory in determining areas may explain the lack of a single effective endemic area method, just as the existence of 22 different species concepts explains the lack of a unifying species concept (see Wilkins 2003).

Biogeography in the Twenty-First Century

Given the fact that much philosophical debate surrounded systematics and biogeography in the 1970s and early 1980s, it is surprising that biogeography is still making so many twists and turns to accomplish an agreed methodology. One would hope that in the twenty-first century we can strive to determine what biogeography is all about. If genetic, ecological, and systematic researchers continue to avoid or ignore each other's work, then biogeography will become even more confusing and balkanized. For example, phylogeography is not a new subject; it belongs to the ideas of early dispersal biogeography, just as area cladistics is a development of pattern cladistics and component analysis. The claims and counterclaims in historical biogeography are due to workers in one field of endeavor not understanding the methods of those in another field. Certainly all biogeographers ask the same question, but that question needs a viable historical answer. Biogeographers need to address the problem of conflicting static and dynamic Earth theories, the differences between species and biotic histories, and the influence experimental sciences have had on a strictly historical science. In biogeography, two critical points are in need of investigation: the defining of endemic areas, and geographical congruence. We believe that the latter has a champion in area cladistics, whereas the former is still in need of investigation and discussion.

Acknowledgments

We would like to thank Gareth Nelson and Dave Williams for their helpful comments during the development of this manuscript.

PART II

PHYLOGEOGRAPHY AND DIVERSIFICATION

Brett R. Riddle and Vicki Funk

Scientists have been asking why certain plants and animals inhabit only certain parts of the world since at least the eighteenth century (e.g., Buffon 1761; see Lomolino et al. 2004). Most of the early works on the subject included explanations ranging from Noah's Ark to transoceanic land bridges. In the nineteenth century, Charles Darwin and Alfred Russel Wallace were leading advocates of the idea that a given taxon or biota had an ancestry at a "center of origin," with subsequent "dispersal over a permanent geography." Their contemporary, Joseph Dalton Hooker, proposed an alternative model that could be viewed as an early version of panbiogeography (Croizat 1958) or, to a lesser extent (since it included no phylogeny), to the phylogenetic biogeography of Hennig (1966) and Brundin (1966) and the "vicariance biogeography" perspective (Platnick and Nelson 1978). Hooker (1867) believed that similarities and differences among floras of the southern continents could be attributed to the emergence and submergence of transoceanic land bridges ("extensions"), leading to fragmentation of an ancestral "continuous extensive flora ... that once spread over a larger and more continuous tract of land ... which has been broken up by geological and climatic causes" (Hooker 1867). Hooker's extensionist view was supported by the geologist Charles Lyell and the botanist Carl Skottsberg, among others, while the "dispersal from a center of origin" view was continued prominently into modern times through the works of William Matthew (1915), George Gaylord Simpson (1940), Ernst Mayr (1942), and Philip Darlington (1957).

In the latter half of the twentieth century, biogeographic approaches that incorporated dispersal as a tractable and important process in taxon and biotic histories (Funk, in press) included the evolutionary biogeography of Philip Darlington (1957), phylogenetic biogeography of Lars Brundin (1966) and Willi Hennig (1966), and a modern version of the latter introduced by Edward Wiley (1981) and continued into the twenty-first century by Dan Brooks et al. (2001). Alternatively, the view that vicariance is a more general and tractable explanation than dispersal became the founding principle of vicariance biogeography (Platnick and Nelson 1978). More recently, vicariance biogeography has developed into cladistic biogeography (e. g., Humphries and Parenti 1999), while Croizat's original panbiogeography and phylogenetic biogeography continue in current literature.

All current methods in historical biogeography, regardless of epistemological and ontological leanings, rely on two scientific advances of the twentieth century, as summarized in Brooks's chapter in this section. First, the arrival of the theory of continental drift and plate tectonics revolutionized the way most biogeographers thought, and did away with the explicitly dispersalist orientation of evolutionary biogeography as the only explanation for current distributions; and second, the advent of phylogenetics as proposed by Willi Hennig (1966) and Lars Brundin (1966) brought new light to the pattern of evolutionary relationships among species over time and space. Of the various types of historical biogeography in use today (see Crisci et al. 2003), only panbiogeography studies do not start with a cladogram of one or more monophyletic groups (although they do incorporate phylogenetic information within their analyses; see Craw et al. 1999), and all of these methods make use of data from continental drift and plate tectonics (see Scotese, this volume).

Briefly, the current discussion between advocates of either cladistic biogeography or phylogenetic biogeography focuses on several fundamental differences in perspective. Cladistic biogeography, which seeks to discover the vicariant pattern in the history of areas, was summarized by Humphries and Parenti (1999), including a summary of component analysis (Nelson and Platnick 1981) using Page's implementation (COMPONENT 1.5; Page 1989). More recent arguments in favor of this approach are available (e.g., Humphries 2000; Ebach and Humphries 2002; Humphries and Ebach, this volume). Phylogenetic biogeography differs in attempting to unravel the full array of historical events associated with the history of speciation and biotic assembly within and among areas. The recently developed method of secondary BPA (Brooks et al. 2001; Brooks, this volume) uses four principles (assumption 0, the missing data protocol, the area duplication convention, and the "Threes Rule" rule) to obtain phylogenetic trees (without homoplasy) that are claimed to depict all area relationships as attributable to either general (e.g., vicariance or geo-dispersal) or one of several taxon-specific events (peripheral isolates speciation or sympatric speciation). A number of recent examples of phylogenetic biogeography are cited in Brooks and McLennan (2002).

The most recent comprehensive review of these and related methods in historical biogeography (Crisci et al. 2003) recognizes some 29 techniques categorized into 9 distinct approaches. To some degree, this large array of methods signals controversy over appropriate approaches to analyzing and interpreting data; for example, see recent exchanges between the cladistic (Ebach and Humphries 2002;

Humphries and Ebach, this volume) and phylogenetic (van Veller et al. 2003; Brooks, this volume) biogeographers. Indeed, it is not a straightforward matter to categorize methods, as evidenced by Crisci et al.'s inclusion of Brooks parsimony analysis (BPA) as a method of "cladistic" biogeography. It seems quite likely, in our view, that this example is related directly to the dynamic nature of modern historical biogeography. As another example of a popular method that is difficult to categorize discreetly, dispersal-vicariance analysis (DIVA; Ronquist 1997) weights vicariant (or predictable dispersal) events as more likely than random dispersal (or random extinction) in the reconstruction of ancestral distributions and area relationships. As a final example, Riddle and Hafner's chapter demonstrates that phylogeography (Avise 2000) cannot in practice be categorized into the "intraspecific" arena as discreetly as might be inferred from Crisci et al. (2003).

While the above discussion shows the presence of controversy within historical biogeography, we believe that these examples also suggest that we are poised at a unique period in the evolution of historical biogeography, where in the not-too-distant future we will see a great deal of momentum toward more agreement than disagreement in sorting through available techniques, leading to the kinds of syntheses between approaches that are signaled by the three contributions to this section. One of the contributions of phylogeography, for example, has been the discovery that the same logic that biogeographers have used to search for a general signal of vicariance among biotas on a global scale might also be important in understanding the diversification and assembly of biotas within continents (Riddle et al. 2000a; Brunsfeld et al. 2002; Calsbeek et al. 2003; Evans et al. 2003) or across oceanic archipelagoes (Wagner and Funk 1995; Heaney 2000). We also find it very interesting that the investigation of diversification by paleontologists in the deep history of life on Earth (see Lieberman's chapter), and phylogeographers within just the past several millions or thousands of years (the chapter by Riddle and Hafner), are converging on exploration of similar underlying attributes (i.e., geo-dispersal) with similar analytical approaches. While phylogenies have been and continue to be used in conservation biology (e.g., Faith 1992; Humphries and Williams 1994), other methods of historical biogeography are also advancing the goals of conservation biology. For example, recent publications such as Hugall et al. (2002) and Moritz et al. (2001) employ the techniques of phylogeography and biogeography for conservation in the Australian wet tropics.

One new area of interest that is just developing in biogeography is that of "supertrees." These phylogenetic trees cover large groups such as spiders (J.A. Coddington, pers. comm.) or the Compositae (V. A. Funk et al., pers. comm.) and allow biogeographers to explore global patterns. Such patterns are developed from extant taxa, but they allow one to infer a biogeographic history for the group that stretches back in time as well. All of these and other new developments are broadening the utility of biogeography and are linking this dynamic research program to more and more fields of research.

The three chapters in this unit explain the rationale behind (Brooks) or expand on (Lieberman; Riddle and Hafner) the modern form of phylogenetic biogeography. As a whole, they discuss the use of this approach on questions ranging from fossils to phylogeography, and from systematics to ecology. Those interested in comparing these chapters with works on cladistic biogeography should read Chapter 4 by Humphries and Ebach.

Dan Brooks's chapter provides an overview of the fundamental differences—from ontological to analytical—between cladistic and phylogenetic biogeography, and proposes a reconciliation and synthesis between systematics and ecology under the latter. He claims that the principal analytical method for phylogenetic biogeography, Brooks Parsimony Analysis (BPA), provides a means by which one can objectively explore the geography of speciation (vicariant, sympatric, or peripheral isolates) and asserts (pg. 135) that "Cladistic and phylogenetic biogeographic methods produce different area cladograms only when dispersal has produced unique or reticulated area relationships." Indeed, Brooks cites a number of recent studies suggesting that reticulated area relationships are actually quite common in the real world, and are to be expected under models of evolutionary radiation such as the taxon pulse or island archipelago speciation (e.g., Wagner and Funk 1995). In the latter case, he discusses the role of phylogenetic biogeography in bringing an explicit evolutionary history component into the MacArthur-Wilson model of island biogeography, providing an opportunity for a new and vital synthesis between systematics and ecology.

Riddle and Hafner's chapter provides a discussion of phylogeography (*sensu* Avise 2000) positioning it between macroevolutionary (historical biogeography, phylogenetic biogeography) and microevolutionary change (demography, population genetics). Phylogeography focuses on biogeography within species and among closely related species, which often involves dispersal and therefore has been rejected by cladistic biogeographers (Ebach and Humphries 2002). More generally, although phylogeography has successfully expanded population genetic theory and questions into the historical geographic realm, it has been less successful at bringing historical biogeography into the realm of fairly recent and spatially reduced taxon and biotic histories. However, as Riddle and Hafner show, the comparative phylogeographic approach is highly compatible with phylogenetic biogeography, providing high-quality distributional and phylogenetic data that can be analyzed in a BPA framework. They argue that an important advantage of using BPA is that the resulting secondary BPA tree provides an objective basis for then returning to a phylogeographic analytical approach to test hypotheses of co-divergence (vicariance or geo-dispersal), peripheral isolates or sympatric speciation, and range expansion.

In his chapter, Bruce Lieberman extends the discussion of phylogenetic biogeography into the realm of paleontology. He stresses the importance of extinction as a factor that must be taken into account in deciphering biogeographic congruence, with the potential for errors arising from extinction increasing in "clades with higher extinction rates or longer durations" (pg. 113). Secondly, he argues that vicariance is not the only historical event that can produce congruent distributions across co-distributed lineages, and he discusses the traditionally overlooked concept of geo-dispersal (congruent range expansion) that might incorrectly be interpreted as vicariance unless assessed analytically as a process with its own distinct signature. This assessment is accomplished by using a modified form of BPA to differentiate geo-dispersal from vicariance in area cladograms. This approach might be especially informative when applied to geographic regions that have undergone cyclical histories of isolation and connection (e.g., continental rifting and collision; Pleistocene sea-level lowering and rising).

Geo-dispersal is only one of the concepts that are interwoven among the three contributions to this section. Each author expresses enthusiasm for future developments (Brooks, for example, indicates that a sophisticated algorithm for implementing BPA is on the horizon); and all have argued that phylogenetic biogeography provides a powerful approach to unraveling the historical complexity of taxa and biotas, regardless of the timescale of interest. While a great deal of controversy remains over methods and interpretation, the road forward promises to be paved more with productive discourse than contentious debate.

We hope these chapters instill a sense of enthusiasm in new as well as more seasoned colleagues interested in the complexity of biological diversification, ranging spatially from the Earth to the most localized habitats, and temporally from the origins of life to the past several thousands of years of climatically and anthropogenically induced habitat changes.

CHAPTER 5

The Past and Future Roles of Phylogeography in Historical Biogeography

Brett R. Riddle and David J. Hafner

> Biogeographic methodology needs to develop analytical techniques in order that complex historical patterns are not concealed by estimates that are very much less complex.
>
> Joel Cracraft, 1988

The Controversial Nature of Modern Historical Biogeography

Of the nine methods in historical biogeography reviewed by Crisci (2001), the fact that most either have been developed or advanced significantly subsequent to the 1988 paper from which the above quote was drawn, testifies to the vitality of this discipline. Yet energetic debate continues regarding the relative value of the different methods. For example, Humphries and co-workers (Humphries 2000; Ebach and Humphries 2002) challenged the legitimacy of both phylogenetic biogeography (as defined by Brooks, this volume) and phylogeography as robust research programs in historical biogeography, drawing rebuttals from advocates of phylogenetic biogeography (van Veller et al. 2002) and phylogeography (Arbogast and Kenagy 2001).

Subsequently, van Veller et al. (2003) and Brooks (this volume) have argued that historical biogeography actually includes two different research programs: *cladistic biogeography*—an *a priori*, event-based exploration of area relationships; and *phylogenetic biogeography*—an *a posteriori*, discovery-based exploration of the geography of speciation. Cladistic biogeography seeks to recover the vicariant backbone among areas of endemism (Humphries 2000). Phylogenetic biogeography is a taxon-based approach to discovering the geographic nature of speciation, and it concerns the history of biotas (Andersson 1996) and vicariance events (Hovenkamp 1997). This distinction results primarily from recognition that "the history of geographical areas is rarely exclusively divergent [and] geological data are overwhelmingly in favour of reticulation as a real phenomenon" (Hovenkamp 1997). In other words, if the biogeographic history of an area includes, for example, episodes of parapatric speciation, post-speciation dispersal, or geo-dispersal (see Lieberman, this volume) in addition to vicariance, the area will exhibit a more complex set of historical relationships with other areas than would result from a single vicariant event, leading to a reticulated pattern of relationships with other areas.

One method, *phylogeography*, is unique in spatial and temporal scope in that it concentrates on "...the geographic distributions of genealogical lineages, especially those within and among closely related species" (Avise 2000). Phylogeography is conceptually positioned between traditionally distinct arenas of macroevolutionary (e.g., historical biogeography, phylogenetic biology) and microevolutionary change (e.g., demography, population genetics; Figure 5.1). It broadens the goals of phylogenetic biogeography by considering the impact of not only vicariance but also dispersal and population genetics on

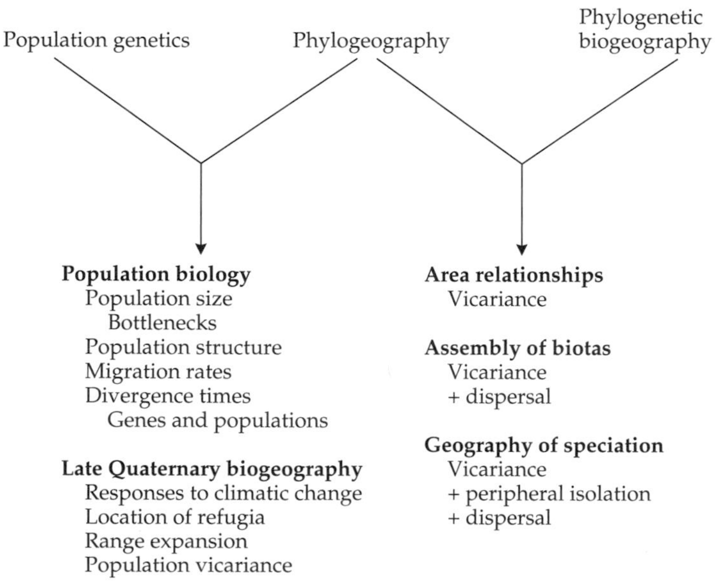

FIGURE 5.1 Depiction of phylogeography as a bridge between traditionally microevolutionary (population genetics) and macroevolutionary (phylogenetic biogeography) disciplines.

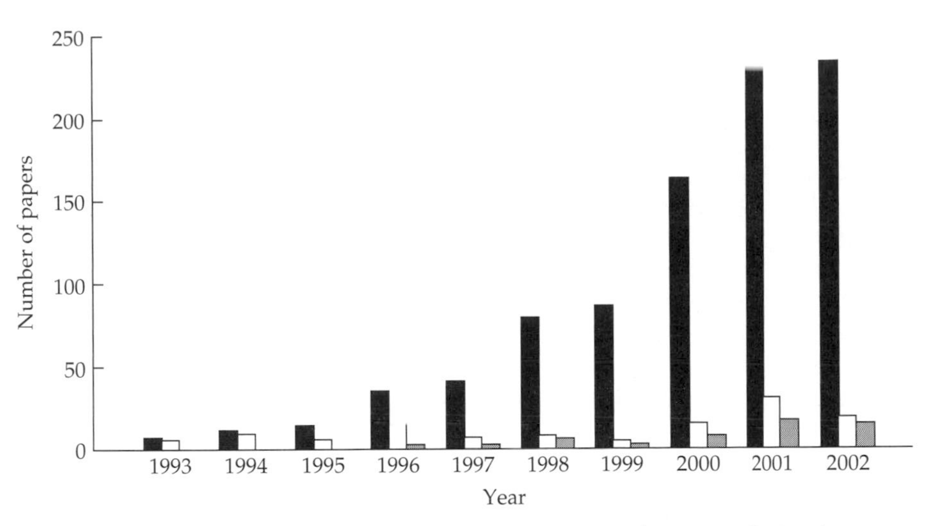

FIGURE 5.2 Summary of Science Citation Index searches showing relative frequencies of studies in which terms describing methods of historical biogeography are used in published papers (within title, abstract, or keywords): phylogeography (black bars), five other methods in historical biogeography (white bars, including combined results for cladistic biogeography, phylogenetic biogeography, panbiogeography, dispersal-vicariance analysis, and Brooks Parsimony Analysis), or comparative phylogeography (gray bars).

the evolutionary and geographic history of genetic lineages (including but not limited to species) and provides the basis for a promising bridge between the often disparate ecological and historical aspects of biogeography (e.g., Ricklefs and Bermingham 2001), as well as between biogeography and Quaternary paleoecology (Jackson, this volume). Although being introduced formally only about 15 years ago (Avise et al. 1987), phylogeography has exploded onto the scene within the last decade (Figure 5.2). This activity is due in large part to the fact that it has been designed explicitly to employ the power of modern molecular genetic methods and analyses. While developed under the premise of being a distinct subdiscipline of historical biogeography (Avise 2000), phylogeography has not received a uniformly warm welcome by other historical biogeographers. For example, Humphries (2000) proclaimed that phylogeography is "the most recent chimera of scenario building" and is responsible for "dispersal scenarios ... being published at a rate faster than ever before," while Ebach and Humphries (2002) suggest that phylogeography resides outside the logical framework established for historical biogeography by Platnick and Nelson (1978).

In our view, phylogenetic biogeography and phylogeography share a strong conceptual overlap. Both approaches are concerned in a general sense with the evolutionary and geographical dynamics of speciation (or gene lineage evolution) and biotic assembly in time and across space, and both deal analytically with problems that arise from reticulation (e.g., between areas, biotas, or populations). Phylogeography has further served as a "bridge" to unify the traditionally quite distinct goals of population genetics and historical biogeography. In this sense, it provides an empirical implementation of the conceptual insights developed by Wiley (1981) into the ways in which phylogeny, population biology, and geography could be used in concert to investigate modes of speciation.

We use title, abstract, and keyword surveys of the Science Citation Index (ISI Web of Science: http://www.isinet.com/isi/) to assess the extent to which phylogeography has become a particularly robust bridge between micro- and macroevolutionary disciplines. We use this exploration as a basis for addressing the nature of perceived conflict between the goals and methods of phylogeography and those of cladistic and phylogenetic biogeography. We argue that phylogeography should strive to establish a framework in which hypotheses of vicariance and dispersal among co-distributed populations and closely related species are developed using phylogenetically based methods such as Brooks Parsimony Analysis, or BPA (Brooks, this volume). But we also anticipate that this synthesis can go further by treating statements in a fully resolved secondary BPA tree as testable hypotheses of dispersal, vicariance, speciation mode, and population history. These hypotheses should then be evaluated using methods of phylogeography, which are designed to sort among alternative historical processes underlying current geographic population structure. We take the relatively recently elaborated subdiscipline of comparative phylogeography (Bermingham and Moritz 1998; Arbogast and Kenagy 2001) as a point of departure for further examining connections between phylogeography and phylogenetic biogeography.

How Well Has Phylogeography Bridged Microevolution and Macroevolution?

Phylogeography fits very comfortably within the arenas of microevolution and macroevolution (Figure 5.3). In its role as a geographical and temporal extension of population genetics, there is little question that phylogeography is developing sophisticated methods within a rigorous statistical and hypothesis-testing framework (e.g., Althoff and Pellmyr 2002; Knowles and Maddison 2002). Most of the recent advances in the rigor of phylogeographic methods are motivated by a desire to extract relatively recent (e.g., late or post-Pleistocene) signatures of historical process (e.g., range expansion, population fragmentation) within assemblages of closely related individuals or populations. For example, one popular method, nested clade analysis (Templeton et al. 1995; Posada et al. 2000) starts with a statistical phylogeny that does not assume *a priori* that genetic affinities are captured by a bifurcating phylogenetic tree (Posada and Crandall 2001). Another popular method, MIGRATE (Beerli and Felsenstein 1999) uses a maximum likelihood approach to infer the

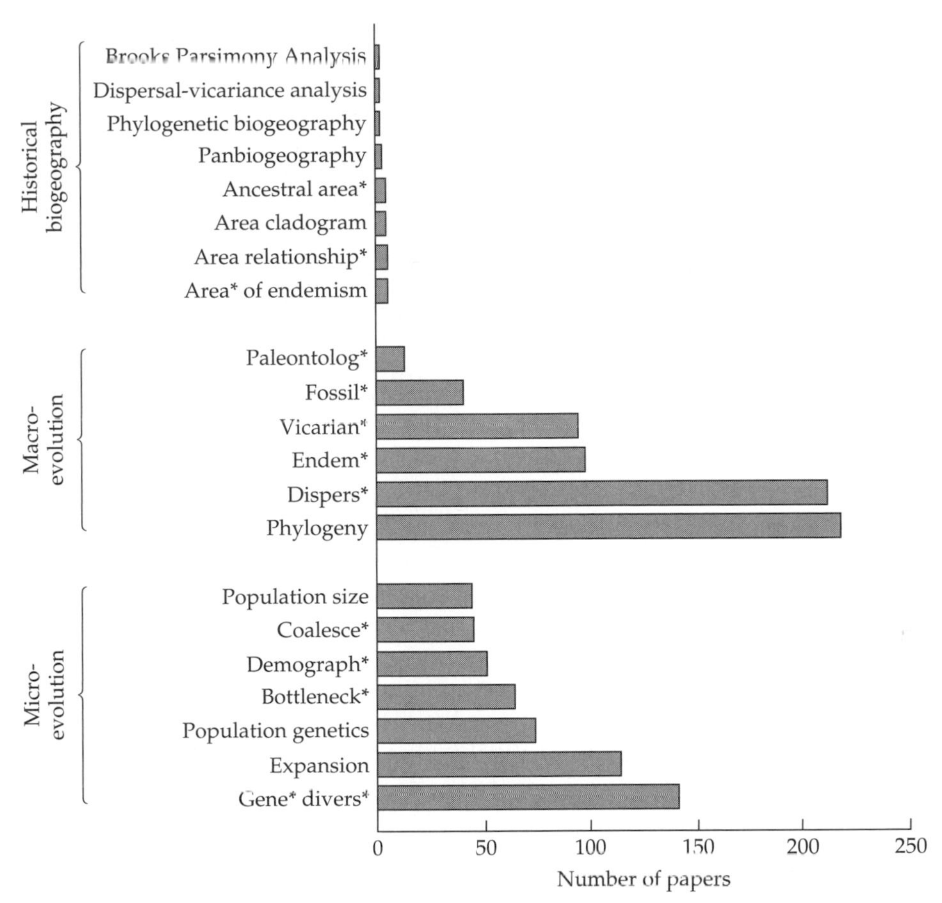

FIGURE 5.3 A demonstration of the frequent use by phylogeographers of the languages of microevolution and macroevolution, but infrequent use of the language of historical biogeography. Science Citation Index searches for combined use of "phylogeograph* and (one of the 21 terms to the left of the bar graph)" together in titles, abstracts, or keywords of published papers.

direction of past migrations between geographically separate populations (e.g., Zheng et al. 2003). When combined with improvements in generating DNA sequence data from more rapidly evolving portions of a genome (e.g., the mammalian mitochondrial control region), these methods are particularly useful in expanding population and ecological arenas into a late-Quaternary time frame. This, in turn, allows exploration of the influence of the glacial-interglacial climatic transformations (Jackson, this volume) on extant populations and biotic assemblages. Yet, while phylogeography does seem to have satisfied its claim to having incorporated both microevolutionary and macroevolutionary perspectives, it seldom has been conducted within the constructs of most other subdisciplines of historical biogeography (see Figure 5.3).

These evaluations of the literature suggest that the phylogeographically driven expansion of biogeography into more recent time frames have only occasionally and secondarily incorporated the concepts of historical biogeography. Perhaps, then, it should not be too surprising that the explosive growth of phylogeography has generated concern among historical biogeographers who reside intellectually within a more time-honored historical biogeography that has analytical roots extending back into the 1970s (Platnick and Nelson 1978) and conceptual roots into the 1950s (Croizat 1958).

Is Phylogeography Really Just a "Chimera of … Dispersal Scenarios"?

Phylogeography does not appear exclusively to be about developing ad hoc or post hoc dispersal scenarios. Notice in Figure 5.3 that vicariance appears frequently (albeit less than half as often as dispersal) within the phylogeography literature. Yet even when vicariance does appear in phylogeographic studies, it may not be approached within a framework advocated by cladistic and phylogenetic biogeographers.

Humphries (2000), representing the view from cladistic biogeography, claimed that historical biogeography "is about classification of areas amongst biological and spatial co-ordinates, and not about dispersal or individual historical scenarios for every group of organisms." Further, the recently presented "Threes Rule" rule states that: "In order to distinguish between general and special patterns of historical association…you must analyze at least 3 co-occurring clades" (Brooks and McLennan 2002; Brooks, this volume). Finally, cladistic and phylogenetic biogeographers would agree that historical biogeography is about "[obtaining] resolved area cladograms that represent historical relationships among areas in which monophyletic groups of taxa are distributed" (van Veller et al. 2003). Yet, from further inspection (Science Citation Index) of 73 of the 94 titles and abstracts from 1993 through 2002 containing "phylogeography and vicarian*," 44 (60%) analyze no more than a single clade. Nevertheless a vicariance hypothesis was supported in 66 (90%) of these studies. Apparently, one of the most important messages from historical biogeography—that a general pattern (vicariance or geo-dispersal) is robustly supportable primarily through rigorous comparison across co-distributed taxa—has not infused the main body of phylogeographic study. Perhaps then, the frequent employment of single taxon studies—which describe at best taxon-specific patterns—as evidence for general patterns, represents a more insidious weakness of current phylogeography, although comparative phylogeography (see discussion below) is a promising move in the direction of resolving this disconnect.

"Dispersal" also has a somewhat different meaning for phylogeographers and historical biogeographers. Regardless of whether one is doing cladistic or phylogenetic biogeography, geographic dispersal is regarded as a process of taxon movement between specified areas. Presumably, there is room in this view of biotic histories to accommodate shifts in geographic ranges within the boundaries of a single area of endemism (e.g., Riddle 1998), but these sorts of "within-area" movements have no analytical standing in either cladistic or phylogenetic biogeography. However, dispersal events such as geographic range shifting, range expansion, and range contraction—without reference to movement among separate areas of endemism—are legitimate and important components of phylogeography, and more generally, biogeography.

Comparative Phylogeography as a Template for Synthesis

Comparative phylogeography (Bermingham and Moritz 1998; Arbogast and Kenagy 2001) appears to have a more explicit alignment with historical biogeography. According to Arbogast and Kenagy (2001), both disciplines share the premise that "the most parsimonious explanation for multiple taxonomic groups that exhibit common spatial patterns of evolutionary subdivision is that they have a shared biogeographic history [explained by] a common set of historical vicariant events." Indeed, vicariance does seem to take on relatively greater significance within comparative phylogeography (Table 5.1a). Notice however, that the comparative framework is still a small subset of the total field of phylogeography (see Figure 5.2).

According to Arbogast and Kenagy (2001), the basis of comparative phylogeography (Figure 5.4) is that "a common set of historical vicariant events has geographically structured a group of ancestrally co-distributed organisms in a similar way." In fact, not all papers using the term *comparative phy-*

TABLE 5.1 ***Search results for Science Citation Index, 1993–2002, frequency of terms associated with historical biogeography***

Term(s)	Associated term(s) and frequency of association (*n*)				
(a) Comparative phylogeography and...	dispers* not vicarian* $n = 8$	vicarian* not dispers* $n = 11$	dispers* and vicarian* $n = 4$		
(b) Phylogeography and...	interspecific $n = 34$	intraspecific $n = 148$			
(c) Phylogeography and cryptic and...	species $n = 38$	speciation $n = 16$	lineages $n = 3$	refugia $n = 4$	vicariance $n = 5$
(d) Molecular and...	biogeograph* $n = 826$	phylogeograph* $n = 404$	biogeograph* and phylogeograph* $n = 129$		

Note: This SCI search was based on titles, abstracts, and key words.

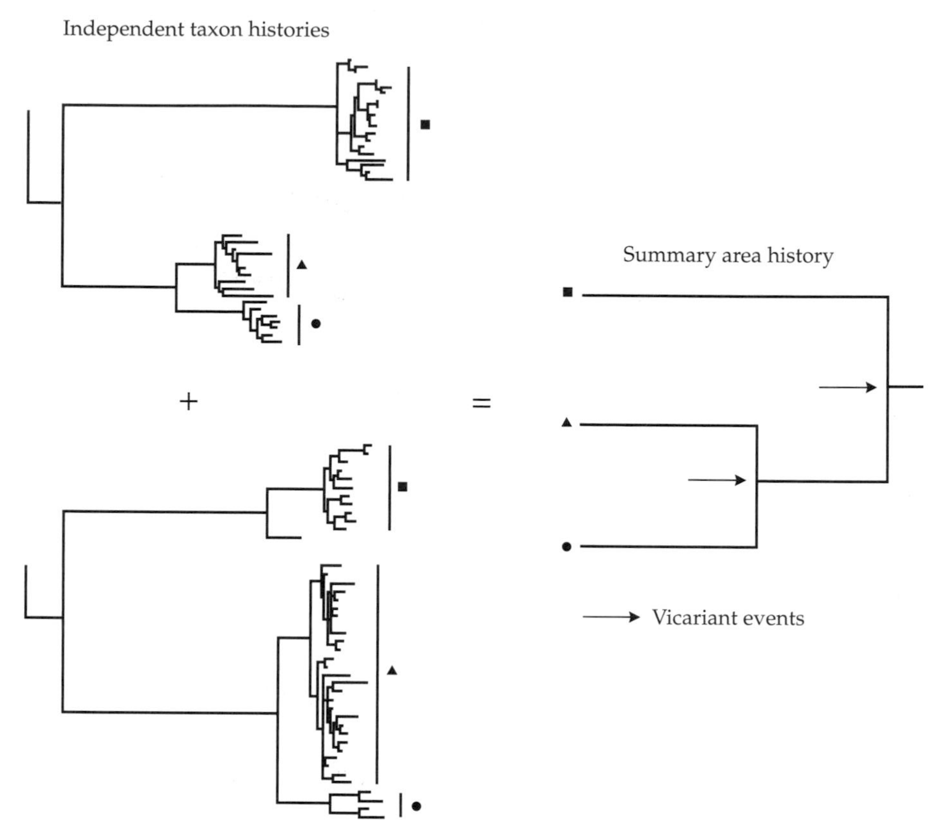

FIGURE 5.4 A depiction of the relationship between comparative phylogeography and historical biogeography, in which phylogroups within taxon phylogenies can be used in the same fashion as species or higher taxonomic units traditionally have, in order to postulate area histories and vicariant events (arrows). Each symbol in the taxon phylogenies depicts a reciprocally monophyletic set of alleles (haplotypes) occurring within and among a set of populations.

logeography overlap appreciably with historical biogeography in this fashion. For vicariant history to be recorded in population genetic structure, a once-widespread, ancestral population must first be fragmented into separate areas. Gene lineage phylogenies embedded within this set of isolated populations (or "cryptic" species) must then have evolved to a state of reciprocal monophyly (i.e., different *phylogroups*, each comprising a set of haplotypes arising from a single common ancestral haplotype subsequent to divergence from other such phylogroups). Perspectives such as Wares' (2002) demonstrate a more inclusive view of comparative phylogeography, one that is focused on biogeographic history as only one of a number of plausible explanations (including ongoing or intermittent extrinsic and intrinsic processes) for genealogical congruence across an assemblage of co-distributed species.

Much of the current growth of analytical methods in phylogeography is concerned with situations where reciprocal monophyly has not been achieved between populations. Althoff and Pellmyr (2002) acknowledged the "phylogenetic" versus "population/demographic" duality embedded within phylogeography, and suggested a sequential approach for examining structure within a single taxon. We expand this perspective by suggesting (beyond) the incorporation of yet another level of complexity into the sequential approach: the comparative analysis of phylogeographic structure within a phylogenetic biogeographic framework.

Fuzzy Borders Between Phylogeography and Historical Biogeography

With the growing interest in pushing phylogeography in the direction of deciphering extremely shallow events in the geographic history of populations, it is perhaps worthwhile to note that whereas phylogeography is heavily skewed toward Pleistocene and Holocene time frames (Figure 5.5), it does not empirically reach its limits there. A number of phylogeographic studies extend back into the Pliocene and Miocene, but then reach a fairly distinct wall that represents a threshold beyond which molecular clock estimates of divergence within extant species and groups of closely related species seldom extend. In a practical sense, this provides a sufficient window of time to examine the geographic histories of closely related species, and indicates that empirical phylogeography has not been delegated exclu-

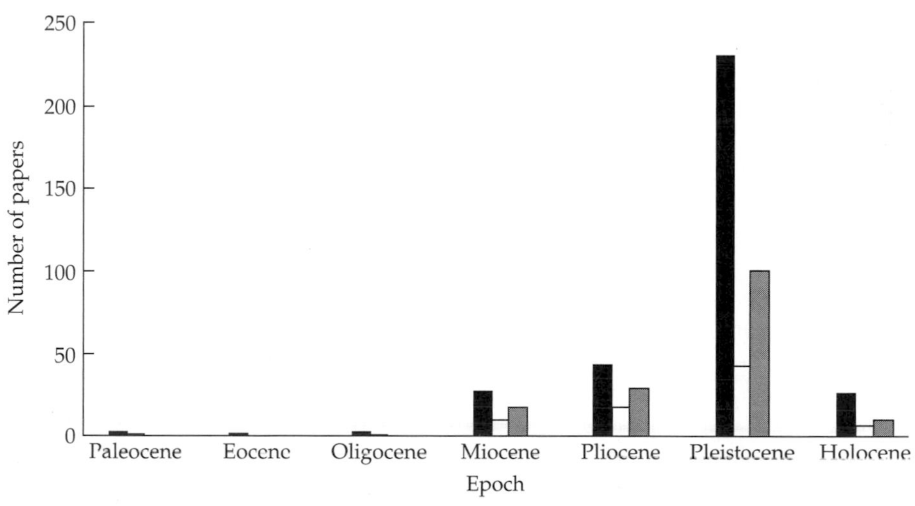

FIGURE 5.5 Summary of Science Citation Index searches showing relative frequency in which specific terms are used in association with Cenozoic epochs in published papers (within title, abstract, or keywords): "phylogeography and..." (black bars), or "phylogeography and vicarian* and..." (white bars), or "phylogeography and dispers* and..." (gray bars).

sively to an intraspecific realm (Table 5.1b). Indeed, one of the strengths of a phylogeographic framework is the frequent elucidation of cryptic geographic and evolutionary structure (Table 5.1c), thereby creating a dynamic linkage between the taxonomically delineated intraspecific and interspecific realms. Avise (2000) has provided case studies that he defines as phylogeographic back into Cretaceous times. In practice, however, there are many studies that consider themselves as molecular biogeography without reference to phylogeography (Table 5.1d), and so the distinction between the two blurs as temporal and spatial scales expand.

Using Phylogeographic Data in Phylogenetic Biogeography

The vast majority of phylogeography studies will sample multiple individuals and localities within and among populations and geographic regions in order to establish empirically the boundaries and evolutionary structure of gene lineages. If the phylogeographic sampling design generates a set of distinct phylogroups, then it has essentially developed a set of operational units for phylogenetic biogeographic analyses (see Figure 5.4). This end point represents a straightforward connection between the two approaches and should not require much further explanation. To this extent, the use of phylogeographic-scale data in phylogenetic biogeography has been implemented in several recent studies (e.g., Taberlet et al. 1998; Zink et al. 2000a; Brooks and McLennan 2001).

One promising method in phylogenetic biogeography is Brooks Parsimony Analysis (BPA). Recent formulations (Brooks et al. 2001; Brooks, this volume; Lieberman, this volume) provide a protocol for extracting information about generalized disjunction events (including vicariance and geo-dispersal) preceding speciation, as well as a variety of events that tend to reduce the generalized disjunction signal (including post-isolation dispersal, speciation of peripheral isolates, and extinction) from an array of taxa and their geographic distributions. This method has been used to analyze a number of data sets (for review, see Brooks and McLennan 2002; Brooks, this volume).

We have recently applied this method to a study of co-distributed vertebrates in the warm desert regions of western North America (Figure 5.6; Riddle and Hafner, in press). The avian portion of this system was investigated previously by Zink et al. (2000a; 2001) using a combination of phylogeographic and dispersal-vicariance (DIVA) analyses; and reanalyzed by Brooks and McLennan (2002) using primary and secondary BPA. Although inferences from those studies were similar in demonstrating a complex history of vicariance and dispersal underlying the assembly of avian communities in these areas, Brooks and McLennan (2002) concluded that BPA was more sensitive in its treatment of different forms of dispersal. Here, we summarize the results of our study for purposes of illustrating potential utility of this approach and refer the reader to Riddle and Hafner (in press) for full details and an evaluation of our results relative to those previous studies.

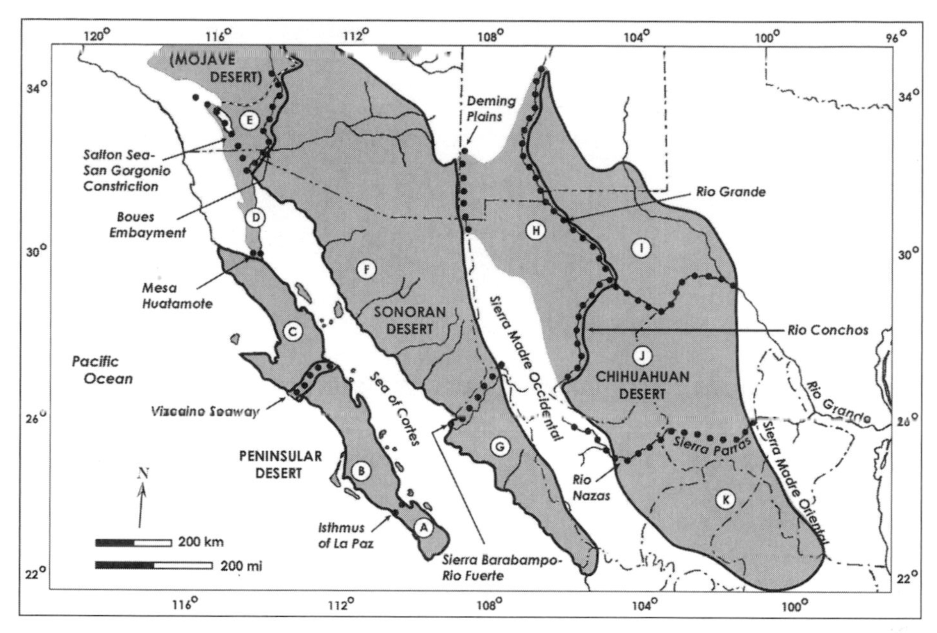

FIGURE 5.6 Southern regional deserts of North America (light shading) with 11 areas of occurrence (A–K) delineated by putative vicariant events. Four postulated areas of endemism identified through Parsimony Analysis of endemicity are indicated by bold outlines: Peninsular South (areas A + B), Peninsular North (C), Continental West (F + G), and Continental East (I + J + K).

Our data set consisted of 9 mammalian, 7 avian, 4 reptilian, 1 amphibian, and 1 cactus cladograms (a total of 64 terminal phylogroups; Table 5.2). In most of these taxon cladograms, terminal lineages represent phylogroups that have been recognized as geographically and evolutionarily distinct lineages through phylogeographic analyses conducted within the past decade; 32 are not recognized as specifically distinct from one or more congeneric phylogroups. The distributions of many of these groups are geographically consistent with a broadly co-distributed, warm-desert assemblage: they have sister taxa in the Gulf of California region and Chihuahuan Desert of the Mexican Plateau, which were separated at one or more times during the late Miocene through early Pleistocene by the Sierra Madre Occidental (Morafka 1977; Riddle 1995; Murphy and Aguirre-Léon 2002). This assemblage also includes the "northern Pliocene vicariant complex," which Grismer (1994) postulated to have been isolated into peninsular and continental areas by northward transgression of the Gulf of California into lowlands of southern California and Arizona.

We began with a set of four postulated areas of endemism (see Figure 5.6): Peninsular South (PS), Peninsular North (PN), Continental West (CW), and Continental East (CE). These four areas represented an amalga-

TABLE 5.2 ***Summary of taxon cladograms, taxa, and character codings used in BPA analyses (see Table 5.3)***

Taxon cladogram	Number of taxa[a]	BPA character numbers[b]	Original references[c]
Ammospermophilus	3	1–5	8; 24
Chaetodipus baileyi species-group	3	6–10	9
Chaetodipus penicillatus species-group	3	11–15	1; 9
Chaetodipus arenarius	3	16–20	8
Chaetodipus nelsoni species-group	3	21–23	9; 24
Dipodomys merriami species-group	3	24–28	8; 24
Peromyscus eremicus species-group	5	29–36	10
Onychomys	2	37–40	7; 11
Neotoma lepida species-group	3	41–45	5
Pipilo fuscus species-group	4	46–52	19; 20; 23
Callipepla squamata species-group	4	53–58	16; 19
Polioptila melanura species-group	3	59–63	14; 15; 19; 22
Toxostoma lecontei species-group	3	64–68	18; 19
Toxostoma curvirostre species-group	4	69–75	17; 19; 21
Auriparus flaviceps	2	76–78	22
Campylorhynchus brunneicappilus	2	79–81	22
Kinosternon flavescens species-group	2	82–84	12
Sauromalus	3	85–89	4
Cnemidophorus tigris	2	90–92	2; 6
Uta stansburiana	3	93–97	13
Bufo punctatus	3	98–101	8; 24
Lophocereus schottii	2	102–104	3

[a] Number of taxa in analysis is an informative subset of total number in the original taxon cladograms. See Riddle and Hafner (in press) for complete data and description of analyses.

[b] Character numbers are given in order of appearance in character matrices (Table 5.3).

[c] 1, Lee et al., 1996; 2, Murphy and Aguirre-Léon 2002; 3, Nason et al. 2002; 4, Petren and Case 2002; 5, Planz 1992; 6, Radkey et al. 1997; 7, Riddle 1995; 8, Riddle et al. 2000a; 9, Riddle et al. 2000b; 10, Riddle et al. 2000c; 11, Riddle and Honeycutt 1990; 12, Serb et al. 2001; 13, Upton and Murphy 1997; 14, Zink et al. 2000a; 15, Zink and Blackwell 1998a; 16, Zink and Blackwell 1998b; 17, Zink and Blackwell-Rago 2000; 18, Zink et al. 1997; 19, Zink et al. 2001; 20, Zink and Dittman 1991; 21, Zink et al. 1999; 22, Zink et al. 2000b; 23, Zink et al. 1998; 24, unpublished data (Riddle, Hafner, and colleagues).

mation of a set of 8 core areas extracted through a Parsimony Analysis of Endemicity procedure from an original set of 11 arid land areas of distribution. The Peninsular South and Peninsular North areas were first postulated through phylogeographic studies to have been separated during the middle Pleistocene by a trans-peninsular seaway between the Pacific Ocean and Gulf of California (Upton and Murphy 1997; Riddle et al. 2000a; Murphy and Aguirre-Léon 2002).

Our primary BPA trees (Table 5.3; Figure 5.7A and 5.7B) postulated two alternative general patterns of area relationships, differing in whether CE is first split from an ancestral western area, or whether the basal split is between

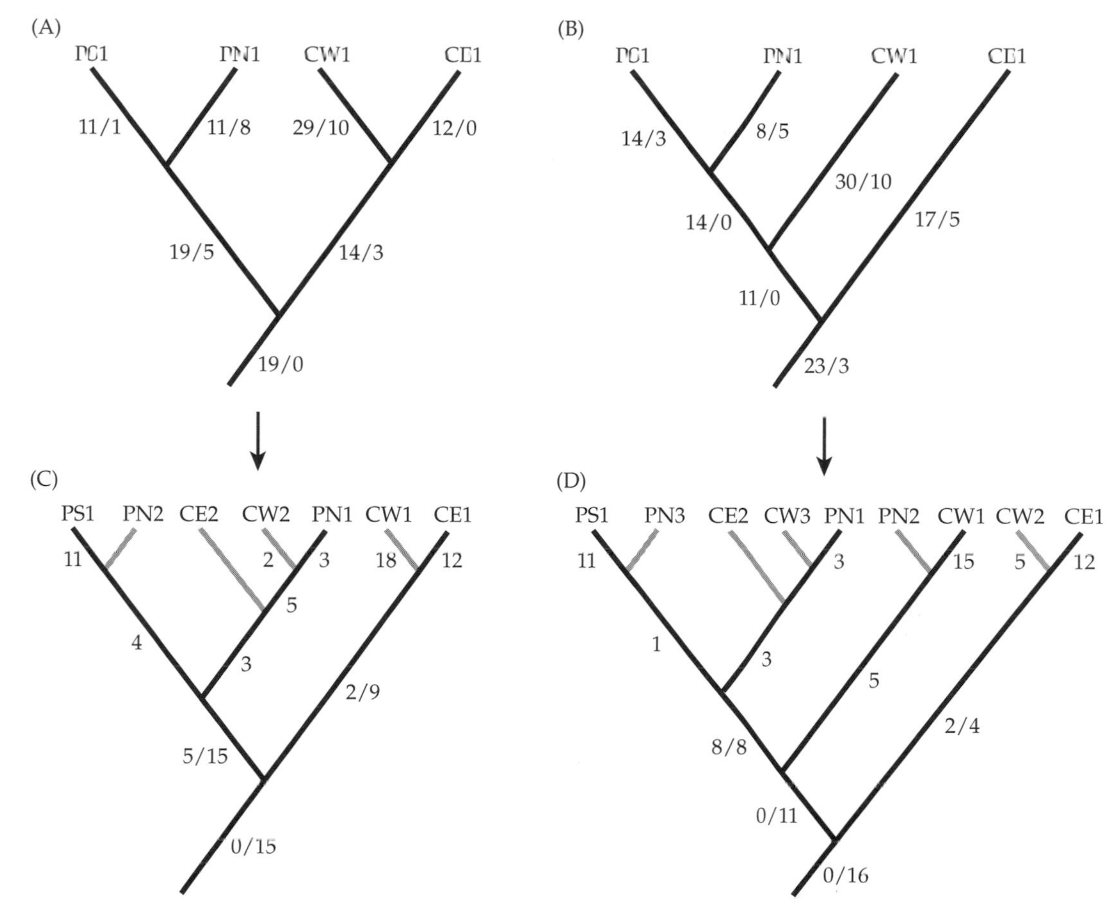

FIGURE 5.7 Results of Brooks Parsimony Analyses. (A, B) Primary BPA trees among four areas (PS1, PN1, CW1, CE1). Numbers on the branches depict [total/homoplasious] character changes. (C, D) Secondary BPA trees among (C) four original and three duplicated areas, or (D) among four original and five duplicated areas. Numbers on the branches depict [terminal taxa] or [terminal taxa/ancestral taxa].

the areas CE and CW versus PN and PS. We attribute this basal ambiguity to the highly reticulate CW area having equivalent numbers of basal sister-group relationships to the east (CE) and to the west (PN and PS) (Riddle and Hafner, in press).

This general pattern is consistent with the late Neogene geological history summarized elsewhere (Morafka 1977; Grismer 1994; Riddle et al. 2000; Murphy and Aguirre-Léon 2002; Riddle and Hafner, in press) and thus with an hypothesis of historical vicariance. However, a total 13 out of 104 characters were homoplasious (arising convergently or in parallel more than once when mapped onto the trees) on the primary BPA tree (CI = 0.72), suggesting

TABLE 5.3 *Presence-absence matrixes for primary and two secondary Brooks Parsimony Analysis (BPA)*

Area code	Area name	Distribution matrix
Primary BPA (Figure 5.7A)		
	Ancestor	00000000000000000000000000000000000000000000000000000000000000000000000000000000000000000000000000000000
PS	Peninsular South	1010110101?????10111???10101111000001???0001110100011010011010101101101???1101101???10101101101011001101
PN	Peninsular North	0110101101?????01101???01101101000001???0001110100011010011010110101111???1011111???01101011011011001101
CW	Continental West	011010001110111?????101011010001011110111110101110110110110111110111011011101101101100011011011010101011
CE	Continental East	00011?????01101?????01101111000011011101?????000011100010101101000111???1011011011101????????000110011???
Secondary BPA (Figure 5.7C)		
	Ancestor	00000000000000000000000000000000000000000000000000000000000000000000000000000000000000000000000000000000
PS1	Peninsular South 1	1010110101?????10111???10101111000001???00011101000110100110101011011010001101101???10101101101011001101
PN1	Peninsular North 1	0110101101?????01101???01101101000001???00011?????????????101011010101100010110110???01101011011011001101
PN2	Peninsular North 2	???1010001101001??????????1010001???101???????????????????????
CW2	Continental West 2	01101?????????????????01101?????????????????0110001011001?????101010110001011011????????01101101???????
CW1	Continental West 1	?????0001110111?????101?????00010111101111101000101100001101111000110000111??????01100011????????0101011
CE2	Continental East 2	???????????????????????01101??011011????????????????????????
CE1	Continental East 1	00011?????01101?????01100011000011011101?????0000111000101011010001100 01011??????101????????000110011???
Secondary BPA (Figure 5.7D)		
	Ancestor	00000000000000000000000000000000000000000000000000000000000000000000000000000000000000000000000000000000
PS1	Peninsular South 1	1010110101?????10111???10101111000001???00011101000110100110101011011010001101101???10101101101011001101
PN1	Peninsular North 1	?????01101?????01101???01101101000001???00011101000110100110101?????1010001011011???01101????????1001101
PN3	Peninsular North 3	??101???????????????????????
PN2	Peninsular North 2	01101??101010110001??????????????01101101???????
CW1	Continental West 1	011010001110111?????101??????????????0111110101100010110 11?????101010110001??????01100011011011010101011
CW2	Continental West 2	????????????????????????????000101111????????0001011??????01111000110000111?????????????????????????????
CW3	Continental West 3	???????????????????????01101??011011????????????????????????
CE2	Continental East 2	???????????????????????01101??011011????????????????????????
CE1	Continental East 1	00011?????01101?????01100011000011011101?????000011100010101101000110001011??????101????????000110011???

a more complex biogeographic history of this biota. The Continental West area shared homoplasies equally to the east with the Continental East area and west with the Peninsular North area, while the latter also shares a large number of taxa uniquely with the Peninsular South area. Both the Continental West and Peninsular North areas appear to be highly reticulated.

Construction of a tree without homoplasy (i.e., such that all characters arise only once when mapped onto the tree) through secondary BPA (Table 5.3; Figure 5.7) required either three (5.7C) or five (5.7D) area duplications. The secondary BPA tree provides a basis for deciphering exceptions to the general history postulated in the primary BPA tree.

In particular, regardless of the precise order of basal separation between continental and peninsular areas, each hypothesis clarifies the highly reticulate nature of both the Continental West and Peninsular North areas. Each of these appears to have played complex roles in vicariance and dispersal, serving both as areas of endemism (particularly Continental West) and as conduits for dispersal at various times during the diversification of this biota. It was also necessary to duplicate Continental East once on each tree in order to resolve a congruent pattern of widespread distributions in the *Dipodomys merriami* species-group, *Auirparus flaviceps*, and *Campylorhynchus brunneicappilus* following a basal divergence in each of these groups across the postulated mid-Peninsular Vizcaíno Seaway barrier.

Testing Phylogenetic Biogeographic Hypotheses with Phylogeographic Data

A traditional historical biogeographic analysis might have difficulty not only in further evaluating evidence for the generalized disjunction backbone postulated in the primary BPA tree, but also in selecting between alternative explanations for the widespread and redundant taxa that remain following the secondary BPA analysis. We believe that a promising approach might be to now return to a phylogeographic perspective to further evaluate hypotheses generated through the BPA analyses.

WIDESPREAD TAXA. Several replicated patterns of widespread taxa are evident. The most common of these (PS + PN, 8 taxa) is interesting because PS also represents an area with a high number of strictly endemic taxa (11). As reviewed above, previous studies have considered the presence of strong endemism in the Peninsular South area to result primarily from middle Pleistocene vicariance. We could postulate at least two alternative histories for the widespread taxa: (1) that they did not respond to either of these vicariance events, or (2) that they were in fact isolated along with the still endemic taxa but have subsequently experienced a post-vicariance range expansion, either from south to north or north to south along the peninsula.

Ancestrally widespread versus recently expanded phylogroups are typically considered in phylogeography as equilibrial versus expanding population sizes, and are therefore amenable to consideration through a population genetics framework (Althoff and Pellmyr 2002; Knowles and Maddison 2002; Zink 2002). Zink et al. (2000a), investigating population structure in one of these widespread taxa, the California gnatcatcher (*Polioptila californica*), provided convincing evidence using the mismatch distribution approach (Rogers and Harpending 1992) that the population genetic signature is consistent with a relatively recent northward range expansion of this species from the southern part of the peninsula. This population architecture is therefore consistent with a scenario in which the California gnatcatcher was in fact isolated and originated as a member of the Peninsular South biota during the postulated mid-peninsular vicariance event, and following erosion of the barrier, subsequently dispersed northward into the Peninsular North area. Similarly, Nason et al. (2002) interpreted the population architecture of a columnar cactus (*Lophocereus*) on the Peninsula to reflect initial isolation in the Peninsular South and subsequent northward expansion, based on hierarchical *F*-statistics.

An intriguing component of this scenario in common between two disparate desert organisms is that we might go beyond considering dispersal in its traditional sense of being an idiosyncratic event (as opposed to vicariance as a generalized, biotic-wide event). If a number of these widespread taxa have population signatures indicating recent northward range expansion from Peninsular South, then this dispersal event would look more like the geo-dispersal (congruent range expansion) model considered by Lieberman (this volume). A similar pattern of geo-dispersal might account for the large number of taxa widespread between Continental West (CW) and the duplicated Peninsular North (PN2) areas.

REDUNDANT DISTRIBUTIONS. Redundant distributions (i.e., more than one taxon within a clade inhabiting the same area) may represent either sympatric or embedded allopatric speciation events, or post-speciation dispersal from another area. Embedded allopatric speciation is a straightforward inference if redundant distributions are still allopatric (e.g., between sister taxa 21 and 22 or 41 and 42 in CW), but differentiating it from sympatric speciation might be more problematic if ranges are overlapping within an area (e.g., between 32 and 35 or 54 and 57 in CW). However, we could again use methods of population genetics within a phylogeographic framework (e.g., nested clade analysis, MIGRATE, mismatch distributions) to evaluate the likelihood of long-term sympatry (consistent with a population genetic signature of historical population equilibrium) versus either embedded allopatric divergence followed by recent population and geographical range expansion, or post-speciation dispersal from another area (consistent with a genetic signature of recent population size expansion).

DISJUNCTION EVENTS. Historical biogeographers have long regarded phylogenetic congruence in area relationships across two or more co-distributed taxa as evidence for a history of vicariance. However, alternative processes that could also explain such phylogenetic congruence include: separate disjunction events occurring at different times or physical locations, which could be falsely inferred to reflect a single event (pseudocongruence; Cunningham and Collins 1994); and the mass movement, isolation, and divergence of entire biotas between areas (geo-dispersal). In order to reject pseudocongruence in favor of true vicariance, fossil and/or molecular data should support the null model of co-divergence at a single point in time and space. Lieberman (this volume) outlines the logic behind evaluating vicariance and geo-dispersal as alternative explanations of geographic congruence.

True vicariance is supportable from molecular divergence calculations assuming that rates of evolution are either constant across a phylogeny, or if not, can be calculated with confidence along each branch (e.g., Sanderson 2002). In either case, one must also have a way to estimate rates using independent dates from fossils or geographic events fixed to one or more nodes on a tree. Another form of complication can arise from variance in ancestral population sizes and gene lineage coalescence times across co-distributed species (Edwards and Beerli 2000; Knowles and Maddison 2002). Even if a suite of co-distributed ancestral taxa were split by a true vicariant event, if each had substantially different population structures or ancestral population sizes, the timing of gene lineage divergence could vary considerably across taxa. Therefore, one should use caution in rejecting a true vicariance hypothesis if estimated times of divergence vary somewhat across taxa in a study, although discrepancies due to differences between gene and population divergence times should become less pronounced as actual timing of events becomes large (Knowles and Maddison 2002). Nevertheless, such estimates can still be useful if applied with due caution.

For example, as an alternative to a single vicariant event separating the Peninsular South from the Peninsular North areas, one could imagine a series of isolating events beginning with Pliocene trans-peninsular seaways or northward marine transgressions about 3 to 4 my BP, followed by a series of seaways or other climatically mediated habitat shifts throughout the Pleistocene up through the latest glacial-interglacial change about 12 ky BP. Available data (Riddle et al. 2000a; Zink 2002) suggest variance in estimates of divergence times but not to the degree that one would expect if isolating events ranged from Pliocene through latest Pleistocene times, and thus do not at this time cause us to reject a hypothesis of a single historical event isolating ancestral biotas between these two areas.

Expanding the Sequential Approach: Combining Reciprocal Strengths of Phylogeography and Historical Biogeography

Over the past three decades, historical biogeography has made great strides toward developing rigorous approaches to addressing the relationship between Earth and biotic history, as well as the relative contributions of vicariance and dispersal in the assembly of biotas and the geography of speciation. Some investigators continue to argue that historical biogeography is strictly about elucidating the formation of general area relationships through a history of vicariant events (Humphries 2000; Humphries and Ebach 2002). Both phylogenetic biogeography (Brooks, this volume) and phylogeography incorporate a more inclusive perspective that seeks to explore both the general (i.e., vicariance and geo-dispersal) and idiosyncratic (i.e., dispersal) signatures of history embedded within biotas. In this sense, phylogeography is more closely aligned conceptually with phylogenetic than with cladistic biogeography.

Humphries and Ebach (2003) argue that one reason for the confusion about what exactly constitutes the realm of historical biogeography is a reluctance of practitioners to fully embrace the shift from a static Earth to a dynamic Earth perspective. Their desire is to make historical biogeography entirely about understanding the distribution of organisms on a tectonically dynamic Earth. We would expand the view of a dynamic Earth temporally to include relatively recent (e.g., late Pleistocene) dynamics, spatially to include intracontinental patterns, and mechanistically to include climatically induced habitat changes (Jackson, this volume). Doing so necessarily challenges us to seek ways to investigate dispersal histories with the same rigor that historical biogeographers have brought to the elucidation of general vicariant histories.

We agree with Humphries and Ebach (2003) about the unfortunate extent to which most practitioners of phylogeography have little experience with the past three decades (at the very least) of conceptual developments in historical biogeography. However, we believe that this relatively new discipline provides a variety of goals and techniques that may be incorporated within a temporally, spatially, and mechanistically expanded view of historical biogeography.

One of our goals has been to present an argument for using the reciprocal strengths of historical biogeographic and phylogeographic methods as a means to discover both vicariant and dispersal events in the history of biotic assembly. Doing so requires treating dispersal and sympatric (or embedded allopatric) distributions analytically rather than considering them to be mere noise. We anticipate a rocky road ahead, with some historical biogeographers saying that this melding of approaches simply cannot be accomplished with sufficient rigor, and perhaps some phylogeographers saying that the relatively new field of comparative phylogeography is sufficiently rigorous not to require synthetic approaches of this kind. However, we do believe that these two approaches in fact offer reciprocal strengths within an expanded conceptual and empirical framework of the fundamental goals of understanding the geographic history of life on Earth.

CHAPTER 6

Range Expansion, Extinction, and Biogeographic Congruence: A Deep Time Perspective

Bruce S. Lieberman

Introduction

Buffon's observations that different regions were inhabited by distinct biotas (Buffon's Law), published in the eighteenth century, established biogeography as an important subdiscipline within biology. Later, it was the recognition by Wallace (1855) and Darwin (1859) that lineages change across geographic space and through time that forged the link between biogeography and evolutionary biology.

My focus here will be on the insights that emerge in biogeography when we unite the deep time perspective of paleontology with the study of biogeography, especially phylogenetic biogeography. This unification is important for three reasons. First, the fossil record is our primary and direct chronicle of the history of life, and consideration of paleontological patterns, especially when these are integrated into a phylogenetic context, allows us to better understand how the Earth and its biota have co-evolved through time. Second, the phenomenon of extinction can influence our ability to retrieve biogeographic patterns. There are several overriding lessons for evolutionary biology manifest in the fossil record, perhaps the clearest of which is the pervasiveness of extinction; indeed, the recognition of extinction was paleontology's first theoretical contribution to biology (Rudwick 1976). It is worth considering how extinction can be factored into biogeographic studies by considering combined paleontological and phylogenetic biogeographic approaches. Finally, phylogenetic biogeographic studies have primarily empha-

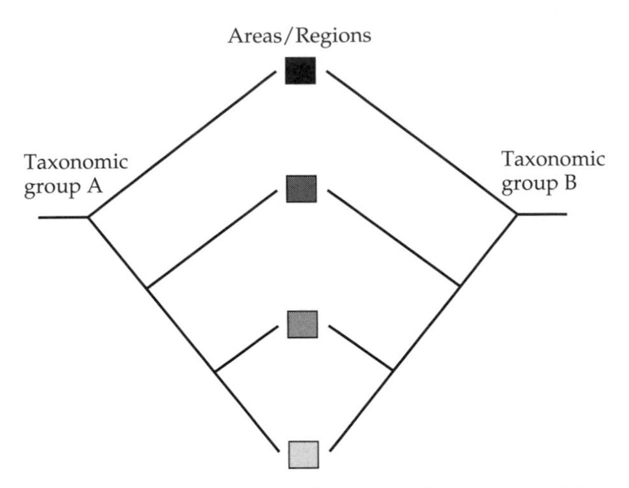

FIGURE 6.1 Two hypothetical area cladograms with congruent biogeographic patterns. The shaded squares represent different areas.

sized vicariance as the fundamental biogeographic phenomenon worthy of study because, it has been argued, this is the only process likely to produce biogeographic congruence, which involves similar patterns of evolutionary relationship across geographic space (Figure 6.1). However, the fossil record is replete with examples of congruent range expansion, or *geo-dispersal*, that can be studied within a phylogenetic biogeographic context (Lieberman and Eldredge 1996; Lieberman 1997, 2000).

In this chapter I will discuss how and why geo-dispersal, originally recognized as a paleontological phenomenon, should be integrated into phylogenetic biogeographic studies. Jackson (this volume) and Vermeij (this volume) consider how a host of other interesting paleontological phenomena can be integrated into ecological biogeographic studies.

Phylogeny and Paleontology

The development of phylogenetic approaches has been important to paleontology because these approaches have increased the rigor of paleontological studies (Eldredge and Cracraft 1980; Smith 1994). One of the best ways to test evolutionary processes "involves the comparison of the patterns of both intrinsic and extrinsic features of organisms predicted from theories of process, with those actually 'found' in nature" (Eldredge and Cracraft 1980: 4). This statement is as true of phylogenetic patterns in the fossil record as it is of phylogenetic patterns in the extant biota. Thus, whenever possible, it would seem worthwhile to consider all available databases describing the pattern of evolution because each of them provides us with information about the history of life.

The fossil record and the data preserved there—extinct taxa with distinctive character combinations—have been shown to have critical relevance for phylogenetic analyses. Gauthier et al. (1988) and Donoghue et al. (1989) have demonstrated that fossil taxa in phylogenetic studies can fundamentally change the recovered evolutionary patterns. The extension of phylogenetic techniques to biogeography also represents an important analytical advance, and provides an explicit means of relating the evolution of groups of species to a series of geological or climatic events. The data of phylogenetic or historical biogeographic studies are evolutionary relationships of groups of species (phylogenies) as well as the geographic distributions of those species.

Extinction and Phylogenetic Biogeography

For reasons analogous to those already described with phylogenetic analysis, it should be expected that including fossil taxa in biogeographic studies would be worthwhile, and to do so might possibly change the biogeographic patterns that are retrieved because they influence the shape of the underlying phylogeny (Lieberman 2000, 2002). By considering the phenomenon of biogeographic congruence this principle can be further elucidated. Such congruence is the primary pattern historical biogeographers try to retrieve between two (or more) clades of organisms (see Figure 6.1). These evolutionary patterns can be related to a series of geological or climatic events, which are predicted to influence groups of organisms in a similar way by causing them to speciate in concert. This pattern of simultaneous speciation across geographic space caused by the emergence of geographic barriers, which subdivide previously fully contiguous regions, is referred to as vicariance. Some authors have argued (e.g., Croizat et al. 1974; Platnick and Nelson 1978; Rosen 1978) that this is the only way that biogeographic congruence can result.

Many factors can cause clades to be incongruent biogeographically, and these make it more difficult to retrieve resolved biogeographic patterns. For example, geographic barriers may affect some groups of organisms but not others, or some groups may speciate in response to barriers while others do not. However, biogeographic incongruence also can arise artificially in extant clades because of extinction. Extant clades may actually be biogeographically congruent but not appear so in a biogeographic perspective based on Brooks Parsimony Analysis (BPA) because of pruning due to extinction (Figure 6.2). Simulation studies described in Lieberman (2002) have shown that biogeographic studies of extant clades that ignore extinct taxa will be predisposed to artificial biogeographic incongruence. Clades with higher extinction rates or longer durations (with more concomitant opportunities for extinction) are especially prone to artificial biogeographic incongruence and thus are less appropriate subjects for biogeographic analysis (Lieberman

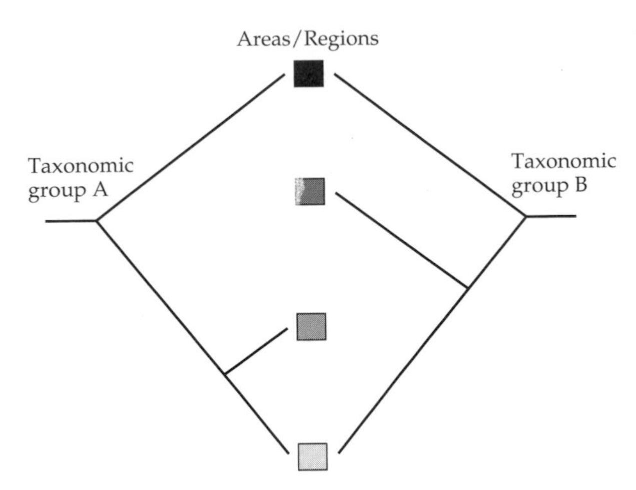

FIGURE 6.2 Now imagine that some of the representatives of the hypothetical clades in Figure 6.1 are extinct. The extant representatives of the same two hypothetical clades are shown. The two cladograms appear biogeographically incongruent, but really they are not. Because each clade has been pruned by extinction, artificial biogeographic incongruence results.

2002). This points out a special role for the fossil record—the repository of extinct taxa—in phylogenetic biogeographic studies. Considering the fossil record of any extant clade allows the inclusion of pruned taxa and diminishes the potential for artificial biogeographic incongruence.

This result is directly counter to the viewpoints expressed in Croizat et al. (1974), Nelson and Platnick (1981), Patterson (1983), and Brundin (1988). They all disparaged the value of the fossil record in phylogenetic biogeographic studies, suggesting that paleontology was a discipline of minimal relevance to phylogenetic biogeography. However, the viewpoint articulated here is consonant with the alternative outlook of Brooks and McLennan (1991, 2002) who advocated a variety of approaches and frameworks for pursuing phylogenetic biogeography. Further, the confounding effects that extinction can have on phylogenetic biogeographic studies of extant biotas had been recognized earlier by Wiley and Mayden (1985) and Brooks and McLennan (1991) who pointed out that clades with a short duration, or minimal extinction, are the best subjects of biogeographic analysis. Brooks (this volume) has also demonstrated that when three or more clades are included in phylogenetic biogeographic studies the negative effects of extinction diminish. Unfortunately, as Riddle and Hafner (this volume) also show, most published phylogenetic biogeographic studies consider only one clade.

The potential for extinction to produce artificial biogeographic incongruence in studies of the extant biota has a bearing on our choice of phylogenetic biogeographic methods. Component analysis (Rosen 1978; Nelson and Platnick 1981) is not designed to deal with any sort of incongruence, and small amounts of incongruence can lead to a loss of biogeographic resolution. Thus, the likely presence of artificial incongruence in extant clades that have experienced some extinction (probably most clades) argues against the use of component analysis in such applications (see additional cogent arguments by Brooks [1990], Brooks and McLennan [1991], and Lieberman [2000]). By contrast, BPA (Brooks 1985, 1990; Wiley 1988a,b; Brooks and McLennan 1991) has little problem dealing with biogeographic incongruence unless it is so severe that it overwhelms the underlying biogeographic pattern. This is another argument for the efficacy of BPA as an analytical tool in historical biogeographic studies in addition to the arguments presented in Zandee and Roos (1987), Funk and Brooks (1990), Brooks and McLennan (1991), Wiley et al. (1991), Enghoff (1995, 1996), and Soest and Hajdu (1997). Roy et al. (this volume) also consider some of the other effects that extinction can have on biogeographic patterns.

Geological Processes and Biogeographic Congruence

It is clear that the phenomenon of vicariant speciation, driven by geological or climatic change, can cause different clades occurring in the same region to show a congruent biogeographic pattern. However, this is not the only way that biogeographic congruence can occur. Congruent range expansion can also occur; indeed, the fact that numerous lineages at various times have expanded their ranges concurrently has been recognized by many generations of paleontologists (e.g., Osborn 1900; Matthew 1939; McKenna 1975; Hallam 1977). "This then is our problem, to connect living distribution with distribution in past time and to propose a system which will be in harmony with both sets of facts" (Osborn 1900:563).

When we speak of range expansion in the sense just enumerated it is not the traditional phenomenon of dispersal (e.g., Humphries and Parenti 1986), which among evolutionary biologists is defined as a taxon's movement over a geographic barrier with concomitant diversification. This traditional dispersal is typically not likely to produce biogeographic congruence in different clades, though it may do so in some cases (e.g., Wagner and Funk 1995; Ricklefs and Bermingham 2001). Rather, different lineages are likely to undergo traditional dispersal independently, and the factors that govern it relate to the individual ecology of each species. Instead, the kind of range expansion referred to here—the kind that produces biogeographic congruence—involves at least three steps (Lieberman 2003a). First, the effacement of geographic barriers due to geological or climatic events causes numerous range expansions or dispersal events (dispersions *sensu* Platnick 1976) involving several different species. The second step involves the re-emergence of barriers in a position within the expanded species range, possibly, though not necessarily, near their original position. The third step involves differentiation and subsequent diversification via allopatry, related to the appearance of the re-emerged barriers. Lieberman and Eldredge (1996) and Lieberman (1997, 2000, 2003a) referred to this three-step process as *geo-dispersal* to distinguish it from traditional dispersal.

Several different clades are likely to show biogeographic congruence when geo-dispersal occurs, but the pattern is not likely to be identical with congruence that has been produced by strict vicariance. This is especially true if a region has been affected by several cycles of barriers falling and then rising. Because the pattern is not identical with vicariance, cladistic biogeographic methods such as component analysis will not be able to retrieve it; yet geo-dispersal represents another clear case of the co-evolution of the Earth and its biota.

Like all ideas in biogeography, a process akin to geo-dispersal may have had its antecedents in earlier works of the eighteenth and nineteenth centuries. For example, Buffon (1761) explained both the similarities and differences between the New and Old World biotas by invoking communication and colonization between the New and Old World at a time when North America and Europe were much less isolated, followed by the increased isolation of North American species, presumably due to the emergence or expansion of the Atlantic Ocean; and finally followed by "degeneration" of North America's biota in isolation. The aspect of geo-dispersal most frequently identified in subsequent works by Lyell, Darwin, and their colleagues is stage one of the process: geographic barriers fall and taxa expand their ranges. The first scientist to definitively identify such a phenomenon (though Buffon, as described above, was an early forerunner) may have been Lyell (1832), who argued that changes in the Earth's physical geography or climate can at various times promote or retard the movement of species (e.g., Lyell 1832, pp. 160, 169). The Darwinian notebooks (Barrett et al. 1987) also contain a few cryptic descriptions of phenomena that may be equivalent to this (Lieberman 2000). Huxley (1870) believed that the herbivorous mammals of Africa "had migrated south (from Europe) once a sea barrier had been removed" (from Bowler 1996, p. 392). Also, Wallace (1876, p. 155) believed that into the southern continents "flowed successive waves of life, as they each in turn became temporarily united with some part of the northern land."

Some subsequent vicariance biogeographers including Rosen (1978), Nelson and Platnick (1981), and Brundin (1988), also identified a phenomenon that is equivalent to what is described herein as stage one of geo-dispersal. Stages two and three of geo-dispersal can also occur if barriers re-emerge after they disappear, or if barriers repeatedly rise and fall. In fact, abundant geological data suggest that geographic barriers rise and fall in repeated cycles. For example, the plate tectonic revolution demonstrated that there have been repeated cycles of fragmentation and coalescence of terrestrial and marine environments occupied by organisms (see Scotese, this volume). These cycles occur on time scales that may or may not be commensurate with those we usually associate with speciation. Overlaid on these, though, there are repeated cycles of sea-level rise and fall, some driven by plate tectonics and some by climatic changes. These and other associated climatic cycles involve repeated events on time scales of tens of thousands to hundreds of thousands and perhaps millions of years. These cyclic phenomena, and their likely concomitant effects on speciation and range expansion, have been described briefly and in general terms by Brundin (1988, p. 356–7), Cracraft (1988, p. 233), Noonan (1988, p. 377), Bremer (1992), Ronquist (1994, 1998a), and Hovenkamp (1997), and in detail by Lieberman and Eldredge (1996), and Lieberman (1997, 2000, 2003a). Vrba (1985) also considered these within a macroevolutionary framework. Finally, Whittaker (this volume) focuses on the role climate plays in producing various ecological biogeographic patterns.

Geo-Dispersal within a Phylogenetic Biogeographic Framework

A crucial issue becomes whether geo-dispersal, a process that produces biogeographic congruence, can be integrated into a phylogenetic biogeographic framework. Although the potential association between geo-dispersal and congruence may seem straightforward, recognition of this process would have limited utility if it could not be studied within a phylogenetic biogeographic framework. Fortunately, one phylogenetic biogeographic method, Brooks Parsimony Analysis, can be readily adapted to study episodes of geo-dispersal in the same way that BPA can be used to study episodes of vicariance. The way this modified version of BPA works is described in detail in Lieberman and Eldredge (1996) and Lieberman (1997, 2000); however, brief discussion is worthwhile here.

The primary modification relative to standard BPA is that a biogeographic study is divided up into two separate analyses: one to retrieve congruent episodes of vicariance among several different clades, and another to retrieve congruent episodes of geo-dispersal. Each uses analysis of a data matrix to provide information about sequences of geological or climatic events, with the results expressed as a tree. The vicariance analysis produces a vicariance tree that makes predictions about the relative sequence of events that fragmented biotas. The closer two areas are on a vicariance tree, the more recently they fragmented or split apart from one another, and thus the more recently they shared a common history. The geo-dispersal analysis produces a tree that provides information about the relative sequence of events that joined biotas. The closer two areas sit on a geo-dispersal tree, the more recently they merged with one another and shared a common history. Both trees can be thought of as in some way analogous to phylogenetic trees; they both provide useful and complementary information about the pattern of shared history. Further, vicariance and geo-dispersal trees can be placed within a geological framework. The pattern of the vicariance tree might be related to continental rifting, and, in the case of marine groups, to sea-level fall, and the pattern of the geo-dispersal tree might be related to continental collision and, in the case of marine groups, to sea-level rise. Moreover, in terrestrial organisms, sea-level rise will cause vicariance and sea-level fall will cause geo-dispersal. For example, when sea level fell as global temperatures cooled during the Pleistocene (see Scotese, this volume), the Bering Strait between North America and Asia became an emergent land bridge that prevented marine organisms from freely and directly moving between the northern Pacific and Arctic Oceans; this encouraged evolutionary differentiation in populations of these species via vicariance. By contrast, geo-dispersal between Asia and North America occurred in a host of large mammals, including *Homo sapiens.* Later, as temperatures warmed over the last 10,000 years, sea levels rose and the Bering Strait once again became a marine environment. Subsequently, wholesale geo-dispersal in marine populations occurred between the northern Pacific and Arctic Oceans. By contrast, the mammal populations now isolated on either side of the Bering Strait could begin to differentiate via vicariance.

The initial steps in the modified version of BPA are similar to those in the standard BPA. First, an area cladogram is required: a phylogeny with the geographic distribution of the terminal taxa substituted for the taxon names. To understand the evolution of a group within a geographic context we also need some information about the ancestral states of the cladogram to see whether the geographic distribution expanded or contracted between any given ancestor and its descendant(s). The algorithm used to determine ancestral states in the modified version of BPA is a form of Fitch parsimony (Fitch 1971), which assumes unordered character transformations. The Fitch algorithm was initially developed in the context of molecular systematic studies and a minor emendation to the algorithm was necessary to make it work in biogeographic studies because it did not allow polymorphisms at ancestral states. This requirement is relaxed when it is applied in the modified version of BPA for the methodological and philosophical reasons described in Lieberman (2000). Essentially, in the modified version of BPA, the final stage of Fitch parsimony (after the downward and upward passes) is dropped. An algorithm that assumes unordered character transformations is then used because, when trying to determine the sequence of transitions between geographic regions as a group evolves, there is rarely a persuasive reason to order movement between regions *a priori.*

After applying this version of Fitch (1971) parsimony, the result is an area cladogram with the biogeographic states mapped to ancestral nodes and terminal taxa. As in BPA, the information about how geographic distribution changes as a group evolves can be converted to a data matrix that can then be analyzed using a parsimony algorithm such as PAUP (Swofford 1998). However, the modified version of BPA involves creating two data matrices. One, the vicariance matrix, is designed to retrieve congruent episodes of vicariance. The other, the geo-dispersal matrix, is designed to retrieve congruent episodes of geo-dispersal. If episodes of vicariance and geo-dispersal are not analyzed separately, the biogeographic signal inherent in the data may be obscured or lost (Lieberman 2000). The best-supported patterns of vicariance and geo-dispersal are what emerge from analysis of the data matrix. If there

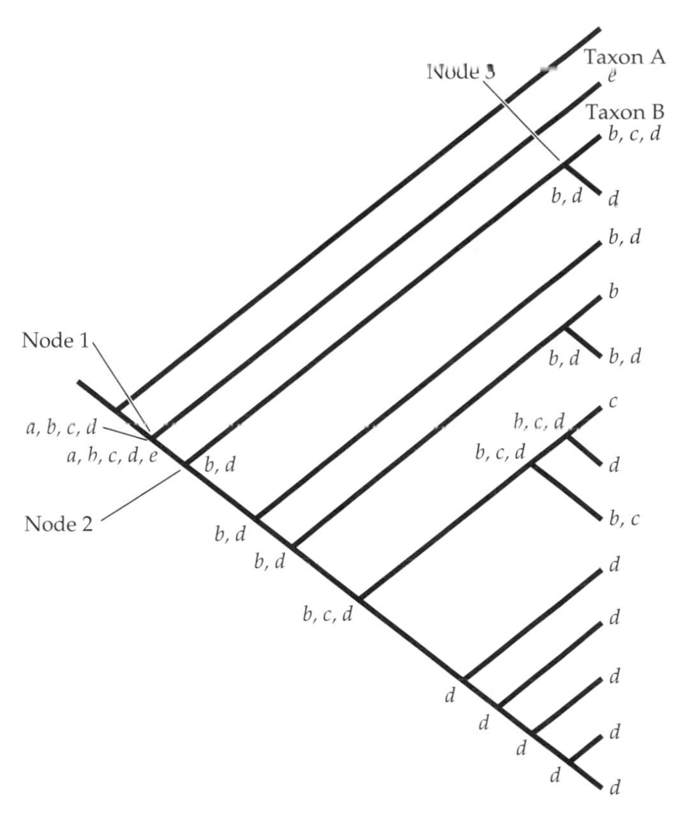

FIGURE 6.3 A hypothetical area cladogram with the geographic distributions of the terminal taxa substituted for the taxon names, and the geographic states optimized to ancestral nodes using the version of Fitch (1971) parsimony described in the text. Labeled nodes and terminal taxa are referred to in the text. The biogeographic state of the root of the tree (*a*, *b*, *c*, *d*) is also shown.

is not a single well-supported pattern, the resulting vicariance and geo-dispersal trees will be weakly resolved or poorly supported by various measures of tree support.

Area cladograms are converted to data matrices in a manner similar to BPA described in Brooks (1985), Wiley (1988a,b), Brooks and McLennan (1991), and Wiley et al. (1991). For example, the rows of the data matrix represent the different areas being studied. The biogeographic states of nodes and terminals are entered into the data matrix along with information that is captured as to how geographic distributions change between ancestral and descendant nodes and between ancestral nodes and descendant terminal taxa using ordered multi-state characters, a procedure described by Mayden (1988) and Lieberman (2000). To illustrate how area cladograms are converted into vicariance and geo-dispersal matrices, refer to Figure 6.3 and Tables 6.1 and 6.2. First, for the vicariance matrix (see Table 6.1), the biogeo-

TABLE 6.1 ***Coding of the vicariance matrix in modified BPA for part of the area cladogram in Figure 6.3***

Areas	Characters				
	1	2	3	4	5
a	1	1	1	0	0
b	1	1	2	1	1
c	1	1	1	0	1
d	1	1	2	1	1
e	1	2	1	0	0
Outgroup	0	0	0	0	0

Note: Characters 1–5 refer to codings for node 1, taxon A, node 2, node 3, and taxon B, respectively. Multi-state characters are treated as additive (ordered).

graphic distribution of node 1 is present in regions a, b, c, d, and e. Node 1 becomes a character in the data matrix and is scored with all "1's" for areas a–e. Character 2 in the data matrix describes the state of taxon A. The transition between node 1 and taxon A involves a contraction in range associated with diversification and thus represents a potential episode of vicariance. Thus, character 2 is coded with all "1's" except for the state of area e, which is coded as "2" to capture the fact that range contraction associated with divergence in this region has occurred. The transition from node 1 to node 2 involves another episode of contraction in geographic range associated with diversification, and thus is another potential episode of vicariance. This is coded as character 3, which has all "1's" except for a "2" in regions b and d. The transition from node 2 to 3 involves no change in geographic distribution.

TABLE 6.2 ***Coding of the geo-dispersal matrix in modified BPA for part of the area cladogram shown in Figure 6.3***

Areas	Characters				
	1	2	3	4	5
a	1	0	0	0	0
b	1	0	1	1	1
c	1	0	0	0	2
d	1	0	1	1	1
e	1	1	0	0	0
Outgroup	0	0	0	0	0

Note: Characters 1–5 refer to codings for node 1, taxon A, node 2, node 3, and taxon B, respectively. Multi-state characters are treated as additive (ordered).

Thus character 4 is simply coded as all "0's" except for a "1" in areas b and d. Finally, the transition from node 3 to taxon B involves an increase in geographic range from present in areas b and d to present in areas b, c, and d. However, this range expansion is not recorded in the vicariance matrix. Instead, this terminal taxon is coded as character 5: 0, 1, 1, 1, 0.

The geo-dispersal matrix (see Table 6.2) is constructed in a similar fashion. Again, node 1, character 1, is coded as present in all regions or all "1's." The transition from node 1 to taxon A involves a contraction in range. This contraction is not coded into the geo-dispersal matrix. Instead, character 2 is coded all "0's" except for a "1" in area e. The transition from node 1 to node 2 again involves a contraction so character 3 for node 2 is coded all "0's" except for a "1," or present, in areas b and d. There is no change in range between nodes 2 and 3, so character 4 is coded the same as character 3. Finally, the transition from node 3 to taxon B involves an expansion in range from present in areas b and d to present in areas b, c, and d. This range expansion is a potential episode of geo-dispersal if congruent events are found in other taxa, and character 5 is coded as 0, 1, 2, 1, 0. Then, an outgroup of all "0" regions is added to each matrix to polarize characters and signify that taxa are primitively absent from the regions of interest.

After the data matrices are analyzed, three items of information can be retrieved. First, there are the two trees that provide information about the patterns of vicariance and geo-dispersal (Lieberman and Eldredge 1996; Lieberman 2000). In addition, the patterns of vicariance and geo-dispersal can be compared. If they are similar, it implies that the same geological or climatic processes producing vicariance have produced geo-dispersal, albeit at different times (Lieberman and Eldredge 1996; Lieberman 1997, 2000). This implies cyclical phenomena such as, in the case of marine taxa, sea-level rise and fall. By contrast, patterns of vicariance and geo-dispersal that are very different potentially imply that unique tectonic events such as continental collisions, which are not cyclical (at least on time scales commensurate with speciation), nonetheless influenced biogeographic patterns.

Although it is clear that theoretically geo-dispersal can occur, an important test of any process is its emergence in pattern analysis. Examples from clades of trilobites suggest geo-dispersal is a well-established phenomenon, and its signal may be as strong as the signal of vicariance (e.g., Lieberman and Eldredge 1996), clear evidence that geo-dispersal has shaped the evolution of biotas (Figure 6.4). This is especially true of times in Earth history with extensive continental collision or repeated episodes of sea-level rise and fall, such as the Middle Devonian. At other times, when continents were mostly separating and there were few episodes of rising or falling sea levels, geo-dispersal was less frequent (Lieberman 1997, 1999, 2003b). In addition, times of excessive geo-dispersal appear to correspond to times of lower speciation rates (Lieberman 1999, 2000). These are all clear indications that Earth history signatures are reflected in corresponding evolutionary and biogeographic patterns.

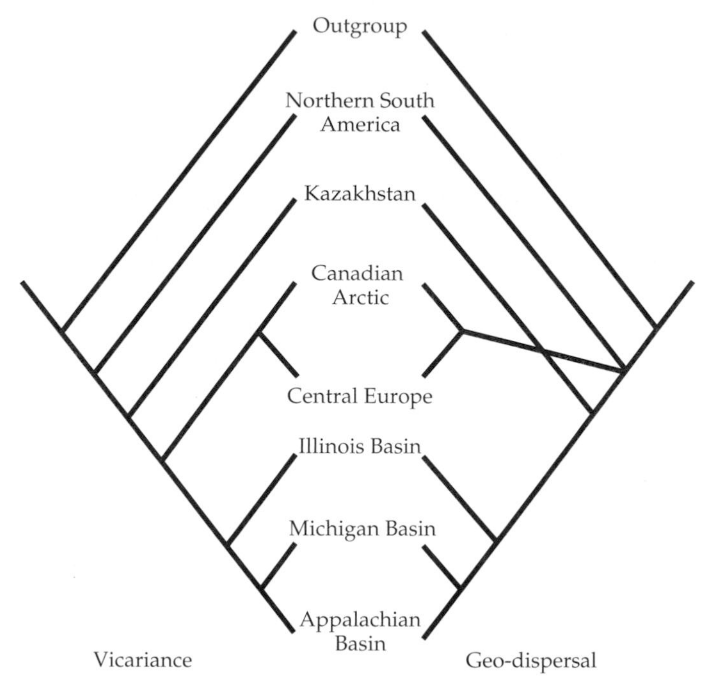

FIGURE 6.4 The most parsimonious vicariance tree and the strict consensus of the most parsimonious geo-dispersal trees for Devonian trilobites derived using the modified version of BPA described in the text (see Lieberman and Eldredge 1996; Lieberman 2000; Brooks and McLennan 2002). These trees were produced by converting several area cladograms of the type shown in Figure 6.3, derived by phylogenetic analysis of Devonian trilobites, to data matrices of the type shown in Tables 6.1 and 6.2, followed by analysis with the parsimony algorithm PAUP (Swofford 1998).

Patterns of molecular phylogeographic differentiation within species also seem to show good evidence for resolved patterns of vicariance and geo-dispersal (Klicka and Zink 1997; Lieberman 2001). This suggests that similar and related patterns of biogeographic differentiation exist at a variety of hierarchical levels.

The modified version of BPA described here unfortunately has not as yet been compared to secondary BPA, which is described in detail by Brooks and McLennan (2002) and Brooks (this volume); however, this will likely prove to be very useful. In any event, it is clear that BPA provides an excellent framework to analyze and study episodes of vicariance and geo-dispersal.

Conclusions

Although geo-dispersal was first integrated into a phylogenetic context in paleontological studies, it is a phenomenon that can be studied in groups that lack a fossil record and in situations when only the extant biota is sampled. The only evidence that needs to be uncovered is repeated patterns of range expansion on area cladograms. Still, the fact that multiple cycles of vicariance and geo-dispersal may shape the evolution of a biota points out an important role for phylogenetic paleobiogeography. The fossil record offers the potential to trace the evolution of groups during the course of several range expansion and contraction events, rather than as the last term in a series of such events. The deficiencies of the fossil record have of course been amply pointed out by various authors at least since Darwin (1859) came on the scene. However, the ability to track several biogeographic events through time would seem to be a strength of the paleobiogeographic approach when it is integrated with the rigorous phylogenetic biogeographic techniques that have already been developed (e.g., Brooks 1985; Wiley 1988a,b; Brooks and McLennan 1991).

Moreover, the key element of phylogenetic biogeography should not be an emphasis on vicariance. Rather, phylogenetic biogeography should be organized around the search for congruence in different clades showing similar changes in their geographic distribution associated with cladogenetic events. These changes can involve either range contraction (vicariance) or range expansion (geo-dispersal). Vicariance should not be dignified as the sole process that produces biogeographic congruence (also see Brooks, this volume). Rather, congruence is a test for uncovering similar biogeographic patterns, just as it is a test for uncovering homologies in phylogenetic analysis. The foregoing leads me to suggest that some promising research areas in phylogenetic biogeography are likely to involve: (1) attempts to study the effects extinction and paleontological incompleteness can have on our ability to retrieve biogeographic patterns in the extant biota and the fossil record; (2) determining the extent to which vicariance or geo-dispersal prevails in different regions and at different times; (3) increasing the integration between paleobiogeographic and biogeographic studies; and finally (4) expanding the integration between phylogenetic and ecological biogeographic studies. Research in the first three areas will help bridge the current gap between paleontology and evolutionary biology using the touchstone of biogeography. Because of the links between paleontology and evolutionary biology, advances in either area afford the opportunity for reciprocal illumination (Lieberman 2000, 2002, 2003c). The fourth area will be particularly relevant in light of the challenges the Earth's biota faces due to the current human-induced biodiversity crisis. Biogeography is a discipline with particular relevance to conservation (see the chapters by Heaney, Rosenzweig, and Sánchez-Cordero et al. in this volume), especially as biogeographers focus on identifying and preserving areas that currently house diversity and that will continue to house diversity in the future. Because of the growing importance of conservation in the twenty-first century, identifying the phylogenetic and ecological biogeographic processes that generate and maintain diversity will become even more crucial.

Acknowledgments

Thanks to Lawrence Heaney, Roger Kaesler, Mark V. Lomolino, Ed Wiley, and two anonymous reviewers for their comments on an earlier version of this paper. Thanks also to Vicki Funk, Brett Riddle, and the organizers of the *International Biogeography Society* meeting for inviting me to participate in a symposium and this volume. Financial support provided by NSF EAR-0106885 and a Self Faculty Award.

CHAPTER 7

Reticulations in Historical Biogeography: The Triumph of Time over Space in Evolution

Daniel R. Brooks

Introduction

Three scientific advances in the latter third of the twentieth century set the stage for transforming biogeography into a dynamic multi-disciplinary field. The first was the publication of the equilibrium theory of island biogeography (MacArthur and Wilson 1963, 1967), which, among other things, established that developments in any area of biogeography needed to be rooted in quantitative, testable methodologies. The second was the acceptance of the theory of plate tectonics and continental drift by geologists (e.g., Dietz and Holden 1970). This breakthrough permitted biologists to recapture a wealth of ideas linking geographic distributions of related species to general distribution patterns on a regional or global scale. The third was the advent of phylogenetic systematics. Hennig (1966) felt that superimposing phylogenetic trees on maps could provide a means of assessing the geographical context of speciation.

Choice of spatial scale greatly influences the types of questions asked and the analytical methods used in ecological biogeography. Macroecologists have suggested that researchers studying ecological associations search for regular patterns of distribution and abundance by expanding the spatial scale of their studies (Brown and Maurer 1987, 1989; Brown 1995; Maurer 1999; Whittaker 2000); recognizing this will involve phylogenetic diversification (Brown and Lomolino 2000; Heaney 2000; Maurer 2000). Brooks (1988) suggested that spatial scale might affect the complexity of phylogenetic patterns in historical biogeography. Specifically, the larger the spatial scale of the study, the more likely we are to find evidence of replicated speciation events, the greater the phylogenetic effects on the diversity examined, the older the origins of the biotas studied, and the more complex the historical explanations for the species composition of those biotas.

This apparent convergence in perspective on the part of some ecologists and some systematists during the late 1980s did not lead to a synthesis at that time. I believe this is partly because it has not been generally appreciated that two quite distinct research programs in historical biogeography have emerged. Cladistic biogeography strives to find a single general pattern of relationships among areas based on the phylogenetic relationships of species living in the areas (Nelson and Platnick 1981; Humphries and Parenti 1999). Its explanatory mechanism for these general biogeographic patterns is a form of allopatric speciation, vicariant speciation, or vicariance. Vicariance occurs as a result of physical changes in the topography or climate of the Earth, fragmenting previously widespread ancestral species into descendants; the possibility of this form of speciation became apparent only with the acceptance of continental drift and associated theories of an evolving geosphere. Vicariance does not depend on any particular biological capabilities, so many different kinds of species (theoretically all the members of a given biota) could be affected by the same vicariant event, giving rise to general or redundant geographic distribution patterns showing congruence between species phylogenies and Earth history.

Methods of cladistic biogeography were developed to seek and find these general patterns. Some have worried that those methods are prescriptions for selectively pruning phylogenetic trees to achieve a desired result (e.g., Simberloff et al. 1980; also Simberloff 1987, 1988) or for selectively interpreting results (Endler 1982). Others objected to the notion that the history of lineage indicates a singular history of place (Brown 1995). They considered this unacceptable, since it discounts dispersal, a phenomenon whose effects could be documented empirically, using, for example, the MacArthur-Wilson model. Cladistic biogeographers noted that, although in its original formulation the MacArthur-Wilson model allowed for within-area speciation, virtually all applications of the model have assumed all species occurring in a given area dispersed there from someplace else—one or more "source areas." Extended to macroevolutionary questions, this meant that all species in all areas evolved "someplace else," or to turn the problem on its head, no species actually evolved anywhere. Result: two diametrically opposed worldviews leading to radically different research programs, including different methods of analysis.

Many biogeographers believe that all historical biogeographers are cladistic biogeographers. This is not so. Parallel to the development of cladistic biogeography has been the emergence and development of an alternative viewpoint. I refer to this as phylogenetic biogeography, in the sense that Hennig

(1966) used the term—that is, the use of geographic distributions and phylogenies to understand the geographic context for speciation. The first phylogeneticist to provide a coherent research program studying the origin of species in this manner was E. O. Wiley, who showed that many significant microevolutionary and macroevolutionary phenomena have correlated phylogenetic and geographic contexts, each producing a distinctive spatial and temporal signature (Wiley 1981; Wiley and Mayden 1985). These signatures can become complex whenever evolutionary events taking place in the same geographical area have historical roots in more than one area. In such cases, the focal area has immediate historical connections with more than one other area, and the historical relationships among the areas are reticulated.

At the microevolutionary level, dispersal and gene flow may produce historically reticulated relationships among geographically localized populations. Dispersal and gene flow are necessary for optimizing trade-offs between local adaptation of populations (Williams 1992) and informational closure of the species as a whole (Brooks and Wiley 1988; Maynard Smith and Szathmary 1995). Dispersal and gene flow are responsible for historically reticulated relationships within clades due to hybrid speciation, and to reticulated relationships among areas due to allopatric speciation by peripheral isolates or to parapatric speciation. Dispersal by different species within clades leading to secondary geographic contact is necessary for speciation by reinforcement and the evolution of isolating mechanisms. Taxon pulse evolutionary radiations (Erwin 1979, 1981, 1985, 1991; Erwin and Adis 1982) produce reticulated area relationships as differentially adapted members of an evolving clade arise in different areas and disperse into areas occupied by older members of the clade—species now exhibiting plesiomorphic ecologies. Biotas produced in this manner have unique histories of assemblage comprising a mixture of residents (species that evolved within the region) and colonizers (species that evolved outside the biota and later colonized it), which results in reticulated historical relationships with other biotas. Colonization sets the stage for evolutionary phenomena driven by interspecific interactions ranging from species exclusion to coexistence by trophic and microhabitat differentiation. Finally, the fusion of land masses through changes in sea levels or tectonic movements or both, called *geo-dispersal* (Lieberman and Eldredge 1996; Lieberman 2000), can merge two or more entire biotas, producing reticulated area relationships on transcontinental scales.

Phylogenetic biogeographers are interested in assigning evolutionary explanations to the observed geographical distributions of species, encompassing vicariant speciation as well as pre- and post-speciation dispersal; in the past, many were also interested in a method of historical biogeography that could both corroborate and falsify hypotheses of vicariance. Advocates of this perspective produced considerable conceptual and methodological development during the 1980s (for a review and references, see Brooks and McLennan 2002), but few recognized that it was an alternative to cladistic biogeography. Two articles make clear the distinction between cladistic and phylogenetic methods in historical biogeography. First, Funk and Brooks (1990) presented an analysis of the "classic case of vicariance" involving the Mesoamerican fish genera *Xiphophorus* and *Heterandria*, casting doubt on the assumption that a single history of vicariance explained the distributions of the members of those taxa. Fitting these two phylogenetic trees to a vicariance model was the basis for establishing the protocols for invoking two auxiliary assumptions in the methods of cladistic biogeography, so this finding also cast doubt on the robustness and independence of the methods of cladistic biogeography. In a second article, Brooks (1990) proposed modifying an existing method of historical biogeography (Brooks Parsimony Analysis, or BPA) specifically to detect and represent not only vicariance events but also events due to dispersal, particularly those producing historically reticulated area relationships.

A Tale of Two Philosophies

Cladistic biogeography and phylogenetic biogeography each have their own quantitative methods. In order to fully understand the philosophical context of the differences between cladistic and phylogenetic biogeographers, we begin at the empirical (or epistemological) level, with an explication of how these approaches treat data, emphasizing the discovery and representation of naturally occurring reticulated area relationships due to dispersal (see also Brooks and McLennan 2002).

The phylogenetic tree shown in Figure 7.1 has been converted into a taxon-area cladogram by replacing the names of the species with their areas of distribution on the tips of the terminal branches of the phylogenetic tree. Given this information, cladistic biogeographers ask, what is the general pattern of area relationships supported by this tree? Cladistic biogeographic methods find two equally well-supported area cladograms (Figure 7.2). Both explain

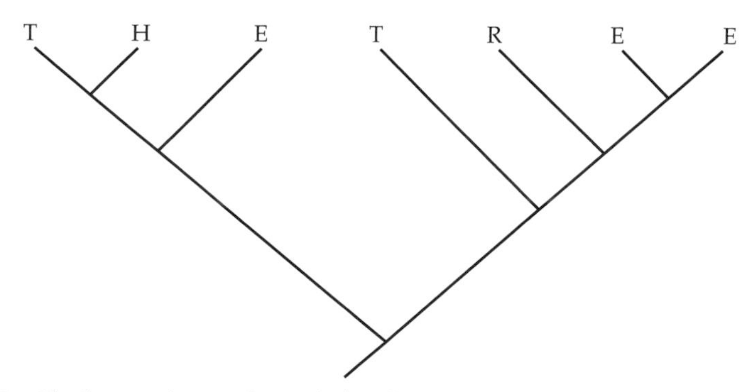

FIGURE 7.1 Phylogenetic tree for a clade of 7 species, converted into a taxon-area cladogram by replacing species names with letters indicating areas of geographical occurrence (T, H, R, and E).

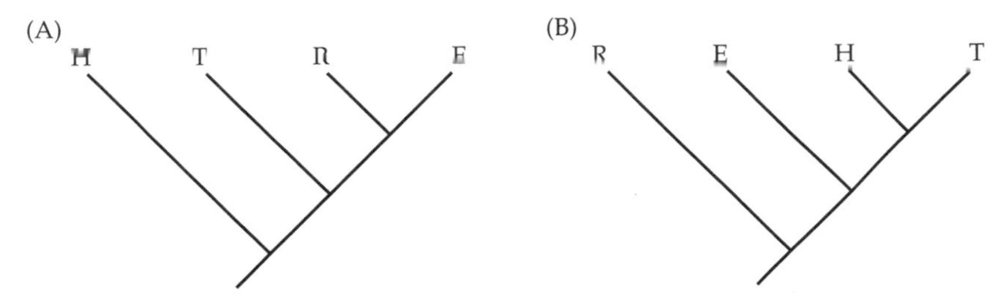

FIGURE 7.2 Two equally parsimonious area cladograms based on the taxon-area cladogram in Figure 7.1, but assuming that each area (T, H, R, and E) has a singular history with respect to the species occurring in it.

the observed distributions by invoking 5 vicariance events, 1 episode of sympatric speciation in a pair of terminal sister species, and 1 episode of lineage duplication coupled with 5 episodes of extinction (lineage sorting events)—or 12 explanations for 6 speciation events. In order to resolve the issue of which is to be the preferred hypothesis, we need more taxon-area cladograms for the same areas.

Phylogenetic biogeographers faced with the same information ask, what are the explanations for the 6 speciation events that produced the 7 species depicted in Figure 7.1? They would assert that it is not possible to make any inferences about general or special area relationships from a singular observation (1 tree). It appears that there are 6 speciation events, including 1 of within-area (possibly sympatric; possibly allopatric on a smaller spatial scale than used in the study) speciation and at least 2 involving dispersal. In order to corroborate those inferences, we need more taxon-area cladograms for the same areas.

Next, we add 2 additional phylogenetic trees converted to taxon-area cladograms (Figure 7.3, both congruent with Figure 7.2A), giving us a total of 12 speciation events. Cladistic approaches produce the area cladogram shown in Figure 7.4A. Reconciling the 12 speciation events to that area cladogram

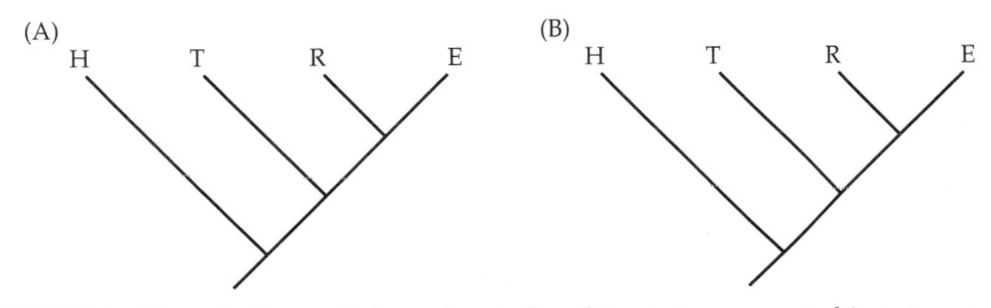

FIGURE 7.3 Two phylogenetic trees for clades of 4 species, converted into taxon-area cladograms by replacing species names with letters indicating areas of geographical occurrence (T, H, R, and E). Note that each taxon-area cladogram is congruent with the area cladogram in Figure 7.2A.

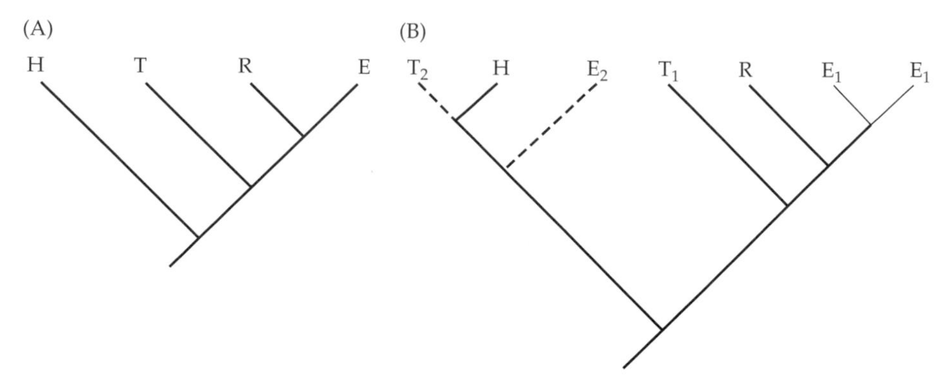

FIGURE 7.4 Area cladograms produced (A) by cladistic biogeographic methods and primary BPA, and (B) by secondary BPA for the three taxon-area cladograms depicted in Figures 7.1 and 7.3. Heavy solid lines indicate general pattern of vicariance; dotted lines indicate dispersal events; thin lines indicate within-area (sympatric) speciation event.

requires 11 episodes of vicariance, 1 episode of sympatric speciation in a pair of terminal sister species, and 2 episodes of lineage duplication coupled with 5 episodes of extinction—in all, 19 explanations for 12 events, none involving dispersal. The phylogenetic method called secondary BPA (Brooks et al. 2001; Brooks and McLennan 2002) produces the area cladogram shown in Figure 7.4B, which requires 9 episodes of vicariance, 1 episode of sympatric speciation, and 2 episodes of speciation by dispersal (peripheral isolates speciation)—or 12 explanations for 12 events, 2 involving dispersal.

Figures 7.4A and 7.4B show a substantial amount of congruence (bold lines on 7.4B). Figure 7.4B differs in depicting areas T and E as having reticulated histories. The areas denoted as H, T_1, R, and E_1 correspond to the best corroborated general distribution pattern. This means that the species occurring in those areas are most parsimoniously interpreted as being the result of vicariance. The species occurring in the areas denoted as T_1 and E_1 represent unique distributional elements, and are thus most parsimoniously explained as cases of speciation by dispersal into areas T and E from area H (dotted lines on 7.4B).

Why the Different Results? Different Optimality Criteria

For cladistic biogeographers, the goal is to find the maximum number of vicariance events. Auxiliary principles, called Assumptions 1 and 2, are invoked to eliminate problematical data from the analysis or to manufacture lineage duplications and extinctions that increase the apparent number of vicariance events supported by the data. In recent years, some advocates of cladistic biogeography have used a variety of models (most notably disperoal vicariance analysis, or DIVA [Ronquist 1997]) to allow exceptions to maximum vicariance, but those are exceptions proposed *a priori* by the investigator, not required *a posteriori* by the data. Phylogenetic biogeographers attempt to document as fully and parsimoniously as possible the geographic context of speciation in co-occurring clades. This includes the possibility that the same area may have been the site of more than one episode of speciation. BPA achieves this through the use of four principles, the most fundamental one of which was formulated by Wiley (1986d, 1988a,b; Zandee and Roos 1987) as

> ***Assumption 0***: You must deal with all species and all distributions in each input phylogeny without modification, and your final analysis must be logically consistent with all input data.

The second principle deals with the issue of what to do with areas that lack a member of a particular clade. Brooks (1981) suggested that clades that had no representatives in an area should be coded as "0" (primitively absent) in the hopes that multi-clade analyses would render those codes as reversals, indicating secondary absence, or extinction. Wiley (1986d, 1988a,b) proposed

> ***The Missing Data Coding Protocol***: All cases of "absence" should be coded *a priori* as "missing."

This principle carries with it a methodological caveat, that analyses using the missing data coding protocol require *a posteriori* interpretations of such data to provide most parsimonious inferences of primitive absence (no member of the clade was ever in the area) or extinction (secondary loss of a member of the clade).

Cressey et al. (1983) noted a third problem with the original formulation of BPA. When a clade contains one or more widespread species or one or more species endemic to the same area, the original formulation of BPA (now called primary BPA) combines the codes for all species living in the same areas, a procedure called "inclusive ORing" (for an example, see Figure 7.5). In certain cases of widespread and sympatric species, inclusive ORing will produce incorrect results detailed in Brooks and McLennan (1991, 2002). Brooks (1990) discovered that inclusive ORing caused problems only when dispersal had occurred bringing 2 or more members of the same clade together geographically, either through post-speciation range expansion or peripheral isolates speciation. He also realized that each such case was a specific falsification of the default hypothesis of vicariance. In order to eliminate the problems caused by inclusive ORing and to represent each falsifier of vicariance explicitly, Brooks proposed

> ***The Area Duplication Convention***: Whenever areas have reticulate histories with respect to the species inhabiting them, assumption 0 will be violated unless those areas are represented as separate entities for each separate historical episode. Duplicate the ambiguous areas until assumption 0 is satisfied.

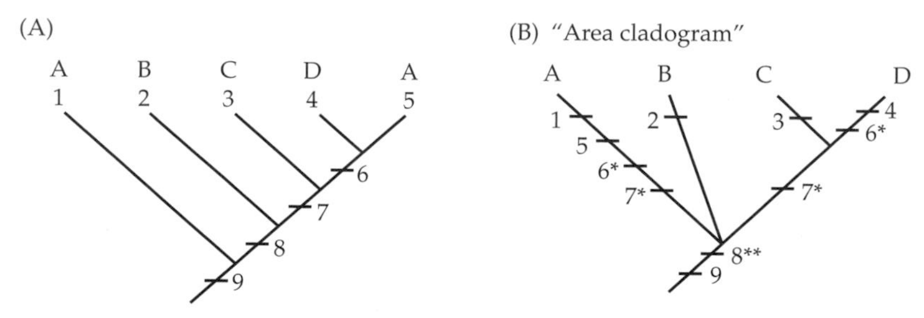

FIGURE 7.5 The problem with "inclusive ORing." (A) Phylogenetic tree for species 1–5 occurring in areas A–D. Coding this taxon-area cladogram for primary BPA involves combining the codes for species 1 and 5, because both occur in area A; this is called inclusive ORing. (B) The primary BPA area cladogram that results from analysis of data that have been "inclusively ORed" produces 3 violations of Assumption 0: characters 6 and 7 each appear twice on the area cladogram, implying that ancestral species 6 and 7 evolved twice (violations 1 and 2); and character 8 appears basal to character 1, implying that ancestor 8 evolved before its sister species (i.e., character 1) (violation 3).

Studies utilizing Assumption 0 and the missing data coding protocol (primary BPA) have confirmed that BPA produces the same area cladograms as cladistic biogeographic methods except when dispersal is more parsimonious than coupled lineage duplication and extinction, in which case the area cladograms produced by primary BPA are more parsimonious. These studies have also shown that under certain circumstances both cladistic biogeographic methods and primary BPA may exhibit internally inconsistent behavior. Cladistic biogeographic methods exhibit internally inconsistent behavior when coupled lineage duplications and extinctions are used to avoid invoking dispersal. The internally inconsistent behavior in primary BPA is due to the effects of inclusive ORing. BPA using the area duplication convention (secondary BPA), which does not include OR data, does not suffer from internal inconsistency (van Veller et al. 1999, 2000, 2001, 2002, 2003; van Veller and Brooks 2001).

The final principle highlights an additional important conceptual difference between cladistic and phylogenetic biogeography. This is best illustrated by an exemplar showing how BPA distinguishes cases of extinction from cases of peripheral isolates speciation.

Adding a fourth taxon-area cladogram (Figure 7.6) to the previous database produces the area cladogram shown in Figure 7.7, accounting for 15 inferred speciation events as 11 episodes of vicariance, 2 episodes of sympatric speciation by sister species that are terminal taxa, and 2 episodes of speciation by dispersal (peripheral isolates speciation). In addition, *a posteriori* explanation of the missing data requires us to postulate 1 episode of extinction (from area R), so there are 16 explanations for 15 events. It is more

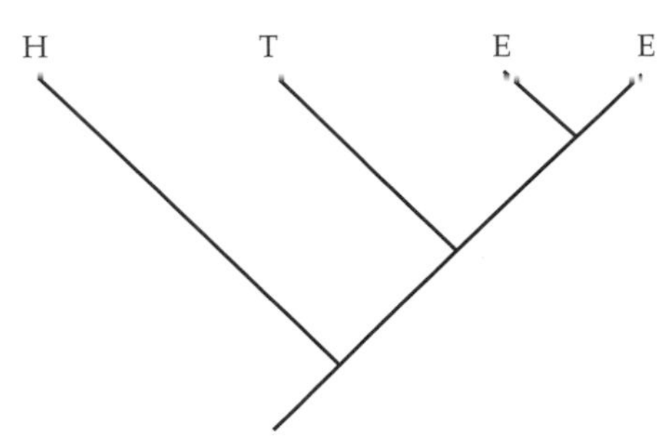

FIGURE 7.6 Phylogenetic trees for a clade of 4 species, converted into taxon-area cladograms by replacing species names with letters indicating areas of geographical occurrence (T, H, and E). This taxon-area cladogram is congruent with the area cladogram in Figure 7.2A, except that it lacks any representative in area R.

parsimonious to explain the absence of a member of clade 4 in area R as the result of a single extinction event than as the result of 3 episodes of parallel speciation by dispersal in clades 1–3. Similarly, it is more parsimonious to suggest that areas T_2 and E_2 are the result of single episodes of peripheral isolates speciation by members of clade 1 than of 3 episodes of vicariance with extinction. Parenthetically, cladistic biogeographic methods produce the area cladogram shown in Figure 7.3A, invoking 13 episodes of vicariance, 2 episodes of sympatric speciation, 2 episodes of lineage duplication, and 6 episodes of extinction—or 24 explanations for 15 events.

Brooks (1981) noted in the original formulation of BPA that, just as any single character might support incorrect phylogenetic relationships due to homoplasy, any single clade might support non-vicariant area relationships;

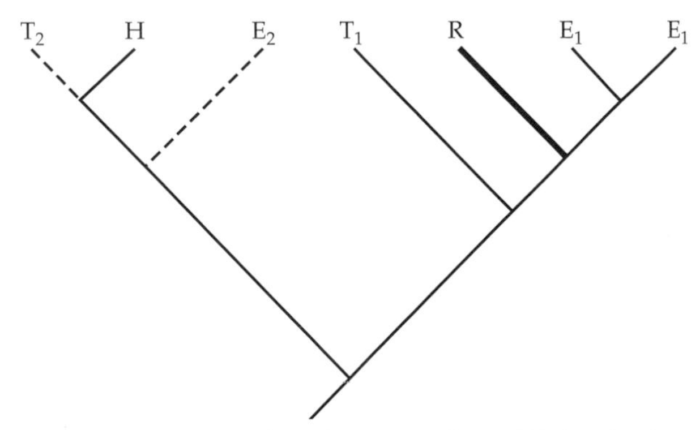

FIGURE 7.7 Area cladogram produced by secondary BPA for the four taxon-area cladograms depicted in Figures 7.1, 7.3, and 7.6. Solid lines indicate general pattern of vicariance (including within-area [sympatric] speciation event in area E); dotted lines indicate dispersal events; heavy solid line indicates secondary loss (extinction) event.

therefore, analysis of multiple clades was necessary to obtain the general pattern. This has been codified (Brooks et al. 2001; Brooks and McLennan 2002) as

> ***The "Threes Rule" Rule*:** In order to distinguish between general and special patterns of historical association based solely on the available data, and in particular to determine whether absence from an area is due to secondary extinction or primitive absence, you must analyze *at least three* co-occurring clades.

In other words, both the general pattern and the exceptions to it must be deduced from multiple examples. For example, Lieberman (this volume) has shown that primary BPA of single clades with missing taxa may produce spurious over- or underestimates of vicariance.

Why Different Optimality Criteria? Different Ontologies

Historical and ecological biogeogeographers in the 1970s and 1980s comprised two different worldviews (ontologies) with different research programs using different methods of analysis, each of which was appropriate for the research program and its underlying philosophy. Cladistic biogeography and phylogenetic biogeography are both characterized by different methods of analysis. Does this mean that they are implementing different research programs justified by different ontologies? Recent discussions seem to assert this to be true (Andersson 1996; van Veller and Brooks 2001; van Veller et al. 2003). Cladistic methods modify the data when necessary to provide maximum fit to the null hypothesis of all-vicariance; this amounts to an *a priori* parsimony criterion. Data that conflict with the null hypothesis must be flawed (see items of error in Nelson and Platnick 1981:417). Assumptions 1 and 2 are essential for eliminating such data from the analysis directly or indirectly by creating lineage duplications and extinctions. The *a priori* hypothesis is not directly tested or falsified, so the research program must be justified by asserting that the world (of historical biogeography) is parsimonious. This is the **ontology of simplicity**.

Phylogenetic methods, by contrast, falsify the null hypothesis of a simple vicariance explanation when the data do not support it, but only to the extent required by the data; this amounts to an *a posteriori* parsimony criterion. Data that appear to conflict with the null hypothesis indicate that the null hypothesis is flawed, providing objective (i.e., *a posteriori*) support for the conceptual framework (Popper 1968); this is the reason Assumption 0 is the fundamental principle of BPA. BPA accounts for flaws in the null hypothesis with each area duplication. Logical parsimony plays a critical *a posteriori* role by determining the minimum number of falsifications of the null hypothesis necessary to explain all the data. Adopting this viewpoint is justified by asserting that the world (of historical biogeography) is not parsimonious—this is the **ontology of complexity**.

Different Philosophies in Historical Biogeography: How Do We Choose?

Cladistic and phylogenetic biogeographic methods produce different area cladograms only when dispersal has produced unique or reticulated area relationships. Thus, if putative area reticulations in real datasets are rare, we could infer that they are atypical enough to be ignored, and we could use the efficiency of cladistic methods to generate simple area cladograms rapidly. How would we determine this? We cannot use cladistic methods because they are designed in such a manner that reticulations can never be found. We can use BPA, however, to ask how common such phenomena are.

All non-reversal homoplasy in primary BPA is the result of area reticulations (this is also the methodological key to implementing the area duplication convention; see below). Published studies that have implemented some version of primary BPA are examples of systems in which substantial numbers of areas can be inferred to have reticulated histories (e.g., Crisci et al. 1991; Ruedi 1996; Van Soest and Hajdu 1997; de Jong 1998; Marshall and Liebherr 2000). The single most extensive database to date showing large amounts of area reticulations is that for the evolution of the biota of the Hawaiian Islands (Wagner and Funk 1995). Equally significant are the fossil data indicating substantial influences of geo-dispersal in deep history (see examples and references in Lieberman 2000), indicating that even general patterns may include reticulate area relationships.

Secondary BPA provides explicit estimates of dispersal-based phenomena, including reticulated area relationships. McLennan and Brooks (2002) noted that BPA-based analyses for birds occurring in 7 areas of endemism in the southwestern United States and Mexico and in 10 areas of endemism in Australia indicated that 35/44 (79.5%) of the areas of endemism identified have reticulated histories; the same is true for poeciliid fishes occurring in 10 areas of endemism in Mesoamerica, and for the arthropods occurring in 17 areas of endemism in the Australian wet Tropics discussed above.

We may also use secondary BPA to assess the evidence in favor of evolutionary radiations involving dispersal and area reticulations. The taxon pulse mode of evolutionary radiations proposed by Erwin (1979, 1981, 1985, 1991; Erwin and Adis 1982) is similar to that proposed by Darlington (1943) and named "taxon cycle" by Wilson (1959, 1961). Both models assume that: (1) taxa and adaptations arise in a "center of origin," and (2) distributional ranges of taxa periodically fluctuate around a more stable, continuously occupied center (Liebherr and Hajek 1990). During expansion phases of the habitat, peripheral patches are colonized. Conversely, during contractions of the habitat, these patches become isolated. Taxon cycles occur over relatively short periods of time ("ecological time") and involve species that disperse actively and colonize new areas without necessarily producing new species, whereas taxon pulses occur over relatively long periods of time ("evolutionary time") and can be thought of as taxon cycles characterized by dispersal leading to speciation by peripheral isolates along a broad advancing front during expansion of suitable habitat (Liebherr and Hajek 1990; Bouchard et al., in press). The historically contingent nature of taxon pulses means that at any given time, different clades comprising a complex biota will likely form a mosaic of different stages in the process; therefore no single clade will likely provide evidence of an entire taxon pulse cycle by itself, especially if there have been extinctions. Subsequently, analysis of co-occurring multiple clades using methods that permit the identification of area reticulations is thus necessary for investigating the possibility of taxon pulse radiations.

Bouchard et al. (in press) performed BPA of 17 areas of endemism in the tropical wet forests of Queensland, Australia based on 15 clades of forest arthropods. Secondary BPA supported 3 general area cladograms, each comprising a mixture of vicariance, peripheral isolates speciation, sympatric speciation, and post-speciation dispersal. These findings indicate a high degree of biogeographic complexity characterized by the following:

- There is no support for a single pattern of area relationships for the 87 species included in the analysis (tested using primary BPA)
- Three independent modules of area relationships are needed to explain the data (different clades support different patterns)
- Vicariance is not the principal mode of speciation
- Within-area speciation (ecological differentiation) is common
- Across-area peripheral isolate speciation accounts for a significant number of events
- All 17 areas of endemism have reticulate histories (uniquely assembled)
- Widespread species occupy between 2 and 10 areas of endemism
- In each of the 3 area cladograms, there is a single clade that becomes associated with other clades through colonization (sequential addition)

These findings suggest that this arthropod fauna has been assembled through a complex history involving taxon pulse radiations of various taxa. Similarly, Marshall and Liebherr (2000) produced a primary BPA for 33 clades of various taxa living in various areas of Mesoamerica. Their area cladogram had a consistency index of less than 55% for informative characters, suggesting that almost half of the speciation events involved dispersal leading to area reticulations.

Finally, phylogenetic biogeographic analysis has begun providing studies that bridge historical biogeography with the MacArthur-Wilson model. MacArthur and Wilson (1963) recognized that species richness on islands is not strictly due to colonization. They hypothesized that "radiation zones" exist in island faunas—adaptive radiation increases with distance from the major source region (the source being an island within the archipelago).

Within-island speciation could thus be a major contributor to species richness. Until recently, however, most research using the MacArthur-Wilson model has neglected the element of within-island speciation. Heaney (2000) produced a series of macroevolutionary predictions derived from MacArthur and Wilson's hypothesis of radiation zones. Islands close to sources of colonizers will not exhibit as much speciation as islands far from a source. This is because colonization on a far island will be a rare event, giving colonizers less chance of experiencing gene flow with parental populations, which would hinder speciation. Small islands will not exhibit as much speciation as large islands, simply because there is not as much opportunity in terms of absolute space and probably in terms of spatial heterogeneity. Thus, there is reason to believe that macroevolutionary aspects of island biogeography complement ecological aspects: (1) within-island radiations should occur most often on large islands that are far away, (2) colonization should occur more often on near than far islands, and (3) speciation should occur more often on far islands than on near islands. Heaney (2000) applied these ideas to Philippine murid rodents and proposed that rare successful colonization of islands from the mainland had been followed by within-island diversification. Similarly, Losos and Schluter (2000) showed that intra-island speciation, rather than colonization, was the primary source of new species of *Anolis* lizards on Caribbean islands larger than 3000 km^2. Finally, Spironello and Brooks (2003) demonstrated, in the case of one clade of insects inhabiting Pacific archipelagoes, a sequence of island colonization, followed by within-island diversification on Tahiti, followed by additional colonization. Their data indicate that islands close to Tahiti have been colonized more often than islands far from Tahiti, as predicted by the MacArthur-Wilson model. They also discovered, however, that colonization of near islands led to speciation relatively less often than did colonization of distant islands. These findings all demonstrate that dispersal has played a significant role in structuring historical biogeographic patterns, and that reticulated historical relationships among areas are common, not rare.

Frontiers of Phylogenetic Biogeography

Deeper Time: Paleobiogeography and Geo-Dispersal

If vicariance has played a major role in the diversification of life on this planet over its entire history, there must have been episodes of biotic expansion between episodes of habitat fragmentation leading to vicariant speciation. Biotic expansion has two components. The first, and most obvious, component involves the active movement of organisms, leading to taxon pulse radiations. The second component, geo-dispersal, is the necessary precursor to area fragmentation producing vicariance events. Geo-dispersal, however, can produce its own general historical biogeographic patterns. Lieberman (2000, and this volume) has developed a method, utilizing BPA and a form of character optimization called Fitch optimization, for distinguishing general distribution patterns due to vicariance from those due to geo-dispersal.

Shallower Time: Phylogeography

Species are subject to two conflicting sets of processes—those tending to split the species into descendant species and those maintaining the cohesiveness of the species as an evolutionary whole (Wiley 1981). Phylogeography aims to uncover the structure of genetic variation within species by reconstructing phylogenetic relationships among geographically localized populations, then examining the effects of relatedness and geography on differences in the genetic structure of those populations (Avise 2000). Lineage diversification and reticulations have generally been represented in separate analyses, although there has recently been a call for the development of methods capable of examining both simultaneously (Althoff and Pellmyr 2002). Applying BPA at the population level requires the use of gene trees for three or more genes (including at least one that exhibits independent assortment) over populations of one species, or of multi-population clades of three or more species. The interpretation of the secondary BPA substitutes "population" for "species" and "gene flow" for "dispersal." Riddle and Hafner (this volume) have begun extending BPA to phylogeographic studies, exploring the possibility that patterns of genetic differentiation among populations within a species can distinguish cases in which a species is found in 2 adjacent areas because it did not speciate when a barrier formed, from those cases due to post-speciation dispersal after the barrier formed.

Technology: Automated Approaches to Generating Area Cladograms

In BPA, different historical biogeographic phenomena are inferred from the analysis *a posteriori* in accordance with the most parsimonious description of the data. This task cannot be simplified by adopting *a priori* costs, probabilities, or weights associated with different historical biogeographic phenomena without decreasing the power of BPA as a tool for discovery. This does not mean that BPA is non-algorithmic, however. BPA can be implemented in two ways using existing computer programs, either by simultaneous analysis of all clades, producing a primary BPA, which is then converted to a secondary BPA by invoking area duplications to account for all dispersals (which appear as non-reversal homoplasy in a primary BPA), or by analysis of all clades by sequential addition of clades using the inclusion/exclusion rule of Hennigian phylogenetic systematics. In the latter case, violations of the inclusion/exclusion rule point to episodes of dispersal, which are accounted for by invoking the duplication rule. For details, see Brooks and McLennan (2003).

TABLE 7.1 *Venn diagram in "message" form of 17 (hypothetical) clades distributed among 8 (hypothetical) areas*

Clade 1	(((TH)(IS))(T(R(EE))))
Clade 2	(((TH)(IS))(R(E(AL))))
Clade 3	(((TH)(IS))((IS)(R(E(AL)))))
Clade 4	(((TH)(IS))((IS)(T(R(EE)))))
Clade 5	(((TH)(IS))((T(HE))(T(R(EE)))))
Clade 6	(((TH)(IS))((T(HE))(R(E(AL)))))
Clade 7	(((TH)(IS))((IS)((T(HE))(R(E(AL))))))
Clade 8	((TH)((IS)((T(HE))(T(R(EE))))))
Clade 9	((T(HE))((R((E(AL)))(T(R(EE)))))
Clade 10	((R(E(AL)))(T(R(EE))))
Clade 11	((T(IS))(R(E(AL))))
Clade 12	(((TH)(IS))(T(RE)))
Clade 13	((TH)((IS)(R(E(AL)))))
Clade 14	(H((IS)(T(R(EE)))))
Clade 15	((T(HE))(T(R(EE))))
Clade 16	(T((IS)((HE)(T(R(EE))))))
Clade 17	((H(IS))(T(R(EE))))

Note: Letters represent areas inhabited by particular species; parentheses represent the phylogenetic relationships among those species.

At present, there is no computer program dedicated to the task of implementing BPA. A novel approach currently under development (Wojcicki and Brooks 2004) is based on text recognition software technology. Each taxon-area cladogram is assumed to represent part of a complex message about the evolution of the members of a group of biotas. Just as a complete message can be reconstructed from multiple fragments, no one of which need be complete, the history of the evolution of multiple complex biotas can be reconstructed from multiple clades, no one of which need represent the complete story. For example, consider 17 (hypothetical) clades distributed among 8 (hypothetical) areas (A, E, H, I, L, R, S, and T). If each of those 17 clades is converted to a taxon-area cladogram and represented as a Venn diagram (Table 7.1), each Venn diagram represents at least part of a message, with the letters representing areas inhabited by particular species and parentheses representing the phylogenetic relationships among those species.

From this perspective, cladistic methods (and primary BPA) correspond to a model of text recognition in which each letter in the alphabet can be used only once (*a priori* parsimony, or the ontology of simplicity). Figure 7.8 depicts the evolutionary message obtained from these data under that constraint. Secondary BPA corresponds to a model in which each letter of the alphabet can be used multiple times as the data demand (*a posteriori* parsimony, or the ontology of complexity). Figure 7.9 depicts the evolutionary message obtained under those conditions.

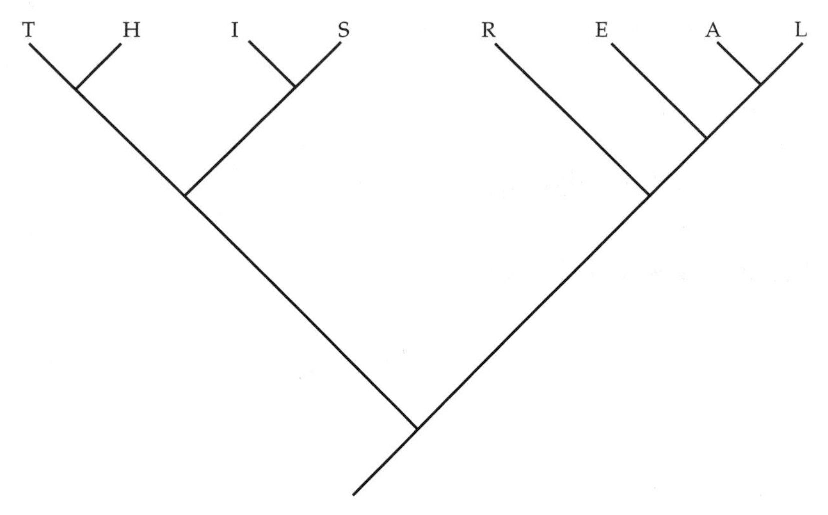

FIGURE 7.8 Area cladogram for 17 hypothetical clades inhabiting 8 hypothetical areas, as depicted in Table 7.1, produced by primary BPA and cladistic biogeographic methods.

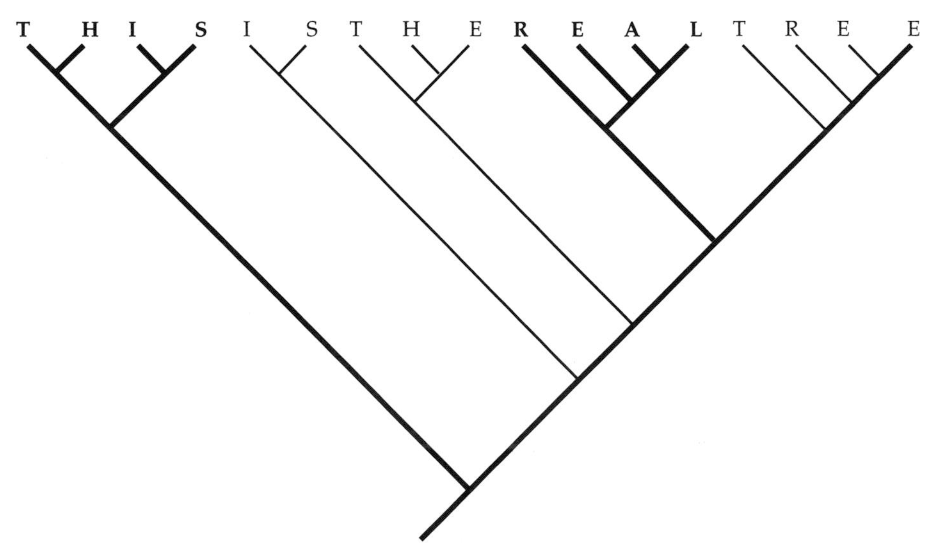

FIGURE 7.9 Secondary BPA area cladogram for 17 hypothetical clades inhabiting 7 hypothetical areas, depicted in Table 7.1.

Conclusion: A Second Chance at a Synthesis

Prominent biogeographers have recently called for a new synthesis in biogeography (e.g., Brown and Lomolino 2000; Heaney 2000; Lomolino 2000a,b; Whittaker 2000). This is an exciting prospect, especially the recognition that historical biogeography is an important element of such a synthesis (Brown and Lomolino 2000; Maurer 2000; Heaney 2000; Whittaker 2000). I believe the synthesis will be aided substantially by the recognition that vicariance produces general patterns, which is not a theory of historical biogeography but part of a methodological basis for it. We are not led inexorably to conclude that all of nature is ordered in a single specifiable area cladogram lacking reticulations. In fact, the same processes producing vicariance can result in redundant patterns of dispersal as well. I suggest that we return to the late 1980s and the rapprochement between historical biogeography and other areas of biogeography—especially macroecology—which should have, but has not yet, happened. Some major elements of common ground between these disciplines can be encapsulated in the following:

- The evolution of the geosphere and biosphere are both temporally and spatially constrained, so they will be historically correlated. That correlation will be complex and non-linear, because Earth and life do not evolve together in a Lamarckian or orthogenetic manner.
- Time trumps space in biological evolution. If the evolution of the biosphere were the result of a single massive dispersal event, followed by nothing but vicariance, there would be one species per area on this planet. If the evolution of the biosphere were the result of a massive number of episodes of sympatric speciation (lineage duplication), followed by a single, parallel massive dispersal event, followed by nothing but vicariance and extinction (lineage sorting), each biota would be a clade. Neither of these occurs anywhere on this planet.
- Increasing the temporal scale increases the spatial scale, but this is not a linear relationship because dispersal can occur in all directions.
- The larger the temporal scale, the more reticulated area relationships there will be. The relationship between temporal scale and historical reticulations of areas will be non-linear, however, again because Earth and life do not evolve together in an orthogenetic manner. It is possible that the only generalization we will be able to derive is that as the temporal scale of a study increases, so will the spatial scale, as well as the chances of dispersal producing reticulated area relationships (for an example, see Figure 7.10). Thus, the relationship between age and area will be a non-linear one mediated by a substantial amount of historical contingency, both biotic and abiotic.

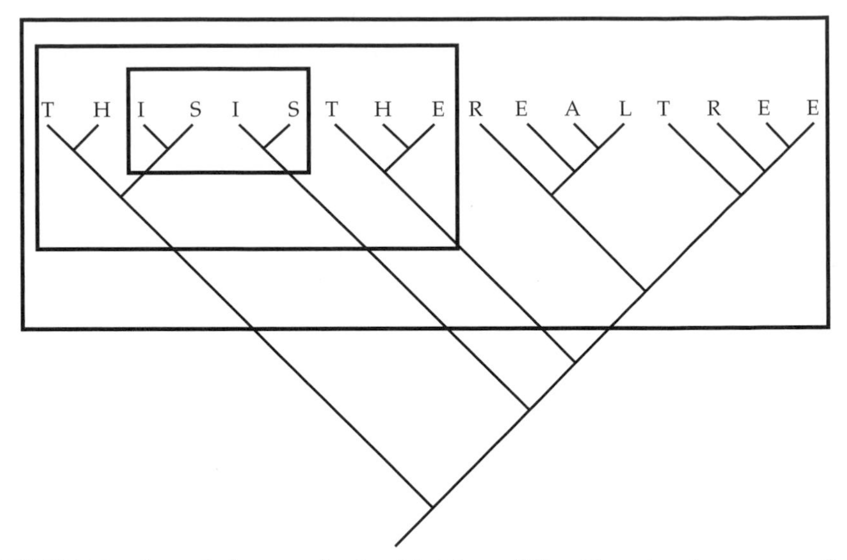

FIGURE 7.10 Area cladogram depicted in Figure 7.8, with rectangles representing heuristic view of increasing temporal scale being associated with increasing spatial scale as well as increasing historical reticulations of areas. The smallest rectangle indicates 2 areas and 2 reticulations, the medium-sized rectangle indicates 5 areas and 4 reticulations, and the largest rectangle indicates 7 areas and 8 reticulations.

If area reticulations are not noise or error, but rather information to be documented and explained, the methodological bases for historical biogeography becomes clear:

- We must discover and evaluate both general patterns and their exceptions, including reticulated area relationships.
- Vicariance is the default explanation for general patterns. This is not a strong explanatory assumption, because other modes of speciation (sequential colonization of islands, or geo-dispersal) may also produce replicated patterns. Does this leave phylogenetic biogeography without a null hypothesis? No. The descriptive null hypothesis is that all distributions of all species in all input phylogenies will correspond to a single set of non-reticulated area relationships. **The explanatory default explanation for such general patterns is vicariance only because it is the only mode of speciation that always produces general distribution patterns**.
- Inductive methods require process models, which should be complex and non-linear with respect to spatial and temporal scaling. None of the currently available methods for cladistic biogeography have integrated reticulated area relationships into their models. The high degree of his-

torical contingency in evolution may mean that no model is capable of capturing sufficient relevant detail, which can be tested by evaluating different models in light of the findings of BPA.

- Hypothetico-deductive approaches require four principles: (1) Assumption 0, (2) Missing Data Coding, (3) Area Duplication Convention, and (4) the "Threes Rule" Rule. At present, these elements are incorporated only in BPA. The high degree of historical contingency in evolution may mean that we will need to analyze many more clades from many more areas than we have today before we can begin to talk about general patterns. We have named less than 10% of the species on this planet, and most of those are known only from their type locality. We have analyzed less than 10% of the named species phylogenetically, and for most of those, the existing phylogenetic hypothesis is the first one ever produced. Finally, of that tiny fraction of the planet's species for which we have even a minimal phylogenetic hypothesis, less than 10% have been analyzed biogeographically, predominantly using methods that do not allow us to detect the effects of dispersal, particularly in producing area reticulations.

The concept of a new and broadly inclusive synthesis in biogeography is not just a good idea for basic research. If there is a biodiversity crisis, and if historical biogeography has anything relevant to contribute to resolving that crisis, it is time for systematists and ecologists to join forces. Ecologists need the descriptive platform provided by historical biogeography in order to understand the complex origins of meta-communities and communities, both their included species and the interactions among those species. Systematists, in order to provide that platform, need first and foremost support for more and faster basic taxonomic inventory work in order to document species, their distributions, and phylogenetic relationships. This should be accompanied by analysis of those data biogeographically using methods that permit us to see all possible historical complexities. Fully implemented, BPA provides a means of discovering arbitrarily high amounts of evolutionary complexity on different temporal scales (and thus evolutionary levels) without sacrificing the rigor of phylogenetics, guaranteeing explicit and accurate representation of the data based on a minimum of *a priori* assumptions. By using BPA, we can determine the general and special elements of geographic distributions based on biological evolution, and then ask if these elements conform to identifiable episodes in the geological history of the areas in question. In this manner, we can begin to understand how the Earth and life have evolved together, even when each evolves independently. And just possibly, from that knowledge will come part of the answer to the question, How can we preserve the biosphere?

Acknowledgments

My deepest thanks go out to Vicki Funk and Brett Riddle for inviting me to speak at the inaugural meeting of IBS. Special thanks to Deborah McLennan, Patrice Bouchard, Ashley Dowling, Vicki Funk, Dominik Halas, Eric Hoberg, Greg Klassen, Arnold Kluge, Bruce Lieberman, Michelle Mattern, Rick Mayden, Craig Mortiz, Bob Murphy, Brett Riddle, Marco van Veller, Ed Wiley, Rick Winterbottom, Maggie Wojcicki, David Zamparo, and Rino Zandee for discussions past and present. Deborah McLennan, Brett Riddle, Bruce Liebeman, and the editors of this volume offered advice that improved this manuscript substantially. Funding for this study was provided in part by an operating grant from the Natural Sciences and Engineering Research Council (NSERC) of Canada.

PART III

DIVERSITY GRADIENTS

Dov F. Sax and Robert J. Whittaker

The study of biological diversity has undoubtedly occupied human thought throughout much of our existence. Many of the works of ancient Greeks and Romans—Aristotle, Pliny the Elder, and others—characterized the diversity of species in the known world. It was not until the age of European exploration, however, that the great scope and variety of biological diversity on the planet was first appreciated (Brown and Lomolino 1998; Briggs and Humphries 2004). These early naturalists were among the first to wrestle with many of the questions associated with biological diversity: how should the components of biological diversity be characterized; why are some places more diverse than others; and what processes have led to the creation of diversity? The characterization and classification of species was most strongly influenced by Carolus Linnaeus, the father of modern taxonomy, who developed the system of Latin binomials that we use to this day to describe species. (Linnaeus also put forward a biogeographical model of how present-day patterns of diversity might have arisen based on the assumption of the biblical flood, but neither the premise nor the model have stood the test of time [Brown and Lomolino 1998].)

Since the time of Linnaeus, the quest to understand biological diversity has largely been a struggle to understand the spatial and temporal variation in the number and kinds of species (although interest in the genetic level has exploded in recent decades). Significant steps were in fact made as early as the eighteenth century by Johann Reinhold Forster and Alexander von Humboldt, who discerned the importance of climate in determining latitudinal variation in diversity. Forster also described the importance of island size (and isolation) in affecting species diversity, which provided a framework for later studies of island biogeography. Of nineteenth century scholars, Alfred Russel Wallace and Charles Darwin, co-discoverers of the theory of natural selection, undoubtedly have the greatest name recognition today. Wallace was perhaps foremost in characterizing geographic variation in biological diversity, and in laying the foundations of zoogeography (Brown and Lomolino 1998). Darwin's *The Origin of Species* also makes much use of biogeographical evidence, devoting two chapters to the topic of geographical distribution and two chapters to the geological record, from which the tremendous dynamism of the Earth and of life upon it was strikingly evident.

Our understanding of biological diversity advanced considerably in the twentieth century. A key development was the emergence of the so-called New Synthesis in the 1930s, when the genetic mechanisms of evolutionary change first became understood. This modern evolutionary perspective was then wed to developments in ecological theory, helping to explain patterns of biological diversity (such as the latitudinal gradient in diversity) that were becoming better characterized as synthetic studies of multiple taxonomic groups began to emerge (for a discussion, see Brown and Sax 2004).

Paralleling the development of the New Synthesis, and the evolutionary and ecological questions that emerged from it, was a growing appreciation of the dynamic nature of biological diversity on all scales of observation. This appreciation was due largely to a number of studies that highlighted temporal change in biological diversity over relatively short periods of time. For example, following the 1883 eruption of the volcanic island of Krakatau, biologists began recording patterns of species recolonization and turnover, providing data that would eventually have a notable role in the development of dynamic theories of island biogeography (MacArthur and Wilson 1963, 1967; Whittaker et al. 1989; Thornton 1996; Whittaker 1998). Also important, was the development of the workhorse paleoecological technique of pollen analysis and its application by the likes of J. Iversen and subsequent workers, who provided dramatic evidence of massive shifts in species distribution and community composition that occurred between the late Pleistocene and early Holocene (e.g., Godwin 1956; Davis 1969; see also Betancourt, this volume).

Of totemic importance in our understanding of dynamic patterns of species diversity, however, was the development of the "theory of island biogeography" by Robert H. MacArthur and Edward O. Wilson (1963, 1967). This work played a key role in providing an explanation of variation in diversity in terms of contrasting rate-limited processes, in this case via the differential effects of colonization (afforced by speciation) and extinction. Island biogeography theory continues to be one of the principal structural models for our understanding of diversity, although it is no longer seen as an adequate model in itself (Brown and Lomolino 1998; Whittaker 1998). Its importance is seen in the change in approach from that which characterized biogeographic studies at the beginning of the twentieth century, when explanations of species diversity patterns were often historically and idiosyncratically based, to those at the end the century, when greater emphasis was placed on the search for general, dynamic models—that is, models that not only recognize the dynamics of diversity patterns through time but which seek out generally consistent (predictable, repeatable) physical or dynamic relationships and mechanisms that regulate life (as described in the chapter by Whittaker).

Many of the dynamic models proposed to explain patterns of species diversity are difficult to test directly; this is largely attributable to the practical and ethical problems inherent in conducting manipulative experiments at geographical scales. As a consequence of this, biogeographers have had to rely upon a range of less direct approaches, including extensive use of comparative and correlative studies, so-called natural experiments, space-for-time substitution, and tests by inference. Many of these approaches suffer from difficulties in distinguishing causation from correlation (but see, e.g., H-Acevedo and Currie 2003) and a variety of other methodological pitfalls.

Another difficulty is knowing when and how to deconstruct (or divide apart) the units of diversity that are being examined. For instance, geographic clines in species diversity can vary dramatically between plant data sets compiled for different growth forms, or between mammals of different body size and/or feeding guilds (e.g. Andrews and O'Brien 2000; Bhattarai and Vetaas 2003). Such differences may be explicable from a consideration of general ecological theory. If we are to develop a predictive and testable science, then we need to develop more sophisticated approaches to modeling species richness.

In this section, Marquet et al.'s chapter explores the idea of deconstructing diversity as a step in the search for a mechanistic understanding of variation in species diversity. They develop a heuristic approach that divides determinants of diversity into three categories: evolutionary patterns; external properties and states of the environment; and internal properties and states of species. By contrasting these types of determinants and looking at the interactions among them, they provide insight into the context-dependent nature of species diversity.

Working from a slightly different premise, the chapter by Whittaker subdivides diversity analyses according to the treatment of geographic area and the diversity metrics considered: on the one hand, considering the island tradition so strongly associated with MacArthur and Wilson (1967) of analyzing and theorizing on the form of species-area relationships, and on the other hand, holding spatial scale constant in order to develop models of the species richness-environment relationships.

One theoretical element that is gaining increasing attention is the body of work on "neutral" or "symmetric" theory, which suggests that many of the diversity patterns observed at ecological and biogeographical scales can be explained without invoking the individual biology of species (Hubbell 2001). Building upon this work, the chapter by Turner and Hawkins explores some of the implications of neutral theory in the context of factors and theories commonly invoked to explain geographic clines in species diversity.

Finally, the chapter by Roy, Jablonski, and Valentine provides an evolutionary and macroecological perspective that stresses the importance of examining other aspects of biological diversity (e.g., morphological diversity) in addition to species richness. Their review shows that patterns of species and morphological diversity are often different at a variety of spatial scales, and thus paves the way for future research that more adequately characterizes all patterns of biological diversity.

Collectively, these chapters suggest a number of key questions and issues that should help guide future research in biological diversity. First, what are the patterns of diversity? Although biogeographers have documented many such patterns, the number of taxonomic groups that are still poorly characterized greatly exceeds those that are well known. Non-species-level patterns of diversity are poorly known for almost all groups and deserve much attention, as these alternate measures of diversity are essential elements of biological diversity. Second, how should we collect, collate, and analyze data on biological diversity in such a way as to avoid methodological pitfalls, ecological fallacy, and tautological reasoning? Certainly the way we collect, manipulate, and analyze data has a large influence on our ability to understand patterns of biological diversity. Third, what are the mechanisms responsible for these patterns? Having a better understanding of the mechanisms should further continue the shift in the study of species diversity from a descriptive stage to a predictive one. Fourth, what are the null expectations of patterns of diversity, and how do these expectations vary with the assumptions we make? Understanding this question may help us to improve the clarity and insightfulness of diversity models.

Finally, and perhaps of the greatest societal relevance, how will diversity change in the future given natural shifts in climate and conditions, and more particularly given the massive changes being driven by anthropogenic activities? Currently, these changes are leading to the extinction of many species on the planet, with many scientists arguing that we are witnessing the beginning of another mass extinction event (e.g., McKinney and Lockwood 1999).

To address these questions, much additional research on biological diversity must be conducted. Ideally, this research will eventually lead to the development of a theoretical framework that can unify our attempts to understand diversity. Accomplishing this will be challenging, but it should also prove to be extremely insightful because, in its fullest development, such a framework would facilitate a predictive understanding of the forces that generate and maintain patterns of diversity. Beyond the development of such a framework, we also need to document and understand how diversity is changing due to current anthropogenic changes in the environment, both among taxonomic groups from bacteria to birds, and across spatial scales from local to global. Recent work suggests that these changes are more complex than had previously been appreciated, and as such they deserve our fullest attention (Sax et al. 2002; Sax and Gaines 2004).

Finally, if we are to continue to make progress toward understanding and conserving biological diversity, then we must find creative ways to test the hypotheses we develop to explain patterns of diversity. To date, few hypotheses in biogeography have been adequately tested. This fact is due in part to the methodological and terminological inconsistencies often present within the field. However, this failure is due principally to the problematic nature of testing hypotheses that operate at broad spatial scales. Some of the best insights to understanding patterns of diversity have occurred when natural or anthropogenic (opportunistic) experiments have been used to circumvent this difficulty; examples include investigations in island biogeography (reviewed in Whittaker 1998), studies of biotic exchanges in the fossil record (e.g., Stehli and Webb 1985), and tests provided by non-native species (e.g., Sax 2001). Future efforts should continue to take advantage of such experiments whenever possible and should continue to develop innovative means of testing our ideas against the real world. This endeavor is not just intellectually challenging; we regard it as having the greatest societal relevance. Hopefully, the final frontiers of biogeography will not be relegated purely to a characterization of species lost, but instead to a better and continually developing understanding of the regulation and maintenance of biological diversity.

CHAPTER 8

Beyond Species Richness: Biogeographic Patterns and Biodiversity Dynamics Using Other Metrics of Diversity

Kaustuv Roy, David Jablonski, and James W. Valentine

Introduction

Biological diversity is difficult or perhaps impossible to measure using a single metric. To most people the term "diversity" is synonymous with "variety," and formal definitions of the term biological diversity try to capture this "variety of life" by including everything from genetic to ecological diversity (Gaston 1996). Yet at present our knowledge of spatial and temporal patterns of biodiversity is based almost entirely on one metric: taxonomic richness. From biogeographic patterns to extinction intensities, all are quantified primarily by using numbers of species or higher taxa. Other measures of biological diversity do tend to correlate with species richness, so that an area with a large number of species is also likely to have, say, higher genetic, morphological, and ecological diversity than an area with few species. However, we also know that the relationship between taxonomic richness and other diversity metrics is often complex and non-linear, and one may not be a good predictor of the other (Foote 1997; Roy and Foote 1997; Wills 2001; but see Williams and Humphries 1996).

In the absence of actual measures of these other facets of diversity such as morphologic, functional, phylogenetic, and ecological relationships, we know little about the spatial and temporal distributions of the "variety of life." More importantly, replacing simple taxon numbers with metrics that attempt to incorporate some measure of differences among species may reveal spatial and temporal trends not evident from taxon counts, and provide richer insights into the dynamics underlying biogeographic and macroecological patterns. In this chapter we use a number of case studies to show how incorporating morphological and functional information (1) changes how we view biogeographic patterns, and (2) helps us better test hypotheses about the processes underlying many types of diversity patterns. We use the morphological and functional examples simply because we are familiar with those metrics, not because we think they are more important than other measures such as genetic or phylogenetic diversity. In our view, the major challenge facing biogeographers is to replace spatial models that primarily incorporate taxon counts with ones that will more fully capture the true and manifold varieties of life.

Quantifying Patterns

Spatial Patterns of Morphological Diversity

Morphological diversity is commonly defined as some quantitative estimate of the empirical distribution of taxa in a multivariate space (morphospace) where individual axes represent measures of morphology (Foote 1997; Roy and Foote 1997; Eble 1998; McGhee 1999; Ciampaglio et al. 2001; Wills 2001; MacLeod 2002). Since most organisms have an indefinite number of potentially quantifiable morphological traits, analyses generally focus on quantifying specific aspects of morphology rather than tackling the impossible task of measuring total morphological diversity (Roy and Foote 1997).

Most species are differentiated on the basis of morphology, and a simple relationship between taxonomic richness and morphological diversity might be expected. Empirical studies of this question are scarce but analyses for birds, bats, lizards, and fishes suggest a general tendency for morphospace volume to increase with taxonomic richness, while nearest-neighbor distances within these morphospaces tend not to vary with richness (Ricklefs and Miles 1994). This pattern holds even for tropical–temperate comparisons. For example, overall morphological diversity of North American mammals, based on skull, dentary, and long-bone traits was higher in tropical assemblages compared to temperate ones, but nearest-neighbor distances showed no significant relationship with species richness (Shepherd 1998); comparable results were obtained by Ricklefs and O'Rourke (1974) in a pioneering study of tropical and temperate assemblages of night-flying moths. Similarly, many of the novel morphologies characterizing tropical taxa of marine molluscs are absent from extratropical areas, and these differences, which correlate roughly to latitudinal differences in richness, have been attributed to differences in predation pressure along latitude, or changes in the physical environment (Vermeij 1978, 1987a; Vermeij and Signor 1992). The tendency for tropical genera of marine molluscs to be species-poor relative to high-latitude genera (Roy et al. 1996) also suggests that tropical species are generally more dispersed in morphospace, but exceptions are known and this is yet to be tested directly.

Despite these rough correlations, morphologic and taxonomic diversity are not so tightly correlated that one can reliably serve as a proxy for the other, and discordances between morphological and taxonomic diversity can raise interesting new questions. For example, a comparison of longitudinal trends of species richness and morphological diversity of Indo-Pacific gastropods of the family Strombidae shows that morphological diversity in the most species-rich regions is no higher than that in regions with half the number of species. In other words, volume of morphospace occupation, as defined by shell shapes, initially increases rapidly with species richness but the rate of increase slows down as more species accumulate (Figure 8.1; Roy et al. 2001; Neige 2003). For the same data, variance in morphospace—a measure of morphological disparity between species—was again not predictable from species richness alone; some areas with relatively few species exhibit disparity values comparable to those with highest species richness (see Figure 8.1). Strombid gastropods are commonly cited as a classic example of the Indo-Pacific species richness gradient, with the Indo-Malayan region having the highest species richness (Briggs 1974 and this volume; Vermeij 1987a and this volume). But when morphology is used as the diversity metric, the longitudinal gradient looks very different from the traditional view (Figure 8.2).

The discordance between taxonomic and morphological diversity is also apparent over evolutionary timescales and is most striking during episodes of intense diversification such as the Cambrian explosion of marine metazoans, when a broad range of morphological variety is established well in advance of peak taxonomic diversity (reviewed in Foote 1997, Valentine et al. 1999, and Wills 2001). These findings suggest that, when analyzed on global scales over millions of years, clades can reach longstanding limits to their morphological diversity—which need not be permanent or absolute—while continuing to accumulate taxonomic richness through time, presumably by increasing the density of morphospace occupation over time. These limits have been attributed to intrinsic factors related to development and to extrinsic factors relating to ecological limitation (e.g., Valentine 1995; Eble 1998; Valentine et al. 1999; Jablonski 2000), but definitive tests of these alternatives remain elusive.

Body size is another aspect of morphology that is of interest from a biogeographic perspective. The expected relationship between species richness and body size has been much debated. For example, some argue that the positive scaling of energy requirements of species with body size should produce an inverse relationship between species richness and body size (Cousins 1989; Blackburn and Gaston 1996). This suggests that average size should increase with latitude for groups that show a strong latitudinal gradient in species richness. On the other hand, the species-energy hypothesis suggests that high-latitude assemblages should have low species richness and smaller body sizes due to the reduced energy availability relative to equatorial regions (Turner and Lennon 1989; Cushman et al. 1993). Empirical analyses have yielded mixed results. Among ectotherms, some groups show

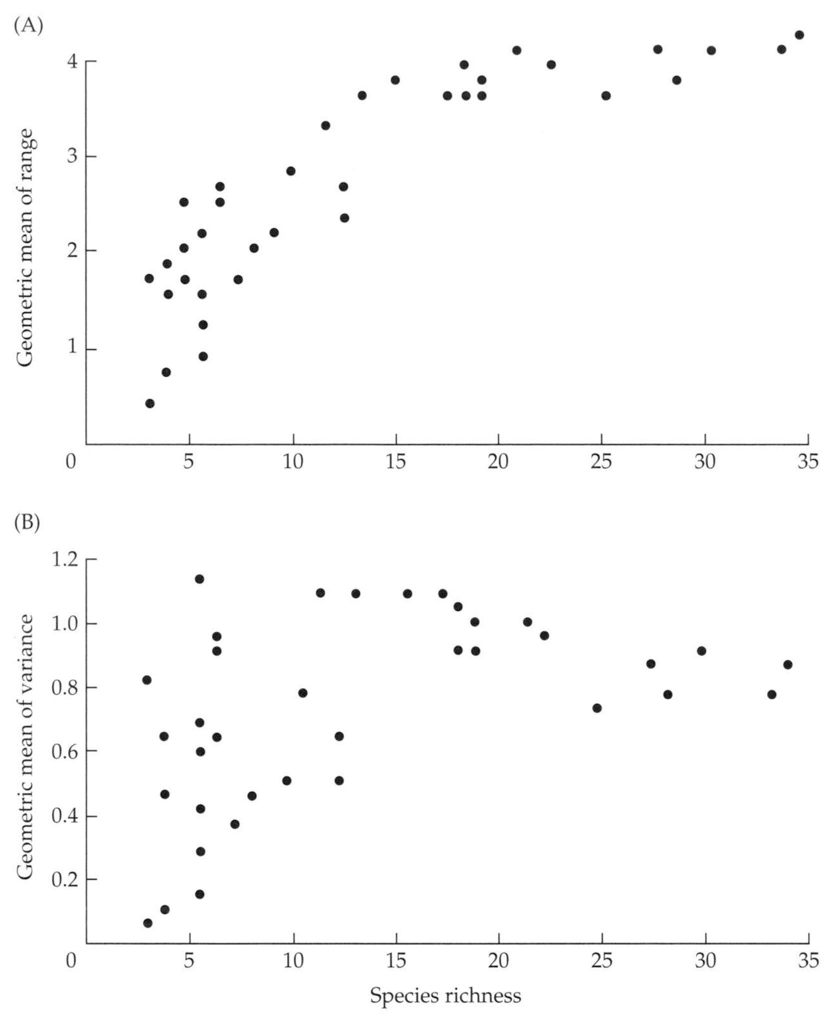

FIGURE 8.1 Relationship between species richness and morphological diversity in strombid gastropods (Family Strombidae). Morphological diversity was quantified using elliptical Fourier analyses of shell shapes. (A) Morphological diversity is defined as the geometric mean of the range of scores on the first six axes of a principal component analysis. This measure correlates with the volume of morphospace occupation. (B) Morphological diversity as the geometric mean of the variance of scores on the same six PCA axes, illustrating the dispersion of species in morphospace. (From Roy, Balch, and Hellberg 2001.)

a positive relationship between size and latitude, while other groups show a negative relationship or none at all (Schoener and Janzen 1968; Cushman et al. 1993; Barlow 1994; Hawkins 1995; Hawkins and Lawton 1995; Kaspari and Vargo 1995; Roy, Jablonski, and Martien 2000; Roy and Martien 2001). A similar situation also holds true for endotherms (McNab 1971). This lack of a consistent relationship between body size and latitude (and, indirectly,

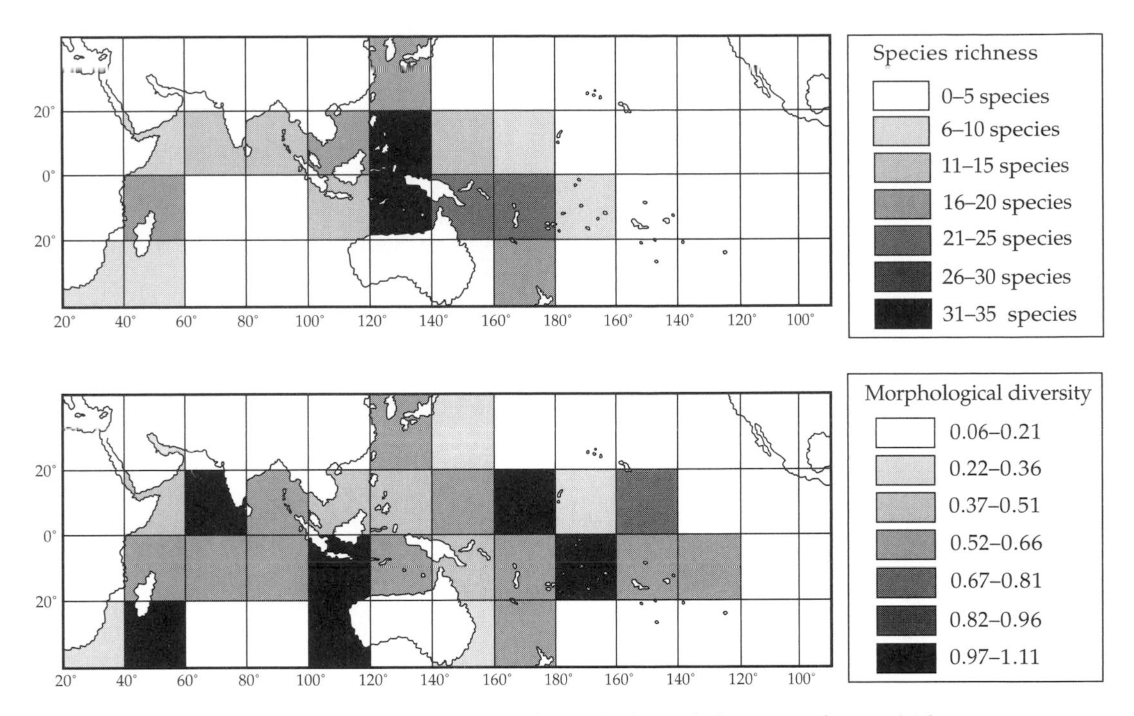

FIGURE 8.2 Spatial patterns of taxonomic and morphological diversity of strombid gastropods across the Indo-Pacific. Morphological diversity represented by geometric mean of the variance of scores on the first six PCA axes as in Figure 8.1B. Note the poor correlation between spatial patterns revealed by the two diversity metrics. (From Roy, Balch, and Hellberg 2001.)

species richness) indicates that biogeographic patterns based on species richness are poor proxies for the biogeography of body size, one of the most important and interesting organismic traits.

Functional Groups and the Latitudinal Diversity Gradient

The latitudinal diversity gradient, with species-rich tropics and depauperate high latitude regions, is perhaps the most striking large-scale biogeographic pattern of all. This pattern is common on land and in the oceans and across many taxonomic groups, but despite a multitude of proposed hypotheses, the processes underlying this gradient remain largely unknown (Rohde 1992; Gaston 2000; Turner and Hawkins, this volume). Part of the problem may be that, despite the progress noted above, we do not know whether high species richness in the tropics necessarily translates into proportionately high morphological, functional, genetic, or phylogenetic diversity. Even clades that show very similar latitudinal trends in species richness may differ strongly in their patterns of functional or morphological diversity along the same gradient. These differences are important since they not only provide a macroecological framework for understanding the dynamics underlying the latitudinal diversity gradient but also, as shown below, can reveal differential responses among functional groups to the same environmental variable. Testing hypotheses about the processes driving latitudinal patterns of diversity using only total species richness of a group obscures these differences.

Marine molluscs, one of the most diverse groups of animals living on the continental shelves of the world's oceans (< 200 m depth), show a strong latitudinal diversity gradient in the Northern Hemisphere (Roy, Jablonski, and Valentine 1994, 1998, 2001). Along the northeastern Pacific margin, both gastropods and bivalves show highest species richness in the tropics, with the number of species declining markedly with latitude (Figure 8.3). A tropical to

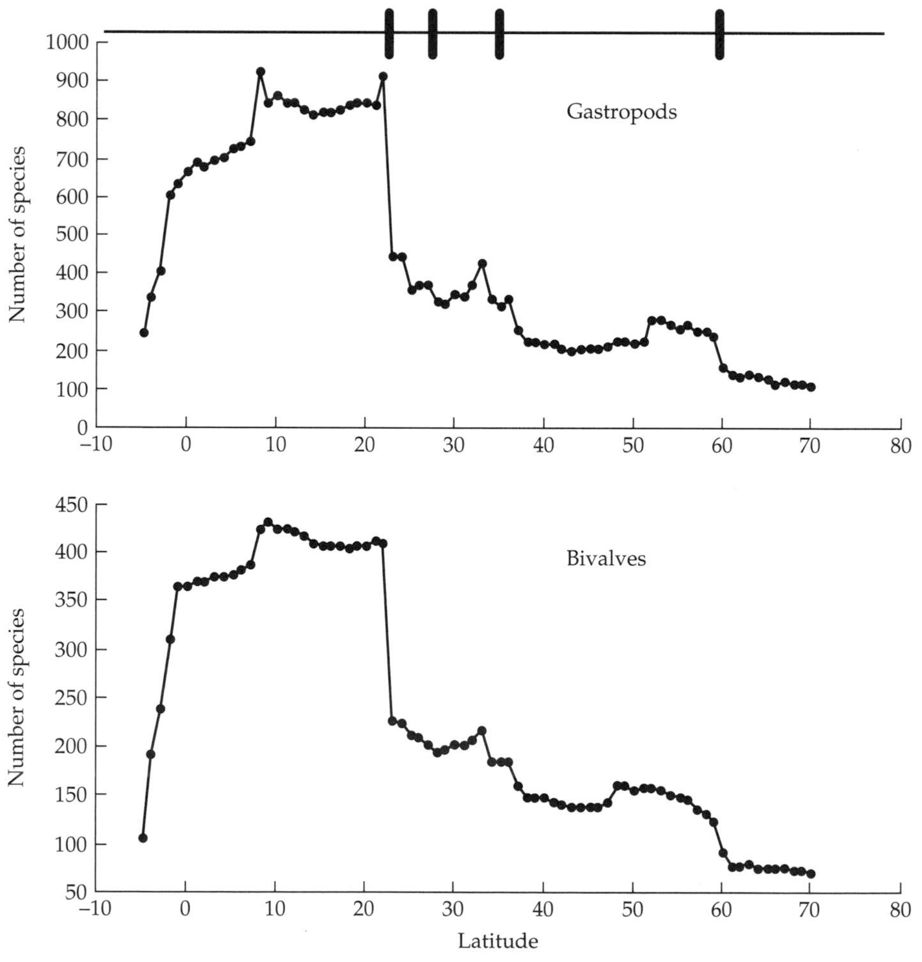

FIGURE 8.3 Latitudinal gradient in species richness of bivalves and gastropods along the northeastern Pacific continental shelf (depth < 200 m). Vertical bars mark the position of the major biogeographic boundaries, defined by species range endpoints.

temperate decline in molluscan diversity is also evident in the Southern Hemisphere, but in this case with a modest reversal of the trend in southern Chile (Valdovinos et al. 2003; Marquet et al., this volume). Functionally, marine bivalves fall into two basic categories, depending on their life positions: infaunal species that burrow or bore into the substratum, and epifaunal species that live at the surface (Stanley 1988). The ratio of the number of infaunal to epifaunal species provides a very simple measure of how total richness is partitioned between these two functional groups. While the species richness of northeastern Pacific bivalves decreases by a factor of four from the tropics to the Arctic, the infauna/epifauna ratio increases along this gradient (Figure 8.4). Infaunal species thus account for a disproportionately larger fraction of total diversity at high latitudes. But such a decoupling between latitudinal trends in functional and taxonomic diversity may not

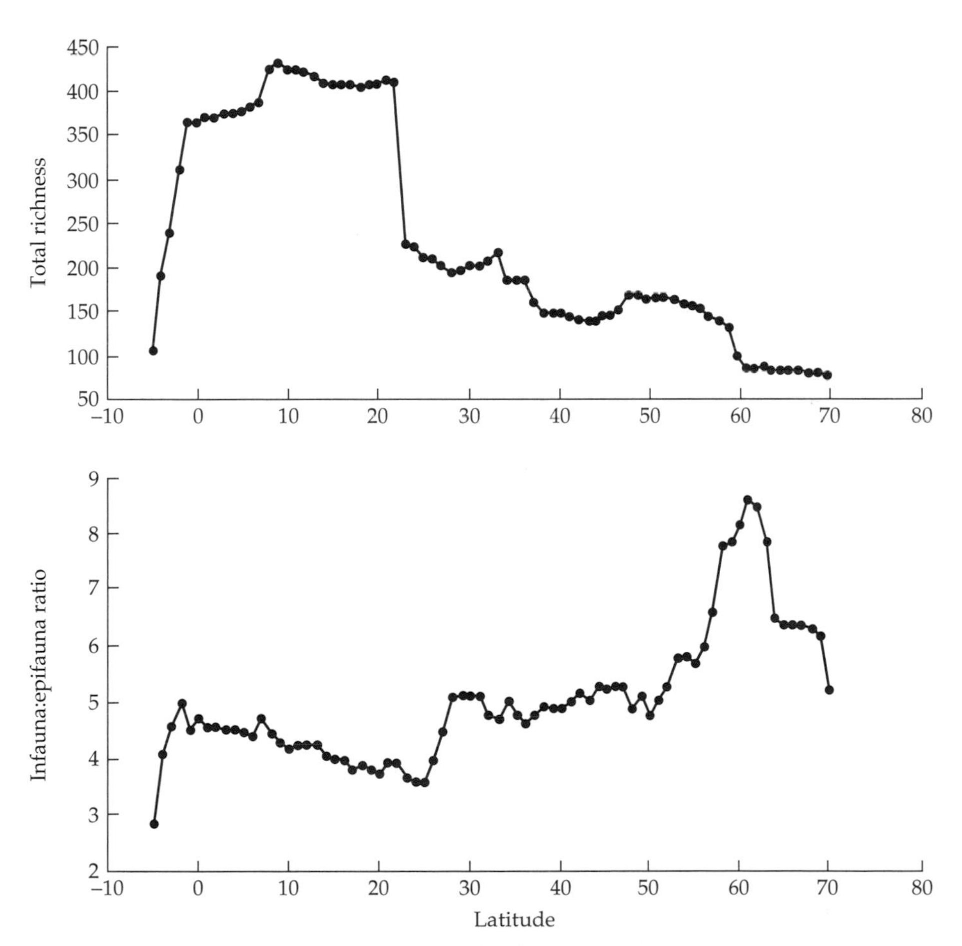

FIGURE 8.4 A comparison of latitudinal trends in species richness and infauna/epifauna ratio of northeastern Pacific marine bivalves. (After Roy et al. 2000.)

have always existed for bivalves. For example, late Jurassic (Tithonian stage, ~145–150 my BP) bivalves also had higher species richness in the tropics than in high latitudes, albeit with a shallower slope (Crame 2002). However, in contrast to today, the infauna/epifauna ratio decreased significantly with latitude during the Tithonian (Crame 1996), a trend that persisted at least until the latest Cretaceous (Jablonski et al. 2000). Thus the bivalves not only steepened their latitudinal diversity gradient over time, but also reversed the latitudinal trend in the proportion of their two primary functional groups. This reversal is probably related to the differential diversification of different clades along latitude (e.g., Crame 2002, and this volume), although the underlying mechanisms remain poorly understood. Without this functional perspective, the late Jurassic latitudinal diversity gradient simply looks like a damped version of the present-day trend in taxonomic richness, although in reality it had a fundamentally different functional makeup.

Bivalves are not the only group where spatial patterns of species richness fail to capture patterns of functional diversity. Gastropods show the expected strong latitudinal gradient in species richness along the northeastern Pacific coast (Roy et al. 1998), but as shown in Figure 8.5, this pattern is decoupled from the latitudinal trend in the ratio of carnivorous to non-carnivorous (C/NC) gastropods (Valentine et al. 2002). The latter is distinctly nonlinear, with lowest values in the mid-latitudes and areas of highest and lowest species richness exhibiting comparable C/NC ratios. This trend in functional diversity is sensitive to the spatial scale over which it is calculated; the decline in C/NC ratio from the tropics to the temperate regions seen in Figure 8.5 essentially disappears at the provincial level (2.1 versus 1.9) even though species richness declines by 20% from 1386 to 1109. As in the case of bivalves, processes driving the spatial patterns of functional diversity in gastropods remain unknown, although the scale-dependence of the pattern suggests that the proximate cause is the differential spatial turnover of the species within the two functional groups (Valentine et al. 2002; Jablonski et al. 2003). Still unclear is whether such differences in distributional patterns reflect ecological controls or historical contingencies such as differential extinctions and/or range shifts in response to late Neogene environmental changes (Roy, Jablonski, and Valentine 2001; Todd et al. 2002).

The latitudinal trend in the C/NC ratio seen in northeastern Pacific gastropods may not be unique to that coast. Taylor and Taylor (1977) documented a qualitatively similar latitudinal trend for northeastern Atlantic gastropods. On the other hand, even though the latitudinal trend in species richness of gastropods along the northwestern Atlantic coast is remarkably similar to that along the northeastern Pacific coast (Roy et al. 1998), the C/NC ratios along those two coasts are markedly different (Roy, Jablonski, and Valentine, unpublished). Data for non-marine groups also show herbivore–carnivore relationships different from that in Figure 8.5 (Gaston et al. 1992; Van Valkenburgh and Janis 1993; Rosenzweig 1995). Clearly, there is a great deal of variation, mostly unknown, in the biogeography of functional groups.

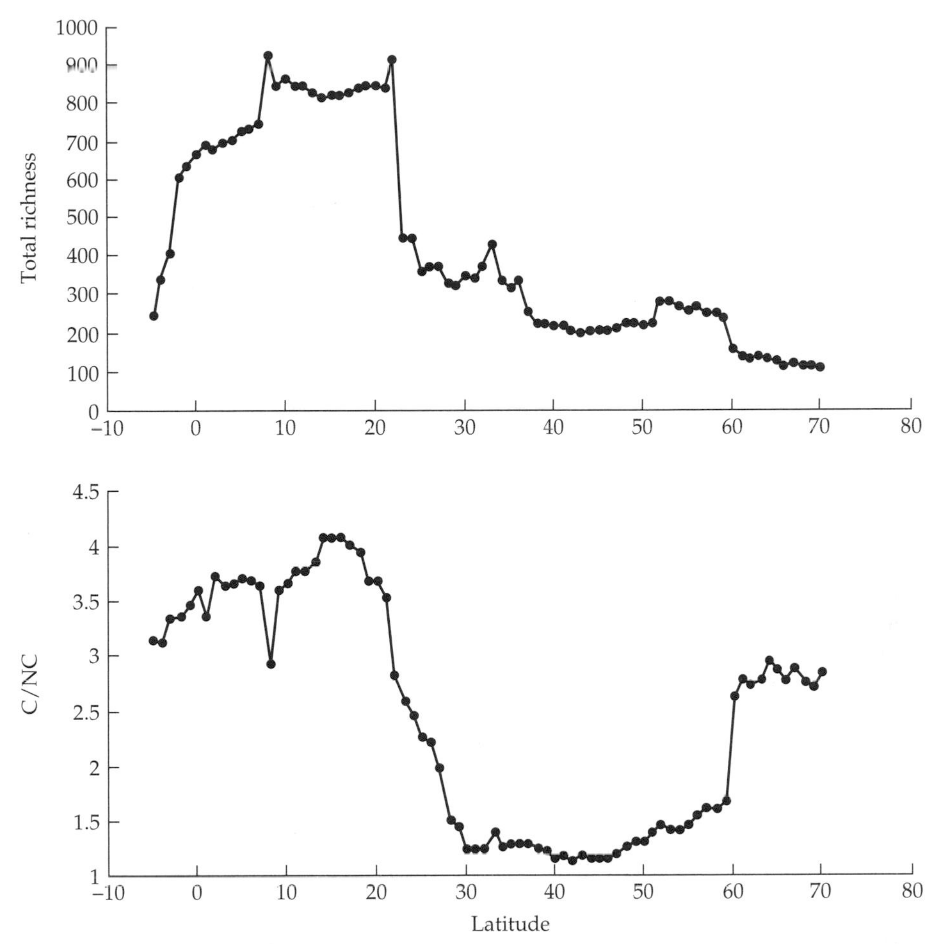

FIGURE 8.5 Comparison of latitudinal trends in species richness and carnivore/non-carnivore ratio (C/NC) of northeastern Pacific marine gastropods. (After Valentine et al. 2002.)

Provinciality and Biogeographic Boundaries

Climatic and oceanographic variables are among the most important forces structuring biogeographic patterns. In some regions, environmental conditions vary abruptly over short distances, whereas in other regions, various aspects of the environment change only gradually over long distances. The effects of such patterns of geographic variation on species distributions and therefore on biotic compositions are well known, ranging from relatively low levels of species turnover in regions of gradual change, to high levels of turnover among many groups in regions of abrupt change. The greatest biotic changes usually correlate with the boundaries of climatic zones and with altitude on land, and depth in the sea. As our experience has been chiefly with marine systems, we draw our examples from the oceanic realm.

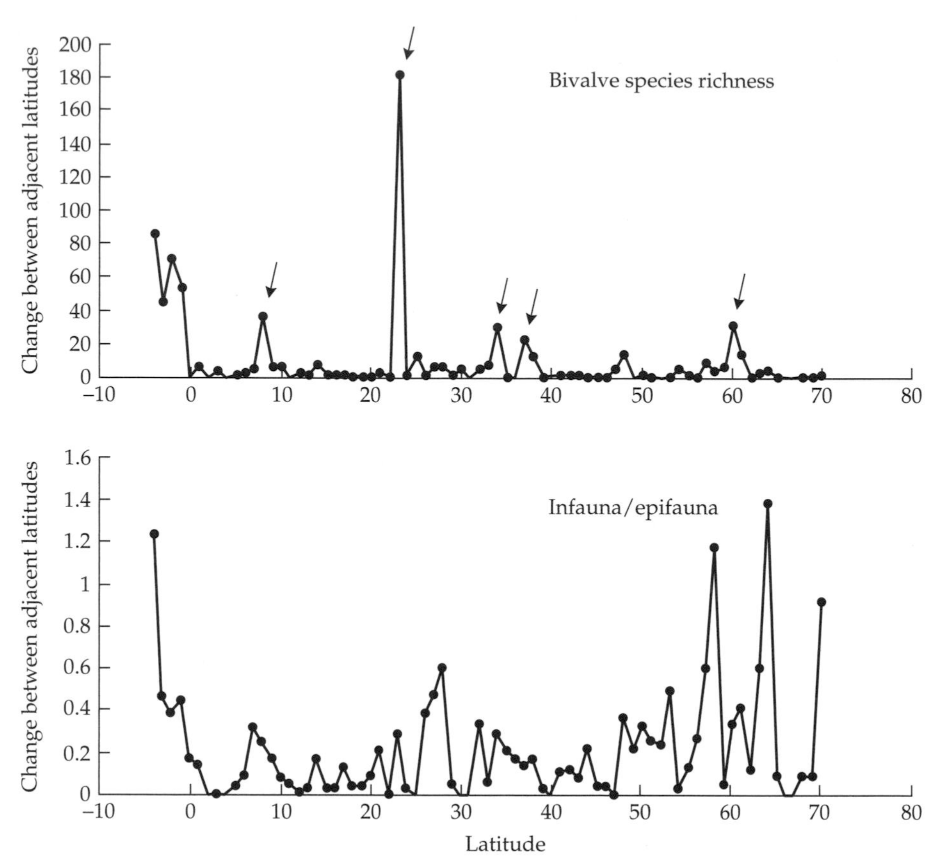

FIGURE 8.6 Spatial pattern of turnover of species richness and functional diversity in northeastern Pacific bivalves. Each point in the upper panel represents the total change in species richness between adjacent one-degree latitudinal bins. Arrows mark the position of major biogeographic breaks representing areas of high species turnover. Each point in the lower panel represents the changes in infauna/epifauna ratio along the same transect. Note the poor correspondence between the two patterns.

The more striking biotic effects of variations in environmental conditions are seen in patterns of provinciality, producing slightly—to distinctively different—faunas along geographic gradients. Provinciality, in the sense of magnitude of faunal turnover, is usually assessed by taxonomic criteria, although as indicated in previous examples, functional criteria are more likely to offer insights into the causal processes. At present, most continental shelves run north and south, crossing climate zones and isotherms, and thus many shelves display north-south chains of biodistributional regions. These regions are particularly well known for molluscs; one of the best-studied regions is the northeastern Pacific continental shelf. Figures 8.3 and 8.6 show a scheme for the main biodistributional regions of this continental shelf, based on molluscan species.

When evaluated in one-degree latitudinal bins, turnover among species of the shelled molluscan fauna of the northeastern Pacific shelf occurs in every bin, with a number of localized peaks (see Figure 8.6; Valentine 1966; Roy et al. 1994, 1998). The locations of high turnover have been taken as the boundaries of formally defined biotic provinces or sub-provinces (see Hall 1964 and Valentine 1966 for reviews); they tend to occur at regions of climatic zonal boundaries. Proceeding offshore, the provincial boundaries that have been studied are recognizable down to depths approximating the outer shelf edge, usually in the region of the thermocline, where the shelf fauna gives way to the cooler- and deeper-water fauna of the continental slope (Jablonski and Valentine 1981). Because temperature change is thus a major correlate of faunal change both latitudinally and bathymetrically, it has usually been considered as the major cause of provinciality, beginning as early as a chart by Dana (1853) and the remarkable biogeographic map by Forbes in Johnston (1856).

However, biodistributional boundaries—peaks in species turnover—also occur in places where local hydrographic conditions, rather than temperature, change abruptly. For example, Gaylord and Gaines (2000) have explored the possibility that a provincial boundary around Point Conception, California (near 34°30′N) is due to transport of larvae in a semipermanent offshore current, thus blocking the dispersal of many species alongshore. They also note that some other biodistributional boundaries are also associated with offshore flows. The local hydrography also results in a temperature change in the area of Point Conception; however, the relative contribution of these two factors to this biogeographic break is not known. Finally, factors other than temperature and oceanography, such as habitat failure owing to deep-water channels or basins that interrupt shallow-water habitats, can also lead to localized turnover in species that can be recognized as a biogeographic boundary.

Clearly, biodistributional units defined from species compositions are of great biogeographic interest, and they may differ by more than one criterion. Factors such as range end-point frequency, level of endemism, taxonomic richness, physical nature of the boundary, and so forth are all useful in defining and evaluating biogeographic units. But a question that has not been much explored is whether functional and morphological diversity patterns also show provincialization, and, if so, whether those divisions coincide with units based on taxonomic criteria. The northeastern Pacific molluscan examples discussed above show that changes in functional diversity are not necessarily concentrated at provincial boundaries marked by clustering of species range endpoints (see Figure 8.6). Thus, a different biogeographic pattern emerges for functional diversity of these molluscs. Similarly, latitudinal patterns of body size of bivalves along the northeastern Pacific coast provide an opportunity to address the effects of provinciality on that important aspect of morphological diversity (Roy, Jablonski, and Martien 2000; Roy and Martien 2001). Surprisingly, at the provincial level the overall size-frequency distribu-

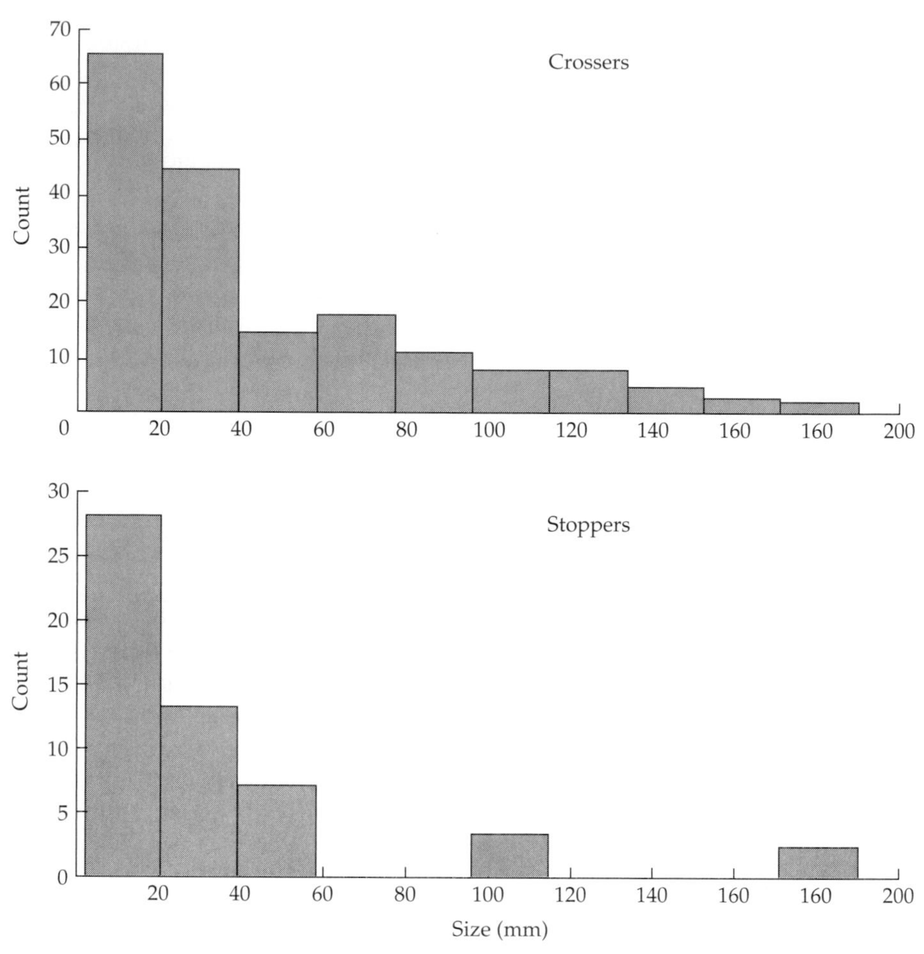

FIGURE 8.7 Body sizes of northeastern Pacific bivalve species across Point Conception, California, a major biogeographic boundary. The top panel shows the size frequency distribution (SFD) of species whose latitudinal ranges span this boundary ($N = 177$). The bottom panel shows the SFD of species whose range endpoints are at Point Conception ($N = 56$). Size is measured as the geometric mean of length and width (see Roy, Jablonski, and Martien 2000). The two distributions are significantly different ($p = 0.02$, Kolmogorov-Smirnov test) and the species ending their ranges at this boundary are smaller (median size 17 mm versus 30 mm).

tion is statistically indistinguishable from the tropics to the arctic, suggesting that provinciality has little effect on regional patterns of body size distributions. Even more remarkably, this similarity exists despite the fact that provincial boundaries show significantly more change in body size than do latitudinal bins away from a boundary (Roy and Martien 2001). In addition, some of the provincial boundaries such as the Californian-Oregonian province boundary (warm- to cool-temperate) are size-selective, with larger-bodied species crossing preferentially (Figure 8.7). Thus the rather monotonous intraprovincial bivalve size frequencies are perturbed at provincial boundaries but are restored within the provinces, implying a replacement of species in size classes that are depleted at the boundary. While this suggests convergent body size evolution between provinces, the underlying evolutionary dynamics remain unknown.

Testing Hypotheses about Diversity Dynamics

Functional Groups and Species-Energy Relationships

Among the hypotheses proposed to explain the latitudinal diversity gradient, the species-energy hypothesis has attracted considerable interest (see Turner and Hawkins, this volume; Whittaker, this volume). This relatively simple hypothesis posits that the latitudinal gradient in species richness is primarily a function of energy availability, with species richness of a region positively related to solar energy input (Wright 1983; Wright et al. 1993; Fraser and Currie 1996; Kerr and Packer 1997). Thus, a positive correlation between some measure of energy and species richness is generally taken as supporting this hypothesis (Fraser and Currie 1996; Kerr and Packer 1997; Roy, Jablonski, and Martien 2000; Francis and Currie 2003). However, the actual mechanisms relating available energy to species richness remain largely unknown (Gaston 2000), so that specific predictions regarding, for example, the slope of the species-energy relationship have been lacking. Energy can be regarded either as an element in trophic dynamics, in which case it is related to the regime of productivity that includes solar radiation and nutrient supply (Valentine 1983), or as generating diversity patterns through metabolic scaling relationships, in which case it is related to temperature *per se* (Allen et al. 2002). While a positive relationship between energy and species richness is known in groups ranging from terrestrial trees and mammals to marine molluscs and reef corals (Wright 1983; Wright et al. 1993; Fraser and Currie 1996; Kerr and Packer 1997; Roy, Jablonski, and Martien 2000; Francis and Currie 2003), the differences in the slopes of the relationship from one group to another have been difficult to interpret. However, a recent theoretical model argues that temperature regulates species richness primarily through the biochemical kinetics of metabolism. This provides a quantitative prediction of how species richness should increase with environmental temperature (Allen et al. 2002). In particular, this model predicts that (1) the natural log of species richness should be a linear function of $1000/T_{env}$, and (2) the slope should approximate –9.0 Kelvin (K) (Allen et al. 2002). These predictions allow us to test quantitatively how different functional groups respond to the same changes of available energy along a latitudinal gradient.

For benthic marine invertebrates, mean sea surface temperature (SST) provides a good approximation of available solar energy. Previous studies have revealed a significant positive relationship between the latitudinal gradient in mean SST and total species richness for both gastropods and bivalves (Roy, Jablonski, and Valentine 2000). However, when these latitudinal trends are partitioned into different functional groups, interesting differences emerge. Gastropods along the northeastern Pacific and northwestern Atlantic coasts exhibit an average slope of approximately –7 K, quite close to the model prediction (Allen et al. 2002). In the northwestern Atlantic, both carnivorous and non-carnivorous gastropods exhibit slopes that are very similar to the overall slope, as do the northeastern Pacific carnivores (Figure 8.8). However, the slope for the non-carnivorous northeastern Pacific gastropods is significantly lower (–2.7 K). Clearly, the latitudinal diversity trend of this latter functional group cannot be explained by the biochemical kinetics model, even though the general prediction of the species-energy theory (i.e., a positive relationship between temperature and species richness) is supported. In addition, while richness is a fairly continuous, linear function of energy input for the eastern Pacific, the western Atlantic patterns are strongly non-linear, with cold-temperate and boreal assemblages of both carnivorous and non-carnivorous gastropods deviating from the predicted slope (see Figure 8.8). Finally, the overall slope for northeastern Pacific bivalves is significantly lower than that of the gastropods and the model's prediction, and within bivalves, the infaunal species exhibit a slope significantly lower than epifaunal species (Figure 8.9).

The differences in the slopes of the temperature-richness relationship between different functional groups raise interesting questions regarding the dynamics of diversity, and help put the discordances between latitudinal patterns of functional diversity and species richness into perspective. Although bivalves do not conform to the relationship found in many other groups, both terrestrial and marine (Allen et al. 2002), they show significant differences related to life habits. Similarly, the non-carnivorous gastropods along the northeastern Pacific are clearly different from carnivorous species along the same coast and from non-carnivorous species along the northwestern Atlantic coast. These differences could stem from historical factors or ecological ones. For example, the deviations could reflect nonequilibrial conditions such as the lack of rebound from selective extinctions during the late Neogene. Alternatively, the deviations may reflect situations where the assumptions of the Allen et al. model, such as a linear relationship between environmental temperature and population density, are violated. Both of these hypotheses are testable using paleontological and ecological data, and the gastropod contrasts are not as surprising when we recall the contrast in C/NC ratio patterns along the east and west coasts of the Americas. Such tests are clearly needed to better understand the dynamics of diversity along latitudinal gradients. In general, these examples also show that tests of species-energy and other hypotheses, using total species richness values, are only crude tests at best; incorporating information about life habits of species reveals patterns previously not detected.

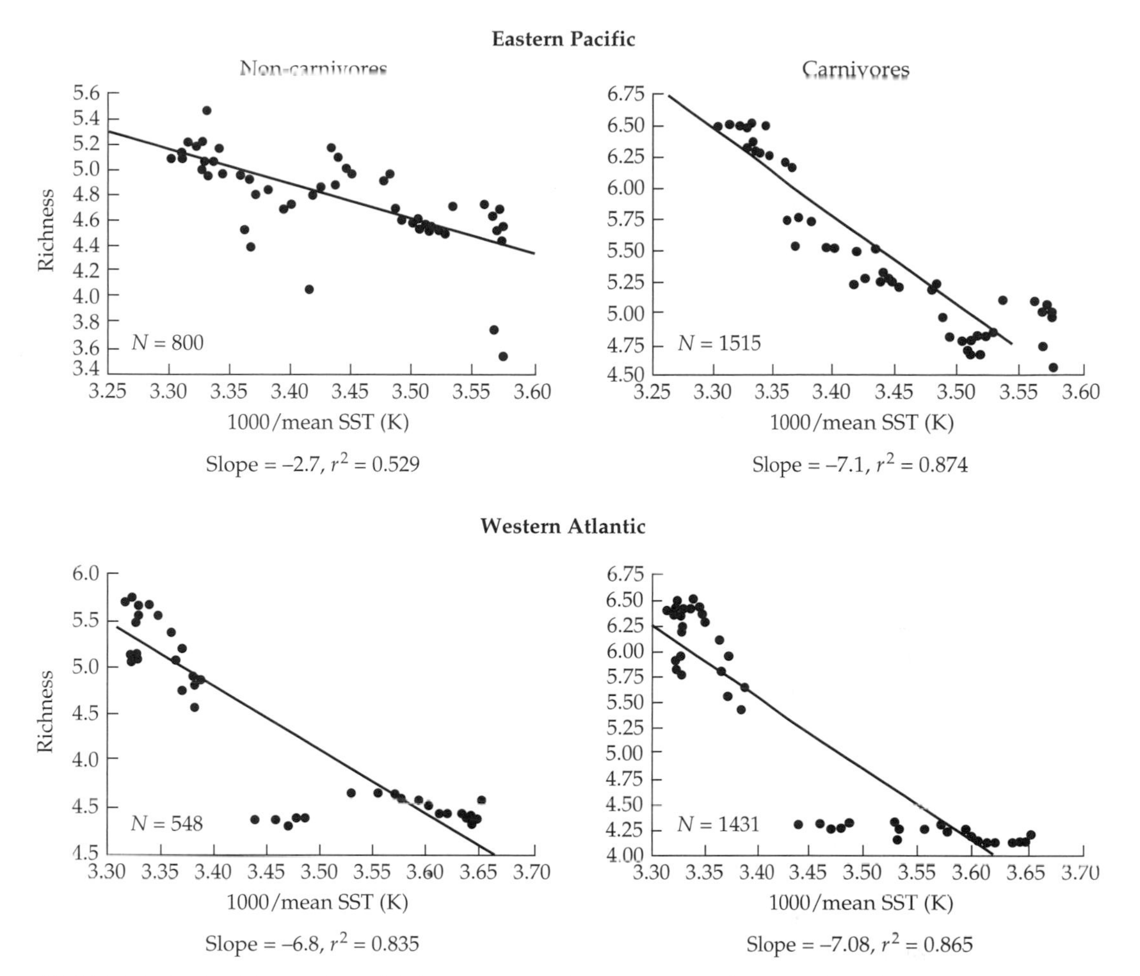

FIGURE 8.8 Test of the relationship between species richness and temperature in marine gastropods along the northeastern Pacific and northwestern Atlantic coasts following Allen et al. (2002). Temperature is defined as 1000/mean SST in Kelvins. The slope of the northeastern Pacific non-carnivorous species is significantly different from the others; see text for details. Species richness and temperature data from Roy et al. 1998.

EXTINCTION SELECTIVITY. Many authors have recognized that the loss of biodiversity involves not only the disappearance of taxa, but of functional groups and morphological variety as well. However, the role of extinction in shaping spatial patterns of biodiversity—functional, morphological, or taxonomic—is poorly known. During the major mass extinctions in the geologic past, one of the most pervasive predictors of genus survivorship was geographic extent (Jablonski 1995), which of course may or may not correlate

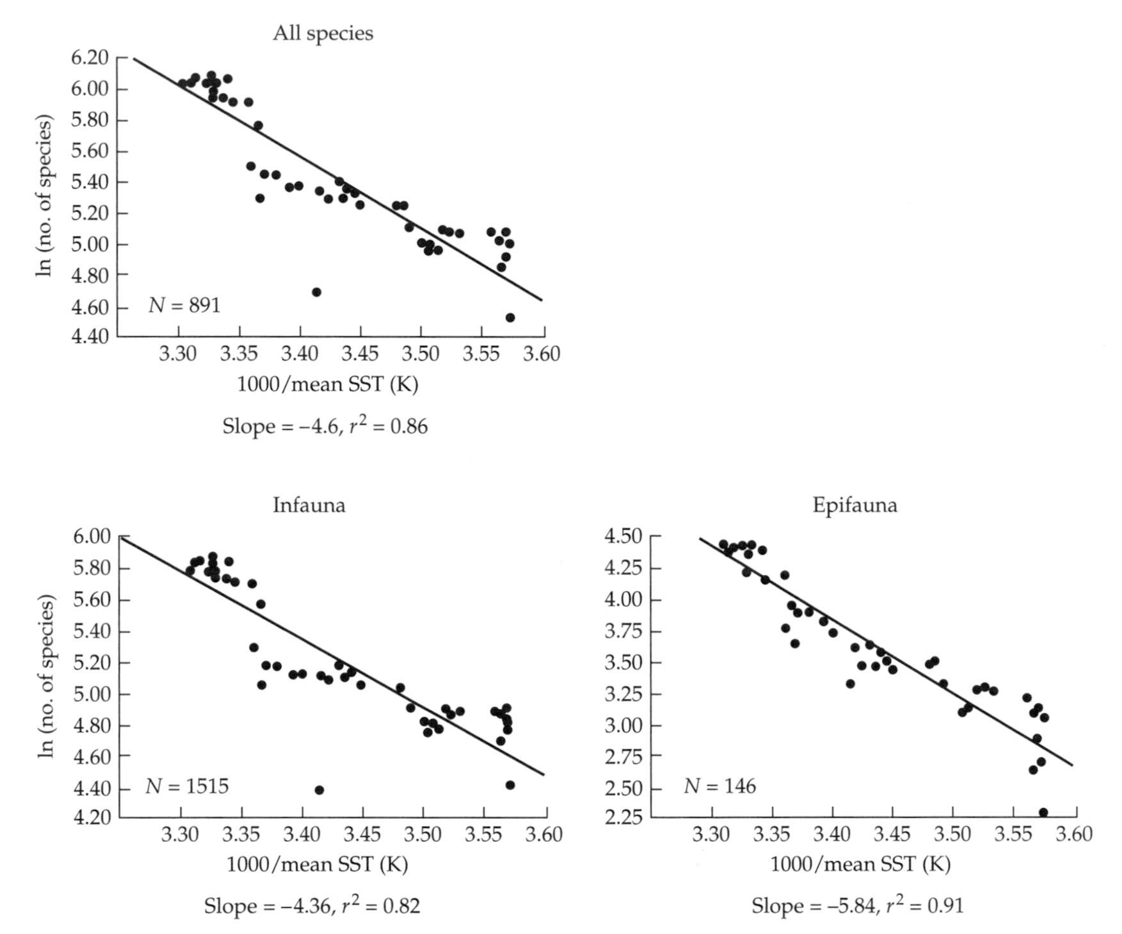

FIGURE 8.9 Same relationship as in Figure 8.8, but for northeastern Pacific bivalves. The overall slope for bivalves is significantly lower than that of gastropods and that predicted by the Allen et al. model. Furthermore, the slopes for the infaunal and epifaunal species are significantly different; see text for details.

with, and thereby promote, other aspects of form or ecology (see also Raup and Jablonski 1993). Thus, the end-Cretaceous mass extinction favors widespread clades but is non-selective with respect to molluscan functional groups, body size, abundance, and many aspects of morphology (Jablonski and Raup 1995; Jablonski 1996; Harries 1999; Lockwood 2003). When other aspects of ecology correlate with geographic distribution, separating true selectivity from hitchhiking effects is difficult. For example, the end-Ordovician extinction apparently selected against high levels of colony complexity or integration in Paleozoic bryozoans, but these factors also correlate inversely with geographic distribution (Anstey 1978, 1986); moreover, among Paleozoic brachiopods extinction selects against "more specialized and higher adaptive morphs having more restricted distributions" (Harper and Rong

2001). Similarly, major extinctions repeatedly removed the most complex forms of planktonic foraminifera. But was the decisive factor morphological complexity, or their narrower geographic distribution? Was it their more specialized position in the water column, or their reproductive ecology? Or, was it a preferred tropical climate zone compared to more widespread, generalist forms (see Norris 1991, 2002)? Such potential covariation also emerges when apparent extinction hotspots tend to coincide with regions particularly rich in endemic genera (Jablonski 1995: 35, and references therein). Extinction selectivities on biotic factors such as body size, abundance, or life habits are actually most pronounced in times of milder extinction intensities than those estimated for the "Big Five" mass extinctions (e.g., in the geologic past, Jablonski 1995; McKinney 1997; Kammer et al. 1998; Smith 2000; in present-day biotas, Gaston and Blackburn 1995; Bennett and Owens 1997; McKinney 1997; Purvis et al. 2000; Cardillo and Bromham 2001; von Euler 2001).

When extinction is truly random, we expect only modest declines in functional or morphological diversity even when a substantial number of species are lost (Foote 1997; Nee and May 1997), although the interaction of low diversity and intense extinction can drastically reduce morphological variety owing to sampling error (e.g., Foote 1996, 1997). Contrary to this expectation, the few analyses performed thus far have found that, even when taxonomic richness is relatively high, morphological variety tends to be truncated rather than thinned during mass extinctions (e.g., Roy 1996; Lockwood 1998; McGhee 1999; see also Petchey and Gaston 2002). But these non-random patterns need not map onto conventional taxonomic or functional groupings. For example, end-Permian and Triassic ammonoid extinctions are not selective with respect to the basic morphotypes within the clade, but tend to leave survivors around the periphery of a multivariate morphospace, which is then filled in again during the evolutionary recovery phase (A. McGowan, pers. comm.). Such vacant portions of morphospace imply some kind of selectivity despite the lack of strong patterns in some of the most obvious morphological or ecological aspects of the victims and survivors; again geography may be crucial here, although we cannot rule out the many ecological factors that remain to be analyzed. Similarly, the loss of phylogenetic diversity (i.e., extinction of higher taxa or total phylogenetic branch length; Purvis et al. 2000) can be greater or less than expected from simple taxonomic extinction intensities depending on the degree and pattern of extinction selectivity: even substantial but random losses will leave phylogenetic diversity relatively intact, whereas more selective extinction can disproportionately reduce phylogenetic diversity (Nee and May 1997; Purvis et al. 2000; von Euler 2001).

Evolutionary and ecological recoveries may be as important in shaping the post-extinction biota as the extinction events themselves (e.g., Erwin 2001; Jablonski 2001, 2002), but here we have even fewer spatially explicit data than for the extinctions themselves. Jablonski (1998) found significant interprovincial differences in the recovery of marine molluscan faunas after the end-Cretaceous (K/T) mass extinctions, but this was strictly a taxonomic analysis and did not address morphological or functional aspects of those faunas (and see Rees 2002 on regional patterns in Permo-Triassic floras). A morphological extension of the differences in K/T recovery dynamics would be worthwhile, particularly because those differences were accompanied by interprovincial variations in the intensity of post-extinction invasions—important for present-day biodiversity—and the invaders were not randomly drawn from the pre-extinction biota (Jablonski 1998). In apparent contrast, although ammonoid cephalopods undergo spectacular episodes of taxonomic loss and recovery (e.g., at the end-Permian and end-Triassic mass extinctions), they show relatively little spatial patterning in their taxonomic and morphological recoveries (Dommergues et al. 2002). This lack of spatial structuring might have been expected from the wide distributions of many ammonoid species and clades and the general absence of a link between their shell morphology and biogeography; more analyses are needed in less vagile groups.

We still do not know the role of selective extinction or recovery during the Neogene or earlier times in sculpting the spatial patterns in functional or morphological diversity seen today. Regional variations in taxonomic extinction intensity have been reported (e.g., Stanley 1986; Todd et al. 2002), but their larger and long-term biotic effects remain poorly known. In the best-studied example, differential extinction and diversification among molluscan functional groups following the uplift of the Panamanian Isthmus drove divergent biotic changes in the tropical Atlantic and Pacific faunas from what had been a single trans-isthmian biota (Todd et al. 2002), and we suspect that comparable differences in extinction and recovery dynamics will prove to be pervasive in the face of the provincialization of the Neogene world as polar refrigeration set in. What is astonishing about the suite of functional divergences in this case is the near-identical taxonomic diversities attained by the two faunas despite their differences in extinction intensities and the differential recovery of disparate functional groups (see Roy et al. 1998).

Present-day extinction risk is clearly not distributed evenly over the globe, as attested by many regional risk assessments and hot-spot analyses. The impact of regional losses in taxonomic richness on the functional or morphological diversity is only now being addressed, however. In a pioneering study, Jernvall and Wright (1998) showed that the expected ecological or functional losses owing to impending taxonomic extinctions in primates differed among regions depending on their respective morphospace occupation patterns and thus were not predictable from taxonomic data alone. Whether this is a common pattern remains to be determined.

Frontiers of Biodiversity Dynamics

A resurgence of interest in biogeography and macroecology combined with wide availability of digital databases is producing increasingly refined data on species distributions, both past and present. These data are being used to test traditional as well as new hypotheses about processes regulating spatial patterns of diversity. However, the vast majority of studies that quantify biogeographic patterns and address macroecological hypotheses to explain those patterns still rely on counts of species or higher taxa. Yet as we discussed above, taxonomic patterns are only one aspect of biodiversity and may often mask interesting spatial patterns of morphological, functional, or phylogenetic diversity. In macroecology, the search for correlations between present-day patterns of species richness and environmental or biotic variables has so far largely ignored the role of functional biology and ecology of individual species or lineages, and hence has limited ability to inform us about the processes actually underlying those correlations.

For example, the species-energy hypothesis for latitudinal and other diversity trends has almost always been tested in terms of correlations of species richness and various energy-related variables. But species' energy use is determined by their functional biology, and differences in energy use should translate into predictable trends in spatial patterns of other aspects of diversity such as body size (Cousins 1989; Turner and Lennon 1989; Cushman et al. 1993; Blackburn and Gaston 1996; Roy and Martien 2001). At least some groups that show a significant positive correlation between taxonomic richness and energy, such as marine bivalves (Roy et al. 2000), fail to exhibit spatial patterns of body size predicted from energy availability (Roy and Martien 2001). Similarly, while marine bivalves and gastropods in the Northern Hemisphere show an overall positive correlation between species richness and energy, the slope of this relationship varies significantly among functional groups. These differences could derive from fundamental differences in macroecological relationships (e.g., relationships between environmental temperature and population density) that reflect differences in the biology of different functional groups, or from historical effects such as differential extinction or recovery during the late Neogene. All of this suggests that general correlations between species richness and energy variables provide, at best, a crude insight into the dynamics of diversity and can mask deviations from expectations that may ultimately be more revealing of underlying mechanisms. Understanding the causes of these deviations is ultimately the key to understanding the processes that relate available energy to biodiversity patterns.

Finally, the role of historical processes in generating present-day species diversity patterns remains seriously understudied. The increasing use of phylogenetic relationships to evaluate the role of historical factors is encouraging in this respect, but a major problem in relying on phylogenies of living species is that it ignores the role of selective extinctions. Yet, there is increasing evidence that extinction events outside of major mass extinctions tend to show selectivities with respect to morphology, ecology, geography, or phylogenetic affinities. Even the "Big Five" extinction events of the geologic past are not entirely random with respect to biodistributional factors. Understanding the role of extinctions during the recent geological past in shaping present-day biodiversity patterns remains a major challenge for biogeography and macroecology. Addressing this would require us to better quantify Neogene extinction patterns for groups with a good fossil record, and to generate better models for how selective extinctions affect topologies of phylogenetic trees; the latter may be the only solution for poorly fossilized groups.

Despite major conceptual advances and the availability of large amounts of data, our knowledge of biogeographic patterns still rests on species counts. Further refinements of such spatial distributional data are undoubtedly important and should be encouraged. However, in our view, a true understanding of the processes underlying diversity patterns requires better information on other aspects of organismal biology and geographic variation in these characters. Incorporating information on morphology, functional biology, and phylogenetic affinities of species in biogeographic databases is not only technologically feasible, but a necessity. Such data would finally produce a biogeography that is truly reflective of the variety of life.

Acknowledgments

We thank the organizers of the Frontiers of Biogeography meeting for the invitation to present these ideas. We also thank Mark V. Lomolino and an anonymous reviewer for comments on a previous version of the manuscript, and the National Science Foundation for financial support.

CHAPTER 9

The Global Diversity Gradient

John R. G. Turner and Bradford A. Hawkins

Introduction

If you tell a non-biologist that you are working on the global biodiversity gradient—explaining that this is the question of why there are so many more species of just about everything in the tropics than at the poles (Hillebrand 2004)—they will probably reply, "Isn't that just the climate?" A biologist, on the other hand, is likely to come up with one or more of the thirty-odd theories for the gradient that have been proposed over the years (see Connell and Orias 1964; Pianka 1966; Rohde 1992; Huston 1994, 1999; Rosenzweig 1995; Turner et al. 1996; Willig et al. 2003; Turner 2004). The gradient (Figure 9.1), which was noted by Forster in 1778 (see Brown and Lomolino 1998) and by von Humboldt in 1808 (see Hawkins 2001), is probably the oldest known fact in biogeography, is much in need of explanation, and is the source of much and sometimes bitter disagreement. We argue here that there has been recent progress in explaining the gradient, and we focus this chapter on what we feel are the most likely contributing factors.

Hubbell's Model of Biodiversity

We start with a recent, controversial idea of what drives diversity. Hubbell's "neutral theory" (*The Unified Neutral Theory of Biodiversity and Biogeography*, Hubbell 2001; Hubbell and Lake 2003) takes the whole question back to basic principles: What happens when individuals are born, live, and die?

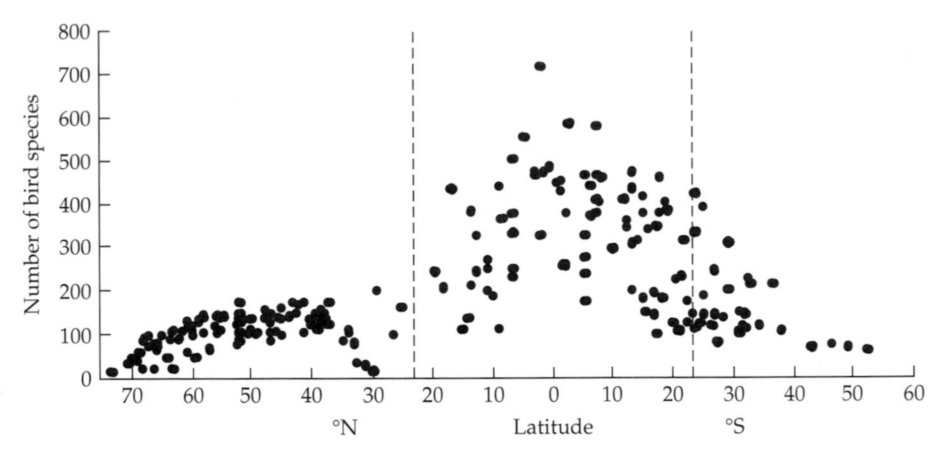

FIGURE 9.1 An example of the latitudinal diversity gradient. Each dot is the species richness of terrestrial birds in one of 206 grid cells, each 48,400 km^2 in size, randomly selected from a larger data set for the birds of North America, northern Central America, South America, Europe, Africa, the republics of the former USSR, and Australia (see Hawkins et al. 2003a). Note the extreme variation in richness within the tropics, which are delimited by the dashed vertical lines.

Imagine a community of identical plants belonging to an arbitrary number of species. Eventually each individual dies, leaving a gap in the vegetation cover that is promptly filled by a propagule from another individual (plants within this dispersal distance constitute a "community"). Because all individuals are the same, there is no variation in competitive ability, and both the deaths and the identity of the species that gets a successful seedling into any particular gap are random. Hence the number of individuals in any species over the course of time increases or decreases purely by random walk. If we follow the fate of any one species it will sooner or later become extinct, a process termed "ecological drift." In time this will happen to all species except one, which will come to occupy the whole habitat. But suppose that at infrequent intervals individual seedlings appear that have a genetic mutation that will result in their descendants becoming a new species. While the majority of such new species, being initially very rare, will themselves rapidly become extinct, a small minority will randomly establish in each community. Thus, while species are slowly and steadily drained away by random extinction, they are also slowly replenished by speciation. (The equilibrium solution is the same as for neutral evolution by genetic drift; Roughgarden 1979.)

A group of communities that can exchange individuals at a much slower rate is now defined as a "metacommunity." The simplest case is an archipelago of separate islands. A species that has originated in one community may thus enter a different community, and species that are locally extinct may eventually recolonize. Over time there will be an equilibrium between the loss of species by ecological drift and their creation by speciation, given by

$$\theta = 2J\nu \qquad \text{Equation 1}$$

where the "fundamental biodiversity number" is θ, J is the total number of individuals (irrespective of species) in the metacommunity, and ν is the rate of speciation (the proportion of individuals in any one generation that sets out on the genetic road leading to a new species). The fundamental biodiversity number is a sounder concept, although annoyingly more abstract, than the actual number of species. How many species we think there are in any one place depends on how big the sample area is, and on how long we keep looking. Subject to these search parameters, species richness can be obtained from θ algorithmically, and therefore we can think of θ as a surrogate for species richness.

Now, if ρ is population density (the number of individuals per unit area of all species combined), and A is the area occupied by the metacommunity, obviously the number of individuals in the metacommunity must be $J = \rho A$, and therefore

$$\theta = 2\rho A\nu \qquad \text{Equation 2}$$

or the fundamental biodiversity depends on the population density over all species, the area occupied by the metacommunity, and the rate of speciation.

It may well be that the assumptions behind this derivation—particularly that all individuals are identical and that all species are equal in competitive ability—are unrealistic enough that the equations are not accurate predictors, and the theory may ultimately be falsified in detail (Clark and McLachlan 2003; McGill 2003). However, the roles of population size and area have long been familiar in the theory of island biogeography (MacArthur and Wilson 1967), which predicts that the lower the immediate rate of extinction, the larger will be the equilibrium number of species (see Whittaker, this volume). Hubbell is not the first to link species richness to productivity, population size, and area (Hutchinson 1959; Brown 1981); moreover, speciation/extinction rates are thought to be sensitive to both area (Rosenzweig 1995) and population size (Connell and Orias 1964). But what is appealing is that these various links arise naturally out of the theory in a straightforward way. As with Brown's (1981) evaluation of the equilibrium theory of island biogeography, even though the real world is considerably more complex than its portrayal in the neutral theory, we believe that the theory provides powerful heuristic value as a way to think about diversity gradients. Equation 2 provides an excellent (if simplified) mental aid in suggesting how population density, area, and speciation rate will jointly influence species richness.

We now ask what controls these three factors over the face of the planet.

The Number of Individuals: Species-Energy Theory

Under the assumptions of the neutral theory, diversity is governed not by the population sizes of individual species but by the total number of organisms. This total population density ρ will be largely dependent on net primary productivity, which governs both plant growth and the biologically available nutrients which power the rest of the food chain; the higher this is, the more individuals can be fitted into a square meter. (Or, productivity governs the total biomass, and given an overall similar distribution of body sizes between species everywhere, the biomass controls the number of individuals.) Many other factors certainly influence animal and plant numbers. However, as we are here interested in very large-scale gradients, we do not initially need to consider factors such as the availability of nutrients, which have a relatively fine-scale or even mosaic pattern; the predominant factor with broad-scale variation that can produce planetary gradients is clearly likely to be climate.

At the global scale, productivity is governed by water-energy balance, often measured by actual evapotranspiration (AET) (Rosenzweig 1968; Lieth 1975; Stephenson 1990); that is, the amount of water driven through the ecosystem by thermal energy. In turn, thermal energy has two sources: the direct input of solar energy to the Earth's surface, and the part of that energy that is stored as the ambient temperature of the air or water in which the organisms are immersed. Both of these sources of energy have a strong latitudinal gradient, declining from the tropics toward the poles, with the gradient in the temperature of air and water being the shallower because of the meridional flux (in which the heated atmosphere and ocean water circulate away from the equatorial regions toward the higher latitudes). Productivity will therefore have a strong latitudinal gradient derived from the gradient in thermal energy. It has rather large and less regular deviations arising from global variation in water availability.

To put it simply, there are more organisms where it is warmer and wetter (which equates broadly with low latitudes), and where there are more organisms there should be more species. This idea has long been known as "species-energy theory" (Wright 1983; Turner 1986; Turner et al. 1987; Currie and Paquin 1987), although this name does not do justice to the importance of water availability. Perhaps confusingly, it is also well known as the "productivity hypothesis."

Productivity and Energy

The productivity hypothesis makes simple and easily testable predictions. If productivity represents the primary constraint on plant and animal diversity, then species richness should be most strongly associated with temperature and rainfall; or with measures directly related to productivity, such as AET or in some cases PET (potential evapotranspiration, the energy avail-

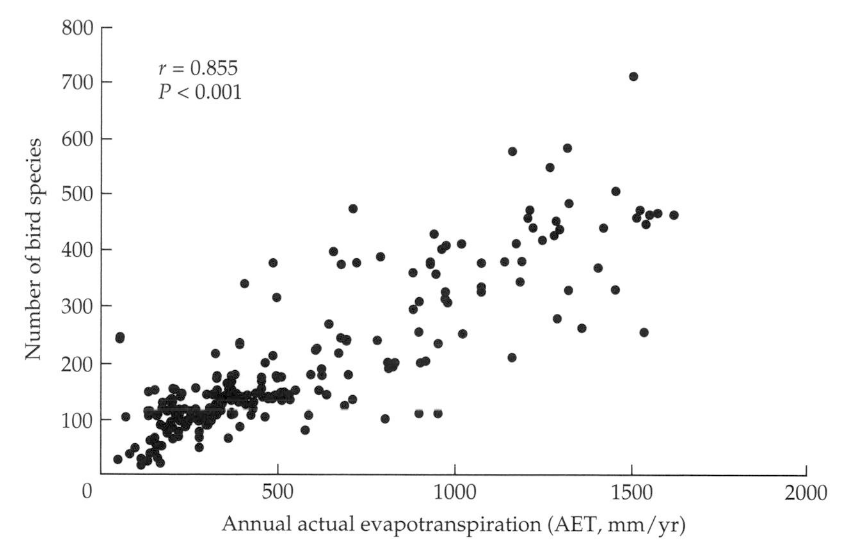

FIGURE 9.2 The bird species richness data shown in Figure 9.1, re-plotted against annual actual evapotranspiration (AET), a commonly used measure of water-energy balance. Despite a strong positive association, 27% of the variation in richness remains unexplained by AET, indicating that additional factors are also influencing bird diversity.

able to move water through the ecosystem); or to productivity itself, which is difficult to measure at the global scale. What evidence exists that productivity, or its climatic proxies, in fact describe broad-scale diversity gradients better than alternative explanatory variables? We believe the evidence is strong and widespread.

In a review of the studies of biodiversity gradients in terrestrial habitats extending over a minimum 800 km linear distance, Hawkins et al. (2003b) found that in 82 of 85 cases, the best explanatory factor, accounting for an average of over 60% of the variance in diversity, was some annual measure of productivity, water availability, or energy (Figure 9.2). Given this weight of evidence, we cannot avoid the conclusion that—unless some even better variable is invented—productivity, energy, or ultimately some aspect of the climate that controls productivity, represents a very powerful predictor of species richness patterns on the continental to global scale.

Climate clearly has a major influence on diversity. Even so, it is not the only factor affecting diversity gradients, even at large scales. Looking at the full ranges of variables that have been found to contribute to statistical models of species richness in selected terrestrial and aquatic environments (sampled in Table 9.1), we find that productivity and its climatic proxies are a consistent component of all analyses, but that alternative variables do usually contribute to the statistical models, sometimes strongly. Obviously, no single variable can account fully for the diversity of all taxa in all places.

TABLE 9.1 *Associations of species richness with various environmental factors in a sample of recent studies*

Species and region	Environmental factors associated with species richness	References
Plants (flowering or vascular)		
Worldwide	AET, productivity, seasonality of temperature, altitude	Scheiner and Ray-Benayas 1994
World islands	AET, area	Wylie and Currie 1993
South Africa: woody	Minimum PET, rainfall, relief	O'Brien et al. 2000
Europe, east Asia, North America: trees	Productivity (climatically modeled)	Adams and Woodward 1989
Zooplankton (foraminifera)		
World oceans	Sea surface temperature	Rutherford et al. 1999
Corals		
(Sub)tropical, worldwide	Sea surface temperature, regional coral biomass, number of islands up-current	Fraser and Currie 1996
Beetles		
North America: *Epicauta*	PET (and solar radiation, temperature, AET)	Kerr and Packer 1999
North America: tiger beetles	PET	Kerr and Currie 1999
Butterflies and moths		
Canada: butterflies	PET, habitat diversity	Kerr et al. 2001; Kerr 2001
North America: swallowtails	PET, altitude, tropical habitats	Kerr et al. 1998
Europe, North Africa: butterflies	AET, rainfall, relief	Hawkins and Porter, 2003b
California: butterflies	AET, relief, altitude	Hawkins and Porter 2003a
Canada: forest moths	PET	Kerr et al. 1998
Molluscs (prosobranchs)		
North American continental shelves	Sea surface temperature	Roy et al. 1998
Fish (freshwater)		
Worldwide	Net productivity, area of river basin, species richness of continent	Oberdorff et al. 1995
Europe and North America	Net productivity, climate, area of river basin, latitude range	Oberdorff et al. 1997
North America: minnows and suckers	AET (whole area), PET (in cold areas)	Kerr and Currie 1999
North America: perch and darters	AET, PET	Kerr and Currie 1999
Amphibians		
North America	PET (and other factors)	Currie 1991
Maine	Climate, relief, tree richness	Boone and Krohn 2000

TABLE 9.1 ***Associations of species richness with various environmental factors in a sample of recent studies (continued)***

Species and region	Environmental factors associated with species richness	References
Reptiles		
North America	PET (and other factors)	Currie 1991
Maine	Climate, relief, tree richness	Boone and Krohn 2000
Birds		
World islands	Productivity, area	Wylie and Currie 1993
West Palearctic	Temperature, AET, PET, relief	Diniz-Filho et al. 2003
Palearctic	AET, PET	Hawkins et al. 2003a
Africa	AET, temperature, relief	Hawkins et al. 2003a
North America	PET, relief	Hawkins et al. 2003a
North America: seasonal residents	Seasonal vegetation index, relief, number of biomes	Hurlbert and Haskell 2003
South America	AET	Hawkins et al. 2003a
Australia	Minimum temperature, rainfall	Hawkins et al. 2003a
Britain: resident species	Temperature, rainfall (negative), habitats	Lennon et al. 2000
Britain: winter visitors	Temperature, summer productivity, relief	Lennon et al. 2000
Britain: summer visitors	Temperature, habitats, latitude	Lennon et al. 2000
Mammals		
World islands: total	PET, area	Wylie and Currie 1993
World islands: herbivores and carnivores	PET, area	Wylie and Currie 1993
North America	AET, minimum temperature, maximum temperature, relief, altitude	Badgley and Fox 2000
South Africa: large mammals	Temperature	Andrews and O'Brien 2000
South Africa: small mammals	Seasonality of temperature, plant richness	Andrews and O'Brien 2000
South Africa: frugivores, insectivores	Minimum PET, seasonality of temperature	Andrews and O'Brien 2000
South Africa: arboreal, aerial	Minimum PET, plant richness	Andrews and O'Brien 2000

Note: The table shows the factors with the strongest associations, not ranked in order of their strength. Usually the most influential factor is some aspect of climate or of productivity (see also Hawkins et al. 2003b). *Relief* is the range between the highest and lowest altitude; *PET, AET, rainfall*, and *temperature* are annual means; *seasonality of temperature* is the variation between the most extreme months; *minimum* and *maximum* are the most extreme months of the year; *habitat diversity* is from satellite scans; *habitats* are the presence of particular kinds of habitat; *climate* is a combination of temperature and rainfall by principal components or other means by which individual variables cannot be separated.

Productivity or Energy?

The terms "energy hypothesis" and "productivity hypothesis" are often used synonymously, but they should not be. There are at least two forms of species-energy theory, "productivity" and "ambient energy" (Hawkins et al.

2003b). The better-known productivity hypothesis (O'Brien 1998; Whittaker et al. 2003) claims that diversity is most strongly governed by the rate at which plants fix energy through photosynthesis (as this governs the growth and performance of the plants, then of herbivores and, through them, all the further links in the food chain).

The ambient energy hypothesis suggests that what controls diversity is some direct effect of climate on the organisms themselves (Turner 1986; Turner et al. 1987; Currie 1991; Hawkins et al. 2003b). For ectotherms, reproduction and feeding are faster and more efficient when it is warm and sunny. The same applies to endotherms, because when it is cold they have to burn energy simply to maintain their body temperature, whereas in warm weather more of this energy can be put to other uses, such as reproduction (Turner et al. 1988).

Which provides the better explanation, productivity or ambient energy? Hawkins et al. (2003b) found that productivity, water availability, or combined measures of water and energy were the best explanations in studies conducted at low latitudes, whereas direct measures of energy (e.g., temperature or PET) were the best explanations at high latitudes, at least for animals; the changeover appears to be at around 50° in the Northern Hemisphere. An obvious explanation is that at low latitudes energy is not limiting on biological activity across broad scales, whereas water availability is. At high latitudes it is not productivity that limits activity, but the temperature, especially in the winter.

This should be particularly clear in small animals, which lose heat faster on account of their higher surface-volume ratio, and among migratory birds at high latitudes where species richness should respond directly to the seasonal changes in climate. Three linked studies of British birds suggest that this might be so (Turner et al. 1988, 1996; Lennon et al. 2000). The species richness of winter visitors should depend on winter temperature only, and of summer visitors only on summer temperature. This is true for the lightest, smallest insectivorous birds in Britain but not for heavier birds, suggesting this direct effect of temperature is confined to species that are particularly prone to heat loss.

However, the effect of freezing is variable. Sometimes winter temperatures seem to have independent influences, suggesting that extreme cold may limit species richness independently of productivity. Usually it is suggested that this reflects the fact that it is difficult to evolve adaptations to deal with freezing (e.g., Ricklefs et al. 1999); but in insects, low temperature may be beneficial, at least in those moderately cold climates where efficient diapause depends on a relatively long spell of consistent refrigeration (Turner et al. 1987, 1996). In birds there is a further interesting effect in winter: they are able to withstand temperatures substantially below freezing at night, provided they can build up enough fat during the day to burn through their nocturnal fast (Newton 1969; Root 1988a,b,c). In winter, temperate zone birds are therefore dependent on the food supply for their survival, and most of this food is the result of summer productivity, now banked as seeds or as invertebrates in diapause. Even at high latitudes, where overall temperature is the best predictor of diversity, this seasonally lagged effect of productivity can be seen in an unexpected correlation between summer temperatures and the winter distributions of birds. Winter visitors, which never experience a British summer, have a diversity that is directly related to the British summer temperature (Turner et al. 1996).

Area

Wright (1983) examined the diversity of flowering plants and birds on 36 islands across the globe, ranging in size from the small Indonesian island of Sumba to the island continent of Australia. Multiplying AET or productivity (per unit area) by the area of the island yielded a good prediction of species richness; the same applies to mammals on islands, worldwide (Wylie and Currie 1993). This looks like confirmation of Equation 2, but unfortunately the number of species is necessarily the inventory for the whole island, so that the area sampled (which strongly influences the count) is thoroughly confounded with the area of the metacommunity. These studies leave the role of area, as defined in Equation 2, untested.

For diversity gradients on continents, however, we cannot yet be certain what the "area" is (Rosenzweig 2003c). In Hubbell's neutral theory, it is the area over which species migrate in the very long term, thus uniting local communities into a metacommunity; or, the distance that a species might spread from its area of origin before it gives rise, somewhere in its range, to another species. Perhaps it is the area of the whole continent, or even of the whole planet. Terborgh (1973) and Rosenzweig (1992, 1995) argued that, because the Earth becomes effectively narrower toward the poles, the tapering of the planet itself might explain the biodiversity gradient. However, this will also predict biodiversity gradients based not just on the Equator but on any or every other great circle (Turner 2004). What is needed is some way of getting the world pointed the right way up, of privileging latitude to distinguish the tropics from the poles. Terborgh and Rosenzweig did this using the contemporary temperature gradient.

As a thought experiment, imagine that arbitrarily defined latitudinal belts are isolated from one another, so that no species are ever exchanged between latitudes. In that case, each latitude belt might be a metacommunity independently subject to Equation 2; the spherical shape alone would produce a tropic-pole gradient in which diversity declines with the cosine of latitude. But even without complete isolation, the shape will still produce a richness gradient provided that the movement of species north-south between belts is much less than the rate of movement within belts, east-west. This is a commonplace observation of horticulture and agriculture, recognized for many centuries. Every gardener knows that to grow tropical or arctic-alpine plants in the temperate zone requires special care, frequently involving greenhouses in which the climate can be artificially regulated. On the other hand, transfers within a climatic zone are easy; temperate Chinese and North American plants grow readily enough in England. When species are expanding or changing their ranges without human assistance, east-west spread is likely to be easier than north-south spread even if climate is not the only factor limiting distribution (see also Woodward and Kelly 2003). While all the other unfamiliar states of the environment to which the species must adapt and evolve change isotropically (to the same degree to all points of the compass), temperature requires little adaptation east-west, but can be a major problem for the spread of species northward and southward. If we accept that there is much faster spread east-west than north-south, then the relevant area for us is the area of the continent at a particular latitude. Because we have no clear idea how deep to make the latitude belt, it is best for statistical purposes to treat it in the limit as infinitely shallow.

It is obvious, however, that the crude widths of the continents do not account for diversity gradients very well, as only South America and Africa taper in the appropriate direction—and then only south of the Equator. Other continents tend to be widest towards the poles or at their middle latitudes (see, e.g., Rohde 1998). The situation becomes even worse when we consider that at very high latitudes America and Eurasia have been periodically connected during the Quaternary glaciations. However, if Equation 2 is valid and productivity and area interact to determine biodiversity, it may be that the major tropic-pole diversity gradient is set by climate and productivity, but that the sizes of the continents—in particular their latitudinal widths—contribute to the details of the gradient. For any group in which the width of a metacommunity is considerably smaller than the landmass, the latitudinal width of the continent (or island) will clearly have very little effect.

We are not aware of studies that have directly examined the effect of the latitudinal width of the continents on diversity in conjunction with climate or productivity (although Turner et al. [1988] did use the width of Great Britain), nor of any using the circumference of the planet with the necessary cosine transformation of the latitude (the predicted gradient is not linear). Several studies have tried to incorporate climate into area-based analyses indirectly by distinguishing "biomes" (Blackburn and Gaston 1997; Ruggiero 1999; Hawkins and Porter 2001), but conclusions are very sensitive to how one defines biomes and how many continents are included in the analysis. Dynamic models like the neutral theory and its more elaborate counterparts depend on the dynamics of individuals, species, or clades, and it is not clear how a biome, in so far as it is defined by the physical structure of its vegetation, can be part of such a model. It is the anisotropy of species migration, not the distribution of biomes, that privileges latitude.

Other Factors

Habitat Diversity

Empirical studies, particularly at much smaller than the global scale, find that other factors often improve the power of the statistical models to account for latitudinal gradients in diversity (see Table 9.1). Some form a fine-scaled mosaic, as indicated by studies showing that species richness of birds and butterflies is correlated with habitat diversity (measured by remote-sensing satellites) (Lennon et al. 2000; Kerr 2001; Kerr et al. 2001). But linking species richness with the number of habitats can border on the tautological; if we take the view (e.g., Lewontin 2000) that a habitat is defined by organisms as the part of environmental hyperspace occupied by a species, then species diversity and habitat diversity must be totally correlated. We obtain lower correlations only because we describe habitats less than perfectly. For plants we define habitats by the presence of particular species constellations ("vegetation types"), so increasing the number of "habitats" must increase the total species list.

If plant richness drives animal richness (Hutchinson 1959), then including more habitats should increase total plant diversity, which then should increase animal diversity. Although a logically appealing idea, there is currently no evidence that broad-scale trends in plant richness directly influence animal richness patterns. The relationship between plant and animal diversity of some more specialized groups can even be negative at the macro scale (e.g., aphids [Dixon et al. 1987]), and even when plant and animal diversity co-vary positively, the association may be driven by similar responses to environmental variables rather than by direct trophic links (Hawkins and Porter 2003a). Due to limited data on broad-scale plant diversity patterns (that include all plants, not just trees), the hypothesis that habitat heterogeneity drives animal diversity via direct links between plant and animal richness has not been well tested. Although finding a relationship between our measures of habitat and the distributions of species has many practical applications (see Rosenzweig 1992), it is not, as far as we can tell, a strong explanation for diversity; it may simply reflect something that is true by definition.

Relief

A factor that consistently favors diversity is roughness of the landscape ("relief" in Table 9.1): crinkly parts of the planet have more species. Switzerland for instance is renowned as having an immense diversity for its area and latitude (von Humboldt was one of the first to report this relationship [Gould 2002a]). This seems to be at least partly independent of the fact that crinkly areas have more habitats, since diversity is often associated with range in elevation more strongly than with more direct measures of habitat heterogeneity (Rahbek and Graves 2001; Kerr et al. 2001; Hawkins and Porter 2003b).

This may reflect the fact that the actual surface area of rough places is greater than the size of the quadrat drawn on the map (O'Brien et al. 2000), but a stronger effect is likely to be the fact that rough places contain an enhanced range of climates, each with the appropriate biota. First, mountain ranges may contain many climatic zones, including everything from tropical forest to Arctic tundra in high tropical mountains. That this is a strong explanation is shown by the change in this effect with latitude—an Arctic mountain is tundra from top to bottom. Thus Rahbek and Graves (2001) found that range in elevation is strongly positively associated with bird diversity in the tropical north of South America, but not in the temperate south. Dividing North America and the Palearctic into northern and southern halves similarly shows that in both regions bird richness is much more strongly correlated with relief in the south than in the north (Figure 9.3). Thus, as the general climate gets cooler, the effect of topographical variability becomes weaker—and even reverses—possibly because the biotic response to temperature gradients becomes weaker. Second, while planar systems (plateaus and plains) have relatively homogenous climates, "crinkly" landscapes contain alternating strips of microclimates along their north- and south-facing slopes, each with biota akin to areas several hundred kilometers both to the north and to the south (Turner et al. 1996).

Relief also may affect the rate of allopatric speciation, particularly when the climate fluctuates over time (Hewitt 1993, 1996; Fieldsa and Lovett 1997). Imagine a group of montane animals or plants that move up and down the mountains as the climate warms and cools. When the mountains have a lot of separate peaks, the populations become isolated during warm periods, when they are likely to become separate species, and can then invade each other's ranges when they move to lower altitudes during cool periods. But it does not follow that contemporary relationships between relief and diversity necessarily arise from patterns of speciation. For example, during the most recent glacial maximum around 18 kya, the Alps were swept clean of species by glaciers, which means that the elevated diversity currently found there represents species that have moved into the mountains following glacial retreat, not species that necessarily evolved in them.

The Niche-Assembly or Structural Theory

The most widely accepted explanation of the diversity gradient over the last fifty years has been the "structural" or "niche-assembly theory"; more species can be accommodated within the resource space of the tropics because ecological niches are narrower and species are more specialized. Adaptation is the key—not simply adaptation to the climate, but adaptation to each other. This connects with a widely held orthodoxy in ecology: that ecological communities are really societies, structured by species interac-

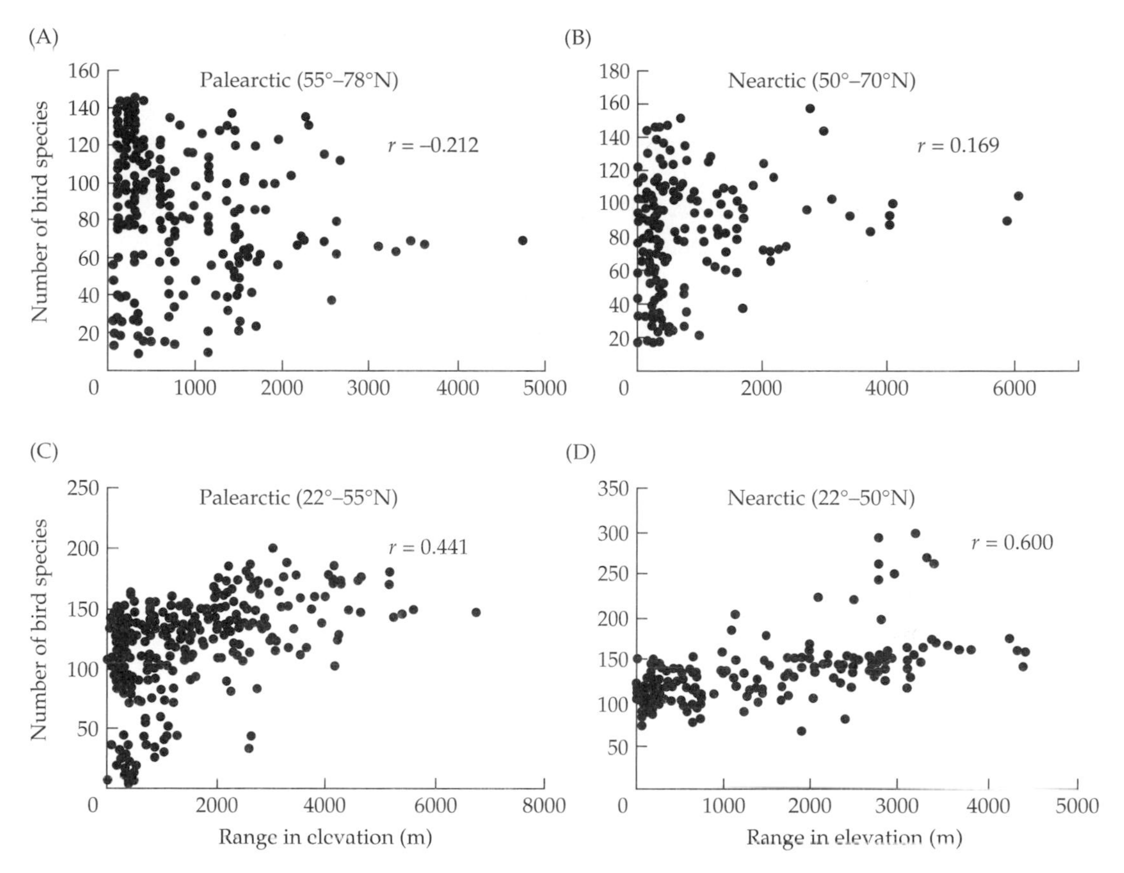

FIGURE 9.3 Relationships between terrestrial bird species richness and range in elevation in the Palearctic (Himalayas excluded) and the Nearctic. Dividing both regions into northern (A, B) and southern (C, D) zones reveals that relief is more strongly associated with species richness when the general climate is warmer. Note that in the Palearctic the relationship is negative in the northern half of the region. All four correlation coefficients are highly significant ($P < 0.001$). (Data from Hawkins et al. 2003a.)

tions. And in contrast with the axioms of neutral theory, species are seen as different in all sorts of ways, particularly in their niches and competitive abilities. There are multiple versions of the niche-assembly hypothesis. The tropics might have more species than a high latitude area with similar resources if resource requirements of tropical species were narrower, or if species could share resources without damaging one another, or if they used somewhat different resources (Figure 9.4). If speciation rates in tropical and high latitude systems are similar, the tropics might contain more species because extinction is lower, or because species spread more readily to become sympatric. How will these things arise from the arrangement of eco-

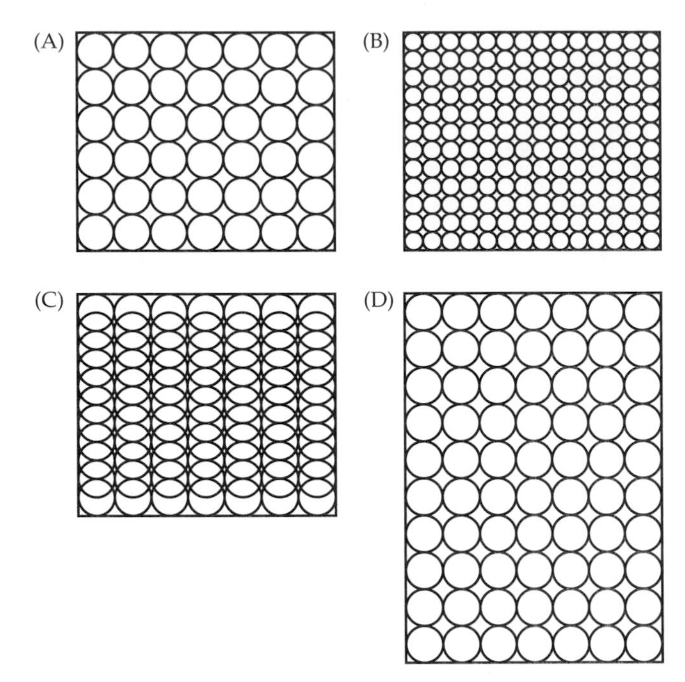

FIGURE 9.4 Three alternative versions of niche assembly theory that might explain why fewer species are found in the ecological space of the temperate zone (A) than of the tropics (B, C, D). The boxes represent the total resource space, the spheres the ecological requirement of species. (B) Tropical species have narrow niches. (C) Tropical niches are the same size as those of the temperate species, but the resource space is shared more because there is less competition and more mutualism. (D) The tropics have a larger resource space than the temperate zone. This alternative corresponds with the assertions of species-energy theory.

logical niches? If species in the tropics evolve so that their requirements for resources are shared more evenly, the distribution of population sizes across species becomes more equable, and rarer species have larger population sizes than high latitude species of equivalent rank. As it is rare species that are disproportionately in danger of extinction, effectively the rate of extinction is lowered, and the equilibrium number of species is increased.

Further, in the tropics species are more likely to be mutualistic and to mutually reinforce one another's populations. The system, like a well-ordered society, behaves less chaotically, so that the random extinction of species is less likely. And it is true that tropical species live in specialized ways that are unknown at higher latitudes (Greenwood 2001).

Explanations proposed for why tropical species evolve as they do are very diverse, although they usually involve some aspect of climate. Widely canvassed is that the stability of the tropical climate is key (Connell and Orias 1964). This means that resources are more reliable and tend to be available year round. Over time, species can become increasingly specialized. At higher latitudes a species that becomes too specialized is at a higher risk of extinction. Therefore tropical species have narrower ecological niches and share ecological space more equably. A more elaborate version of this theory adds a difference in the rate of allopatric speciation (Dynesius and Jansson 2000); at high latitudes, species are selected not only to be generalists, but also to have high individual vagility. Alternatively, we suggest that if the inter-generation variance in climate (seasonally or annually) is lower in the tropics, fluctuations in population size will also be lower, substantially decreasing the risk of extinction for the rarest species. Climatic variability does feature as a predictor of species richness in some empirical analyses, but not often, perhaps because it is not always measured.

But how do we know which is the cause and which is the effect? Are there more species because niches are narrower, or are niches necessarily narrower because there are more species? Unfortunately, testing many niche-assembly theories requires data that do not yet exist or are virtually impossible to quantify satisfactorily (niche breadths have to be measured for comparable samples of tropical and high latitude species through a sample of surrogate measures of niche breadth; see Roy et al., this volume), and the various theories tend not to make strong and clear predictions by which they can be tested. This leaves them as a set of plausible just-so stories that are very difficult to incorporate into quantitative models of diversity.

Speciation Rate, Migration, and Evolutionary History

It is not quite correct to say that correlation does not imply causation; type II errors excepted, a significant correlation always implies causation (Sokal and Rohlf 1969). But it does not necessarily imply direct causation. Thus the strong association between biodiversity and climate (see Table 9.1) might vindicate the species-energy theory, or it might indicate only that more species are adapted to lower latitude belts, the correlations arising because each species occupies its appropriate place in the climate gradient (Turner 1986). The correlation could be explained not by ecology, but by evolution. What processes might lead to the evolution of more species adapted to low latitudes?

Speciation Rate

Speciation might be higher in the tropics, leading to more tropical-adapted species. In warmer climates generation times are shorter, and of course shorter generation times lead to faster evolution of all sorts, including speciation (Rohde 1992, 1998). It is also likely that sister species originating in allopatry become sympatric only after their niches have differentiated somewhat, usually by becoming narrower. Shorter generation times and more rapid evolution will result simultaneously in both greater per area diversity and narrower ecological niches.

However, there is a drawback with this argument (encountered also in controversies over neutral evolution theory); the speciation rate in Equations 1 and 2 is per generation, not per unit time. And if speciation is in equilibrium with extinction, which should also be reckoned per generation (as it is in the neutral theory), the two effects cancel out. It is not obvious that latitudinal variation in speciation rate will necessarily account for the planetary diversity gradient.

Effects of Glaciation and Recent Climate Change

British biogeographers in particular have liked the idea that the species richness gradient in the temperate zone reflects simply a delay in full recolonization after the retreat of the glaciers (e.g., Ford 1945; Mathews 1955). This is almost certainly untrue. The pollen record shows that recolonization was initially rapid and then stabilized (Silvertown 1985), and exotic species introduced into new areas spread rapidly to occupy the latitude span they had in their native continent (Sax 2001). Latitudinal gradients have existed for a long time (see chapters by Roy et al. and by Crame, this volume), and during periods of climatic change "habitat tracking" is rapid (Jackson, this volume). The richness of butterflies in England underwent a decided dip during the nineteenth century (Figure 9.5), corresponding to a substantial deterioration of the British summer temperatures of about 0.8°C, which lasted from around 1870 to 1920 (Manley 1974).

From the current regression of butterfly richness on temperature (Turner et al. 1987), about half of the drop should be attributable to climate change; in the two areas sampled, industrialization may account for the rest (Fryer and Lucas 2001). The late twentieth century rise in local species richness appears to result from global warming: 36 species of British butterflies have recently expanded their ranges (perhaps to be discounted by 11 rare species, where the expanded range may be an artifact of better censusing), 1 species has had a net contraction equal to its expansion, and 12 species have contracted (Asher et al. 2001). In Europe, birds and butterflies have been rapidly extending northward (Parmesan et al. 1999; Thomas and Lennon 1999). However, it is hard to say whether these changes simply represent habitat tracking by climatically adapted species, or whether they illustrate the mechanism behind the species-latitude gradient itself.

Alternatively, the repeated extirpation of species by massive ice sheets in the northern Nearctic and Palearctic might account for some lowered diversity (Clarke and Crame 2003), although finding a historical signal has been difficult (Currie and Paquin 1987; Turner et al. 1988; Adams and Woodward 1989; Oberdorff et al. 1997; but see Oberdorff et al. 1999) and sometimes contentious (Currie and Paquin 1987; Latham and Ricklefs 1993; Francis and Currie 1998; Ricklefs et al. 1999). Using the approximate time when areas became exposed for recolonization by the retreat of the ice (a more refined measure than simply distinguishing glaciated and non-glaciated areas),

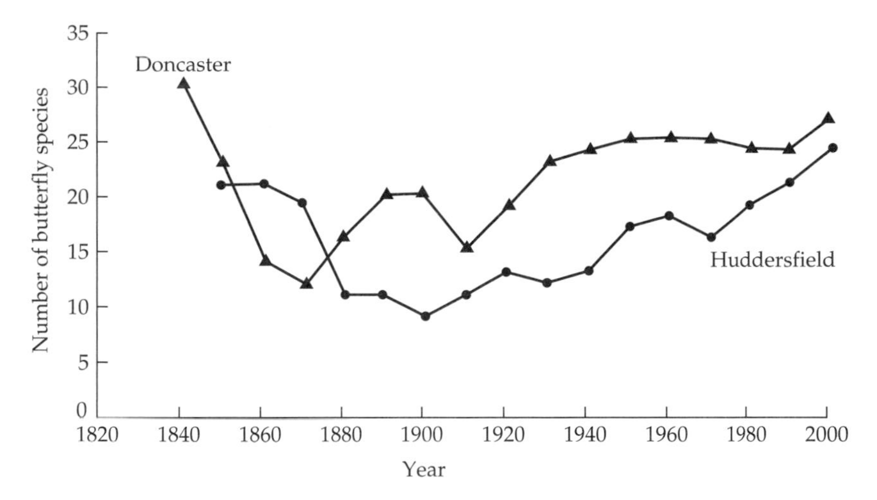

FIGURE 9.5 The decline and recovery of butterfly species richness in 1000-km^2 areas around Huddersfield and Doncaster in Britain from 1830 to 2000. The decline and initial recovery corresponds with a period of cool summers in Britain from 1870 to 1920 (the decline seems to start earlier at Doncaster, which is at a lower altitude and less industrialized); the late twentieth century increase is likely attributable to global warming. (Data from Rimington 1992 and Fryer and Lucas 2001; most recent point for Doncaster from Asher et al. 2001.)

Hawkins and Porter (2003c) found that PET explained 76% to 82% of the variance in richness of birds and mammals currently inhabiting the most recently glaciated part of North America, but the ice-free age accounted for an additional 8% to 13%, primarily due to effects operating at large spatial scales. Thus, although contemporary climatic conditions have the strongest apparent influence on the diversity gradient, a historical signal persisting thousands of years is still detectable.

Migration between Climatic Belts

Species probably do spread between climatic belts, although perhaps rather rarely (Rosenzweig 2003c). Could "tropical overspill" then produce the global gradient (Rosenzweig 1992)? If species spread northward and southward at the same rate, migration will indeed increase diversity (per area) all over the planet, but it will not in itself produce a diversity gradient. In order to generate the observed gradient, we have to suppose some or all of the following (see also Turner 2004):

1. *Speciation rates are higher at lower latitudes, or, higher productivity results in higher tropical richness.* Further, the difficulty of adapting to new climates inhibits spread equally toward the poles and the tropics. The steepness

of the existing gradient is then maintained by the inhibition of spread to foreign latitudes, but because an equal percentage of species spread north and south, numerically more tropical species will enter higher latitudes than higher latitude species colonize the tropics: there will indeed be a net "tropical overspill." It smoothes out the diversity gradient, but also causes the tropics to be the net source of evolutionary novelty for the rest of the planet—the great powerhouse of evolution seen by the classical evolutionary biologists (Jablonski 1993; Clarke and Crame 2003).

2. *Adaptation to cold is slower than adaptation to heat, or, adaptation to frost is harder than to the absence of frost* (Latham and Ricklefs 1993). This means species spread toward the tropics faster than they spread poleward. There will be "temperate overspill" that generates the observed gradient, and the tropics will be the net recipient of evolutionary novelty. However, Sax (2001) showed that exotic introductions tend to spread poleward rather than toward the tropics.
3. *Originally, no species were adapted to frost* (which requires physiological resistance to ice formation in tissues), and the planet has progressively cooled. Higher latitudes have been deprived of their species, and the difficulty of adaptation is substantially delaying their replacement; those clades that have adapted have yet to produce enough species (Latham and Ricklefs 1993). On the other hand, if the planet was warming, and if it is indeed also hard to adapt to warm climates, the gradient should be in the reverse direction. The planet has progressively cooled since the Eocene (Clarke and Crame 2003), but has substantially warmed since the last glaciation. Which time scale is appropriate?

Theories, and Theories

So, where do we stand with respect to understanding the largest pattern in ecology and biogeography? We have distinguished three major types of theory for the biodiversity gradient: species-energy theory, niche-assembly theory, and a variety of theories that chiefly invoke evolution and history. All may be correct, and historical contingencies may be significant in both energy and niche-assembly models.

Species-energy theory has been popular for the last 10 years, and very unpopular for the previous 30. It was proposed in something like its modern form by Hutchinson (1959), alongside what we would now call niche-assembly theory. While his niche-assembly theory was then widely developed, the odd thing is that Hutchinson's adjacent suggestions about productivity and extinction seemed to be largely ignored by cohorts of ecologists. When rediscovered in the 1980s, these same suggestions met with vigorous opposition. The theory has been parodied (e.g., Willig et al. 2003) either as a flawed deterministic (not, as it is, stochastic) model (Rohde 1992), or by over-simplification: "More energy equals more biological activity, greater evolutionary speed, and therefore more speciation" (McGlone 1996; see also Turner 1992).

Faced with a choice between cause and effect (diversity regulates niche widths versus niche widths regulate diversity), many of us look first to such rich Darwinian concepts as natural selection and, above all, adaptation, which are indeed powerful and often compelling explanations. Niche-assembly theories are thoroughly Darwinian: they use the theory of adaptation by natural selection to build up a view of ecological communities as living wholes, greater than the sum of their parts. But if Darwinism explains much, it does not necessarily explain everything (Gould 2002b).

We can naturally take the view that the world is more complicated than the description provided by the neutral theory of biodiversity. Indeed it is. The point of such a very simple model is not that it captures all aspects of reality, but that it tries to isolate the crucial aspects and show us how they fit together (Equation 2). It has the strength of making rather clear predictions that can be tested and found wanting (or not wanting). A problem with niche-assembly models is not that they are wrong, in the sense that adaptation and natural selection do not occur (they do), but that they generate predictions that are too vague to test with empirical data. As with random genetic drift, it is not a matter of whether "ecological drift" occurs (Turner 1992); depending as it does on the birth and death of individuals, and the fact that population sizes are finite, it is inevitable. The question is how many further processes occur which distort or obscure the patterns it produces.

The profitable strategy for macroecologists is to build their explanations of biodiversity "from the bottom," to start with the simplest theory, the neutral-species-energy model, and to test this theory rigorously against the data (Whittaker and Field 2000; Hubbell 2001). Great insight will then come from building the neutral theory upwards into the more elaborate, rich, and complicated structural theories. As theories are developed, it will be very important to make firm predictions from them, and then to gather empirical data that will test these predictions to destruction.

So far we have been able to show that of the three factors that the neutral theory suggests will be major determinants of large scale patterns of biodiversity—productivity, metacommunity area, and speciation rate—there is strong support for the energetic element of climate. There is, however, an urgent need to incorporate area into diversity studies that also include climatic and productivity variables (Turner 2004). We need also to develop better databases on the abundances and distributions of non-woody plants; trees, which are much better studied, may be atypical in that the large amount of dead wood they contain makes them partly supra-terrestrial fossils, subject to the dynamics of deposition and decay as well as birth and death.

A complication in distinguishing the explanatory power of productivity and ambient energy, even when AET or energy-water variables seem to provide the best explanation, is that in the far north, plant growth is limited by temperature and light rather than water. Therefore, we cannot rule out the possibility that any strong association between animal diversity and energy operates via the indirect effects of energy on plants, rather than via direct physiological effects on the animals themselves. This is particularly serious when productivity is estimated from energy and water variables, producing a set of factors that are so tightly intertwined it is virtually impossible to distinguish among them. However, some progress may now be possible by using remotely sensed vegetation data to provide estimates of plant productivity that are computationally independent of water and energy variables (Kerr et al. 2001; Hurlbert and Haskell 2003; Hawkins and Porter 2003b). By including both types of variables simultaneously, and using predictive causal models (e.g., structural equations) that allow one to distinguish direct and indirect effects of climate on diversity, it may be possible to tease apart the roles of productivity and ambient energy.

Although spatial covariation between paleo- and contemporary climates will make it a challenge to fully disentangle their effects, the inclusion of phylogenetic information in diversity analyses promises to be the single most important improvement in analyses of the global diversity gradient. As quantitative phylogenetic comparisons among biotas (e.g., Cardillo 1999; Kerr and Currie 1999; Qian and Ricklefs 1999; Poulin and Guégan 2000) expand to large taxonomic groups and over larger parts of the Earth, our ability to evaluate the role of history should improve.

Finally, the current warming of the global climate should now provide us with an important, if undesirable, natural experiment testing species-energy theory. So far investigations have concentrated on the less predictable movement of individual species, although climatic models have been used to predict changes in the richness patterns of birds in Britain (Lennon et al. 2000) and of vertebrates and trees in the United States (Currie 2001).

In closing, we believe that progress is being made rapidly, and we strongly disagree with those who claim that we have no idea what drives the global diversity gradient. Biogeographers have some very good ideas, and some of them are very well supported. We believe that we are getting close to answering the largest scale pattern in ecology and biogeography.

Acknowledgment

We thank Richard Field for his critique of an earlier version of this chapter.

CHAPTER 10

Diversity Emerging: Toward a Deconstruction of Biodiversity Patterns

Pablo A. Marquet, Miriam Fernández, Sergio A. Navarrete, and Claudio Valdovinos

Introduction

Life is about diversity, heterogeneity, and novelty. The understanding of how these properties interact, emerge, are maintained, and eventually destroyed is at the core of the research programs of biogeography, ecology, and evolution. Diversity (or more properly, biodiversity), although present at all levels of organization, is most commonly measured by ecologists and biogeographers as the number of species found at a particular point in time or space (species richness). For more than two centuries ecologists, paleontologists, and biogeographers have been quantifying the number of species found in almost all imaginable places on the Earth, from the deep ocean to mountain tops, from soil to canopies, and foremost, along gradients in the physical environment, such as those along latitude, longitude, altitude, salinity, aridity, and depth (Brown 1988). We can certainly be assured that diversity changes and that it does so in a nonrandom way. However, our understanding of the principles underlying those changes has remained elusive. Thus, although the task of enumerating species at a given location might be simple (given the required time and funding), understanding the processes and underlying mechanisms is not.

Why are there more species of trees in a 0.1-ha plot in the Peruvian Amazon than in a plot of similar area in the temperate forest of Chile? To answer questions involving differences in diversity between places or along geographic gradients, it is tempting to resort to changes in the external environment that are associated with available energy, area, or temperature; or to changes in the intensity of ecological interactions such as predation, competition, or disease; or to historical reasons that might have affected dispersal, speciation, or extinction rates (Hutchinson 1959; Pianka 1966; Schall and Pianka 1978; Diamond 1988; Rohde 1992; Rosenzweig 1995; Brown and Lomolino 1998; Waide et al. 1999; Gaston 2000). A good example in this regard is the latitudinal pattern in species diversity, for which around 25 or so hypotheses have been proposed (e.g., Brown and Lomolino 1998; Roy et al., this volume; Turner and Hawkins, this volume; Whittaker, this volume).

In this chapter, we argue that part of the difficulty in achieving a grand unified theory for patterns in species richness resides in the nature of the response variable itself. Species richness is an aggregated variable that subsumes, in a single number, the variety of life at the species level found in a particular place and time. Further, because each species counts the same when computing richness, all species are in practice considered equivalent (see also Lomolino 2000a,b), but of course, they are not. We all know that the only variety that the concept of species richness captures well is that contained in Latin binomials. Diversity indices (e.g., Shannon's *H*) go a step further by weighting species by their abundance. But these indices still assume that, other than their shear abundance, species are essentially equivalent. We do not attempt to trivialize the concept of species richness; our aim is simply to show that in order to understand patterns in species richness we need a deeper understanding of the quality and nature of the units that are being measured.

In this chapter, we will try to deconstruct richness patterns. We use the word *deconstruction* in its etymological sense, as a "turning to the roots" of what is being measured, or "disaggregation," to make apparent what is hidden. As shown here, deconstructing biodiversity patterns or examining other expressions of biodiversity (see also Roy et al., this volume) can sometimes open easier paths because the biological attributes of species are more tangibly related to their ecological and evolutionary implications. Interestingly, deconstruction is usually done but not pursued consciously as a methodological strategy to analyze richness patterns, for we usually do not work with richness per se but with richness of a particular group of organisms. This widespread practice usually goes unnoticed and is taken to reflect our own taxonomic expertise. On the contrary, we claim that deconstruction should be consciously performed as a methodological strategy for at least four reasons:

1. To understand the causes underlying patterns in species richness
2. To reconcile seemingly disparate explanations of richness patterns that emphasize an overriding importance to a given environmental factor, such as area or energy (e.g., Rosenzweig and Sandlin 1997; Rhode 1998; Chown and Gaston 2000)
3. To make apparent the need to overcome current methodological restrictions on the way richness data are collected and analyzed
4. To restate the question of richness in a comparative framework

As is usually the case in science, most of the time we realize that the wheel was already in place when we discovered it. In this sense we pay due recognition to the work of Michael Huston, who in his book stressed (1994:2):

> One of the central premises ... is that biological diversity can be broken down into components that have consistent and understandable behavior. The other central premise is that the various components of biological diversity are influenced by different processes, to the extent that one component may increase, while another decreases in response to the same change in conditions.

In this context, our attempt here is to emphasize some key issues and develop these ideas further, showing both their theoretical foundations as well as their empirical appeals.

The Deconstructive Approach to Species Richness

The basic assumption behind deconstructing patterns in species richness is that assessing diversity as richness does not adequately characterize the way in which species differ from each other, and which cause them to respond in different ways to changes in the environment. Further, as we show below, by decomposing richness into smaller subsets of species that are internally homogeneous, in terms of sharing a particular attribute, we can gain better insight into the causes that underlie richness changes in time and space. As pointed out by Huston (1994), the more types of organisms that are included in a sample of species richness, the more mechanisms are likely to influence the observed richness.

A simple example can help illustrate our point. In Figure 10.1, we plot the relationship between area and the total number of species for 11 islands within the archipelago of the Sea of Cortés (Cody and Velarde 2002), as well as the same relationship on the same islands for different groups of organisms distinguished according to taxonomic affiliation (i.e., a taxonomic deconstruction). Note that we use only those islands for which complete inventories of all taxa under study have been conducted. We would like to make three obvious points about this figure. First, as observed anywhere in the world, the number of species that different groups attain in a given area differs by more than one order of magnitude (e.g., plants versus mammals). Second, while in general species richness in different groups increases with area, it does so at very different rates. Third, the amount of variance in species richness explained by area varies among groups, from 75% in land birds to 24% in tenebrionid beetles (Figure 10.2).

The differences in the effect of area on richness likely result from differences in dispersal abilities or opportunities, energetic demands, and in general, from differences in the characteristics that define a group of organisms as different from one another in the first place, and which determine that they have different readings of the environmental text. However, we can continue

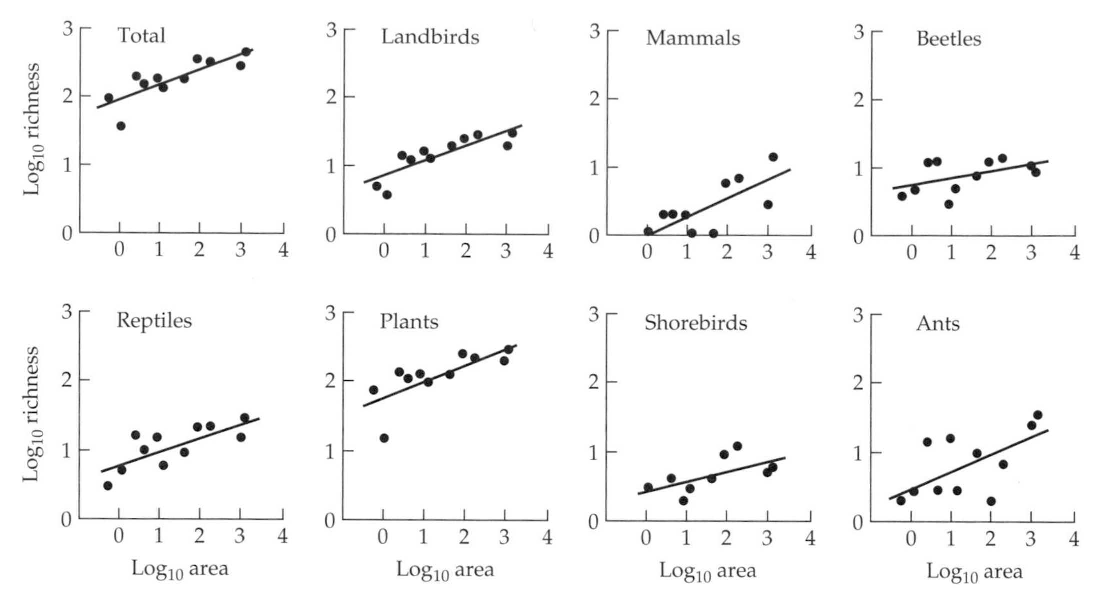

FIGURE 10.1 Relationship between richness (number of species) and area on 11 islands in the Sea of Cortés.

this deconstruction even further, for even within taxonomic groups (e.g., plants) we can distinguish different kinds of organisms characterized by a suite of life history traits that will likely cause them to respond differently to changes in their external environment. We could also deconstruct total species richness on the basis of physiological attributes of taxa (such as endotherms versus ectotherms) or other taxonomic distinctions (such as vertebrates versus invertebrates), or on ecological attributes such as food web position (decomposers, producers, primary consumers, and secondary consumers).

What is the right way to deconstruct richness patterns and where or when should we stop deconstructing? We will postpone discussion of these issues until the end of this chapter; meanwhile we would like to emphasize several important conclusions derived from the preceding example. First, species richness can be decomposed in many different ways. Second, depending on the criteria used to make this decomposition, we will likely arrive at different answers regarding the importance of a given environmental factor such as area (see the above example). Third, deconstruction opens the way to a comparative approach to richness patterns by allowing one to ask why there are more species of one type than of another in a given place or time, and why some groups of species respond in different ways to the same environmental gradient. Notice that this is possible because the data in this example are derived from the survey of several groups of species on the same set of islands. This might not be a necessary requirement for deconstruction, but it certainly might constrain its full application.

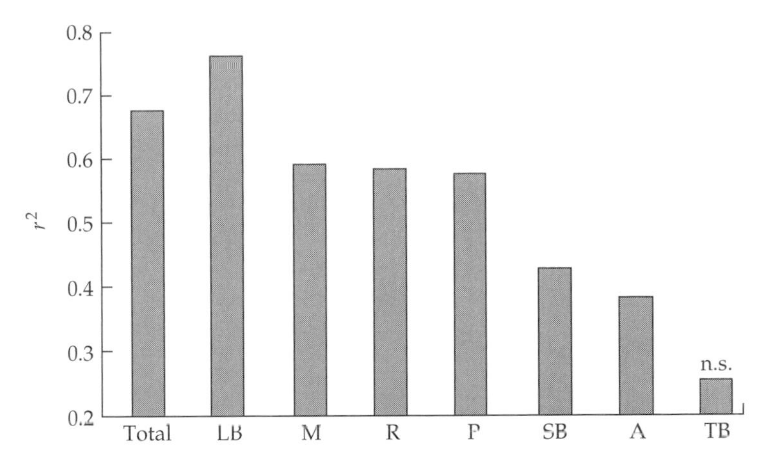

FIGURE 10.2 The importance of area in affecting richness of different taxa expressed as the amount of explained variance in number of species (r^2). LB, landbirds; M, mammals; R, reptiles; P, plants; SB, shorebirds; A, ants; TB, tenebrionid beetles.

At the risk of being redundant, thus far we have argued that to analyze how the total number within a group of species (e.g., vascular plants) changes across an environmental gradient, assumes that all species are equal or ecologically equivalent, but they are not. On this point, we concur with Lomolino (2000a,b) in that non-random variation in species characteristics is of fundamental importance in understanding biogeographic patterns. To further illustrate the use of deconstruction, we show a hypothetical example in Figure 10.3. The traditional method of analyzing patterns in richness treats all species as equivalent and assesses the relationship between the response variable (i.e., total richness, S_T) and an explanatory variable (i.e., environmental factor, E). However, this procedure masks the real complexity of the pattern, for a large amount of error variance in the relationship between S_T and E can be explained by further disaggregating S_T in species groups (S_1, S_2, and S_3) that show qualitatively different responses to changes in E. Thus there is a diversity of diversity patterns. A recent paper by Bhattarai and Vetaas (2003) illustrates this point (Figure 10.4). These authors studied the elevational gradient in plant species richness in the Himalayas and tested for the effect of several environmental variables upon total species richness, as well as across five groups of species defined on the basis of their life form (e.g., ferns, herbs, climbers, trees, and shrubs). As shown in Figure 10.4, the relationship between species richness and potential evapotranspiration (PET) changes, depending on life form (see also Arroyo et al. 1996 for a similar analysis along a latitudinal gradient). The pattern is unimodal for shrubs and trees, positive and monotonic for climbers, and negative and monotonic for ferns, while herbs show no pattern. Thus in order to answer the question of why the total number of plant species increases nonlinearly with PET, one should first

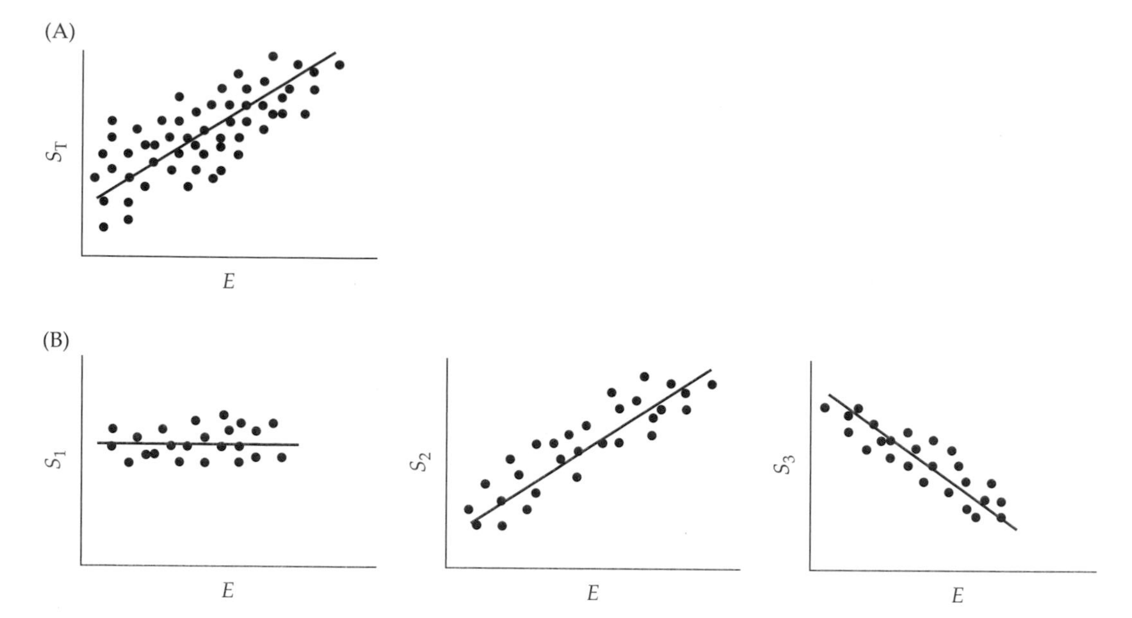

FIGURE 10.3 Schematic representation of the deconstructive approach. (A) An illustration of how total richness (S_T) relates to environmental factor E. (B) Total richness can be decomposed or dissected by analyzing how different groups of species (S_1, S_2, and S_3) respond to the same factor E.

answer the question of why trees, shrubs, and climbers increase in richness with PET, but herbs and ferns do not.

As we have already emphasized, and will continue to develop further, comparing and analyzing ecological systems in terms of numbers of species does not capture adequately the heterogeneity of the system and, in some situations, it might obscure instead of foster, understanding. Richness variation across geographic gradients can be meaningfully decomposed so as to shed light upon the underlying causes. Most current explanations of the geographic variation in species richness consider species as equivalent and emphasize the causal role of the external environment (e.g., area, productivity, water availability), but these explanations, as well as the mechanisms that they invoke, fail to recognize that the processes that control species diversity at any particular location (i.e., migration, extinction, and diversification) are not independent of the life history or physiological and ecological attributes of a species, for these attributes determine what aspects of the outside world are relevant to them. Thus further understanding of the dynamics of species diversity requires an expanded conceptual framework where differences among species are emphasized and explicitly considered in order to understand species richness patterns. Although it might seem old-fashioned to speak of deconstruction when unifying theories are becoming trendy (e.g.,

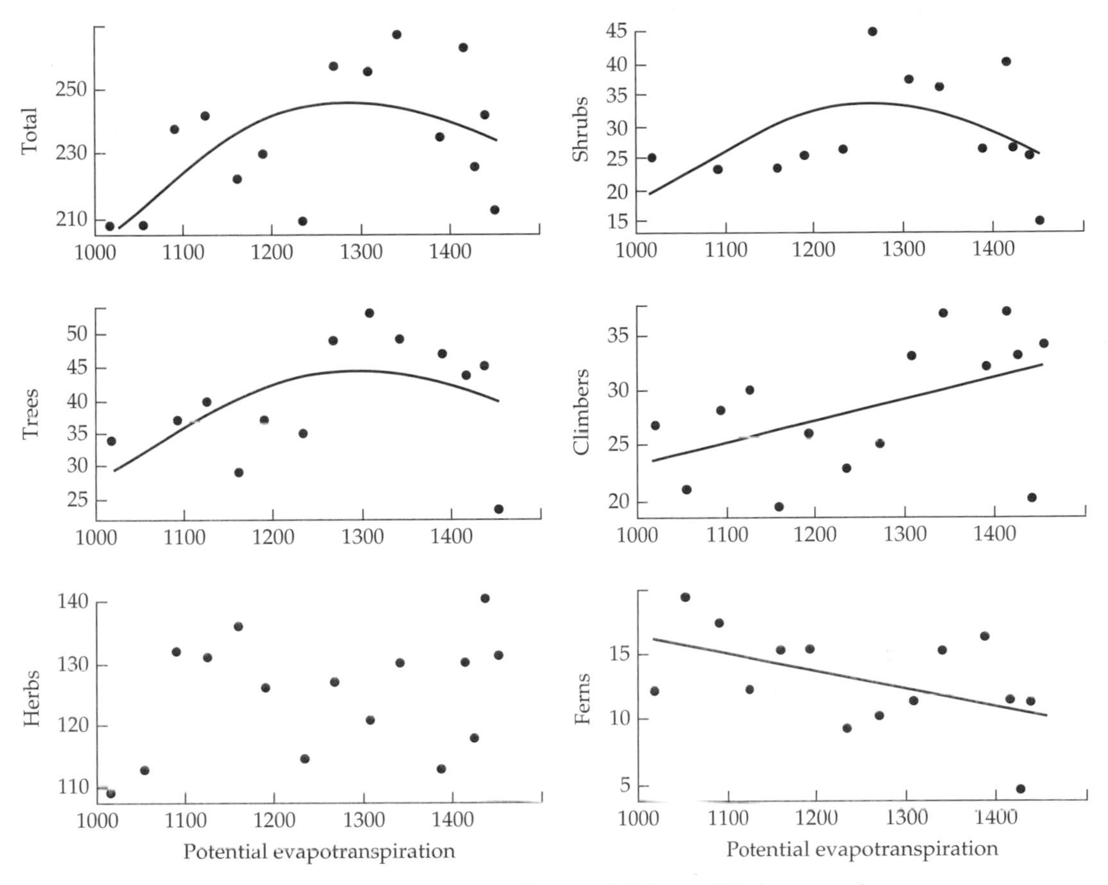

FIGURE 10.4 Relationship between species richness of different life forms and potential evapotranspiration (PET). (After Bhattarai and Vetaas 2003.)

Hubbell 2001), we think it is a logical imperative to properly qualify the complexity of patterns before unification is attempted. Deconstruction capitalizes on the differences among organisms and explores how patterns change depending on the criteria used to disaggregate richness.

At the most basic level, deconstruction is possible because the number of species in a given time and place emerges from the interaction between an organism's strategies and attributes and the environment wherein it exists and has been embedded. This, in turn, affects the extinction and diversification dynamics of the lineages to which it belongs. In Figure 10.5, we illustrate this approach by depicting an explanatory domain for species richness patterns as the space contained within a triangle formed by three major axes: one, accounting for the internal properties and states of organisms (i.e., their physiological, ecological, and life history traits); another, depicting the external properties and states of the external environment wherein the organisms are embedded

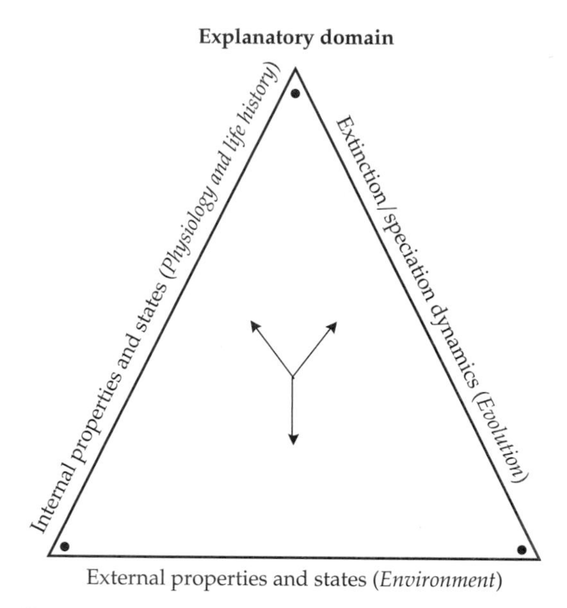

FIGURE 10.5 Conceptual model delimiting the explanatory domain for species richness patterns. The black dots indicate that most current explanations of richness patterns have privileged explanations along one single explanatory axis (see text for discussion).

(e.g., temperature, precipitation, productivity) including the effects of other co-occurring species (e.g., competitors, predators, parasites); and a final axis depicting extinction/speciation dynamics that largely result from the long-term interaction between organisms and their environment. Within this explanatory domain, patterns emerge as the result of the interaction among the characteristics of species, their environments, and their evolutionary dynamics, emphasizing that species differences do affect the patterns we observe, thus rendering possible the use of a deconstructive approach. Eventually, deconstruction could be performed along each of these three axes.

For the sake of simplicity, in this contribution we emphasize deconstruction with a focus on how traits of organisms affect patterns in richness. We present two case studies that illustrate potential interactions among these axes. The first example makes the point that the persistence of species in a particular landscape depends on the interaction between the attributes of the species and those of the environment wherein the species is embedded. The second example illustrates that the differential persistence of species with particular life history traits along environmental gradients can give rise to the differential proliferation of taxa, and thus to the emergence of diversity patterns. In this case, it is shown that only by deconstructing richness patterns according to life history traits is it possible to further understand the emergence of richness gradients.

Interaction between Species Traits and Environmental Characteristics

Landscapes are not homogeneous. In fact, we can discern a landscape pattern, represented, for example, by the distribution of attributes such as patch size, patch shape, and connectivity; and a landscape dynamic, represented by the rate of change in landscape pattern. Because species are not equal in terms of their sensitivity to landscape patterns and dynamics, persistence is a function of species attributes. Typically, species with restricted dispersal will be more sensitive to connectivity, while wide-ranging species might be more sensitive to amount of habitat. In recent contributions (Keymer et al. 2000; Marquet et al. 2003), a rigorous treatment of this idea was attempted. Using a simple metapopulation model that represented the pattern of patch occupancy by a given species, Keymer et al. (2000) coupled this model to a dynamic landscape model that included patch dynamics, where the proportion of available habitat patches changed dynamically. In this model, the landscape dynamic is represented by the rate of patch creation (λ) and destruction (e), where $\tau = 1/e$ gives the mean lifespan of a patch that can be colonized, and $s = \lambda/(e + \lambda)$ gives the expected amount of available patches in the landscape (Figure 10.6). The species dynamic is given by the rate of propagule production (β) and the intrinsic species-specific extinction rate in patches (δ). Following Keymer et al. (2000), the non-spatial version of this model or mean field approximation can be written as

$$\frac{d}{dt}p_o = e(p_1 + p_2) - \lambda p_o$$

$$\frac{d}{dt}p_1 = \lambda p_o - \beta p_1 p_2 + \delta p_2 - e p_1$$

$$\frac{d}{dt}p_2 = \beta p_1 p_2 - (\delta + e)p_2$$

As is typical of patch occupancy metapopulation models of the Levin's type, metapopulation persistence can be assessed by the reproductive number (R_0) (Marquet and Velasco-Hernández 1997), which in analogy with its use in epidemiology (Anderson and May 1991) measures the number of secondary colonizations that an occupied patch will accrue during its lifetime, or in our case, the "infective" properties of a particular life history. Clearly, persistence will be achieved whenever $R_0 > 1$, at which point a species invading a landscape will persist and reach a positive equilibrium abundance. As shown by Keymer et al. (2000), this model can be expressed as

$$R_0 = \frac{\beta s}{\delta + \tau^{-1}}$$

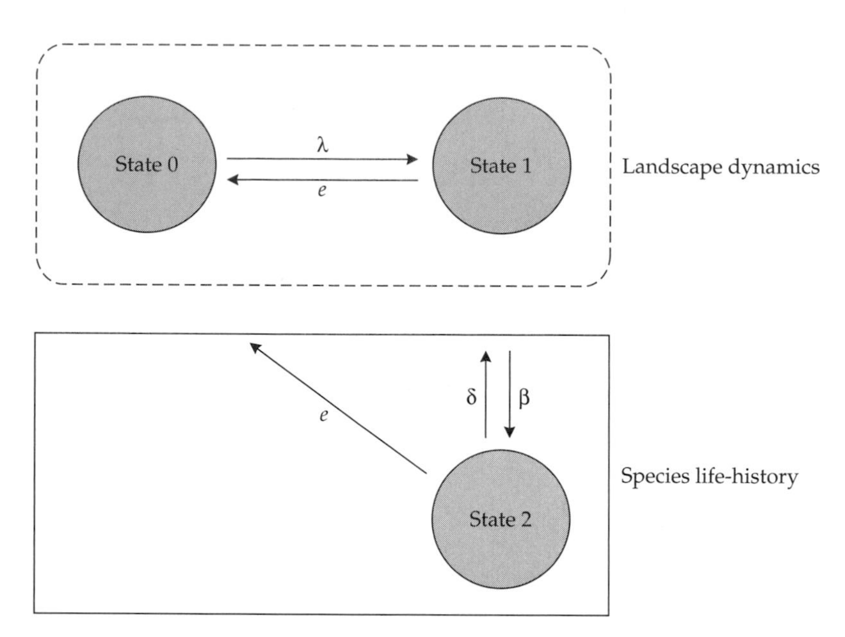

FIGURE 10.6 Schematic representation of the model developed by Keymer et al. (2000). State 0 = destroyed patch; State 1 = habitable empty patch; State 2 = habitable occupied patch (see text for discussion).

As shown by this expression, persistence depends both on the life history of the species inhabiting the landscape (given by β and *e*), as well as on the properties of the landscape itself, which in this case are represented by the amount of available habitat (*s*) and its average lifetime (τ). Thus, persistence of species depends on the interaction between external (environmental) and intrinsic (life history) factors. In a multispecies case, those species whose life history traits allow them to achieve R_0 greater than 1 will persist and will become more represented in a given landscape (i.e., they will attain higher richness). Similarly, along a gradient in landscape characteristics (i.e., *s* and/or τ), a gradient in richness will emerge as a result of the interaction between species traits and landscape features.

Of Species Traits, Environmental Contexts, and Speciation–Extinction Dynamics

As shown above, the characteristics of the environment do affect species persistence. Thus, environments with different combinations of biotic and abiotic attributes will be dominated by species showing different sets of adaptive strategies, some of which will do better than others and will tend to predominate by increasing persistence (i.e., reducing extinction rate). This would result in a progression in the dominance of different strategies along a sequence of environments, or an environmental gradient. The ability of a species to persist is a necessary cause, although not a sufficient one, for further species proliferation via speciation. In this context, it is expected that both extinction and speciation rates will not be random, but will be correlated with several species traits (e.g., Owens et al. 1999) as has been shown for traits such as body size (e.g., Maurer et al. 1992), geographic range (e.g., Cardillo et al. 2003; Jablonski and Roy 2003; Jones et al. 2003), fruit type (Smith 2001), sociality (Muñoz-Duran 2002), type of larvae (Jablonski and Lutz 1983), habitat use (Crame 2002), and growth form (de Queiroz 2002), among others. Ecological requirements related to habitat or resource use are correlated with speciation and extinction rates within a lineage (Rensch 1959; Eldredge 1979; Jackson 1974). In particular, generalist species capable of using different resources in alternative environments (such as a lineage might encounter through time), or specialist species whose resource patches are abundant, widespread, and persist through time, are subjected to less directional selection, and their lineages subjected to low speciation and extinction rates, as exemplified by large African mammals (Vrba 1980, 1987). Dispersal mode and ability also affect macroevolutionary dynamics. For example, weak dispersal in marine gastropods that lack planktotrophic larvae results in narrow geographic ranges, and consequently, in higher rates of extinction, as well as of speciation. In contrast, species with planktotrophic larvae and high dispersal abilities have low extinction and speciation rates and, correspondingly, large geographic ranges (Valentine and Jablonski 1983). A similar but opposite pattern has been documented for herbs as compared to woody plants (Levin and Wilson 1976). Thus, characteristics of species can be associated with their differential speciation and extinction or, more generally, species-sorting (Vrba and Gould 1986; Vrba 1989). In this scenario, and with everything else being equal, species richness along a large-scale environmental gradient is expected to be dominated by different proportions of species with different traits. If species-sorting were to be compounded by traits promoting higher speciation or lower extinction rates in some environments, then gradients in diversity would be observed. As explained below, these mechanisms might account for contrasting gradients in the richness of marine invertebrates of the southeastern Pacific (Valdovinos et al. 2003; Astorga et al. 2003).

Valdovinos et al. (2003) analyzed the diversity and distribution of 629 shelled molluscs along the Pacific South American shelf, from 10°S to 55°S, including only those known to live in waters more shallow than 200 m, corresponding approximately to 95% of all described mollusc species for this continental shelf (Valdovinos 1999). Strong latitudinal changes in mollusc species diversity exist along the Peru–Chilean shelf for all major mollusc taxa studied and especially for the most diverse group within molluscs, the prosobranchs (Figure 10.7A). However, the change in species diversity is not

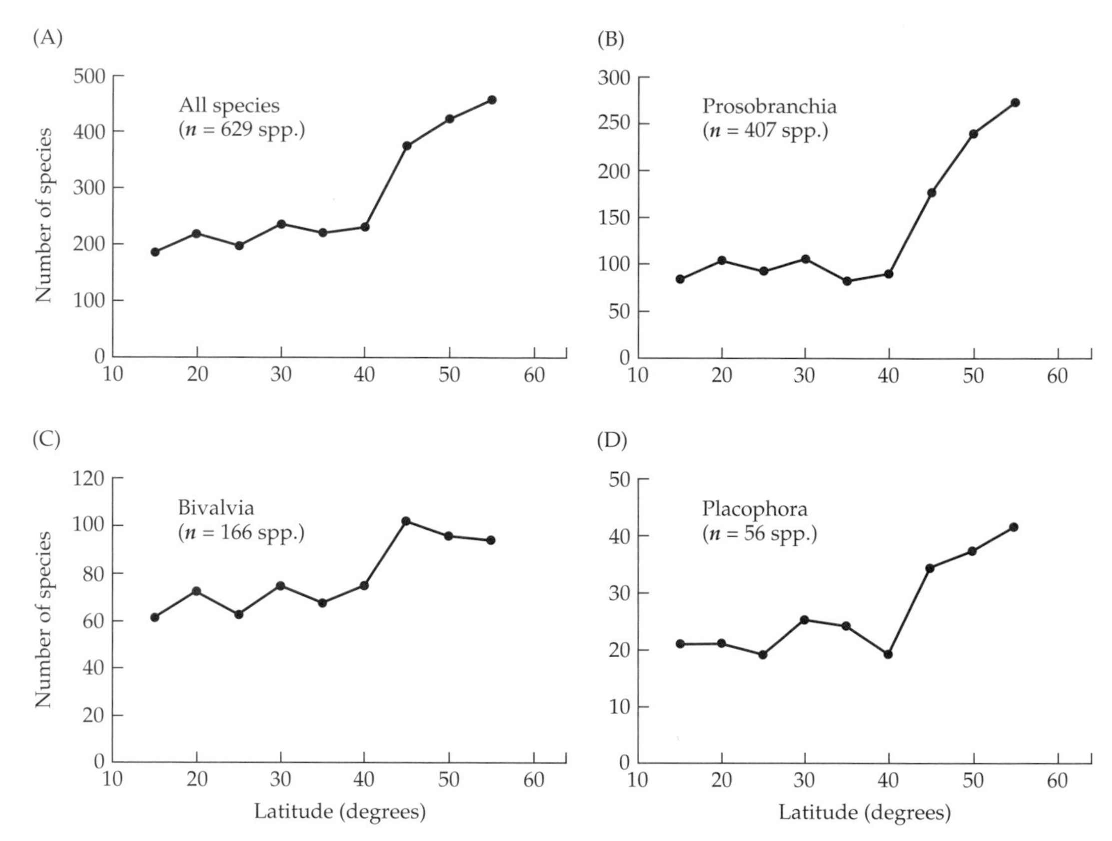

FIGURE 10.7 Relationship between total number of mollusc species and latitude in the southeastern Pacific. (After Valdovinos et al. 2003.)

monotonic across latitudes. Diversity of prosobranch species per latitudinal band remains relatively low and constant around a value of 100 between 15°S and 40°S, and then sharply increases to the south, reaching around 300 species per band in the area around Cape Horn (Figure 10.7B). This general pattern is similar for all taxonomic groups (Figures 10.7B–D). Thus along the Peru–Chilean coast, mollusc species richness does not follow the typical decline toward the poles that is observed in several other marine and terrestrial groups (Brown and Lomolino 1998; Roy et al. 1998; Rhode 1999).

Discordant latitudinal diversity gradients have previously been documented for other marine taxa in different regions of the world, for which different causal factors have been proposed (Thorson 1965; Gaines and Lubchenco 1982; Santelices and Marquet 1998; Clarke 1992). However, the present case is particularly intriguing, since Roy et al. (1998) have shown that prosobranch species richness in the northeastern Pacific increases toward

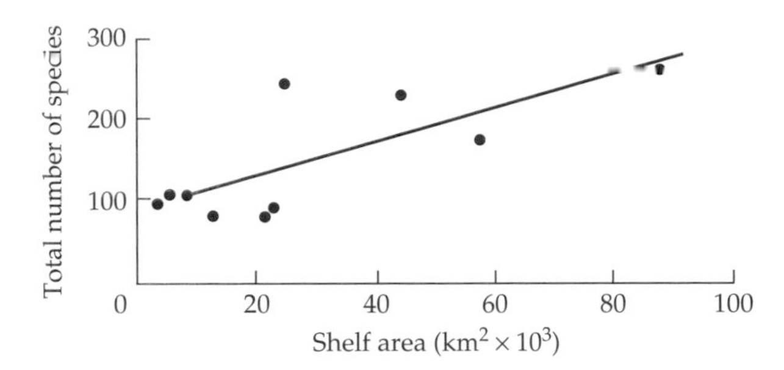

FIGURE 10.8 Relationship between total number of mollusc species and shelf area in the southeastern Pacific. (After Valdovinos et al. 2003.)

tropical latitudes; this pattern is positively correlated with sea surface temperature, thus lending support to the energy input hypothesis (Currie 1991). On the contrary, Valdovinos et al. (2003) show that shelf area, but not temperature, explains a significant portion (59%) of the variance in species richness (Figure 10.8). This significant relationship was produced by the large increase in area south of about 42°S to 45°S, a region dominated by large archipelagoes, fjords, and convoluted interconnected channels (see Castilla et al. 1993). Thus, while the solar-energy input hypothesis (Currie 1991) does seem to be a good explanation for diversity patterns in the Northern Hemisphere, it clearly does not account for mollusc diversity in the Southern Hemisphere. This analysis suggests that toward high latitudes along the southeastern Pacific, there may be a shift in the relative importance of the factors controlling diversity, with available shelf area playing a more prominent role. Valdovinos et al. (2003) hypothesize that the observed trend in mollusc diversity has been the result of higher diversification of molluscs at higher latitudes (south of 42°S) triggered by the use of discrete refugia that might have enabled taxa to survive repeated glacial advances over the past 40 million years. This, coupled with fragmentation and isolation of mollusc distributional areas as a consequence of range shifts, could have actually favored the diversification of species in southern latitudes (Crame 1997, and this volume), conforming a pocket of high taxonomic diversity. This fragmentation of distributions and consequent allopatric speciation created a taxonomic diversity pump (Valentine 1984; Clarke and Crame 1997). As predicted by this model, there is a higher proportion of congeneric species south of 42°S (3.3 species per genus) than on the central coast of Chile (1.6 species per genus), suggesting that the increased diversity was produced by local radiation and not by recent colonization of species from other areas, such as New Zealand or the Antarctic Peninsula (Powell 1973; Dell 1990; Crame 1997).

From a deconstructive perspective, it is tempting to ask whether all species are, on average, equivalent in terms of their response to changes in area. If this were the case, no further knowledge regarding causal factors could be obtained by deconstructing the pattern since under deconstruction this pattern would stay the same. Available data suggest that this is not

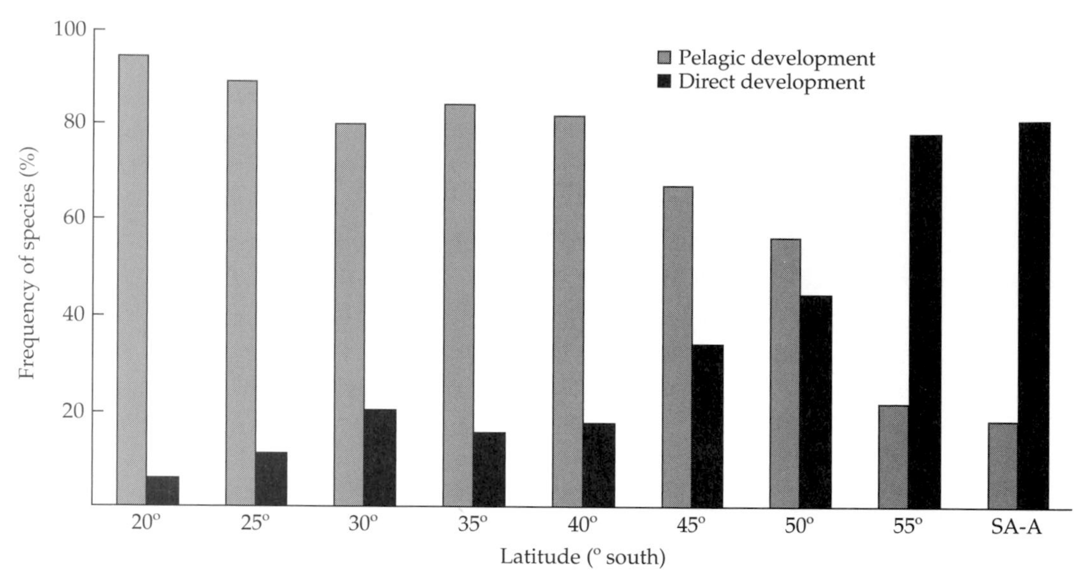

FIGURE 10.9 Relationship between latitude and frequency of mollusc species with pelagic versus direct development in the Southern Hemisphere. (After Gallardo and Penchaszadeh 2001.)

the case, for this latitudinal trend in species richness is paralleled by a trend in mollusc mode of development, which suggests that deconstructing along this life-history axis could increase our knowledge of the causes behind richness patterns. This phenomenon, which has been known for a long time, has been dubbed Thorson's rule. Basically, this rule states that, as latitude increases, most marine invertebrate species tend to show a direct mode of development, while as latitude decreases, species tend to show indirect development with a well-defined planktotrophic phase (Thorson 1946, 1950). Although originally proposed for the northeast Atlantic Ocean, the Mediterranean, and the Persian Gulf (Thorson 1936, 1950; Mileikovsky 1971), Gallardo and Penchaszadeh (2001) have extended Thorson's rule to include marine gastropods in the southeastern Pacific (Figure 10.9). This latitudinal gradient in developmental mode lends further support to the mechanisms proposed by Valdovinos et al. (2003) to explain the higher diversity of mollusc species toward high latitudes in the southeastern Pacific.

Species with restricted dispersal (i.e., brooders or direct developers) are more likely to suffer range fragmentation and become isolated as a consequence of repeated glacial advances and climate change, thereby enhancing the probability of speciation. Astorga et al. (2003) recently proposed this link between gradients of developmental strategies and diversity. Having compared latitudinal diversity patterns for brachyuran and anomuran

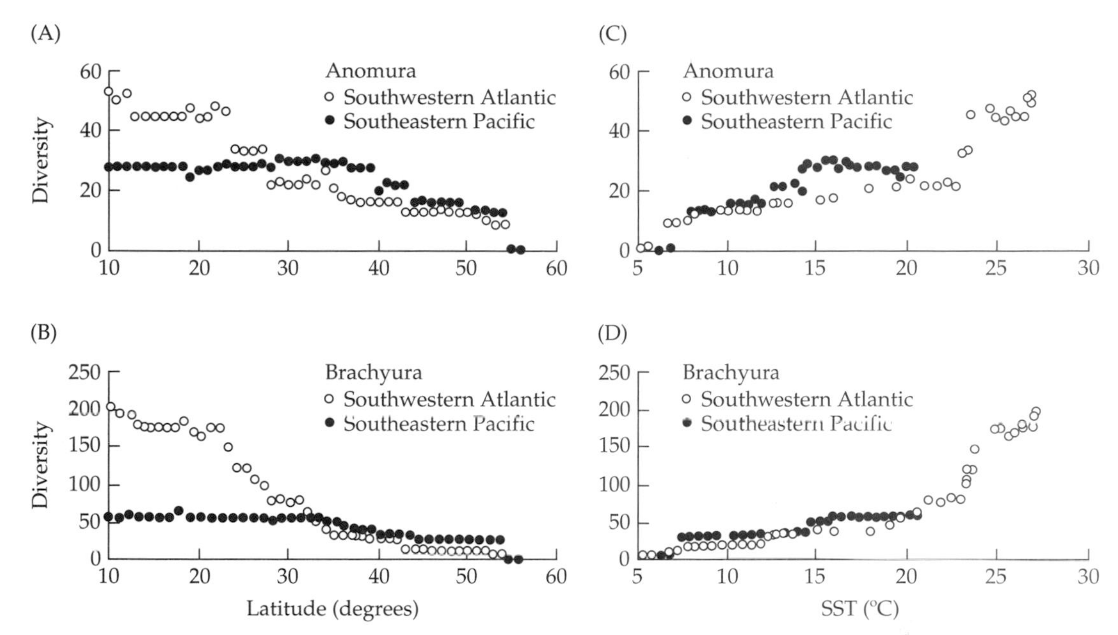

FIGURE 10.10 Relationship between number of species, latitude (top two panels), and sea surface temperature (SST; bottom two panels) for anomuran and brachyuran crustaceans in the southeastern Pacific and southwestern Atlantic. (After Astorga et al. 2003.)

crustaceans, which share the same mode of development (planktotrophic) in two oceans (southeastern Pacific and southwestern Atlantic), Astorga et al. found that both groups of species show the typical decrease from tropical to high latitude areas (Figure 10.10). Further, these authors showed that sea surface temperature correlates positively and significantly with species richness (see Figure 10.10) in contrast with the finding of Valdovinos et al. (2003) for mollusc species, and demonstrates that a decrease in richness is usually found for species with planktotrophic development, whereas anomalous or inverse patterns are usually found in groups with direct development.

It is clear that at least in this case, and likely in most cases, deconstruction not only provided better insights on the determinants of species richness gradients, but helped to reconcile divergent hypotheses and patterns regarding its generation. Diversity patterns depend on the interaction between species properties and the environmental contexts wherein they are embedded, highlighting the complexity inherent in attempts to explain patterns in species richness. As this example makes clear, the deconstruction of richness patterns is essential to understanding the causes that underlie species richness patterns in marine environments (see also Gaines and Lubchenco 1982; Roy et al. 2000).

Frontiers: Toward a Relational Theory of Biodiversity

In his presidential address to the American Society of Naturalists, Hutchinson (1959) appointed Santa Rosalia as the patroness of evolutionary studies. Given the current state of the art in our understanding of *Why are there so many kind of Animals?*, a skeleton encrusted in a stalactite of whom little is reliably reported seems an adequate choice. Our attempt in this chapter has been to elaborate on the importance of the components of species richness based on early ideas put forward by Huston (1994). Although we cannot claim devotion to Santa Rosalia, we think that this recognition is essential to free her from her cave and her encrusted state.

Throughout this chapter we have advanced what we called a deconstructive approach to the analysis of species richness patterns. By this we mean the analytical strategy of disaggregating species richness into smaller subsets of species that share a particular characteristic such as mode of development or other phylogenetic, ecological, or life-history trait. We claim that this procedure allows a better understanding of changes in richness in space and time to the extent that it makes explicit how the environment interacts with the biological attributes of species, subsequently giving rise to richness patterns. Ideally, we should start with the most comprehensive dataset possible, then deconstruct it and pose the question of richness in comparative terms, or why a group with trait x has more species than one with trait y (where x and y are alternative ecological, physiological, or life-history traits). This method is in contrast to the usual approach of measuring the number of species of a given taxon, for example along a spatial gradient, and comparing the numbers of the same taxon at different positions along the gradient. When measuring species number in this manner, species are treated as equivalent, and comparisons are relative to the same taxa; the observed differences are thought to be driven by changes in an external environmental factor affecting all species equally. In this context, the ultimate causes of richness changes are taken as external and independent of the species themselves (because they are all treated as equivalent). However, we should realize that the response variable—the number of species—is internally heterogeneous, in such a manner that it can be decomposed in different groups of species, each affected in different ways by the same environmental factor. Therefore, what is causing the pattern is neither external nor independent to the species being considered, but resides in the interaction between species traits and the external factor being investigated.

At a more fundamental level, the deconstructive approach is a relational approach to diversity, asserting that the number of species in a particular place or time depends on the relationships that exist between the environment and the characteristics of species— interaction that, in the long term, is expressed in extinction/diversification dynamics (see Figure 10.5). As emphasized in our examples, we cannot speak of the effect of an external environmental factor without making reference to a particular type of species in comparison with others, for each type specifies its own relevant environment and interacts differently with it.

In theory, deconstruction can be iterated down to single species, because each species is unique. However, before reaching this level, we will likely lose the pattern (or the statistical power to detect the pattern). Similarly, deconstruction can be performed in many different ways, but to do so requires knowledge of how much deconstruction is enough and what attributes should be used to carry out deconstruction. We do not have a definitive answer to these questions. Suffice it to say that deconstruction should be guided both by theory and by empirical knowledge of the species under analysis and should stop, at least in theory, whenever internally homogenous and equivalent sets are achieved (i.e., they fulfill the symmetry assumption; see Hubbell 2001), because for these sets, further deconstruction would be impossible to carry out. If our interest is to test for the effects of energy availability, then a sensible way to carry out deconstruction would be by disaggregating richness on the basis of energy metabolism, distinguishing endotherms from ectotherms, body size, or diet. However, this procedure will ultimately be constrained by data availability.

Notwithstanding these limitations, we think that deconstruction can help to test and refine current hypotheses on the factors controlling species richness to the extent it focuses on the interaction between an environmental factor and the biological attributes of species, which most of the time have well-known consequences upon species distribution and abundance. As emphasized by Huston (1994:69), acknowledging that organisms are functionally different within ecological systems allows the partitioning of richness into different components that might be influenced by completely different processes. However, in contrast with Huston (1994), we do not see this as a problem that can be solved by carefully selecting the group of species to be analyzed but as an opportunity to disentangle, through deconstruction, the mechanisms that underlie richness patterns. In this regard, deconstruction should be a common methodological strategy to be fostered in order to reduce the complexity and increase our understanding of the pattern itself.

For the sake of simplicity, we have treated the interaction between the organisms and the environment as static. However, recent theoretical work suggests that this interaction is dynamic and involves a two-way process whereby the organisms, while themselves affected, also affect their local environment by modifying biotic and abiotic components. This operates through a process whose ecological consequences are reflected in the modification, creation, maintenance, and destruction of habitat for other organisms (i.e., engineering) (see Jones et al. 1994) and in the modification of the sources of natural selection acting upon themselves (i.e., niche construction) (see Lewontin 1983; Odling-Smee et al. 1996; Laland et al. 1999; and Kitchell 1990). This means that the axes of the explanatory domain depicted in Figure 10.5 are not independent but interact with each other creating a dynamic process whose end result is the emergence of diversity. These propositions, as well as our deconstructive approach, are built on the fact that species are not equivalent, and they stress that further development in our understanding of ecological, evolutionary, and biogeographic patterns and process should focus on the relational aspects underlying the interaction among organism properties, environmental contexts, and resulting evolutionary dynamics.

The deconstructive approach is not limited to the analysis of species diversity patterns. In fact, it can be applied to any ecological, evolutionary, or biogeographic pattern based on the analysis of a large number of species or ensembles. Examples include patterns in body size distributions (e.g., Marquet and Cofre 1999; Gaston et al. 2001), species abundance distributions (e.g., Magurran and Henderson 2003), the analysis of patterns in geographic ranges (e.g., Gaston 2003), and in general, the patterns usually analyzed by macroecologists (e.g., Brown 1995; Blackburn and Gaston 2003). However, there are two major challenges that prevent this approach from becoming widespread and useful. The first is related to the availability of data, for deconstruction requires knowledge of the ecological traits and life-history characteristics of the majority of the species under analysis, because these traits provide the axes for deconstruction. The second relates to the development of statistical procedures, on the one hand, to assess the relative contribution of the different axes used for deconstructing patterns and, on the other hand, to deal with the issue of non-independence among them. Because the axes for deconstructing a pattern are multiple, it is essential to have a statistical procedure to identify useful axes in terms of how much new information they provide. However, besides these problems, strategic deconstruction is a useful approach to understanding biodiversity patterns, to discovering new ones, and to refining our expectations on how they will change in the face of global changes in the environment.

Before closing, a final digression is in order. As noted by some reviewers, we managed to write a paper on species diversity without discussing two major recent approaches that have made great advances in this field: macroecology (e.g., Brown 1995; Brown et al. 2003), and the statistical-mechanical, symmetric, neutral, grand unified theory of biodiversity (Hubbell 2001). From a deconstructive perspective, Hubbell's symmetric theory could be called a trophically deconstructed symmetric theory of biodiversity because its domain of applicability is restricted to trophically similar species, which are envisioned as equivalent in terms of each having on average equal fitness, as well as experiencing the same per capita demographic processes. Thus this theory assumes that once trophic deconstruction is performed, no further deconstruction is possible because we arrive at a set composed of equivalent entities. While we think that this theory holds great promise for the scientific advancement of the fields of ecology and biogeography, it is not clear why it should be restricted to trophically similar species. It is somehow a contradiction for a unified theory to be so restrictive in its application. Further, even trophically similar species may differ in many significant ways that render them non-equivalent, violating the symmetry assumption, and paving the way for further deconstruction. However, it should be emphasized that Hubbell's theory identifies a criterion that might be useful as a stopping rule for the deconstructive process: symmetry. Thus, deconstruction should stop when species are grouped into categories that satisfy the symmetry assumption.

Finally, macroecology envisioned as the research program whose aim is the search for general law-like principles underlying the emergence of ecological patterns, structure, and organization (e.g., Brown 1999; Marquet 2002; Brown et al. 2003) takes, at first glance, a completely opposite perspective from deconstruction to the extent that it aims to find the minimum set of principles (e.g., metabolic scaling; Brown et al. 2003), whose application would put all species on a common ground because it cues on attributes that are general to all living systems. In this context, macroecology holds the promise of coming out with a truly unifying perspective on the diversity of life: one where the deconstructive approach will be meaningless. However, because deconstruction capitalizes on the differences among organisms and explores how they affect the emergence of ecological patterns and structures, it complements unified theories by pointing out the functionally important and different pieces of the ecological puzzle that should be eventually unified.

Acknowledgments

This contribution was made possible by the support of Project FONDAP-FONDECYT 1501-0001. We are grateful for the invitation made by the organizing committee to the first meeting of the *International Biogeography Society*, which resulted in this chapter. Several people provided us with valuable discussion and disagreement. In particular we thank Michael Rosenzweig for pointing out to what he calls the "dimensionless pattern of diversity"; to James H. Brown and Clive Jones for sharing with us a good taste for wine and valuable thoughts on the organisms–environment interaction; and to Dov F. Sax and Mark V. Lomolino for encouraging us to get our ideas in print. P.A.M. acknowledges support from the Santa Fe Institute through an International Fellowship. This is Contribution No. 2 to the Ecoinformatics and Biocomplexity Unit. Finally, we thank Mark V. Lomolino, Lawrence R. Heaney, and two anonymous reviewers for their constructive comments and criticisms.

CHAPTER 11

Dynamic Hypotheses of Richness on Islands and Continents

Robert J. Whittaker

Introduction

The central goal of biogeography is to describe and understand the distribution of diversity across the planet and the development of these patterns through time. I restrict attention herein to spatial patterns, in which respect some clear, emergent patterns of diversity can be considered to be well established and robust (Brown and Lomolino 1998). These include geographical gradients of species richness from the poles to the equator (low to high species richness) and the increase in species richness with increasing area. Other patterns that were thought of as general—such as the decline in richness with altitude in the tropics—are turning out to be less robust, with evidence for certain taxa (notably mammals) indicating that a mid-elevation peak may be more general (Heaney 2001; Lomolino 2001; Li et al. 2003).

The fluidity of our knowledge of patterns is nothing compared to the turmoil in theoretical frameworks over the last half century. As a simple classification, the major theories for the explanation of diversity patterns can be classed as historical, dynamic, and artifactual. The distinction between historical and dynamic hypotheses is based on the probability of recurrence of a particular state or form. Historical, or time-bound knowledge, refers to the analysis of complex states having very low probabilities of being repeated (i.e., states of low recoverability) (Schumm 1991). For instance, plate tectonics theory is undoubtedly important to understanding patterns such as the global distribution of mammal families, but the sequence and direction of continental movements does not follow a predictable pattern or cycle (Scotese, this volume); thus, key transforming events have very low probabilities of being repeated. Historical theorizing about spatial patterns in diversity can be encapsulated by the dictum "the past is the key to the present." Dynamic, physical, or time-less knowledge refers to the analysis of states having a high degree of probability of being repeated. The search for dynamic hypotheses of species richness can thus be likened to the search for the general laws of biogeography. By this I mean governing relationships operating through time and space in a broadly consistent fashion. These in turn produce predictable patterns, from which historical contingency supplies the deviations. The third class of argument holds that at least some of the patterns we perceive may be artifacts of the sampling design of our "natural experiments."

I will focus on two approaches to the study of species richness, offering a brief synthesis of the state of the field, and where we might be heading. The first approach, strongly associated with the study of islands, concerns variation in richness with increasing area. The second is concerned with spatial variation at the macroscale across large landmasses. Dynamic hypotheses have been formulated in respect to both traditions, but framed in very different terms.

Some years ago I reached a point in thinking about island biogeography where I had become highly critical of the dynamic equilibrium model of MacArthur and Wilson (1963, 1967): it just didn't capture enough of the action in the Krakatau system to satisfy (for details, see Bush and Whittaker 1991, 1993; Whittaker 1995). At the same time, I had become convinced that the grand cline of richness was essentially a climatic gradient (O'Brien et al. 1998, 2000). I faced a paradox: how to reconcile my non-equilibrium island thinking with an essentially equilibrium theory of diversity patterns across continents. In thinking further about this, I reached the conclusion that the key was to account for the scale of analysis (Whittaker 1998, 2000; Whittaker et al. 2001, 2003). Hence, scale features as an important part of the present assessment, in which I consider in turn the two traditions and their associated dynamic hypotheses.

Island Ecology and the Species-Area Relationship

MacArthur and Wilson's (1963, 1967) dynamic equilibrium model postulates that the richness of a taxon on an island is the product of forces leading respectively to the gain and loss of species, resulting in a continual turnover of the species present (Figure 11.1). Their model was generated to account for apparent regularities in the form of species-area relationships (SPARs), invoking varying rates of gain and loss as a function of isolation and area, respectively. In the famous graphical model, MacArthur and Wilson show immigration declining exponentially and extinction increasing exponentially

$\Delta s = M + G - D$

where

s = number of species on an island

M = number of species successfully immigrating per year

G = number of species added per year by local speciation (radiation)

D = number of species dying out per year

FIGURE 11.1 The first formal statement of the dynamic model, by MacArthur and Wilson (1963, p. 378). Key to equation slightly abridged.

as an initially empty island fills up to reach its equilibrial richness value, shown by the intersection of the two curves. The immigration rate curve flattens with increasing isolation, while the extinction rate flattens with increasing area, thereby generating a family of curves that provide unique combinations of richness and turnover for each combination of area and isolation. Evolution is subsumed in the model with the assumption that, as an island becomes ever more isolated, new forms are increasingly likely to arise as a result of *in situ* radiation rather than immigration. The model was clearly intended as a general model covering all forms of islands (Whittaker 1998).

Their theory stimulated a research program concerning the form of the SPAR, involving all forms of insular habitat and many different taxonomic groups. This research can be characterized first as examining how SPARs change as a function of a group of subsidiary variables (e.g., isolation, elevation, habitat heterogeneity, regional richness); and second, as based on or sometimes testing the assumption that the pattern is the dynamic outcome of the above processes, conditioned by variation in environment. Central to the first agenda was the recognition of different categories of SPARs, such as the distinction between samples (within contiguous blocks of habitat) and isolates curves and that there should be systematic differences between them (Wilson 1961; MacArthur and Wilson 1963, 1967; Scheiner 2003; Tjørve 2003).

Progress in this area of island ecology has been hindered by problems of inventory, multi-collinearity of variables, and of insufficient attention to issues of scale (Whittaker 1998, 2000). For instance, the finding that isolation has no statistical power in explaining richness variation of birds across a series of isolates is trivial if the islands are isolated by distances of a few hundred meters to a few kilometers; as far as these organisms are concerned, isolation displays no meaningful heterogeneity across this range of values. The same can apply to plants (e.g., Morrison 2002b). Similarly, for a different taxon at the same scale, or for the same taxon across a different range of distances, isolation may be highly significant.

Problems of inventory are most apparent when analyzing species turnover through time on islands (e.g., Whittaker et al. 2000). This is unfortunate, as analyses of species turnover are arguably the best way of testing the MacArthur-Wilson model against competing explanations for SPARs (Gilbert 1980). Alternative hypotheses for the basic pattern of increase in species number with area include:

1. The random placement hypothesis (if individuals are distributed at random, larger samples will be richer)
2. The habitat diversity hypothesis (larger islands equate to more habitats—a part of the MacArthur-Wilson explanation; see quote below)
3. The equilibrium hypothesis (i.e., the MacArthur-Wilson dynamic)
4. The incidence function hypothesis (minimum area requirements lead to more species on larger islands)
5. The disturbance hypothesis (small islands suffer disproportionately from disturbances and hence lack disturbance–intolerant taxa; Whittaker 1998)

Of these, only hypotheses 3 and 5 imply turnover; indeed, only the MacArthur-Wilson model predicts a pattern of stable species number with predictable turnover rates and with variation in the form of the SPAR as a function of isolation.

Evaluation of the Dynamic Model of MacArthur and Wilson

I will take four key MacArthur-Wilson postulates in turn. First, that the number of species on an island represents the outcome of opposing rates of immigration (afforced by speciation) and extinction. This, of course, has to be true.

Second, they assumed that most islands are at, or near, equilibrium most of the time. This is debatable. In particular, highly disturbed islands (either by human or other agency) and the most remote islands, may be rarely in line with their theoretical equilibrium value. Examples include (1) Hawaiian *Rhyncogonus* weevils, which show a negative SPAR but a positive age-richness relationship (Paulay 1994); (2) the doubling of plant richness on many remote islands as a result of human introductions (Sax et al. 2002); and (3) Ricklefs and Bermingham's (2001) phylogenetic analyses of small land birds in the Lesser Antilles.

Third, the theory demands that, where equilibrium conditions apply, the equilibrium represents a dynamic outcome, with measurable turnover occurring. This has also been challenged (e.g., for birds on remote islands) (Lack 1969, 1976; Walter 1998).

Fourth, underpinning the dynamic model, immigration should decrease predictably with distance, and extinction with area (as larger islands have greater habitat diversity and a larger resource base). As plotted in relation to richness of the system, these rates were envisaged to be hollow curves (MacArthur and Wilson 1967). Subsequent research has shown, problematically, that immigration rates can vary with island area as well as isolation, both for passive and active dispersers (Buckley and Knedlhans 1986; Johnson 1980; Lomolino 1990; indirect support from Whittaker and Jones 1994; Shilton et al. 1999), leading to a higher immigration rate for larger islands. It has also been found that isolation may influence extinction rates, with populations on near islands being rescued from extinction by supplementary migrations from the mainland (the "rescue effect"; Brown and Kodric-Brown 1977). If both island area and distance can simultaneously influence immigration and extinction, then the neat predictive power of the MacArthur-Wilson model is significantly diminished. Although it is extremely difficult to measure these rates directly (apart from within "model" systems), data from a number of island systems suggest just such two-way effects on immigration and extinction rates (e.g., Toft and Schoener 1983; Manne et al. 1998; Whittaker 1998).

Spatial and Temporal Variation in the Capacity for Richness on Islands

MacArthur and Wilson (1967:8) wrote of area:

> Our ultimate theory of species diversity may not mention area, because area seldom exerts a direct effect on a species' presence. More often area allows a large enough sample of habitats, which in turn control species occurrence. However, in the absence of good information on diversity of habitats, we first turn to island areas.

I should like to restate the problem, moving a little away from the initial area/isolation focus, and ask, What determines the equilibrium point, or species saturation point, or capacity for richness, S, of an island? First, there is a dependence of biological activity, and hence population mass (and, in consequence, of S) on the resource base of the island (i.e., the energy regime, water regime, nutrients, and shelter). Second, S will vary as a function of the variety of habitats, and for animals, the availability of appropriate food resources and breeding sites. These factors determine the array of niches that can be supported and which particular species from within a pool may be able to sustain a breeding population. The first set of factors constitutes the first order controls on productivity and thus on richness. The second set of factors structures the distribution of productivity amongst populations of different species, and has a modifying effect on S. Thus we might think of the two sets as first order and second order "capacity rules" (defined by Brown [1981] as "the physical characteristics of environments which determine their capacity to support life").

MacArthur and Wilson (1967) devoted attention to the carrying capacity, K, for particular species, and what makes for a good island colonist. But if we are concerned only with S and not with compositional structure, then we only need focus on translation of number of individuals into number of species within a given taxon. MacArthur and Wilson tied their theorizing to Preston's log-normal species-abundance distribution. There are alternative species-abundance models but, while important (Hubbell 2001), the differences between them are not crucial at this level of argumentation. Empirically, MacArthur and Wilson knew that in any ecosystem, a high proportion of species occur in low numbers (i.e., are rare). Hence, relatively slight alterations of critical rates (birth, reproduction, mortality) in an isolate should raise or lower the overall species number measurably (which I think is the sort of mechanism Brown [1981] had in mind by his term "allocation rules"). Sink species—those with negative growth rates—will go extinct on a remote island and not be replaced with a new propagule for many years. By contrast, in near-source habitat patches, sink species may be sustained for long periods by immigration of surplus individuals from optimal habitat nearby. Hence, increasing isolation influences the overall S of the island by influencing the frequency of arrival of propagules. We can think of this as modifying the allocation of energy among species by truncating the long tail of rare species in the system. Really remote oceanic islands, where speciation is the dominant process for raising species number, should have no sink species, a shorter tail of rare species, and thus lower overall richness.

In developing their model, area represented a good surrogate for the resource base/habitat diversity component, encapsulating those aspects regulating productivity and overall numbers of organisms (i.e., all aspects bar the modifying influence of the proximity/isolation effect). MacArthur and Wilson, by focusing on the fixed variables of area and isolation, heightened the attention on biotic turnover as a population process, and simultaneously downplayed the significance of variations through time in the overall carrying capacity of the system. Table 11.1 highlights some of the ways in

TABLE 11.1 *How might the capacity for richness of an island vary through time?*

Factors leading to an increase in richness	Factors leading to a decrease in richness
Primary succession: increasing biomass, system complexity and niche space, driven in part by species colonization, which enables further colonization	Late successional stages, if passed through simultaneously across an isolate, might see competitive exclusion of suites of earlier successional species (e.g., Wilson and Willis 1975; data from Barro Colorado Island, Panama)
Arrival of a keystone species (e.g., possibly fig trees)	Arrival of a "superbeast" (e.g., first vertebrate predator) may over-predate naïve prey species
Humans may increase habitat diversity, introduce new species, manipulate ecosystems to raise productivity, etc.	Humans may clear habitat, introduce new pest/predator or highly competitive species, degrade habitats, etc.
Area/habitat gain (e.g., through coastal erosion, deposition, uplift, sea-level fall)	Area/habitat loss (e.g., through coastal subsidence, sea-level increase)
Moderate disturbances might open up niche space (e.g., for non-forest species on an otherwise forested island)	Major disturbance (e.g., hurricane or volcano) may wipe out much standing biomass, massively reducing populations
Increasing climatic favorability: increasing biological activity, increasing NPP	Climatic deterioration: lower NPP
Evolution on remote islands—a similar effect to primary succession	[*In situ* co-evolution is unlikely to generate reductions in richness]

which abiotic, biotic, and human influences might lead to fluctuations in species numbers that are, at least in part, independent of area and isolation.

Recognizing the dynamism of many island environments, we may posit that different taxa from the same island system or the same taxa from different island systems may be better explained by different concepts, equating to the four corners of Figure 11.2 (Table 11.2). First, island data showing both the equilibrium richness and the homogeneous pattern of turnover postulated by MacArthur and Wilson (1963, 1967) can clearly be considered consistent with the dynamic equilibrium theory. Second, dynamic, non-equilibrium patterns (i.e., exhibiting turnover but not attaining equilibrial richness prior to further alterations of carrying capacity) support the "disturbed island" ideas of Bush and Whittaker (1991, 1993). Third, some data suggest a static equilibrium, at least for secular timeframes, whereby the system maintains a static composition, unaffected by minor environmental fluctuations and rejecting potential colonist species, as argued for some bird data on oceanic islands by Lack (1969) and Walter (1998). Fourth, where the system appears to be essentially stable in composition on secular timescales—but at a richness out of line with the carrying capacity predicted by the MacArthur-Wilson theory—we might term this a "static non-equilibrium," first exemplified by Brown's (1971) mountain-top island mammals (but see Lawlor 1998; Grayson and Madsen 2000; Beever et al. 2003).

This conceptual framework recognizes the dynamic equilibrium condition, but embraces those systems that do not comply with all of the properties of the MacArthur-Wilson model. Even so, it remains incomplete. One failing is that it does not recognize that, even within a single taxon or growth form, the scale at which different species interact with their environment may be radically different. The Krakatau system provides illustration. The large fruit bat *Pteropus vampyrus* has lately been present intermittently on the Krakatau islands as a colony of several hundred individuals (Tidemann et al. 1990; Whittaker and Jones 1994). *Pteropus* gather in large roosts on islands but typically fly to mainland sites to feed (and sometimes vice versa), and their nightly range can be as much as 70 km (Dammerman 1948). The colonies are semi-nomadic rather than reliable seasonal migrants, and Krakatau must provide just one roost site used by the same colony within a larger geographic area. We may speculate that their intermittent use of the Krakatau roost could reflect regional variation in food supply, human interference at other roost sites, and avoidance of periodic ash falls on Krakatau, rather than an immigration/extinction dynamic as envisioned in island theory. A similar picture can be painted for the large fruit-pigeon *Ducula bicolor*, which has been described as roaming in large flocks between offshore islands in the region (Dammerman 1948). Arguably, the behaviors of these particular species, both of which are important seed dispersers and thus have important roles in plant colonization, fall outside the notions of residency demanded by island theory. Continuing this theme, analyses of five tree species have shown extensive gene flow to be occurring between mainland and Krakatau populations, in one case indicating the populations to be

TABLE 11.2 *Exemplification of classification of studies of island richness and turnover*

Dynamic, non-equilibrial	Static, non-equilibrial
1. Krakatau, plants (Whittaker et al. 1989; Bush and Whittaker 1991) 2. Krakatau, butterflies (Bush and Whittaker 1991) 3. Bahamas, plants on small islands (Morrison 2002b)	7. Krakatau, reptiles (several species introduced by people; only two species have been lost, both related to habitat losses) (data in Rawlinson et al. 1992) 8. Great Basin mountaintops of North America, small mammals (Brown 1971) 9. Lesser Antilles, small land birds (examined over an evolutionary timeframe) (Ricklefs and Bermingham 2001)
Dynamic, equilibrial	**Static, equilibrial**
4. Krakatau, birds (a reasonable fit although they show clear successional structure in assembly and turnover) (Thornton et al. 1993 and Bush and Whittaker 1991) 5. Mangrove islets of Florida Keys, arthropods (Simberloff and Wilson 1970) 6. British Isles, birds on small islands (Manne et al. 1998)	10. Krakatau, terrestrial mammals (no recorded extinctions to date) (Thornton 1996) 11. Bahamas, ants on small islands (Morrison 2002a) 12. Oceanic island birds (Lack 1969, 1976 [but see Ricklefs and Bermingham 2001]; Walter 1998) 13. Canadian woodlots, plant data (if successional effects are removed from the analysis) (Weaver and Kellman 1981)

Note: Different taxa from within a single island group (Krakatau) support different models; moreover, in different contexts, data from the same taxa (terrestrial vertebrates, birds, invertebrates, plants) have been interpreted as supporting different models.

panmictic (i.e., as if involving random mating within a single extensive population) (Parrish 2002).

For some species of birds, bats and plants then, it may be reasonable to think of Krakatau not as a sink into which very occasionally individuals may migrate, but as merely one node in a network of sites. While the direction of movement of propagules may largely be inward to Krakatau (for some species at least), a two-way exchange of genetic information between mainland and island populations can be envisaged. Such a scenario is suggestive of metapopulation dynamics, although the dynamics may not necessarily conform perfectly to existing metapopulation models (see e.g., Freckleton and Watkinson 2003). On the other hand, Krakatau under-samples some groups (e.g., tree species with large, winged, wind-scattered propagules; those with no particular dispersal specialisms; and those dispersed by large terrestrial mammals) (Whittaker et al. 1997). Even among generally highly dispersive taxa like birds, some species (e.g., pittas and hornbills) are thought unlikely to colonize across such a large stretch of open sea. In short, even within a single taxon, 40 km of ocean may be an absolute barrier to some groups but trivial to others.

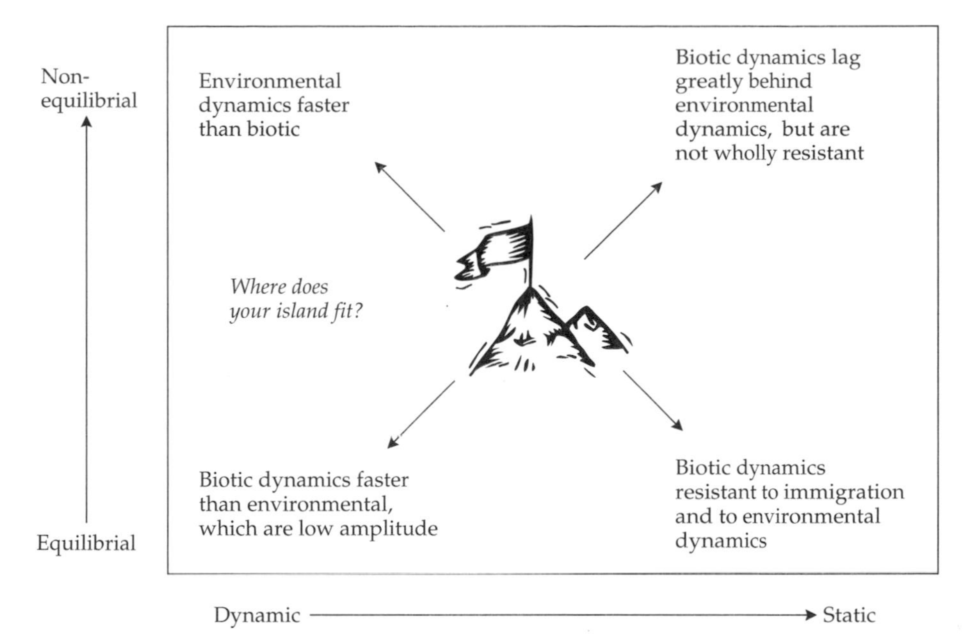

FIGURE 11.2 A representation of the conceptual extremes of island species turnover. The dynamic equilibrial condition corresponds to MacArthur and Wilson's (1963, 1967) theory; the static equilibrial equates to, for example, David Lack's (1969) ideas on island turnover of birds; the static non-equilibrium equates to James H. Brown's (1971) work on non-equilibrium mountaintops; and the dynamic non-equilibrium equates to Bush and Whittaker's (1991, 1993) interpretations of Krakatau plant and butterfly data. Considering a single taxon, different positions in this diagram may correspond to different islands or archipelagoes, while different taxa in the same island group may also correspond to different positions (see Table 11.1). (Modified from Whittaker 1998, 2000.)

There are, moreover, important hierarchical links between different trophic levels. For example, animal dispersal of plant seeds and animal pollination strongly influence colonization and persistence of particular plant species. On the other hand, the availability of food plants and habitat dictate to varying degrees of specificity the ability of the island to support groups such as butterflies, frugivorous birds, and frugivorous bats. Such hierarchical links structure the island assembly process, in the Krakatau case producing non-monotonic changes in the rates of immigration and extinction (Bush and Whittaker 1991, 1993).

Complicating the issue of assigning islands to a place in Figure 11.2 is the impact of humans. For instance, human introductions to a set of 13 oceanic islands/archipelagoes ranging from Pitcairn Island and Cocos Island to New Zealand and Hawaii have approximately doubled the size of island floras with few resulting native losses (Sax et al. 2002; and see Lockwood, this volume). At first glance this appears to invalidate the idea of those islands initially being at dynamic equilibrium (above). But humans have also greatly altered habitat availability, reducing the area of forests, yet in the process creating new habitats and opening up resource space to introduced species. Human actions have also massively increased immigration rates. So this failure might be reasoned away as consistent with alterations in resource space and immigration, and with a greatly elevated new equilibrium number, as the islands move from one SPAR curve to another (see Rosenzweig, this volume). Moreover, we know many island forms, if not yet extinct, are classified as in danger of extinction (i.e., their loss may be subject to some time lag but is inevitable in the fullness of time). These counter-arguments are reasonable enough, yet it is a challenge to sustain the case that the richness of remote islands really represents a dynamic equilibrium condition featuring predictable rates of turnover (Brown and Lomolino 2000; Heaney 2000).

The Form and Biological Meaning of Species-Area Relationships

Much of the work stimulated by MacArthur and Wilson (1967) has focused around the form of SPARs and has taken the approach of finding the transformation that generates the best linear fit, typically involving both axes being log-transformed [i.e., $\log S = C + z (\log A)$, where S = species number, A = area, and C and z are constants, with z describing the slope in log-log space]. MacArthur and Wilson (1963, 1967) seized upon the apparent systematic difference between z values for islands ($z = 0.20$ to 0.35) and non-isolated sample areas on continents (or within large islands) ($z = 0.12$ to 0.17). Much rests on the generality of these ranges, and in his review, Williamson (1988) reported z values ranging as follows: for islands, 0.05 to 1.132; for samples, –0.276 to 0.925; and for habitat islands, 0.09 to 0.957. The wider range and overlap in values between different categories of SPAR reported by Williamson likely reflects scale effects in the range of values of area and of other causal variables in the original studies. Some scientists (e.g., Rosenzweig 1995 and this volume) hold that the MacArthur and Wilson (1967) generalizations remain about right. Nonetheless, we should be wary, especially in conservation science, of making unsubstantiated assumptions about the properties of SPARs, as they certainly can range beyond the limits typically assumed.

A variety of regression approaches have been developed to take account of other variables that may improve the variance explained. One of the more interesting approaches was Wright's (1983) linkage of species and energy, whereby he estimated total energy availability as the product of energy and island area. In illustration, he was able to account for 70% of the variation in plant richness for 24 islands worldwide—a significant improvement on area across the same data set, and an approach worth further development.

There has been a recent resurgence of interest in SPARs, with Rosenzweig (1995) arguing that there are at least three different "scales" of SPARs (islands, within-province, between-province) and Scheiner (2003) distinguishing six types of SPAR depending on the way in which sample data points are configured across a system of isolates or non-isolated patches of habitat. Tjørve (2003, p. 833) has perhaps controversially argued that our approach to model fitting for SPARs should "be based on the recognition of biology rather than statistics"—that is, we should examine the fit of theoretical (mechanistic) models against empirical data sets rather than setting out to find the best straight-line fit through the data. Lomolino (2000c) has also questioned the approach of using data transformation in search of the best linear fit. He argues that there may be good ecological reasons to posit more complex scale-dependent relationships, and that untransformed SPARs should exhibit a sigmoidal form, with a phase across low values of area where species numbers scarcely increase, followed by a rapid increase with area and a subsequent flattening as the number of species approaches the richness of the species pool. The existence of a "small island" effect appears well established (Lomolino 2000c, 2002; Lomolino and Weiser 2001), but Williamson et al. (2001, 2002) contend that there is no evidence of the upper asymptote whereby the log-log species-area curve flattens off across the largest islands.

It is noteworthy that the form and biological meaning of SPARs remains the subject of so much debate and confusion. One important reason is that each empirical data set has its own unique combination of variation in contributory environmental variables such as area, elevation, climate, habitat type, isolation, disturbance histories, etc. Each of these variables has relevance, but does not necessarily have significance over all scales within the empirical data gathered. Perhaps unsurprisingly, few SPAR analyses have attempted to tackle such threshold responses in respect to the complex array of potentially important, interacting variables. It has thus proven difficult to identify general patterns and to develop general models through this approach.

Where Next?

I will close this section with a few remarks on what I think are the more interesting ways to go with respect to dynamic island ecological hypotheses. First, we have paid insufficient attention to scale effects and need to develop a better grasp of the scales over which particular groups of organisms respond to their environment, paying attention to nonlinearities and threshold effects. Allied to this endeavor, we may need to rethink the objectives of our analyses of SPARs. Second, there is scope to develop more sophisticated non-equilibrium dynamic models capturing physical environmental dynamics and the structure and lags in the ecological responses. Third, more complex "successional" models are required to account for the hierarchical links across taxa in the assemblage and turnover of island biotas. There is an applied relevance to all of this (for example, see Rosenzweig, this volume), and a need to update the advice provided to conservation scientists on the dynamics of SPARs.

Island biogeography is a far richer sub-field than this brief review has indicated, and some of the most exciting work is being done on themes such as assembly rules, nestedness, evolutionary change, and conservation (e.g.,

Wagner and Funk 1995; Grant 1998; Whittaker 1998; Lockwood, this volume; Vermeij, this volume), topics which regrettably lie outside my present remit. Island biogeography has been championed as a testing ground, or natural laboratory, from which we have gained many key insights. I concur while cautioning that, in my view, building a general diversity model out of the complexities of the SPAR literature will be a hard task. To do so is to engage with that most complicating of phenomena: variation in the amount of space.

Spatial Patterns in Richness and the Grand Clines

My second theme is spatial variation in richness between areas, for which the touchstone has long been the so-called latitudinal gradients in richness. I say "so-called" because, for terrestrial systems at the macroscale, they are two-dimensional. Richness can vary as much longitudinally as latitudinally (see O'Brien 1993, 1998; Crame, this volume), a simple but crucial consideration (e.g., Hawkins and Diniz-Filho 2002). These gradients are thus much better described as geographic gradients in richness. In addition to dynamic hypotheses, there is a long tradition of historical theorizing in respect to these gradients and, more recently, several authors have warned that artifacts arising from the spatial structure of the data might be misleading (e.g., Colwell and Lees 2000; Lennon 2000).

Dynamic hypotheses seek to explain geographic patterns in richness as an outcome of climatic regulation of ecological processes (Whittaker et al. 2001, 2003). This research program has been hindered by a paucity of high-quality species range and climatic data, problems of multi-collinearity and a general puzzlement about how to handle spatial autocorrelation (Lennon 2000; Lomolino 2001; but see Hawkins and Diniz-Filho 2002; Diniz-Filho et al. 2003), and, once again, by insufficient attention to scale.

Scale

Given our knowledge of species-abundance distributions within communities and of SPARs, it should be self-evident that to examine spatial patterns in richness, we have to control sampling area. This can be done statistically using SPAR regression, or it can be done in sampling (Williamson 1988). The latter is to be strongly preferred because environmental variables exhibit biologically meaningful heterogeneity at different scales to one another. Hence, statistical approaches to removing area's signal may in cases be fatally confounded by co-variation between area and other causal variables. Analyses that do not control area are thus liable to mislead, and should generally be avoided.

Spatial scale has varying meanings within ecology. A crucial distinction is that between (1) the geographical extent of a study system (i.e., the space over which observations are made, such as a hillside, state, continent, or the Tropics); and (2) the grain or focus, being the contiguous area over which a single observation is made or at which data are aggregated for analysis (1 m^2 quadrat, 1 ha plot, or 1° grid cell). Extent and grain/focus are neither equivalent nor substitutable, an obvious point frequently neglected in ecological reviews of diversity theory (Whittaker et al. 2001, 2003). A study that uses 0.1 ha plots remains a fine grain study even if the extent is pantropical. With such a widely distributed dataset, climatic variables will emerge as significant, but we should also expect other variables that exhibit measurable heterogeneity principally at fine scales of analysis (e.g., soils, slope, aspect) to be more evident than, for instance, in a study of regional to continental extent using 10,000 km^2 grid cells.

Table 11.3 proposes a linkage between the spatial scale (grain) of analysis, the typical diversity phenomena examined, the more prominent explanatory variables, and the temporal scale over which processes operate to shape the pattern (Willis and Whittaker 2002). In practice, if we are aiming to

TABLE 11.3 *A hierarchical framework for diversity theory*

Spatial scale	Diversity phenomena	Predominant environmental variables	Temporal scale at which processes occur
Local	Species richness within communities and within patches	Small-scale biotic and abiotic interactions (e.g., habitat structure, disturbance by fires, storms)	~1–100 years
Landscape	Species richness between communities; turnover of species within a landscape	Changes in soils; topography; altitude; drainage	~100–1000s years
Regional	Species richness patterns across large geographical areas within continents	Water-energy dynamics; area effects (e.g., peninsula effect)	Last 10,000 years (i.e., since end of last glacial)
Continental	Differences in species lineages across continents	Aridification events; glacial /inter-glacial cycles of the Quaternary; mountain-building episodes (e.g., Tertiary uplift of the Andes)	Last 1–10 million years
Global	Differences reflected in the biogeographical realms (e.g., distribution of mammal families across continents)	Continental plate movements; sea-level change	Tens of millions of years

Source: From Whittaker et al. 2003, adapted from Willis and Whittaker 2002.

Note: This table aims to identify the main diversity phenomena of interest at particular spatial scales of analysis, some of the more prominent explanatory variables, and the temporal scales over which their variation is particularly prominent.

develop "dynamic" theories, then we are seeking general mechanisms and processes that operate through space and time, and which are therefore not constrained to a discrete period of time. Yet some patterns, such as the distribution of zoogeographic and phytogeographic realms, can only be understood as the outcome of evolutionary and tectonic processes operating over many millions of years (e.g., Brown and Lomolino 1998; Roy et al., this volume; Scotese, this volume). Similarly, the distributions of pockets of endemicity and of disjunct species at a continental scale typically reveal the signal of climatic changes spanning the Quaternary and tail end of the Tertiary. Within continents, we also observe predictable geographical patterns in the distribution of richness as a function of the differential overlap of species ranges of varying size; these patterns can be linked to climatic controls over the capacity for richness (O'Brien 1993, 1998; Whittaker et al. 2001). At finer scales, we can detect a far greater array of abiotic and biotic factors, increasingly tractable to an experimental ecological approach.

Climatic Gradients in Richness

That climate might provide the fundamental explanation for macroscale patterns in species richness is an old idea (O'Brien 1998; Lomolino et al. 2004). Contemporary authors commonly root their work to Wright's (1983) species-energy theory, developed as an extension of the island approach to SPAR, and which he operationalized using estimates of the rate at which resources available to a given taxon were produced across entire islands. Wright's chosen metrics reflected this definition, but many subsequent authors have interpreted "energy" in a more limited sense (e.g., as the thermal energy regime). Yet if we begin with plants, it should be immediately apparent that the energy regime, while important, is insufficient. We need to capture the dynamic interaction between water and energy (O'Brien 1993, 1998; O'Brien et al. 1998, 2000).

Why begin at the macroscale and with climate? First, climate exhibits significant variation at the macroscale and, while climate is modified by vegetation cover, it can be considered as an independent variable for present purposes. Second, species ranges within continents are directly or indirectly controlled by climate, as are the major vegetation/biome types within which species find their preferred habitat space. At the macroscale, what we record as geographic patterns in species richness arises from the differential size and overlap of species ranges. Third, climate controls macroscale variation in biogeochemical processes, primary productivity, and biological activity in general.

In examining terrestrial richness patterns, there is a strong logic to beginning with plants as the point of energy capture, and within plants, trees. Trees are long-lived and are the dominant, largest elements of the most productive terrestrial ecosystems, for which climatic controls might be anticipated to be most directly evident. Macroscale models limited to the temperate northern latitudes were developed in the late 1980s by Currie and Paquin (1987) and Adams and Woodward (1989) based on actual evapotranspiration (AET, a derived variable indicative of the water balance) and on varying net primary productivity (NPP). Around the same time, O'Brien (1993) developed a climate-based model for Southern African woody plants based on Thornthwaite's minimum monthly potential evapotranspiration (PEMIN) and annual rainfall (PAN). O'Brien (1998) subsequently developed a protocol to translate this two-variable regional model into an "interim general model," which successfully predicts the general pattern of richness, so far as it is known, for both tropical and temperate regions and for both longitudinal and latitudinal patterns of richness.

O'Brien's (1993, 1998) argument is that woody plant species richness tends to be greatest where the potential amount and duration of biological activity (beginning with photosynthesis) is greatest, decreasing as the climatic potential for biological activity decreases. The climatic capacity for richness is theorized to be a function of climate-controlled water-energy dynamics. In this model, richness and photosynthesis both increase as a linear function of the availability of liquid water (although there is an upper limit to this, e.g., where you have a lake) and a parabolic function of energy regime (heat/light). The parabolic function means that the dependent variables (richness, photosynthesis) and liquid water only exist within a portion of the full range of energy values: where it is too cold, water is present as ice; and where it is too hot, as a gas. Empirically, the variables selected in O'Brien's water-energy model capture an element of seasonality in the energy variable PEMIN. For southern Africa, the model accounts for 79% variance across a grid of 25,000 km^2 cells (O'Brien 1993). Moreover, adding topographic relief produced a three-variable model accounting for 86%, 87%, and 82% variance at species, genus, and family levels, respectively (O'Brien et al. 2000). These findings suggest the existence of a first-order "capacity rule" (*sensu* Brown 1981), wherein the geographic pattern of variation in richness is a by-product of climate-controlled water-energy dynamics (O'Brien et al. 1998, p. 496).

It is not yet clear how the model extends to other plant growth forms. Trees are orders of magnitude larger than forbs, for example, and even "other woody plants." The space and resources taken up by these respective growth forms must on average scale accordingly (Niklas and Enquist 2001). Thus, the number of individuals that can be assembled per unit area is much larger for these other synusia (i.e., a more or less distinct layer composed of plants of a similar life-form) than for trees, and we can anticipate this translating into increased richness as trees give way to fully vegetated but lower-stature vegetation types. Hence, in crossing climate gradients from forested to treeless vegetation types, the richness patterns of other synusia can be expected to vary in a fashion not simply extendable from the woody plant model. Support for this argument is provided by analyses of altitudinal richness gradients in Nepalese plants (Bhattarai and Vetaas 2003), in which a water + (energy-energy^2) model provided the best fit for trees, but not for other growth forms.

Francis and Currie (2003) have produced alternative models based on a measure of water deficit combined with either temperature or potential evapotranspiration (PET). Their analyses of family level plant data do not differentiate among different growth forms. They demonstrate that a consistent climate-richness relationship pertains globally and that, taking the data for one continent out of the analysis, a good fit can be generated for that continent using the data from the rest of the world.

Collectively, these studies do not rule out an important role for historical contingencies in explaining the relatively small amount of residual variation in macroscale richness patterns (e.g., Hawkins and Porter 2003c), but they do indicate that the great geographical patterns in richness can be considered climatic gradients (Whittaker et al. 2001, 2003; Turner and Hawkins, this volume).

On Linking Mechanisms

Within the dynamic MacArthur-Wilson model, the mechanism determining variation in richness is seemingly simple. In contrast, it is a challenge to tie dynamic hypotheses of richness at the macroscale to a single mechanism. Rather, the "water-energy dynamics drives photosynthesis, drives subsequent biological activity, drives the capacity for richness" argument (O'Brien et al. 1998) invokes, by default, numerous linking mechanisms that operate across space and through time to generate predictable patterns in richness. I have reviewed some of these mechanisms elsewhere (Whittaker et al. 2001, 2003), and here merely list a few invoked in relation to the high richness of tropical rain forest:

1. Disturbance and turnover along the lines of the intermediate disturbance hypothesis (Connell 1978; Clinebell et al. 1995)
2. High productivity linked to high rates of forest turnover (Phillips et al. 1994)
3. Direct cycling of nutrients at high rainfall in tropical forests (Clinebell et al. 1995)
4. Density-dependence (e.g., in the pest-pressure hypothesis) (Janzen 1970; Condit et al. 2000)
5. Specialist mutualisms involving seed dispersal and pollination relationships
6. Regeneration niche differentiation

While interesting, none of these themes has produced a general, predictive model. In so far as these mechanisms have relevance to the "grand clines," they rely upon systematic differences in climate between high and low diversity regions (i.e., they provide linking mechanisms rather than the fundamental driving force).

The argument connecting climate and tree species richness articulated by O'Brien et al. (1998) invokes a positive, monotonic species richness-productivity relationship (SRPR). Yet, some argue that the general form of the SRPR may be humped (Rosenzweig and Sandlin 1997). NPP is so difficult to measure directly across large areas that we are reliant upon inference or modelling efforts to establish the form of SRPR over large extents at the macroscale. Despite claims to the contrary (Mittelbach et al. 2001 [Appendix A], 2003), I am aware of no evidence of humped SRPR for trees at this scale (Whittaker and Heegaard 2003). Indeed, this appears to be another area in which relationships vary in form from the macro-to-local scale (Whittaker et al. 2001, 2003; Chase and Leibold 2002).

Climate Gradients in Richness of Animals

Climate variables also have, directly or indirectly, strong descriptive power for macroscale patterns of animal richness (e.g., Turner et al. 1988; Currie 1991; Fraser 1998; Boone and Krohn 2000; Aava 2001; van Rensburg et al. 2002). The models currently developed appear to vary as a function of the groups selected and the climate regimes and vegetation types encompassed, in turn reflecting ecologically structured differences between, for example, large and small-bodied mammal species (relating to habitat, thermal, water, and food requirements; e.g., Andrews and O'Brien 2000). Thus macroscale species-richness patterns of animal taxa across continents may not mimic tree richness in simple fashion and may not show a simple positive correlation with NPP (see Aava 2001; Balmford et al. 2001). Nonetheless, as Turner and Hawkins (this volume) show, several studies have reported strong relationships between water-energy models and animal richness. Initiatives to develop improved dynamic models of richness for animal taxa should prove rewarding (e.g., H-Acevedo and Currie 2003).

Where Next?

I believe we can develop dynamic models of some generality and predictive power. I expect to see most advances in the near future coming from top-down models, beginning at the macroscale with the climatic pattern, and modifying macroscale variables (e.g., topography, surface water availability). Once the macroscale is successfully captured, it is then possible to downscale the approach to incorporate the landscape-local scale patterns and processes. Such models would be valuable research tools for both basic ("pure") and applied (conservation) purposes. Alternative, macroecological approaches that employ allometry in developing mechanistic models also hold considerable promise (Allen et al. 2002) and have contributed important insights as to the importance of water-energy interactions at the level of the stand and indeed of the individual (Enquist et al. 1999). Developing macroscale global models for more taxa requires systematic data on species' ranges, climate, and other environmental variables. Rapid advances are being made in converting existing data to digital form. The resulting analyses are demonstrating that the basic explanation of the grand cline—as von Humboldt suspected in the nineteenth century—is climate (O'Brien 1998; Francis and Currie 2003; Turner and Hawkins, this volume). Having reached a stage where simple models can account for 70%–80% (or more) of the pattern, we can now resolve the details of which variables provide the "best" model and how to develop more complex models accounting for any remaining predictable variation at the macroscale.

Some readers may be surprised at my neglect thus far of issues such as multi- collinearity, autocorrelation, and geometric constraints (e.g., Colwell and Lees 2000; Lennon 2000). These are serious issues, requiring consideration in analysis and interpretation (see Grytnes and Vetaas 2002; van Rensburg et al. 2002). They also require consideration in the design of sampling programs. For instance, analyses that treat disjunct species distributions as solid blocks (as do the original papers at the center of the debate on geometric constraints) are liable to smuggle in artifacts and biases that can be avoided by using data sets derived from systematic species range maps based on actual distribution records. At a given scale of analysis, such data allow us to map patterns in species richness, allowing species ranges to have disjunct distributions within our data sets, as of course species ranges do in the natural world. I believe we recently have made considerable advances in understanding these issues and can now specify protocols to reduce the danger of smuggling artifacts into our analyses (e.g., Grytnes and Vetaas 2002; Hawkins and Diniz-Filho 2002; Diniz-Filho et al. 2003; Zapata et al. 2003).

Toward a More Complete Theory of Diversity: Reconciling Divergent Perspectives

The development by O'Brien (1998) of an interim general model based on southern Africa assumes that tree richness across this region can be considered close to being in climatic equilibrium. The IGM is then used to predict the capacity for richness and, following O'Brien's (1998) logic, significant residual variation then points to (1) missing environmental variables, and (2) historical legacies in other regions of the world (e.g., differential impacts of Pleistocene climate cycles). So, to the apparent paradox: Is it in fact unreasonable to think of equilibrium at such a coarse subcontinental scale of analysis, when evidence (e.g., from island biogeography and of "undersaturation" in communities) suggests that non-equilibrium richness often pertains at fine scales? I contend that it is possible to reconcile these perspectives if we allow for the patchy nature of species ranges and as long as species exhibit sufficient flexibility in ranges in response to climate change (Whittaker et al. 2001). First, we should recall that species richness patterns at the macroscale are a function of differential overlap of species ranges. Species range boundaries are more or less continuous envelopes containing far more spaces than objects due to, for example, the existence of suboptimal patches of habitat, the presence of interacting species, and temporal environmental variability—meaning that not all suitable habitats are occupied at any one time, and that species distributions exhibit disjunctions at all scales of analysis down to that of the individual plant or animal. Three examples illustrate the flexibility of ranges alluded to: (1) Kullman's (1998) palaeoecological analyses of rapid range responses by trees in Scandinavia at the end of the last Ice Age, (2) Sax's (2001) demonstration that climate gradients have formed up swiftly among non-native species following human transportation of species between land masses, and (3) H-Acevedo and Currie's (2003) demonstration that seasonal changes in species richness of North American birds are underpinned by consistent relationships to climate variables. Moreover, we know from palaeoecological analyses that species respond individualistically in adjusting their range boundaries to changing climatic conditions (e.g., Birks 1989), often via jump dispersal at low densities (Kullman 1998). Hence, it is possible to reconcile the occurrence of undersaturated patches of habitat at fine scales within landscapes across which climatically predictable, macroscale patterns of richness occur.

Some biogeographers express concern at the notion of a biogeography without history, arguing that those seeking explanations of a contemporary nature are dealing in pattern seeking and are capable only of building correlative models. This is an issue that MacArthur and Wilson (1967) recognized in their seminal monograph. They argued some 36 years ago that the historical tradition of the subject had failed to encourage generalizations. They set out a very bold agenda—to forge a new biogeography connecting patterns in the distribution of life with a universal dynamic model expressed at the species and population level. Some authors have implied that their theory ignores evolution. But this is not strictly true, as evidenced by, for example, Wilson's (1961) earlier work on the taxon cycle, which forms a part of MacArthur and Wilson's 1967 monograph and of their body of theory. I see a close analogy with the dynamic (climatic) models of richness: their role is to capture the predictable, revealing the imprint of history in the unpredictable or residual variation. In addition, the vagaries of history are abundantly evident as soon as you move from questions of "how many" to questions of relatedness ("how different"), and to analyzing changes in the range of a taxon through time (see, for example Roy et al., this volume). MacArthur and Wilson's bold hope was that, if we focus on the currency of the species, patterns in richness might be predictable as a function of a general dynamic process, common to all evolutionary ecological systems irrespective of the degree of biological distinctiveness of the particular (island) systems involved. Clearly, immigration-extinction dynamics are important; it is just that the rates of these processes do not always behave in accordance with the requirements of their theory and, thus, its predictive capacity is diminished.

As others have remarked, it can be as instructive to understand where and why theories fail as it is to see them operate successfully. The latest attempt to follow MacArthur and Wilson's lead is provided by Hubbell's (2001) Unified Neutral Theory, which is remarkable in showing how far it is possible to go in accounting for variations in species diversity by reference to a similar, basic dynamic, and setting aside all the detail of niche theory and competitive interactions, historical biogeography, etc. (see Turner and Hawkins, this volume). It may well prove rewarding to explore the predictive capacity at different scales of analysis of linking the Unified Neutral Theory to a climate model.

We should be prepared to be bold and to seek dynamic models capable of capturing the first order variation in richness. This need not mean ignoring history. First, and crucially, for such a dynamic model to have generality it must be operational through time as well as space, and so the conditions for its operation must show a degree of continuity that can be specified and which we can test for using palaeoecology, or by making use of "natural experiments" (e.g., H-Acevedo and Currie 2003). Second, by developing successful dynamic models that capture the first order pattern in our data, we can isolate the residual variation wherein the imprint of historical idiosyncrasies (time-bound knowledge) can be discerned and explored (Whittaker and Field 2000; Hawkins and Porter 2003c). The challenges involved are considerable, especially so in geographical contexts that defy the easy application of the scale framework developed in Table 11.3, for example, within archipelagoes possessing a mix of the geological and biological features of oceanic and continental islands (cf. Heaney 2000). However, I think we can claim to have made progress and to have at least a basic understanding of some of the more prominent biogeographic patterns, not least the great geographic/climatic gradients in richness across the Earth's continents.

Finally, we must never ignore scale. While I advocate that the approach of holding scale constant within an analysis, and then moving on to the next scale, holds greater promise than the further refinement of theories and models of species-area curves, I recognize a role for both traditions. If we are to continue developing biogeography as a predictive science, we must seek dynamic, time-less components wherever possible and wherever they stand up to testing. To avoid blind alleys, we should be careful to state our assumptions, specify our theoretical constructs and our models in plain terms, and be perpetually wary of circular arguments. Our broader goal should be to build a coherent, conciliate body of diversity theory incorporating the time-less and the time-bound.

Acknowledgments

My attendance at the International Biogeography Society meeting was enabled by a grant from the Royal Society. The ideas expressed in this chapter owe much to stimulating discussions with many colleagues, and I should particularly like to thank James H. Brown, Richard Field, Brad Hawkins, Mark V. Lomolino, Eileen O'Brien, Dov F. Sax, and Kathy Willis. I thank the following for constructive comments on a draft of this paper: Lawrence R. Heaney, Mark V. Lomolino, David R. Perault, and an anonymous reviewer.

PART IV

MARINE BIOGEOGRAPHY

John C. Briggs, Brian W. Bowen, and Michael A. Rex

The field of marine biogeography may have begun with Charles Lyell (1830–1832), who was the first to recognize geographic provinces for marine algae. However, it is Edward Forbes who deserves the title "Father of Marine Biogeography." His *Report on the Mollusca and Radiata of the Aegean Sea* (1843) included the important observation that "parallels in depth are equivalent to parallels in altitude" and reported the discovery of an azoic or lifeless zone at the greatest depths (see Lomolino et al. 2004). He subsequently published the first worldwide compendium on marine biogeography (Forbes 1856): a map of the distribution of marine life, together with descriptive text, in Alexander K. Johnston's *The Physical Atlas of Natural Phenomena*. This seminal work describes 25 provinces, 9 "homoizoic belts," and 5 depth zones. His posthumous publication, *The Natural History of European Seas* (Forbes 1859, finished by his friend Robert Godwin-Austin) concluded that: (1) animals created in one biogeographic province were apt to become mixed with emigrants from other provinces; (2) individuals tended to migrate out from their center of origin; and (3) provinces, like species, must be traced back in time in order to understand their origins.

James Dwight Dana, a contemporary of Forbes, served on the *United States Exploring Expedition to the South Seas* (1838–1842). Hired as a geologist, he was also interested in the distributions of corals and crustaceans. Dana (1853) separated the surface waters of the world into zones based on isocrymes (lines of mean minimum temperature), and made the prescient observation that "The cause which limits the distribution of species northward or southward from the equator is the cold of winter rather then the heat of summer or even the mean temperature of the year."

Ludwig K. Schmarda (1885) was the first to unite marine and terrestrial zoogeography, in *Die geographische Verbreitung de Tiere*. Additional landmarks in the late nineteenth century include Arnold Ortmann's *Grundzuge der marinen Tiergeographie* (1896) and Philip L. Sclater's treatise on the distribution of marine mammals (1897). Four decades later, Sven Ekman initiated the modern field of marine zoogeography with *Tiergeographie des Meeres* (1935), the first worldwide survey of marine animals. Ekman's revised English edition, *Zoogeography of the Sea* (1953) was widely circulated and highly influential. In this book, Ekman defines marine faunal regions in terms of endemic species, endemic genera, and their evolutionary relationships. The emphasis on taxonomic or phylogenetic relationships would become one of the most powerful movements in twentieth-century biogeography.

While Ekman's book was guiding generations of marine zoologists, there were concurrent tremors in the field of geology that would eventually yield upheavals in the field of biogeography. Between 1915 and 1929, Alfred Lothar Wegener published four editions of *Die Entstehung der Kontinente und Ozeane*, including an English edition (*The Origin of Continents and Oceans*). These works created considerable controversy, but the theory of continental drift was not widely accepted until the discovery of the mechanism that drives continental movements. In 1960, Harry H. Hess proposed that new seafloor extrudes from the mid-oceanic ridges. Robert S. Dietz named this process seafloor spreading, and suggested that old seafloor is absorbed beneath deep ocean trenches and along young mountain ranges (see H. W. Menard's [1986] history of the development of the plate tectonic theory).

The plate tectonic revolution had a gradual but decisive effect on biogeography. Previously, the historical relationships among biogeographic provinces were analyzed within a framework of stationary continents. To explain the distribution of terrestrial groups under this flawed assumption, implausible cross-oceanic landbridges and sunken continents were proposed by nineteenth- and twentieth-century zoologists (see Figure 2.5 in Brown and Lomolino 1998). Released from this constraint, marine paleontologists were among the first to take advantage of the new framework of plate tectonics, resulting in four important works in the 1970s: F. A. Middlemiss and P. F. Rawson (eds.), *Faunal Provinces in Space and Time* (1971); N. F. Hughes (ed.), *Organisms and Continents Through Time* (1973); A. Hallam (ed.), *Atlas of Paleogeography* (1973); and J. Gray and A. J. Boucot (eds.), *Historical Biogeography, Plate Tectonics, and the Changing Environment* (1979).

Another important but controversial result of the plate tectonic revolution was the rise of vicariance biogeography (see Parts I and II of this volume). The advent of protein electrophoretic technology in the 1970s allowed researchers to examine genetic relationships on a broad scale, including tests of vicariance hypotheses for marine organisms (Rosenblatt and Waples 1986). By 1990, studies of the geographic distribution of gene genealogies, under the umbrella of phylogeography (Avise 2000), provided additional tests of biogeographic hypotheses. Based on calibrated mutation rates ("molecular clocks"), the age and permeability of biogeographic boundaries can be evaluated with an objective genetic yardstick. DNA sequence divergence (Knowlton et al. 1993). These and other molecular studies in marine biogeography have demonstrated that ecological traits (e.g., habitat preferences), geography, and oceanography shape the biogeographic patterns of many marine biotas (e.g., marine fishes; see Bernardi et al. 2000, and Rocha et al. 2002). These studies routinely uncover evidence of cryptic evolutionary lineages—patterns only occasionally observed in the better-studied terrestrial realm. Marine organisms show separations that are much more recent than the ancient (Cretaceous or older) partitions predicted by vicariance biogeography (Grant 1987; Banford et al. 1999). Previous hypotheses on dispersal models for marine organisms received broad support from DNA sequence comparisons (Bowen and Grant 1997), and ultimately provided important insights on the role of vicariance and dispersal.

Eventually, the controversy between vicariance and dispersal schools subsided, and most marine biogeographers now acknowledge the importance of both mechanisms (Cunningham and Collins 1994; Lessios et al. 1999; see also Part II of this volume). Furthermore, there is increasing evidence that mechanisms apart from strict allopatric speciation, such as sympatric and parapatric speciation, are important in the marine environment (Briggs 1999c; Rocha 2003a).

The technology of ocean exploration has always lagged behind that of terrestrial biology, as indicated by the publication of comprehensive marine surveys (Ekman 1935; Briggs 1974), a full six to ten decades behind the terrestrial appraisals of Alfred Russel Wallace (1876). However, by the end of the twentieth century, marine biogeographers had finally advanced to keep apace with, and sometimes lead their colleagues working in the terrestrial realm. Important books on marine biogeography were published during the latter decades of the century, including those by John C. Briggs, *Marine Zoogeography* (1974); Geerat Vermeij, *Biogeography and Adaptation* (1978) and *Evolution and Escalation* (1987); S. van der Spoel and A. C. Pierrot-Bults (eds.), *Zoogeography and Diversity in Plankton* (1979); Oleg G. Kussakin (ed.), *Marine Biogeography* (1982, in Russian); and A. C. Pierrot-Bults et al. (eds.), *Pelagic Biogeography* (1986). Several other important contributions, dealing with both terrestrial and marine distributions, have been published in recent years, notably by Hallam (1994), Briggs (1995), and Brown and Lomolino (1998).

As the contributions in this volume indicate, marine biogeography has emerged as a rigorous discipline—one that often defines the frontiers of biodiversity studies in general. Indeed, any of the three chapters in this Part could have been seamlessly inserted in other sections of this volume. Geerat J. Vermeij contrasts island life in the marine and terrestrial realms, demonstrating how the comparative approach to the study of island diversity can illuminate several very important aspects of biogeography and evolution. John C. Briggs summarizes the evidence for a center of speciation in the East Indies Triangle and emphasizes the distinction between evolutionary process and biogeographic pattern. The center of origin hypothesis mandates that conservation priorities include protection of the regions where successful speciation takes place. J. Alistair Crame presents a comprehensive analysis of latitudinal and longitudinal diversity gradients using contemporary distributions and fossil data. The causes of such gradients are closely linked to the processes of speciation and extinction. Collectively, these contributions demonstrate the scope and vitality of current research in marine biogeography.

In addition to the topics covered in these chapters, the frontiers of marine biogeography include studies of the deep sea below the continental shelves, which is Earth's largest and most recently explored ecosystem. Its permanent cold and darkness, enormous pressures, and scant food resources present some of the most extreme environmental circumstances found anywhere in the biosphere. Yet late nineteenth- and early twentieth-century expeditions found that life exists throughout the deep sea even at its greatest depths. The development and use of much more effective sampling gear during the 1960s revealed that the deep seafloor actually supports a highly diverse invertebrate fauna (Hessler and Sanders 1967). The complete shift in our perception of the deep sea from the sterile realm envisioned in Forbes's azoic hypothesis to the discovery of a rich benthic community is one of the most dramatic reversals in the history of biogeography. Species diversity in the soft sediment habitats that dominate the deep seascape is now recognized to vary spatially at local, regional, and global scales, and temporally at ecological to geological scales (Stuart et al. 2003). The reproduction and population dynamics of deep-dwelling species are coupled to patterns of primary production in surface waters (Gage and Tyler 1991). Just as in other environments, the deep sea experiences periodic large-scale catastrophic disturbances, including massive submarine landslides, widespread anoxic events, volcanic eruptions, high-energy currents, and bolide impacts (Rex et al. 2000; Levin et al. 2001). The realization is growing that the deep sea is not a single ecological system, but a complex of distinctive interrelated habitats including hydrothermal vents, cold seeps, oxygen minimum zones, trenches, seamounts, canyons, methane hydrates, and ferro-manganese nodule fields. Given the pace of discovery and the fact that most of this vast environment remains unexplored, we can expect to encounter many new and surprising deep-sea habitats and their associated faunas.

In the late twentieth century, advances in engineering and biotechnology brought the farthest reaches of the ocean into the realm of historical biogeography. These, coupled with other recent advances including genetic surveys of plankton (Bucklin and Wieb 1998), pelagic fishes (Graves 1998), and hydrothermal vent fauna (Karl et al. 1996; Black et al. 1997) provide exciting new appraisals of evolutionary and phylogeographic hypotheses of biological diversity in general. The advent of molecular genetic studies, in particular, has added a powerful line of inquiry to the field of marine biogeography, allowing new evaluations of the boundaries between provinces that are defined primarily by species distributions.

Studies of the geographic distribution of genetic lineages, under the umbrella of phylogeography (Avise 2000), allow at least two advantages for testing biogeographic hypotheses:

1. The age and permeability of biogeographic boundaries can be evaluated with an objective yardstick: DNA sequence divergence. Previously such evaluations were the exclusive domain of paleontology. With calibrated mutation rates, the divergence between regions separated by oceanic and geographic boundaries can be evaluated with an independent line of evidence for comparison to paleontological dates and geographic evidence (Knowlton et al. 1993).
2. Under the vicariance school of biogeography, models based on rare dispersal events were dismissed as "unscientific" because they did not fit readily into a testable format (Nelson and Rosen 1981). Molecular genetic studies, however, demonstrate how DNA sequence comparisons can document rare dispersal events, at least for the cases of successful colonization (Bowen and Grant 1997). Genetic studies have reinforced the position, based on numerous field observations, that dispersal events are a significant component of biogeographic processes (McDowall 1978; see also the chapters by Brooks and by Lieberman).

We note in closing that marine biogeography has always prospered at the scientific crossroads of other disciplines, and will continue to benefit from advances across a broad range of disciplines, including ecology, genetics, physiology, climatology, geography, phylogenetics, and paleontology.

CHAPTER 12

Island Life: A View from the Sea

Geerat J. Vermeij

Introduction

Biologists have long loved islands. To our mostly mainland sensibilities, the terrestrial life forms of islands—many of them now ravaged by extinction—are unfamiliar and even bizarre. Influential theories—chief among them the equilibrium theory of island biogeography (MacArthur and Wilson 1967), in which an island's diversity (species number) is held to represent a balance between immigration and extinction—were inspired by comparative studies of terrestrial life on islands. Evolutionary biologists have looked to islands as models for how diversification (or "adaptive radiation") takes place, but they too have concentrated on denizens of the dry land. In short, islands have been viewed as tidy, self-contained systems in which ecological processes and their evolutionary consequences become transparent in the way that island air is clear without the obfuscation of continental haze (see chapters by Heaney, Whittaker, and Rosenzweig, this volume).

But islands are, quite literally, tidy places for seashore life as well. The water barriers that isolate islands from other lands may work differently for marine species than they do for terrestrial ones, but they nonetheless prevent dispersal of shallow-water marine bottom-dwellers to islands, and render marine island ecosystems partially isolated from their counterparts around larger land masses. Marine biogeographers have, of course, been well aware of the distinctiveness of marine insular biotas, but their investigations of these biotas have remained largely descriptive. Everyone knows that islands support fewer marine species than continents, and that many species lacking planktonic dispersal stages cannot reach islands; however, few marine-oriented biologists have asked how the marine life on islands differs from that near mainlands, and even fewer have thought about how islands as seen from the perspective of sea creatures differ from islands as we humans and other land-adapted species might perceive them.

These are the questions to which I address myself in this chapter. Drawing on my own experiences with island biotas around the world and on my reading of a highly dispersed taxonomic and monographic literature, I argue that marine islands, like their terrestrial counterparts, are characterized by lower diversity and generally lower productivity than waters near larger land masses. A crucial difference between marine and terrestrial islands, I shall show, is that major high-level predators in the sea are little constrained by dispersal across broad stretches of ocean, whereas terrestrial top consumers (other than humans and their associates) rarely disperse to islands. As a result, marine clades do not display the radical ecological shifts on islands that terrestrial clades often do.

The Terrestrial Perspective

In order to place my remarks into perspective, I shall first summarize how islands for land-adapted species differ from larger, less isolated land masses, which for convenience I refer to as continents or mainlands. Islands and mainlands differ in three interrelated aspects of their terrestrial ecosystems: diversity, productivity, and clade membership of functional groups. Islands typically have fewer species than comparable mainland environments. A striking exception occurs among land snails, which are dramatically more diverse on isolated islands lacking natural enemies than on topographically comparable larger land masses where mammals, predatory gastropods, and other consumers of land snails are common (Solem 1973; but see also Heaney, this volume). Rates of production of the most prolific members of island ecosystems are lower than those in comparable continental settings because of metabolic restrictions associated with small population sizes. Finally, ecological roles—herbivore, top predator, canopy tree, and the like—are filled by species belonging to different clades on islands and mainlands.

The best-known ecological characteristic of islands is the low diversity (number of species) of the biota. This dearth of species compared to mainlands arises for one or more of three reasons: (1) a low rate of immigration, or successful establishment, of species dispersing from elsewhere; (2) a low rate of speciation (species formation) in established island clades; and (3) a high rate of extinction of island populations. Dispersal, evolution, and extinction therefore affect the diversity of species. These factors depend on the size of the island, on the productivity of the island's ecosystems, and on the distance between the island and potential source regions for immigrants.

Although we have grown accustomed to think of islands as home to spectacular radiations of endemic species—Hawaiian drosophilid fruit flies, Macaronesian tree composites, Galápagan groundfinches, Jamaican and

Cuban lizards and land snails, New Zealand moas, and Malagasy lemurs and tenrecs, to name just a few—these radiations are in reality modest in terms of species numbers and net rates of species formation. The living fauna of Madagascar contains just over 100 mammals, representing four independent and temporally separate over-water invasions from Africa during the last 70 million years (Yoder et al. 2003). The dozen or so groundfinch species (*Geospiza*) of the Galápagos (Grant 1986) are the net product of at least 14 million years of evolution in islands over the Galápagos hotspot (Werner et al. 1999; Hoernle et al. 2002). Although the 700 species of living endemic Hawaiian drosophilid flies may all have evolved on the current Hawaiian Islands, which have a collective age of 3 to 5 million years (Price and Clague 2002), the high speciation rate in this group—98 species in 9 invasions over the last 1 million years on the island of Hawaii (Carson et al. 1970), certainly the result of intense sexual selection and differentiation (Carson 2003)—may be more the exception than the rule among endemic insular clades.

As Sax and colleagues (2002) point out, human-assisted species introductions of plants and birds have seldom reduced, and have often significantly enriched, the number of species coexisting on islands. Taken together, these findings indicate that low rates of immigration and speciation contribute importantly to observed low species numbers on islands.

Populations of invaders become extinct on small islands at high frequencies (Schoener and Schoener 1983), but the rate of extinction of well-adapted endemics may be low. Evidence from fossils and from molecular sequences implies that West Indian and New Zealand island endemics have persisted for thousands, to hundreds of thousands of years (Ricklefs and Bermingham 2001; Worthy and Holdaway 2002). Current rates of extinction of island endemics are extremely high because of human activity, notably extensive habitat modification and the introduction of plants, animals, and pathogens from continents. Islands, before human contact, evidently had stable, long-lasting endemic elements that, once they had adapted to the constraining conditions of island life, were relatively resistant to extinction. The low diversity of islands therefore derives from limits on speciation and dispersal and on frequent extinction among recent immigrant populations.

Colonization of islands by land species is limited by: (1) the size of the source area and source population; (2) the ability of dispersing units to survive the journey; and (3) the receptivity of the island biota to the establishment of new arrivals. All else being equal, the likelihood of successful dispersal and establishment should rise as the size of the source population increases. With a larger source population, more individuals will arrive, and the likelihood that the arriving population is large enough to resist stochastic extinction is greater. This is why large-bodied species, with smaller population sizes, are less likely to disperse and establish successfully than smaller species. Likewise, species that depend uniquely on one or two other species are unlikely to disperse and to become established on islands. This category includes pathogens, parasites, and mutualists specialized to single hosts or requiring a succession of unique host species in a complex life cycle. Host species able to disperse to an island are, therefore, less likely to be as heavily parasitized, or to carry as many parasitic species, as those in the source population (see Wolfe 2002; Mitchell and Power 2003; Torchin et al. 2003). Finally, species without resistant stages are unlikely to disperse across hostile habitats. This is why mammals do badly, whereas lizards, weedy flowering plants, ferns, land snails, and opportunistic ants—all groups that are capable of prolonged suspended animation as seeds, spores, eggs, or patient, fasting adults—do well as dispersers from mainlands to islands (Wilson 1961; Heatwole and Levins 1972; A. R. Smith 1972; Diamond 1974; Yoder et al. 2003).

These limits on species formation and dispersal, in turn, are related to a second, less frequently noted characteristic of island ecosystems, namely, low productivity. Because of their small size, islands cannot sustain huge populations of any species. The species most affected by this constraint of size of the economy in which they operate are those whose per capita or collective demand for resources, measured in units of power, is high, and whose productivity is therefore also high. They include large trees, weedy herbs, warm-blooded vertebrates (especially large-bodied species), and social insects with large colonies. Burness and colleagues (2001) have quantitatively described the relationship, recognized earlier by Bakker (1980), between maximum body size of top carnivores and herbivores, and the effective size of land masses. With decreasing land area, islands support smaller top consumers. Warm-blooded animals, with their higher energy budgets, are more affected than cold-blooded species. I am unaware of explicit comparisons of terrestrial primary productivity between island and physically similar mainland ecosystems, but the relationship between maximum power and the size of ecosystems leads me to suggest that primary productivity, and the productivity of the most powerful island endemics, is lower than that in comparable continental biomes. In short, sustainable demand is dependent on the size of an economy, which in turn dictates productivity. At the high end of performance, therefore, island endemics do not measure up to their mainland counterparts.

Support for this hypothesis comes from the observation that large-bodied invaders with high demand often experience sharp declines in body size as they adapt to island conditions. Size decrease of this type has occurred in the evolution of insular elephants, hippos, deer, and sloths (see, e.g., Foster 1964; Lomolino 1985; Brown and Lomolino 1998). Flightlessness has emerged as a common means of island endemics to reduce metabolic demand relative to mainland ancestors (Diamond 1991; McNab 1994; Dudley 2000). Size increases have often characterized rodents, lizards, turtles, raptors, owls, and pigeons—all animals that on mainlands occupied subordinate positions relative to non-dispersing mammals. The metabolic and productivity characteristics of endemic terrestrial species on islands therefore arise in part from the dispersal filter, which favors species that have permanently low demand or that can suspend activity for extended periods, and from the constraints on the population sizes of high-demand species in small economies.

As a consequence of this relationship between ecosystem size (often expressed as area) and maximum demand, the species that fulfill the most influential roles on islands belong to clades different from those with top roles on the mainland. On today's continents, top predators and herbivores tend to be mammals, whereas on islands these niches are more often filled by birds, reptiles, and in the case of the smallest islands, spiders and insects. Among land plants, island forest trees belong to the Compositae or other secondarily woody taxa, whose continental ancestors are members of the understory or of the community of opportunistic weeds. Even on islands the size of small continents—South America during the Cenozoic, Australia and Madagascar today—functional roles are carried out by animals that differ architecturally and historically from those on larger mainlands. Armored glyptodonts were effective herbivores in South America, whereas elsewhere the role of large herbivore was not occupied by an externally armored vertebrate after the Cretaceous, when ankylosaur dinosaurs partially filled that role.

All of these factors—low diversity, low productivity, and assumption of new roles for clades—make insular assemblages quite different from their mainland counterparts. Although selection on islands favors greater power among competitive dominants much as it does on continents, the maximum achievable power is more constrained on islands, with the result that continental species able to disperse and to establish themselves (today usually with the aid of humans) often outcompete locally adapted species either by using more resources more rapidly, or by eating the natives. These facts were already well known to Darwin (1859) and have been amply confirmed in hundreds of subsequent studies. It is thus the small size and isolation of island economies that accounts for the unique composition and performance characteristics of island biotas of land-adapted species.

Marine Island Biotas

Like their terrestrial counterparts, the marine biotas of islands are consistently less species rich than mainland biotas. All shallow-water environments—coral reefs, intertidal rocks, sandy beaches, mudflats, mangroves, kelp beds, and seagrass meadows—are affected by this impoverishment. A few examples will illustrate this point. At least 3000 species of molluscs—most of them less than 5 mm long—inhabit a 295-km^2 reef system on the west coast of New Caledonia, a large continental island in the southwestern Pacific (Bouchet et al. 2002). By contrast, the molluscan faunas of the geographically related Hawaiian Islands (about 1000 species; Kay 1979), the Pitcairn group (about 200 species; Paulay 1989; Preece 1995), and Easter Island (115 species; Rehder 1980) are small. Indonesia, western Melanesia, and the Palau Islands in southwestern Micronesia support 9 or more species of seagrass; most islands east of the Marianas and Samoa lack these productive plants altogether (Tsuda et al. 1977). Some 30 mangrove species occur together in the richest forests of southeast Asia; none exists on islands east of Samoa (Woodroffe 1988; Woodroffe and Grindrod 1991). There are at least 300 symbiont-bearing coral species in Indonesia and nearby parts of Melanesia and Micronesia; just 58 species occur in the Pitcairn Islands in southeast Polynesia (Paulay 1989). The tropical western Atlantic (including the Caribbean and the mainland coast of eastern Central and South America) supports a fauna of some 400 species of crab (Forest and Guinot 1962; Coelho and Ramos de Araùjo 1972). The isolated islands of St. Paul's Rocks (8 species; Holthuis et al. 1980), Ascension Island (about 65 species; Manning and Chace 1990), and St. Helena (12 species; Chace 1966) harbor much smaller numbers of crab species. In the south-temperate to sub-Antarctic zones, oceanic islands typically lack intertidal barnacles and mussels, and have at most one or two widespread species of crab (Knox 1960; Holthuis and Sibertsem 1967; Beurois 1975; Newman 1979). Finally, the tiny island of Rockall in the North Atlantic is home to just two molluscs and no barnacles, echinoderms, crabs, or hermit crabs (Moore 1977).

Marine islands present a fascinating contradiction. On the one hand, even the most isolated islands have relatively few endemics in their biotas, implying that dispersal is frequent enough to prevent divergence of island populations from those in source areas; on the other, dispersal is sufficiently rare or unreliable that many species cannot establish or maintain viable island populations. Kerguelen, in the southern Indian Ocean, which at 1200 km from the nearest land is the world's most isolated island, has a fauna of 130 molluscs, few if any of which are endemic (Cantera and Arnaud 1984). Only one of 20 crab species (5%) and about 42% of the 113 molluscs at Easter Island are endemics (Garth 1973; Rehder 1980). I know of four monotypic genera endemic to islands: the gastropods *Nodochila* (Columbellidae) at Easter Island and *Cyrtulus* (Fasciolariidae) in the Marquesas, the bivalve *Pascahinnites* at Easter Island, and the crab *Euryozius* (Carpiliidae) on South Atlantic islands. There appear to be no marine endemic genera with more than one species. As Kay (1984) has noted, local evolution on marine islands has involved modest radiations of species within relatively widespread genera, such as the limpets *Patella* in the Macaronesian Islands (Weber and Hawkins 2002) and *Cellana* in the Hawaiian Islands (Kay 1979). The vast majority of island marine endemics are single species belonging to widespread clades. In short, speciation must be considered a rare phenomenon in the marine biotas of islands.

Ocean Barriers to Marine Dispersal

The character of island endemics, and of island biotas as a whole, is profoundly influenced by the effectiveness and selectivity of the dispersal barrier between islands and their source biotas. For terrestrial organisms, the ocean represents an entirely hostile barrier because none of their life stages profits directly from it. The situation is different for shallow-water, bottom-dwelling marine species, and especially for near-shore swimmers like fish. For them, deep ocean basins offer at most a porous barrier, an environment in which one or more life stages can float, swim, and often feed. In order to investigate ocean barriers for marine species, we need to know where they are, how they affect potential dispersers, and which categories of species are least likely to be constrained by them.

The first step in investigating the nature of barriers is to find out where the barriers are. This might seem too obvious to merit comment or analysis, but the identification and position of barriers have proved to be more difficult to determine than by merely pointing to a stretch of open ocean. For a given island or archipelago, moreover, there may be several barriers to dispersal, not all of which work in the same way. For example, the Hawaiian Islands receive many species from the west—from the region of southern Japan, the Ryukyu Islands, and the island arc stretching south from Japan through the Bonin (Ogasawara) and northern Mariana Islands to Guam—but they may also receive species from Polynesia to the south and east (Vermeij et al. 1984; Vermeij 1987b). St. Helena in the South Atlantic has a strong South African element in its biota, but there are also species with relatives at other more tropical South Atlantic islands such as Ascension and the islands and banks off the coast of Brazil (E. A. Smith 1890; Chace 1966; Briggs 1974; Manning and Chace 1991), and still others seem to have arrived from the Indian Ocean. Marine species in Bermuda are almost all of Caribbean origin, rather than coming from closer-by North Carolina. Islands off the west coast of tropical America receive dispersers from both the mainland American coast and from islands in the central Pacific. For example, of the 34 crab species at Clipperton Island, almost half (16 species, 47%) belong to Indo-Pacific taxa, and the rest are from western America (Garth 1965).

Two lines of evidence are available for the identification and ranking of dispersal barriers. The first is climatic and oceanographic. We can ascertain where and how rapidly ocean currents carrying dispersing individuals flow, or ascertain wind patterns that affect surface drift. Although these data are important and useful, they cannot stand alone as evidence about the nature of ocean barriers because species differ in the mode of dispersal and in the depth of water in which they disperse. Consequently, a second and more powerful line of evidence comes from the species themselves. Molecular markers now allow us to identify which populations are genetically most closely related to the island population in question. Phylogenetic analysis likewise can identify the closest living relatives of island endemics. This permits us to identify the most likely source of the island population or species, and therefore specifies which barrier was crossed successfully. If many populations and species are studied in this way, we can rank the barriers according to permeability or effectiveness.

The position and effectiveness of barriers change through time. As ocean currents strengthen or weaken, as land masses move relative to one another, as islands grow and shrink according to falling or rising sea levels, and as species themselves evolve dispersal-related characteristics, islands become the targets of different groups of invaders, all of which leave their mark on the island's biota. They are influenced by the physical history and geography of Earth's surface as well as by evolutionary events in the species themselves. Like their larger counterparts along mainland coasts, island biotas accumulate species over time, their composition reflecting the integration of history of the barriers that isolate the biotas from others. For example, limited fossil evidence indicates that oceanic barriers may have been somewhat less effective in the past than they are today. Pleistocene fossils recovered from raised terraces in the Hawaiian Islands, and representing populations living during interglacial intervals when sea levels were higher than at present, chronicle the presence of western Pacific species no longer occurring in the islands today (such as the strombid gastropod *Gibberulus gibberulus*, the turbinid gastropod *Turbo setosus*, and the cone snail *Conus geographus*, among others) as well as extinct endemic taxa derived from western Pacific immigrants (such as the strombid *Canarium mutabile ostergaardi* related to *C. m. mutabile*, and *Conus kahiko* related to *C. chaldaeus*) (see Ostergaard 1928, 1939; Kosuge 1969; Kay 1979; Kohn 1980). Similarly, in the Azores, Miocene deposits laid down during high sea stands contain many European taxa belonging to clades not currently represented in the Azorean fauna (Zbyszewski and da Veiga Ferreira 1962). The barriers between the islands and their source biotas may thus be less porous today than they were during warmer intervals when sea levels were higher. The reasons for such changes in the effectiveness of barriers are unknown. One possibility is that currents transporting dispersing individuals were stronger; another is that island shores were more productive and therefore less hostile to small immigrant propagules.

There are as yet few studies of how specific barriers filter potential immigrants. In my studies of ocean barriers among islands in the western and central Pacific, I found that sand-dwelling snails were only half as likely as snails living on hard bottoms to cross the 600 km of ocean between the continental coasts of eastern Indonesia and northern New Guinea and Palau, the nearest Micronesian archipelago (Vermeij 1987). Barriers between Palau and Guam, Guam and the Hawaiian Islands, the northern Marianas and the Hawaiian Islands, and the Line Islands and island outposts of the tropical eastern Pacific are unselective with respect to substratum type (Vermeij et al. 1984; Vermeij 1987). Hard-bottom gastropods with occluded or narrowly elongated shell apertures difficult for predators to enter, are, as a group, less affected by oceanic barriers than are those whose apertures are broadly open. Predatory gastropods, notably tonnoideans and neogastropods, are far less affected by barriers than their herbivorous or suspension-feeding counterparts (Kay 1967, 1991; Vermeij 1987b). Finally, among rocky-shore species, gastropods specialized for life on the upper shore are often strongly affected by barriers between mainland coasts and those of nearby oceanic islands, but among oceanic islands they are little affected (see also Kay 1967; Vermeij 1972, 1987b; Reid 1985, 2001). Even so, insular endemics are frequent among high-shore gastropods (Vermeij 1972).

Most biologists associate high powers of dispersal in bottom-dwelling invertebrates with the presence of larval stages capable of living and feeding in the plankton for long periods. This perception is accurate for the tropics, but not for temperate regions. Nearly all of the gastropods at Kerguelen in the southern Indian Ocean, and many species in the Aleutian Islands in the North Pacific, have very broad distributions despite the absence of planktonic stages (Arnaud 1974; Cantera and Arnaud 1984; Vermeij et al. 1990). The presence of rafting or floating algae probably accounts for the broad distribution of species lacking planktonic stages in these regions.

The Biotic Environment of Marine Islands

Direct comparative studies of marine communities on islands and mainlands have not targeted the intensity of competition, predation, or parasitism. Even if such studies existed, their relevance to the evolution of island endemics would be questionable because human interference has profoundly altered the biotic environments of islands and mainlands alike, often in ways that cannot be easily predicted or inferred. In the absence of direct comparative measurements of biotic interactions, we can let the phenotypes of island endemics do the talking. In particular, it is instructive to compare the predation-related or competition-related aspects of form in island endemics with the same traits in close relatives with mainland or geographically widespread distributions. The phenotype reflects selective regimes operating in those parts of the population where the density and number of individuals are highest. It can therefore tell us what the biological environment, as reflected by adaptation, is like on islands and mainlands.

Although systematic studies remain to be done, preliminary evidence indicates that marine gastropods endemic to part or all of Polynesia—a collection of isolated oceanic islands comprising the eastern outposts of the vast Indo-West Pacific (IWP) region—have shells with generally reduced antipredatory shell armor compared to their nearest relatives with more widespread distributions in the IWP (Vermeij 1978). Additional evidence from the IWP and elsewhere generally confirms this impression. Comparisons summarized in Table 12.1 are gleaned from taxonomic monographs and from my own observations on the morphology of species in my own collections and in museum holdings. In the absence of cladistic analyses, I have relied on my own assessments of relationships and on those of taxonomic experts. Of 27 species pairs for which I was able to make comparisons, 23 (85%) show reduced armor in species endemic to eastern Polynesia. In the endemic category I include species found only in the Hawaiian

TABLE 12.1 ***Morphological comparisons of Polynesian endemic gastropods with widespread Indo-West Pacific relatives***

Polynesian species	Widespread species
Turbo (Marmarostoma) sandwicensis, Hawaii, **s**	*T. (M.) argyrostomus*
Astralium milloni, Polynesia, **h**	*A. rhodostoma*
Casmaria erinacea kalosmodix, Polynesia, **h**	*C. e. erinacea*
Drupa aperta, Hawaii, **a**	*D. clathrata*
D. elegans, Polynesia, **h**	*D. arachnoides; D. ricinus*
D. morum iostoma, southeast Polynesia, **s**	*D. m. morum*
D. speciosa, Polynesia, **h**	*D. rubusidaeus*
Menathais armigera (*affinis* form), southeast Polynesia, **h**	*M. armigera*
Morula praecipula, Easter, **h,s**	*M. uva*
Nassa tuamotuensis, Polynesia, **s**	*N. serta*
Purpura harpa, Hawaii, **h,s**	*P. persica*
Euplica loisae, Easter, **h,s**	*E. varians*
Zafrona consobrinella, Easter	*Z. striatula*, **s**
Caducifer englerti, Easter	*C. decapitata*, **s**
Benimakia marquesana, Polynesia, **s**	*B. rhodostoma*
"Peristernia" ustulata, Hawaiian form, **s**	*"P." ustulata*
Niotha candens, Marquesas, **h**	*N. albescens*
Vasum armatum, Polynesia, **h**	*V. turbinellus* group
Harpa gracilis, Polynesia, **h**	*H. amouretta*
Mitra nubila hawaiiensis, Hawaii, **s**	*M. n. nubila*
"M." ancillides, Polynesia, **h**	*"M." nivea*
Neocancilla papilio langfordiana, Hawaii, **s**	*N. p. papilio*
Strigatella flavocingulata, Easter, **s**	*S. scutulata*
"Ziba" verrucosa foveolata, Hawaii, **s**	*"Z." v. verrucosa*
Conus pascuensis, Easter, **h**	*C. miliaris*
Hastula lanceata, Hawaiian form	*H. lanceata*, **s**
Terebra argus brachygyra, Hawaii	*T. argus*, **s**

Note: Boldface letters refer to the characteristics by which the Polynesian endemics listed in the left column differ from the more widespread taxa in the right column. **a**, broader or less occluded aperture; **h**, higher spire; **s**, less sculpture.

Islands, only at Easter Island, only in the Marquesas Islands, or only in southeastern Polynesia (Society, Tuamotu, Austral, Gambier, Marquesas, Pitcairn, and Easter Islands). These endemics differ from more widespread relatives by having a higher, more protruding spire; finer, weaker external shell sculpture; a larger, less occluded shell aperture; or some combination of these traits (Figure 12.1).

Gastropods on tropical western Atlantic islands (Fernando de Noronha, Trindade, and Ascension) similarly differ from more widespread relatives. The population of *Leucozonia nassa* at Fernando de Noronha has less sculpture than do other populations from the Caribbean and Brazil (Vermeij and Snyder 2002). *Thais meretricula* from Fernando de Noronha and Ascension is externally almost smooth, whereas the related West African *T. nodosa* typi-

FIGURE 12.1 Shell morphology illustrating the features summarized in Table 12.1. The upper left shell, *Drupa aperta*, illustrates "**a**, broader or less occluded aperture"; the upper right shell, *Drupa clathrata*, does not show this feature. The lower left shell, *Nassa tuamotuensis*, illustrates "**s**, less sculpture"; the lower right shell, *Nassa serta*, in comparison, has defined ridges.

cally has well developed knobs. *Stramontia bicarinata*, a species endemic to both Ascension and St. Helena, has weak sculpture and a tall spire compared to its Caribbean and Brazilian relative *S. rustica* (Vermeij 2001a).

These same relationships hold on temperate islands. Shell thickness, one of several measures of resistance to shell breakage, is notably low in populations of periwinkles (*Littorina* spp.) and dogwhelks (*Nucella* spp.) on the Aleutian Islands relative to those of Japan and mainland Alaska (Vermeij 1992). Moore (1977) noted that the population of *Littorina saxatilis* on Rockall Island is thin-shelled and wide-apertured. Rock lobsters of the genus *Jasus* in Australia, South Africa, and New Zealand often have ridges on the abdomen that provide resistance against forces pulling these crustaceans out of the crevices they normally inhabit (Pollock 1990). Ridges are less well developed in populations on isolated southern islands, including Amsterdam in the southern Indian Ocean and Tristan da Cunha in the South Atlantic (Pollock 1990).

One interpretation of the morphology of these insular endemics is that selection due to predators breaking or entering gastropod shells is reduced relative to the mainland. This interpretation is supported by observations on insular-endemic species of high-shore, neritid gastropods from Easter Island, the Hawaiian Islands, and the South Atlantic islands of Fernando de Noronha, Trindade, and Ascension (Vermeij 1978). Repaired injuries in these species are either infrequent or (in most cases) absent, whereas they are common (up to 40% of individuals in some populations) in neritid species that are more widespread. Similarly, Aleutian populations of *Nucella* and *Littorina* show low frequencies of shell repair when compared to snails from other, less isolated parts of the North Pacific (Vermeij 1992). Low frequencies of repair indicate a low intensity of selection from shell-breaking predators (see Vermeij 1982, 2002b).

Another line of evidence for reduced competition on islands comes from the distribution of gastropods with a labral tooth. The labral tooth is a blunt or sharp, downwardly projecting tooth or spine on the edge of the shell's outer lip in some predatory gastropods. In the cases where its function has been experimentally and observationally investigated, this structure speeds up predation on hard-shelled prey and may protect the predator's proboscis or foot when the predator inserts part of its body between the moving skeletal parts of its prey (Vermeij 2001b). On many mainland coasts in the warm-temperate and tropical zones, muricid gastropods with a labral tooth comprise 5%–17% of muricid species assemblages. Island faunas in Polynesia and the Atlantic contain 0%–3% muricids with a labral tooth. Only in the Galápagos, an archipelago unique for its high, continent-like productivity, do muricids with a labral tooth form a higher proportion (6%) of the muricid fauna. Tooth-bearing members of other families, such as the Buccinidae and Fasciolariidae, occasionally occur on islands, but I know of only one instance (out of about 60 independent acquisitions of the labral tooth) in which the evolution of the labral tooth from a toothless ancestor took place in an insular setting. The single instance is "*Buccinum*" *cinis*, a buccinid known only from the Galápagos and Cocos Islands in the eastern Pacific (Vermeij 2001b). Given that mechanisms which speed up predation reduce the risk that the predator itself falls victim to competitors and its own predators, I interpret the pattern of distribution and evolution of the labral tooth to mean that the intensity of competition and the intensity of selection for rapid means of dispatching prey, are reduced in marine island ecosystems.

A second interpretation, not necessarily in conflict with the first, is that the morphology of island endemics reflects lower rates of growth and therefore lower productivity. In many snails, slow growth is associated with a tall spire and with reduced relief, whereas fast growth is accompanied by a lower spire and stronger external ornament (for review and discussions, see Vermeij 1993, 2002a). Similarly, among island populations of suspension-feeding bivalves in the tropical Pacific, slow growth is indicated by inflated (strongly convex) valves. Flatter-shelled relatives are confined to more planktonically productive continental shores (Vermeij 1990). Atolls and raised, high limestone islands are generally surrounded by clear oceanic waters of low productivity, mainly because there is little runoff of land-derived nutrients from these islands. Only in the closed lagoons of some atolls can productivity be high (see Salvat 1969, 1971; Paulay 1990, 1996). Unfortunately, these productive environments are highly ephemeral because during times of low sea level, such as during glacial episodes in the Pleistocene, the lagoons disappeared, with the result that island shores became uniformly unproductive (Paulay 1990, 1996).

Two important observations clash with the interpretations that marine islands are unproductive places relatively safe from predators. The first is that most islands support lush communities of benthic primary producers—corals and coralline red algae in the tropics, kelps or other large seaweeds near and below the low tide line on temperate islands. Although no measurements of benthic primary productivity are available, descriptive accounts of Rockall Island (Moore 1977), Amsterdam and St. Paul Islands (Beurois 1975), and other south-temperate islands (Knox 1960) emphasize the abundance (if not the high diversity) of algae. Many of these same islands (but not Rockall) support large suspension-feeding barnacles (and sometimes mussels) in these low-shore habitats.

The second, perhaps even more counterintuitive, observation is that the top marine predators of fish, invertebrates, and primary producers in the sea tend to have enormously broad distributions and great powers of dispersal, with the consequence that they are frequently encountered on even the most isolated islands. Although I first noted this fact on the basis of impressions of the wide distributions of mollusc-eating and coral-eating predators (Vermeij 1978), more systematically collected data confirm the pattern. For predatory fish, I tabulated data on maximum body length and geographic range of species in three groups of fish occurring on the Great Barrier Reef and the Coral Sea, a region of high diversity whose fish fauna is relatively well known and representative of "mainland" settings (all data are from Randall et al. 1990). I divided species into those with a broad range (here defined as encompassing the entire IWP region, or an even larger region) and those with a narrower distribution. Fish in the latter category often have enormous distributions, but are not found in the Hawaiian Islands of southeastern Polynesia.

Two of the three groups I examined fit the pattern. Among tropical sharks, the 15 narrowly distributed species are significantly smaller (median length 107 cm) than the 20 species encompassing at least the whole IWP region (median length 320 cm; difference $p < 0.001$, Mann-Whitney U-Test). For strictly marine moray eels (Muraenidae), 8 narrowly distributed species (median length 43 cm) are smaller than the 20 more wide-ranging species (median length 80 cm; difference $p < 0.05$). I found no difference between narrowly distributed and wide-ranging epinephelid groupers (26 and 16 species with median lengths of 45 and 41 cm, respectively). Even in this group, however, by far the largest species (the 270 cm long *Epinephelus lanceolatus*) has a broad IWP range.

In her analysis of IWP mantis shrimps (stomatopods), Reaka (1980) similarly found that large-bodied species, including highly aggressive competitors as well as powerful shell-smashers, have broader distributions than smaller-bodied species. Although smaller species may be undersampled and therefore appear to have smaller ranges than they actually do, the important conclusion that large, powerful species are very widely distributed is incontestable.

The largest, most powerful shell-breaking predators in the sea all have gigantic ranges. The spiny puffer *Diodon hystrix* (maximum length 71 cm) is circumtropical, as is the ray *Aetobatus narinari* (disc width 250 cm). Distributions encompassing the entire IWP characterize the bonefish *Albula neoguinaica* (100 cm), the molar-toothed lethrinid *Monotaxis grandoculis* (60 cm), the wrasses *Cheilinus undulatus* (229 cm, weighing 190.5 kg) and *Coris aygula* (more than 100 cm), the crabs *Carpilius maculatus* (carapace width 15 cm) and *C. convexus* (11 cm), and the box crab *Calappa hepatica* (about 8 cm), among others. Spiny lobsters (genus *Palinurus*) are found on all IWP islands, although some of the insular forms are regarded as endemic species.

The same feature—enormous geographic range—applies to major consumers of corals. These consumers include the snails *Coralliophila violacea*, *Quoyula madreporarum*, and *Drupella cornus*; the crown-of-thorns seastar *Acanthaster planci*; and species of the puffer genus *Arothron*. The giant parrotfish *Bulbometopon muricatum* occurs east to the Line Islands, though not in southeast Polynesia or in the Hawaiian Islands.

The fact that many large and powerful consumers have very broad distributions and amazing powers of dispersal does not necessarily mean that island biotas are consistently subjected to intense selection by these animals. Paulay (1989) has noted that many species are rare and occasionally absent on islands, and that many populations are extirpated there from time to time. It may be that island populations of widespread species rely for their persistence on frequent recolonization ("rescue," *sensu* Brown and Kodric-Brown 1977) by dispersal from less isolated regions where population sizes are larger. From the perspective of victim species, then, marine islands provide an environment of potentially intense but variable selection due to consuming species. If this interpretation is correct, the chief differences between terrestrial and marine islands therefore lie in the dispersal abilities of consumers with high per capita demand. Among land organisms, dispersal abilities are notably low; for marine species, they are very high.

This difference, in turn, implies another. On land, the dominant competitors and top consumers in mainland biotas belong to clades that do not disperse easily to islands. Island dominants therefore tend to belong to other clades, which on continents occupy subordinate roles. This kind of clade replacement does not occur on marine islands, where the dominants belong to the same clades as do the dominants in mainland marine biotas.

Frontiers of Marine Island Biogeography

We can learn a great deal about terrestrial island biotas and about marine island biotas by making comparisons between the two and by noting how these biotas differ from their mainland counterparts. Despite all the attention that biologists have paid to islands, this comparative approach is in its infancy. Given that habitat fragmentation is one of the dominant human effects on the biosphere, an increasing number of biotas will take on island-like characteristics, both on land and in aquatic environments. The comparative study of islands should, therefore, be a major preoccupation of biogeographers and evolutionary biologists in the coming years.

What form should such studies take? During the last 40 years, diversity—the number of species, or some variant that also takes abundance into account—has been the chief phenomenon of interest. Now that the prefix *bio-* has been added to the word diversity, conservation biologists and politicians have created an additional market for research that emphasizes numbers of species. Although diversity may provide a convenient summary of other, less easily quantifiable phenomena in biogeography and ecology, I think we are long overdue for a change in emphasis from the abstract, database-friendly notion of diversity to a richer understanding of biological variety, in which species names and abundances are associated with and complemented by ecological roles, capabilities, adaptations, histories, interactions, and ecosystem-level properties (see chapter by Roy, Jablonski, and Valentine, this volume). The latter include productivity, nutrient retention and enhancement, invasibility, resistance to disturbance and extinction, and reactions to invasion and species loss.

All processes that affect the composition of island and mainland biotas—speciation, extinction, and invasion—are selective in the sense that some clades, ecological types, reproductive syndromes, and geographic groups are more vulnerable than others. Not only do we need to characterize patterns of selectivity in terms that are biologically meaningful to the taxa involved, but we also need to identify, characterize, and compare the barriers that separate islands from mainlands. In order to assess the chances of survival or recruitment of a species on an island, we need to distinguish between source and sink populations—between those that can "export" individuals and those that rely for persistence entirely on recruits from elsewhere—and we must understand how the increased insularization of the biosphere transforms many source populations into sinks. Basically, we must learn how barriers work.

As I noted earlier, barriers come and go, and become more or less porous. Neither islands nor mainlands reflect long-term stability, but at this point we know nothing about how or whether islands and mainlands differ in the stability or composition of their biotas. This ignorance is part of a larger void: we have barely begun to investigate how the biotic environments of islands differ from those of mainlands. Because the biosphere is becoming more fragmented, predictions about the ecological and evolutionary future of species will critically hinge on such an understanding of the biotic environment of isolated fragments as compared to larger, more continuous regions. As I noted earlier, the architecture and physiology of organisms may offer the most reliable guide to the evolutionary conditions that prevailed before the advent of human interference. We can and should, of course, study the biotic environments of contemporary islands, but we must remember that all these environments are profoundly altered by humans and the habitat modification and invading species humans have brought with them. The adaptations of species are, quite precisely, integrated responses to the environments with which their bearers had to grapple. I am, therefore, an ardent proponent of interrogating the species that lived and evolved on islands, and of asking how these species differ from relatives or ancestors in mainlands. Altogether, then, this is a plea for doing the hard work of putting the ecological and evolutionary context back into studies of island and mainland biotas. Flesh and blood is, after all, the stuff of life. Stripping it away leaves skeletons—names, distributions, and abundances—that are so devoid of biological meaning that the patterns and explanations fail to capture the very phenomena that motivated our interest in the first place. It is time to put these skeletons back in the closet and to reveal ecological and evolutionary reality in all its splendor.

CHAPTER 13

A Marine Center of Origin: Reality and Conservation

John C. Briggs

No matter how strange or rigorous the inanimate or noncompetitive environment may be, the evolutionary products of such environments are not as thoroughly refined by competition, and are thus not as well prepared for widespread success, as are the products of highly competitive associations.

Hobart M. Smith, *Evolution of Chordate Structure*, 1960

Introduction

After more than 50 years of argument, and despite the accumulation of much pertinent evidence, some marine biologists are still skeptical about the reality of centers of origin. The question of existence or nonexistence is an important one to settle. Should we think of the marine system as dynamic in an evolutionary sense, being driven by the production of dominant species from certain areas? Or do successful species come from almost anywhere so that we cannot conceive of the ocean having an evolutionary order? The controversy has revolved primarily around the Indo-West Pacific (IWP) and the status of its high diversity center in the East Indies.

FIGURE 13.1 The East Indies Triangle, a center of evolutionary radiation lying within the Indo-West Pacific.

The East Indies Center

One of the most interesting features of the IWP is that, despite a basic homogeneity caused by the occurrence of many wide-ranging species, there are great differences in species diversity (richness) among the various parts of the region. The majority of tropical marine families have their greatest concentration of species within a comparatively small triangle formed by the Philippines, the Malay Peninsula, and New Guinea (Figure 13.1). As one leaves this East Indies Triangle to consider the biotas of the peripheral areas, one finds a notable decrease in species diversity that is negatively correlated with distance.

The controversy over the East Indies began with Ekman (1953) and Ladd (1960). Ekman referred to the Indo-Malayan region as a faunistic center from which other subdivisions of the IWP recruited their faunas. But Ladd, who studied Recent and fossil molluscs, thought that the original faunistic center was located among the islands of the central Pacific and that prevailing winds and currents had carried species into the East Indies. Ladd's hypothesis received support from studies by McCoy and Heck (1976) and Heck and McCoy (1978, 1979). Another theory was introduced by Woodland (1983), who concluded from his work on the fish family Siganidae that the high diversity of the East Indies was due to an overlap of species from the Indian Ocean and the western Pacific. Woodland's conclusion was supported by Donaldson (1986) through the latter's work on the fish family Cirrhitidae.

Rosen (1984), who studied the paleogeography of corals, decided that species formation took place largely around the islands of the central Pacific. Subsequent dispersal to the East Indies supposedly allowed the accumulation of a high level of diversity. Jokiel (1989, 1990), also a coral biologist, described the transport of reef corals on floating pumice from the outer islands toward the west. Later, Jokiel and Martinelli (1992) developed a "vortex" model to show how diversity could accumulate through the mechanism of peripheral species being carried to the East Indies by ocean currents.

Kay (1990), who worked on the molluscan family Cypraeidae, noted that speciation had occurred in the Indian Ocean and central Pacific. For this reason, she did not recognize the East Indies center and instead supported Ladd's accumulation hypothesis. Pandolfi (1992), after investigating two coral genera, concluded that relatively derived species, and the greatest endemism, existed on the periphery, and therefore much of the diversity must have come from the outer areas. But Veron (1995), in his comprehensive book, *Corals in Space and Time*, observed that the East Indies and other regions extending west to the Red Sea had probably acted as long-term net sources of species diversity, and that the more peripheral regions were probably net sinks. Veron also stated that there was no evidence to show that peripheral species were the most derived.

In their work on the biodiversity of reef fishes, Ormund and Roberts (1997) emphasized the substantial amount of endemism that can be found in peripheral areas. Because speciation is active in such areas, they inferred that the production of new species in the East Indies was relatively minor. Paulay (1997) suggested that the East Indies was both a center of origin and a center of accumulation. Palumbi et al. (1997), from his work on mitochondrial DNA (mtDNA) in sea urchins, suggested that new species might arise anywhere in the IWP. Randall (1998), who reviewed the zoogeography of the shore fishes of the Indo-Pacific, suggested that the high diversity of the East Indies was due to local speciation as well as to the immigration of species from peripheral areas. Benzie (1998), who worked primarily on tridacnid clams (family Tridacnidae), saw evidence in the genetic structure of these animals that indicated species moving into the East Indies from outer regions must have played an important role. He also felt that the high genetic diversity of the clams was partially due to an overlap of the Indian Ocean faunas with those of the western Pacific. Wilson and Rosen (1998), in their chapter on the paleontology of IWP corals, argued that an appropriate model should include speciation in the center, accumulation of species from other areas, and function of the center as a refuge for species survival. On the other hand, Lessios et al. (1999), who worked on the phylogeography of sea urchins, preferred the hypothesis of an original circumtropical common stock that had become fractionated.

Most of the critics of the center of origin hypothesis were convinced that the high diversity of the East Indies was due, at least in part, to the accumulation of species from elsewhere. Others have pointed out that speciation occurs everywhere in the IWP. Why should speciation in the East Indies be more important than that which takes place in the peripheral areas? And what is there to prevent marginal species from moving into the diversity center? Also, why is diversity not increased by an overlap of western Pacific and Indian Ocean faunas? The answers to these questions are inherent within our current knowledge about six coincident biogeographic patterns (Briggs 1999a) and the results of a widespread sampling program conducted by Mora et al. (2003).

Six Patterns

The first and most well known of the coincident patterns is that of species diversity. As noted above, species diversity decreases from the East Indies outward.

The second pattern is that of geographic variation in the geological age of certain genera and species. The East Indies is inhabited by the youngest hermatypic (reef-building) coral genera, and the average age increases with distance from that area (Stehli and Wells 1971). Some authors have argued that the age contours illustrated by Stehli and Wells are statistical artifacts, because older genera will have more species and a wider distribution by chance alone (Vermeij 1987a; Jokiel and Martinelli 1992). However, Veron (1995) corroborated the generic age pattern of Stehli and Wells (1971). In so doing, Veron pointed out that the average age in peripheral regions was the product of a small number of early Cenozoic genera and that it was not attributable to older genera having more species. Similar gradients have been reported for the geological age of barnacle and crinoid genera, and ostracod species (Briggs 1999a).

The third pattern is that of dispersal tracks that radiate outward from the diversity center. By determining the evolutionary relationships within certain families, some authors have inferred historical routes of dispersal from the East Indies. Examples may be found among the fish families Pomacentridae (Allen 1975) and Callionymidae (Fricke 1988), and the gastropod family Cypraeidae (Foin 1976). Myers (1989) depicted the likely routes of colonization, taken by the inshore fish fauna as a whole, as reaching the outer parts of the western Pacific from the East Indies.

The fourth pattern is that represented by the geographic regression of phylogenetic relationships in various clades of marine organisms. Such patterns are almost always related to dispersal tracks, but the nature of the connection is often not easy to deduce. In the past, some researchers have maintained that, as a given phyletic line extends into new territory, relatively advanced (apomorphic) species will evolve at the periphery. Others were convinced that the more apomorphic species were characteristic of the center and that their more primitive (plesiomorphic) relatives were displaced horizontally or from shallow waters to greater depths. In the IWP, the latter pattern has been consistently displayed so far. Examples are the fish family Gobiesocidae (Briggs 1955), the gastropod family Cypraeidae (Foin 1976), the fish family Callionymidae (Fricke 1988), the fish genus *Cynoglossus* (Menon 1977), mangroves (Chapman 1976; Ricklefs and Latham 1993), and sea grasses (den Hartog 1970; Specht 1981).

The fifth pattern is that created by extinctions. In the IWP, when many disjunct patterns representing various shelf species are examined, in almost all cases a distributional gap occurs in the East Indies. The most common type of disjunction demonstrates an anti-tropical or anti-equatorial pattern, in which the relict populations occur to the north and south of the central tropics, but not within them; however, there are few cases that show east-west disjunctions. Various theories have been proposed to explain these patterns. The tropical extinction hypothesis seems to best fit the evidence for most shallow water species (Briggs 1987, 1999b). The drama of extinction often makes its debut in the East Indies, gradually spreads outward, and plays itself out among small populations far from their center of origin.

The sixth pattern is that of geographic variation in the level of genetic diversity within individual species or clades. Data from mtDNA in *Echinometra* sea urchins demonstrated a decreasing gradient of diversity that paralleled the decline in species diversity across the Pacific (Palumbi 1997). Information from *Tridacna* clams showed routes of gene exchange consistent with the hypothesis of dispersal from the East Indies (Benzie and Williams 1997). An investigation of mtDNA sequences in shrimp belonging to the genus *Penaeus* yielded data indicating an origin and maximum diversity in the IWP, with an historic radiation to achieve its present circumtropical range (Baldwin et al., 1998). An mtDNA analysis of olive Ridley sea turtles (*Lepidochelys* spp.) indicated the general East Indies area as the place of origin for their interoceanic radiation (Bowen et al. 1998).

The foregoing six biogeographic patterns, when considered together, appear to provide strong support for the center of origin hypothesis. In contrast, the theory that the high diversity in the East Indies is caused by an overlap of the Indian and Pacific Ocean faunas is not consistent with these patterns. The fauna of the Indian Ocean immediately to the west of the East Indies is virtually the same as that of the western Pacific immediately to the east (i.e., the great majority of species found in the diversity center extend outward into both oceans). Although there are some closely related (geminate) fish species separated by the Malay Peninsula (Randall 1998), their number (52) is so small compared to the total fish fauna (about 4,000 species) that they are comparatively insignificant. In order for an overlap to notably enhance diversity, the two faunas would have to be significantly different.

Despite the existence of evidence to the contrary, the accumulation theory has continued to be supported. In writing about the history of fishes on coral reefs, Bellwood and Wainwright (2002) noted that if high-diversity reefs were the site of origin of fish lineages, then we should expect to find basal (plesiomorphic) taxa still living there. Instead, their study found that such forms were located in deeper waters or in peripheral locations. They noted that the peripheral endemics and the many marginal and non-reef species probably played a continuing role in the generation of fish diversity on coral reefs. Furthermore, they suggested that some of the morphological and ecological characteristics of extant coral reef fishes originated in non-reef habitats and were later transported onto the reefs. Such inferences are consistent with the accumulation hypothesis, but the necessary evidence is lacking.

Another example is that of Santini and Winterbottom (2002), who examined the distributions of species belonging to 13 genera of coral reef organisms (mostly fishes). Using data generated from the present geographic patterns and phylogenetic relationships of the various species, the authors claimed to be able to correlate diversity with a series of tectonic events beginning with the separation of New Zealand in the mid-Mesozoic and continuing through the early Cenozoic. Marine species, however, have an average duration in the fossil record of only about 4 million years (Sepkoski 1999), and most families of coral reef fishes are only about 50 million years old (Wood 1999). Although their data would only be applicable to the later Cenozoic, the authors concluded that the present high diversity of the East Indies center was the result of the tectonic rafting of endemic and marginal species from outer areas.

Most recently, Mora et al. (2003) reported the results of a broad-scale survey of reef fishes that encompassed the entire Indian and Pacific oceans. They analyzed the distributional data for 1,970 species from 70 locations. The species belong to 13 families that are well known in a taxonomic sense. When the geographical ranges of the individual species are plotted, the range midpoints for both the latitudinal and longitudinal ranges clustered to form a unimodal curve with the peak located in the East Indies. Such results would not be expected from either the overlap hypothesis or the accumulation hypothesis. Moreover, the authors found that local speciation seemed to have played a minor role, given that only about 2% of the species in any community were endemics. Their results demonstrated that speciation in and dispersal from the East Indies played the major part in assembling communities throughout the Indian and Pacific Oceans.

Geological History

In a geological sense, the East Indies as a center of origin has a relatively brief history (see Scotese, this volume). Wilson and Rosen (1998) traced the development of reef corals in Southeast Asia and found very little development in the East Indies until the early Miocene. Kay (1990) found that the earliest cowries (Cypraeidae) occurred in the Paleocene of Pakistan and India. The earliest coral reef fish assemblage has been found in the Eocene of Monte Bolca in Italy (Bellwood 1996). It is now apparent that parrotfishes (family Scaridae) were part of that community (Streelman et al. 2002). The parrotfish clade that inhabits modern coral reefs arose about 42 my BP. The fossil and Recent distributions of the fish family Triacanthidae were found to be concordant with a Tethyan origin and a subsequent colonization of the Indian and western Pacific oceans (Santini and Tyler 2002).

In their study of fossil molluscs, Piccoli et al. (1987) noted that, in the Tethys Sea, the area of greatest species diversity extended from Europe and North Africa to India. Beginning in the mid-Eocene, fossils from Java indicated the beginning of a build-up in the East Indies. By the early Miocene, there was a substantial increase in the molluscan diversity of Java with some species found as far east as the Fiji Islands. By the late Miocene, the diversity center in the East Indies was evidently well established.

Thus, it seems clear that the previous location of the tropical diversity center was in that part of the Tethys Sea that extended from Europe and North Africa to India. Two principal events were probably responsible for the eastward migration of the center. First, in the mid-Eocene the global temperature began to decline and the tropics began to undergo a latitudinal contraction. Second, the early Miocene collision between Africa and Eurasia eliminated the Tethys Sea and established the Mediterranean. The tropical biota that was trapped in the Mediterranean gradually disappeared as the climate cooled. A contributing factor was probably the continuing fusion of India to Asia, with the formation of the Arabian Sea and the Bay of Bengal. The northern parts of these two seas, with mud and sand bottoms, cannot support coral reef communities, although their infaunal diversities may be high.

The Geography of Speciation

The remaining problem for the center of origin hypothesis has to do with the geographical distribution of speciation. In an early work on the subject, Briggs (1966) observed that dominant, advanced species apparently came from certain favorable centers and, because we know that speciation also takes place in areas peripheral to such centers, we should recognize that two kinds of evolutionary change are taking place—one that may be successful in terms of a phyletic future and another that is generally unsuccessful. Over the short term, successful evolution may be defined as the process that takes place when a new species expands from its area of origin and establishes itself in new territory. Unsuccessful evolution takes place when a new species remains more or less confined to its place of origin and finds itself in an evolutionary trap. In the long term, successful evolution has the potential to develop into extended evolutionary lines. The products of unsuccessful evolution may persist for long periods of time if they exist in locations that are protected from invasion by newer species or other environmental disturbance. Because successful evolution evidently occurs in high diversity centers of origin, dominant species that have been able to spread over large geographical areas probably arose in such places. Small, isolated populations are capable of rapid speciation, but they possess limited genetic variation and are vulnerable to genetic drift and inbreeding depression. Such changes appear to be detrimental to long-term success.

The manner in which speciation takes place has been the subject of considerable argument, but most biologists would agree that there are three basic modes. Within the allopatric mode, where speciating populations are separated by an extrinsic barrier, there are two types: (1) vicariant speciation, which may occur when a new barrier arises to isolate parts of a formerly continuous population, and (2) dispersal speciation, which may occur via dispersal when propagules manage to overcome an existing barrier to succeed in colonizing a new territory. Examples of vicariance are often found where fluctuations in sea level have caused the making and breaking of land barriers. The occurrence of geminate species separated by the Malaya Archipelago, a reflection of historic low sea levels, has been noted by Randall (1998) and several other authors. In other parts of the IWP, a glacial-stage land barrier was formed between New Guinea and Australia, and a sill formed across the mouth of the Red Sea. The appreciable endemism now found along the coast of northwestern Australia and in the Red Sea may be attributed to these blockages.

Although vicariance has been an important impetus to speciation, dispersal has been no less so (see chapters by Brooks, Lieberman, and Riddle and Hafner, this volume). The major part of the western Pacific is underlain by the huge Pacific Plate. As the Plate moved westward, it created a series of island arcs between New Zealand and The Aleutians. The Plate also contains many sites of intraplate volcanism where island chains were created above hot spots or mantle plumes. The result of these tectonic activities is an enormous expanse of scattered islands and archipelagoes that extend more than one-third of the way around the world. Islands so created are obligated to pick up their shallow marine (and terrestrial) biotas via dispersal.

Parapatric speciation takes place when two populations occupy essentially non-overlapping areas but are in continuous contact along a mutual border with no extrinsic boundary to separate them. Often, the border will be marked by a narrow hybrid zone. This kind of speciation is considered to be important in oceanic species where only a partial separation can occur. Although it has been traditionally associated with planktonic organisms, parapatric speciation may also be important in the benthic biota. In a review of sibling marine species, Knowlton (1993) noted the occurrence of parapatric populations along continuous coastlines. Her data included 133 cases of sibling relationships representing 91 genera in 9 phyla. In approximately 20% of these cases, parapatric distributions were identified.

Recently, the process commonly known as sympatric speciation has been given a new name: "competitive speciation" (Rosenzweig 1995). This new designation is appropriate since it refers to the natural condition under which the process takes place; that is, the absence of extrinsic barriers means that speciation has probably taken place in the presence of gene flow. Consequently, allopatric speciation depends on isolation opportunities (vicariance or dispersal) while competitive speciation depends on ecological opportunities.

How can ecological opportunities arise? There are many well-documented cases of host-switching among insects. But Palumbi (1994) reviewed genetic divergence and reproductive isolation in the marine environment and observed that evidence for the action of selection in increasing reproductive isolation in sympatric populations is fragmentary. On the other hand, Lazarus et al. (1995) examined fossil material that demonstrated competitive speciation in two large foraminiferan populations. In addition, Gosliner and Draheim (1996), who worked on the biogeography of opisthobranch gastropods, found no vicariance of sister taxa in the IWP, but did find a marked sympatry.

For contemporary populations without fossil records, speciation mode is probably best determined by comparing the geographic patterns of closely related (sibling) species. This assumes that such new species will have had very little time to spread from their place of origin. In her survey of sibling species among the marine shallow-water invertebrates, Knowlton (1993) found that competitive speciation may be very important. Of the total of 133 cases, the great majority demonstrated sympatry. In marine fishes, Briggs (1999c) illustrated patterns of sympatry in five sets of sibling species. Two of the sets involved three siblings, each with almost completely overlapping ranges. In each example, the more restricted member of the sibling group was primarily confined to the East Indies while its sister species were more widespread. Shine et al. (2002) reported a case of apparent sympatric speciation involving two sibling species of sea snakes in the East Indies. Evidently, competitive speciation does take place in the marine environment and, in some cases, it does so in the putative center of origin.

After reviewing the modes of speciation in the IWP, I concluded that all three basic modes were active and, while competitive speciation may not be as common as the other modes, it was important to find it apparently taking place in the East Indies (Briggs 1999c). A severe handicap inherent in the vicarianist philosophy is that it recognizes only vicarianism as being of evolutionary importance (but see chapters by Lieberman and Brooks, this volume). Nevertheless, vicariance is only one type of allopatric speciation. Allopatric speciation by dispersal is occasionally recognized by vicarianists, but it is not considered to be an important process (Santini and Winterbottom 2002), and competitive speciation has been almost entirely ignored.

As many recent systematic works have demonstrated, widespread geographic patterns, consisting of a mosaic of closely related species or subspecies, can often be traced to the previous existence of a broadly distributed ancestral species. The presence of such a mosaic, with its evidence of vicariant speciation events, has led many investigators to conclude that this mode of speciation has been of predominant evolutionary importance. Yet such conclusions do not usually take into consideration the place of origin for the ancestor or the reasons for its widespread success. Wide-ranging species, having evolved in the competitive cauldron of the diversity center, may undergo secondary speciation in peripheral areas, but the latter processes are unsuccessful in producing continuing phyletic lines.

Most speciation events are not concentrated in the center of origin (Briggs 2003b). In fact, the enormous areas of lower diversity that exist outside the center certainly produce more species than do the comparatively limited areas of the centers. Speciation in small, isolated populations is apt to proceed at a faster than normal rate. The difference is that successful species tend to be produced in the centers rather than elsewhere. All newly formed species do not have equal potentials. Those that are produced in the diversity centers by large populations with high levels of genetic variation are the ones that disperse outward to keep the system going.

The Centrifugal Mechanism

The East Indies Triangle covers a very large geographic area allowing sufficient space and structural variety so that any of the speciation modes could be accommodated. Once a new species is formed, what are its chances of success? Some years ago, William L. Brown (1957) proposed his theory of "centrifugal speciation." In essence, he predicted that species produced by large, central populations were the ones that would succeed in the long run. Three years previously, Mayr (1954) published his peripatric theory in which he proposed that the ultimately successful species should come from small peripheral populations. Mayr argued that geographical speciation was the almost exclusive mode among animals, and that small, isolated populations played an important role in creating new phylogenetic lines (the founder principle). Mayr's peripatric theory became widely accepted and, for the ensuing 40 years, Brown's centrifugal concept was almost completely ignored.

The peripatric theory received additional support when Eldredge and Gould (1972) published their hypothesis of "punctuated equilibria." They noted, "The central concept of allopatric speciation is that a new species can arise only when a small, local population becomes isolated at the margin of the geographical range of its parent." More recently, Gould and Eldredge (1993) made the more direct observation that the speciation events ultimately responsible for rapid punctuational changes took place within such small, isolated populations. Finally, Gould (1994) acknowledged that the theory of punctuated equilibria was little more than Mayr's peripatric theory translated into geologic time.

Following the original publication by Eldredge and Gould (1972), Brown (1987) published a critique of punctuated equilibria pointing out that the original examples failed to support the theory. In his article, Brown observed that a more descriptive term for peripatric speciation would be "centripetal speciation," for it would describe the movement of a peripheral daughter species if it did supplant a central parent species. At the same time, he emphasized the lack of solid, or even strongly suggestive, evidence to support the idea of peripatric (centripetal) speciation. Although there have been some reports of fossil sequences that apparently demonstrate this kind of speciation, the evidence has been equivocal. As was pointed out by Frey (1993), the peripatric mode had previously been regarded as the dominant type of animal speciation while the centrifugal process had largely been ignored.

The centrifugal hypothesis is useful because it supplies a mechanistic explanation for the manner in which centers of origin operate. Large central populations with adequate genetic resources are necessary for the production of successful species. Once such species are formed initially—by allopatry, parapatry, or sympatry—they will proceed to move outward into new territory. As they do so, they will dominate and replace other species in accordance with the centrifugal prediction. On the other hand, if a small peripheral population speciates and moves in to replace the large central population, it would be a confirmation of the centripetal process. So far there is little or no evidence that this has ever happened.

The accumulation hypothesis implicitly assumes that peripheral species have the capacity to move into the diversity center. Not only is this genetically improbable, but there is empirical evidence to show that such movements are very unlikely. The IWP as a whole is a highly diverse biogeographical region. It is bounded on the east by the deepwater East Pacific Barrier and on the west by the Old World Land Barrier. Thousands of IWP species have been able to transcend these barriers and establish themselves in the neighboring tropical regions. But exceedingly few—probably fewer than 100—have performed reciprocal migrations. The two boundaries, which have been described as "one-way filters" (Briggs 1974), illustrate the fact that it is very difficult for marine species to colonize areas that possess a greater species diversity.

The Reality of an Evolutionary Center in the Marine Tropics

Although the center of origin hypothesis was first stated by Darwin in 1859, it has been the subject of much contention and reinterpretation since that time. For more than 50 years, the theoretical center of origin in the Indo-

West Pacific Ocean has been the focus of more argument than that of any other proposed area. Recent research has made the distribution patterns in that part of the world much better known. This knowledge offers an opportunity to settle the question of the reality of an evolutionary center in the marine tropics.

The East Indies Triangle appears to be the place of origin for a series of dynamic systems that extend across the entire Indo-West Pacific. These systems are apparently maintained by a continuous outflow from the East Indies. The best known of these is the diversity gradient, but others are equally important. Average generic age increases with distance from the East Indies, dispersal tracks of individual animal groups extend outward, the East Indian species tend to be the more advanced (apomorphic), extinction patterns appear to originate in the East Indies, and a gradient of lessening genetic diversity extends outward.

A continuous flow of species replacement from the East Indies over the past 10 million years would create and maintain the systems that have been identified. The fact that these six systems exist and operate as they do makes sense only within the context of the center of origin hypothesis. Successful species that are produced in the center must be able to dominate peripheral relatives that are less fit in a genetic sense.

If one wished to disprove the East Indies center of origin hypothesis, it could be done by finding evidence against the six patterns that provide its support; that is, if one could demonstrate diversity gradients that increase with distance from the center, generic age gradients that decrease with distance, dispersal tracks that extend inward, phylogenetic patterns with plesiomorphic species in the center, widespread extinction patterns that do not originate in the East Indies, or a gradient of increasing genetic diversity that extends outward, then the hypothesis would be called into question. Also, in regard to the discovery that the great majority of fish species have their ranges centered in the East Indies (Mora et al. 2003), a critic would need to show that most species are not centered in that area.

Recent discoveries by paleontologists have an important bearing on the center of origin concept (Briggs 2003). We know now that there is an historic onshore-to-offshore replacement process, and that high-diversity tropical settings are sources of evolutionary innovation. Furthermore, taxa such as species, genera, and families are most successfully produced under high diversity conditions. In other words, "diversity begets diversity." Both of these findings have direct relevance to the operation of centers of origin. Therefore, we can now perceive the center as playing an important role in vertical as well as horizontal distributions. Over time, the onshore to offshore replacement sequence has had a cumulative effect in the deep sea.

It is important to note that elements of the tropical biotas have been able to extend their ranges by means of latitudinal dispersal. Fish families in high latitudes are generally older than those of the tropics. Flessa and Jablonski (1996) determined that tropical bivalve faunas were significantly younger than non-tropical ones. In a more detailed study, Crame (2000a; see also his chapter in this volume) concluded that the steep latitudinal gradients of the youngest bivalve clades provided additional evidence that the tropics have served as a major source of evolutionary diversification. As a result, the tropics were seen as a species pool supplying bivalve taxa to the higher latitude regions. In view of the information now available about the East Indies, it seems that the successful tropical origins have probably been taking place in that center.

Frontiers of Evolution and Conservation in the Marine Realm

We can learn much more about the operation of the evolutionary process through the study of biogeography. By this, I mean biogeography in the context of research progress in a variety of more circumscribed disciplines (systematics, physiology, genetics, paleontology, climatology, geophysics) and with an appreciation for the geographical and ecological settings within which the evolutionary change takes place. In the marine environment, the evolutionary importance of centers of origin has now become apparent (Briggs 2003b, and this chapter). So far, most of the evidence for the centers of origin and their evolutionary influence is based on fishes, and to some extent on molluscs and corals. I suspect that similar patterns will be found in other phyla, but no one will know until more research is done.

There is no longer any doubt, however, about the existence of a primary center of origin in the East Indies. It is the source of a sequential series of successful species that extend across the entire Indo-Pacific, and often into other tropical regions and higher latitudes. As long as it continues to function, the center will produce species that are increasingly better adapted and more competitive. It is the continuous arrival of such species that permit peripheral ecosystems to maintain their diversity. Evolution, as it copes with changes in the physical and biological environments, functions most efficiently in centers of origin where populations possess relatively high levels of genetic variation. If such a center is damaged to the extent that its operation is compromised, the production of improved species will diminish, and evolution will be less able to cope with changes in the environment. If this happens, one can predict ecosystem changes resulting in the loss of much of our present species diversity.

Perhaps even more important, the presence of an intact center of origin means that well adapted species will be available for damage repair. Using the example of coral reefs, many of them have been severely affected by such anthropogenic disturbances as freshwater runoff due to deforestation, pollution, destructive fishing practices, and coral mining. In cases where the cause can be removed, the damage can eventually be repaired by the influx of species that had arisen in the center. That is, damage to the peripheral pattern is repairable in the long run as long as the source is protected. However, if the source is degraded, peripheral damage is likely to be persistent, if not permanent. We need strategies not only to protect biological diversity, but also the evolutionary processes that sustain it (Moritz 2002).

In a recent article, Roberts et al. (2002) chose biodiversity hotspots using two primary criteria: species diversity and endemism. The difficulty with this approach was that these two criteria do not correlate well (Briggs 2002; Hughes et al. 2002). Almost all reef areas that harbor relatively large numbers of endemics are located on the fringes of the IWP, many consisting of small oceanic islands. Typically, these are places with low species diversity but with a high endemicity. Studies of historic patterns of extinction and replacement suggest that the endemic species found in such places are phylogenetic relicts or they have been derived from relicts that have speciated in isolation (see chapter by Heaney, this volume). In each case, they usually exist in small populations that are apparently on their way to extinction.

Mora et al. (2003) identified about 90 fish species that are apparently confined to the inner part of the East Indies. These were considered to be neoendemics that were recently derived and had yet to expand their ranges. These, and the many invertebrate neo-endemics, may be considered to represent the evolutionary future of the tropical marine fauna. As they spread out, they will displace many of today's common, widespread species. Future human generations will be able to appreciate these new species and their descendants *provided* we protect the East Indies.

Although the East Indies Triangle is relatively small compared to the vast extent of the IWP, it covers a large geographic area including all of the Philippines, the Malay Peninsula, the large islands of Borneo, Sulawesi, Sumatra, Java, New Guinea, and thousands of smaller islands. It also includes the northern edge of the Great Barrier Reef. Much of Malaysia and the Philippines support dense human populations and consequently many of the coral reefs and their surrounding habitats have been degraded. (In the Philippines, less than one-third of the reefs are in a healthy condition; Gomez et al. 1994). Aside from the reef structures, overfishing is evident in many areas so that subsistence and commercial fisheries are no longer productive. Without additional protection, the damaged reefs and the overfished populations will not be able to recover.

Conclusion

When one takes an overview of distributions in the IWP, it is possible to find patterns in various stages of development, as might be expected in a dynamic evolutionary system. In the East Indies there are many apparently new species that do not occupy large areas; indeed, older ones extend farther out in various directions and a few are circumtropical. Still older species—some the products of subsequent speciation—tend to occur in peripheral areas. In terms of a phyletic future, species formed in peripheral locations are generally unsuccessful. At the same time, extinction patterns also develop, many beginning in the East Indies, with the final demise often taking place far from the center. Such general biogeographic patterns may be highly modified depending on the reproductive mode and dispersal ability of the group concerned.

In recent years, much research has been devoted to the identification and characterization of conservation hotspots that need immediate attention. Although the research has been helpful, the result in the marine environment has been the discovery of numerous hotspots scattered over the world ocean, mainly in the tropics. Certainly it has been worthwhile to call attention to those places, but the lack of a priority system presents a quandary. The informative article by Roberts et al. (2002) does place an emphasis on the East Indies because of its high diversity, but it does not provide a basis for making choices. If we want to protect the place of highest diversity and also the process that produces that diversity, the choice becomes obvious. The East Indies possesses the greatest species and generic diversity, and it also functions as the oceanic world's primary center of evolutionary radiation.

The hotspots that are located peripheral to the East Indies represent the pattern but not the process. This is not a suggestion that we ignore the pattern but, in terms of priority, it should be secondary. To set aside significant portions of the East Indies as reserves or marine parks will be an enormous undertaking requiring international cooperation. If marine biologists from many nations can agree that this is *the* greatest conservation challenge, it will be an important first step. The East Indies has not been completely neglected. The Nature Conservancy and other organizations have projects in the area, but none of them approach the large-scale effort that is needed. Success in protecting diversity in this region depends on whether we in the United States can persuade our government to take a leadership role.

Acknowledgments

I wish to thank J. C. Tyler and E. A. Hanni for their comments on the manuscript, and the University of Georgia for continued access to its databases.

CHAPTER 14

Pattern and Process in Marine Biogeography: A View from the Poles

J. Alistair Crame

The major puzzle in biodiversity is, for me, the problem of the origin and maintenance of richness, and the ecological structures that that implies.

M. H. Williamson, 1997:14

Introduction

At what may be called the dawn of the modern era of biogeography (*sensu* Brown and Lomolino 1998), in the late 1960s F. G. Stehli and his co-workers published a series of maps of the distribution of marine organisms; this series of maps was to have a profound effect upon our view of global biodiversity (Stehli et al. 1967; Stehli 1968). This was, at first sight, somewhat ironic because the primary objective of these maps was to pinpoint latitudinal gradients in taxonomic diversity through the geological record and thus the approximate former positions of geographical poles. These in turn could be used to test hypotheses of polar wandering and continental drift.

Although this innovative procedure was soon surpassed by rapid advances in paleomagnetism and other geophysical techniques, these basic maps remained as one of the first attempts to look at the distribution of living marine organisms on a global scale. Groups covered included turtles, planktonic foraminifera, hermatypic corals, cypraeid and strombid gastropods, and bivalves (Stehli 1968; Stehli and Wells 1971). The latter group was of particular importance because of their sheer abundance and widespread distribution from the tropics to the poles (Stehli et al., 1967, figures 2-9). Stehli and his group went on to use these maps to suggest how large-scale biodiversity patterns might be underpinned by global variations in evolutionary rates (i.e., rates of speciation and extinction) (Stehli et al. 1969, 1972).

The contoured plot of bivalve species diversity derived by Stehli et al. (1967, figure 3) showed a number of particularly striking features. Maximum diversity values occurred in close proximity to the equator, and from there a regular decline in numbers of species to both the North and South Polar Regions. In fact, there was approximately an order of magnitude difference in the number of taxa between the equator and the poles. Furthermore, the plot showed that tropical high diversity was partitioned into two distinct centers: one that was termed "Indo-Pacific" and the other "Atlantic" (Stehli et al. 1967:458). For the first time, we were beginning to appreciate the complex nature of geographic variation in taxonomic diversity in the marine realm.

It is the intention of this paper to review how our understanding of large-scale taxonomic diversity patterns in the marine realm has developed since the pioneering studies of Stehli and his associates. The primary emphasis here will be on establishing what we now know about the nature of both latitudinal and longitudinal gradients and the processes that are thought to underpin their formation. Although the discussions will include deep-sea biodiversity patterns, there will be no detailed review of the nature of depth (i.e., onshore-offshore) gradients. For recent reviews of this important subject area the reader is referred to the papers by Gray (1997) and Levin et al. (2001).

Methodologies and Scale

One of the most important methodological advances in biogeography in recent years has been the development of high-speed computers (Brown and Lomolino 1998). Compared with 35 years ago, it is now very much easier to assemble large data sets and subject them to sophisticated statistical analyses; this is particularly so given the advent of powerful relational databases and geographic information systems (GIS) (Markwick and Lupia 2002). Such technology has added a new rigor to the study of taxonomic diversity gradients and transformed it into a much more analytical procedure.

We also know a great deal more now about plate tectonics and the precise timing of those key geological events that have helped to shape the modern world (Hallam 1994; Cox and Moore 2000; Scotese, this volume). These in turn have sharpened our perceptions of the disjunct distributions and episodes of vicariance that underpin the burgeoning school of cladistic biogeography and phylogeography (Brown and Lomolino 1998). While the role of cladistic techniques in the analysis of latitudinal diversity gradients is not readily apparent (Pielou 1979), their use in the study of longitudinal patterns may be crucial. This is particularly so in tropical regions where a number of studies of Indo-Pacific corals is beginning to reveal a series of significant east-west disjunctions (Rosen 1988; Pandolfi 1992; Veron 1995). It would appear that a once-homogeneous fauna was disrupted by a series of Cenozoic tectonic and climatic events to produce a pattern of overlapping, stepwise species ranges. Some of the most derived taxa occur on the margins of the province, well away from the Indonesian high diversity focus (Pandolfi 1992; see also, below).

Biodiversity phenomena are very obviously scale-dependent and in any study it is important to match both patterns and processes at the appropriate scale (e.g., Aronson and Plotnick 1998; Lyons and Willig 1999; Gaston 2000). Using a nested hierarchical scheme, Ricklefs and Schluter (1993) suggested that there were six critical levels at which diversity could be considered: local population, metapopulation, landscape (i.e., habitat dynamics and selection), region, province, and globe. Levin et al. (2001) employed a similar scheme but simplified it down to just three levels: local (patch dynamics on a 1–10 m^2 scale), regional, and global. The emphasis in this review is very much on the global scale, with only occasional reference to smaller (regional) patterns and processes.

It is also important to emphasize that any biodiversity pattern can be heavily dependent upon the sampling strategy used to obtain it. The recent introduction of terms more widely used in landscape ecology has helped to clarify three basic components of species richness in any given area. First, there is grain (the minimum scale sampled—often the size of an individual quadrat or dredge); then lag (the inter-sample distance); and finally extent (the total area from which samples are collected or the maximum scale) (Palmer and White 1994). In the past, grain has often been confused with extent, although the two can of course have very different diversity values.

How Widespread is the Latitudinal Gradient from the Tropics to the Poles?

Patterns on a Regional Scale

The latitudinal gradient is easiest to detect at the broadest of geographical scales. For example, if samples are grouped into natural biogeographical regions (or latitudinal bins of 5°–10°extent), then it is possible to demonstrate strong gradients in both hemispheres for groups such as the bivalves (Stehli et al. 1967; Roy et al. 1994; Flessa and Jablonski 1995; Crame 2000a,b), gastropods (Roy et al. 1994, 1998; Crame 1997, and references therein), coastal fishes (Rohde 1992), asteroids (starfish) (Price et al. 1999), planktonic foraminifera (Bé 1977; Rutherford et al. 1999), benthic foraminifera (Buzas and Culver 1999) and pelagic copepods (Woodd-Walker et al. 2002). Further pelagic gradients in the northeastern Atlantic have also been recorded in ostracods, decapods, and euphausiids (Angel 1997); furthermore, in the northwestern Atlantic there is a cline in bryozoans (Clarke and Lidgard 2000). It should be emphasized that these are essentially shallow-water gradients delineated at the regional or gamma-diversity scale (*sensu* Margurran 1988). They are by no means uniform in style, having in common only the feature that the highest diversity values occur between 0°–30°and the lowest between 60°–90°(see below).

Some of the strongest regional latitudinal diversity gradients are undoubtedly those that pass from coral reef into non-reef environments. Although taxonomic diversity is by no means uniform across any one reef tract, it is certain that these tropical features contain the highest concentrations of shallow-marine invertebrate, vertebrate, and plant taxa anywhere in the world (Briggs 1996; Gaston and Williams 1996; Reaka-Kudla 1996; Paulay 1997). Of course, there are some exceptions to this "rule" and it is possible that certain peracarid crustaceans, such as the amphipods, may be more diverse in the polar regions than the tropics (De Broyer and Jazdzewski 1996; Myers 1997). It is also striking how benthic algae show a regular diversity gradient along the Atlantic coast of Europe but a reverse gradient along the Pacific coast of South America (Santelices and Marquet 1998).

Patterns on a Smaller Scale

A number of recent studies have suggested that if samples are taken on significantly smaller spatial scales then it is much harder to detect a latitudinal gradient. This is particularly so of the macrobenthic taxa associated with soft substrates, where an early study by Thorson (1957) found no evidence of an equatorial-polar gradient (Clarke 1992). In the shallow marine realm, level-bottom muddy substrates can indeed be compared over very considerable distances and they are beginning to suggest that there is no obvious relationship between within-habitat (or alpha) diversity and latitude. For example, using standardized sampling and analytical techniques, Kendall and Aschan (1993) could find no major differences in the infaunal diversity values obtained from a series of sites in Spitzbergen (78°N), the North Sea (55°N), and coastal Java (7°S) (see also Kendall 1996). Some surprisingly high soft-sediment diversity values have been calculated from southern Australia (38°–40°S) (Poore and Wilson 1993), and rarefaction values obtained for gastropod-bivalve-isopod assemblages in the Weddell Sea (70°S) are very similar to those from tropical latitudes (Brey et al. 1994; Arntz et al. 1997).

Intriguing as these results may be, it is important to ask whether the sample species richness measured within any given habitat or assemblage (i.e., the grain) is actually representative of that habitat or assemblage (i.e., the extent). Gray (2000) suggested that we should in fact be looking at four scales of species richness: point species richness (a sampling unit); sample species richness (several samples from a site of defined area); large area species richness (including a variety of habitats and assemblages within an area of given size); and biogeographic province species richness. Thus, even if Kendall and Aschan's (1993) diversity values give a reliable estimate of sample richness, can we necessarily infer from them that they are reliable estimates of large area or biogeographic province richness (or the species richness of any defined habitat) (Gray 2000)? The true total number of species in any given habitat, large area, or province may only be calculable with the aid of some form of species accumulation curve and the determination of an asymptotic value (e.g., Colwell 1997; Paterson et al. 1998).

Ellingsen (2001) and Ellingsen and Gray (2002) carried out a detailed investigation of taxonomic diversity patterns within the soft-sediment macrobenthos of the Norwegian continental shelf. They collated data from 101 separate sites (56°–71°N) that comprised some 68,000 individuals belonging to 809 species. Calculating sample richness from the mean of five grab samples (point richness: 0.5 m^2 each), they could find no evidence for a latitudinal cline in alpha diversity. When the sites were grouped into five large areas it could be shown that there was considerable latitudinal variation in sample richness but neither mean sample richness per large area nor gamma diversity varied with latitude. In a comparable detailed study of the taxonomic diversity patterns of 535 bryozoan species at 304 continental shelf sites in the North Atlantic, Clarke and Lidgard (2000) could find no obvious trend in alpha diversity from 0°–55°N; thereafter diversity does fall sharply, but the paucity of samples from 55°–90°N precluded any firm conclusions about the nature of high-latitude trends. It is also clear that sample richness varies widely within all of Clarke and Lidgard's (2000) analytical sample bins (i.e., large area richness). The latter study has been complemented by López Gappa's (2000) compilation of the distribution of bryozoans on the continental shelf of Argentina (35°–56°S; coast to 50°W). Plotting the distribution of some 246 species on 1° × 1° grid cells, this author was able to show that the highest within-habitat diversity values occurred in the highest latitudes (48°–56°S). Although the more northern sites are affected to at least some extent by both facies and collecting biases, this strong latitudinal trend is in marked contrast to what would normally be expected.

Patterns in the Deep Sea

It is also possible to sample at a smaller geographical scale (but over a considerable latitudinal range) in the deep sea. Here, the sampling is essentially from level-bottom substrates but, because of considerable variation in sediment grain size, bottom water oxygen levels, and bottom currents, it is not necessarily strictly within-habitat (Levin et al. 2001; Ellingsen and Gray 2002). In a pioneering study based in the deep (500–4000 m) Atlantic Ocean,

Rex et al. (1993) examined diversity patterns exhibited by some 214,000 individuals from 97 epibenthic sled samples (77°N–37°S). Measuring species diversity by rarefaction, they established highly statistically significant latitudinal gradients for bivalves, gastropods, and isopods in the North Atlantic and somewhat less well-defined patterns in the South Atlantic (Rex et al. 1993, figure 2). Here, there seemed to be a complex pattern of interregional variation with comparatively low values from Walvis Bay being matched by exceptionally high values from the Argentine Basin.

As striking as these latitudinal gradients in the deep sea appear to be, their existence has not gone without question. Much of the criticism has centered on the rarefaction methodology, as it is known that the estimate of the number of species (ES_N) for any given number of individuals can be strongly affected by small sample sizes. In certain circumstances this can lead to a gross over-estimate of species richness because of the uncertain effects of dominance and an assumption of random rather than aggregated distributions (Gray 1997). Gray (2000) noted the large degree of variance in sample level richness with latitude in Rex et al.'s (1993) data, which casts doubt over the validity of the patterns produced; this was especially so in the Southern Hemisphere.

To counter criticisms made of their initial study, Rex et al. (2000) reanalyzed their North Atlantic data set using species diversity indices based on both the richness and evenness components of taxonomic diversity rather than just rarefaction alone. They found both species richness (S) and the Shannon-Wiener diversity index (H') to be highly correlated with latitude in all three taxa, but evenness (J) was less so. There was a significant correlation of J with latitude for isopods but only a weak correlation in gastropods and none at all for bivalves (Rex et al. 2000, figure 2). These authors concluded that the latitudinal gradients for all three taxa are more strongly affected by species richness than evenness.

Some doubt has been cast on the Rex et al. (2000) data set, principally because the bivalves and gastropods still show a considerable degree of variation in the 0°–40°N latitudinal range. It is only the exceptionally low values from the Norwegian Sea that make the latitudinal gradients statistically significant (Gray 2001; Ellingsen and Gray 2002). Nevertheless, it is rather striking how very similar diversity patterns are shown in the deep-sea North Atlantic by a completely different taxonomic group, the benthic foraminifera. In this case the low values obtained from the Norwegian Sea are backed up by an independent data set from the western central North Atlantic and Arctic basin (Culver and Buzas 2000). The benthic foraminifera data set was compiled from some 93 sites in the 2000–4000 m depth range, spanning 60°N–72°S. Diversity values for each of these sites were calculated in terms of species richness (S) and α (the parameter of Fisher's log series, which should be independent of sample size, N). Statistically significant latitudinal gradients for both of these values were found in each hemisphere, although for any given latitude North Atlantic values were considerably lower than South Atlantic values (Culver and Buzas 2000, figure 2). It is particularly interesting to see that the south polar regions again exhibit considerably higher estimates of within-habitat diversity than the northern ones (*cf.* Brey et al. 1994; Arntz et al. 1997).

To complete this section on deep-sea patterns it should be noted that neither nematode nor polychaete diversity appears to show any simple relationship to latitude in the North Atlantic (Lambshead et al. 2000; Glover et al. 2001). Using both rarefaction and species counts per unit area, it can be demonstrated that diversity in both these taxa increases significantly from 10°–56°N, before falling sharply in the Norwegian Sea. Such a trend could well be linked with a known increase in surface productivity (and thus of organic flux to the deep sea floor) away from the equator into higher latitudes (Lambshead et al. 2002).

A Latitudinal Gradient of Between-Habitat Diversity?

Even if there were no latitudinal gradient of within-habitat diversity, we could perhaps explain the regional gradient entirely on the grounds of between-habitat (or beta) diversity. In one of the first major reviews of species diversity patterns on a global scale, MacArthur (1965) argued that evolution in the tropics acted much more to fit additional species along environmental gradients by habitat differentiation than to pack them more tightly within any one habitat type. Whereas the number of species within any given habitat should reach saturation levels relatively quickly, those within a larger region would continue to grow by increasing differentiation between habitats. Eventually, all new diversity would be between-habitat (MacArthur 1965).

We might expect *a priori* that beta diversity would be particularly high in the tropical shallow marine realm because of the sheer three-dimensional complexity exhibited by coral reefs. Between the latitudes of approximately 30°N and 30°S, and at water depths essentially <40 m, reef-building corals and associated framework organisms have generated a vast array of both hard- and soft-substrate types. These in turn have undoubtedly facilitated the proliferation of large numbers of closely related suprabenthic (especially fish), benthic, and cryptic taxa (Newell 1971; Sorokin 1995; Reaka-Kudla 1996; Kohn 1997; Wood 1999). Nevertheless, there have been very few quantitative demonstrations that higher rates of beta-diversification actually occur in the tropical marine realm.

In an investigation of the soft-substrate macrobenthic assemblages from 25 lagoon sites on Great Astrolabe Reef (Kadavu Island, Fiji), Schlacher et al. (1998) demonstrated that 42% of species (a total of 211 species) were restricted to just one site and 80% were restricted to no more than four sites.

They concluded that there were high levels of "spot endemism" within the lagoon and high rates of community turnover between each of the sites. Using a new taxonomic similarity index, Δ_S, Izsak and Price (2001) found high levels of beta diversity in the echinoderm assemblages occurring at 16 sites on one small island (Pula Wé) in Sumatra. On a much larger scale, Price et al. (1999) divided the Atlantic asteroid (starfish) fauna into 26 regions and established that the highest levels of beta diversity were consistently associated with the shallowest-water sites. These authors suggested that, on a regional scale, there was a positive correlation between beta diversity and habitat heterogeneity. This was also the conclusion of Ellingsen and Gray (2000) in their very detailed investigation of taxonomic diversity patterns within the Norwegian continental shelf fauna. Using four different measures of beta diversity, they were able to demonstrate a much stronger correlation of beta diversity with environmental heterogeneity than with latitude.

Clarke and Lidgard (2000) carried out a comprehensive study of latitudinal variation in beta diversity within the North Atlantic bryozoan fauna. Using a modified version of Whittaker's statistic, β_W, they were able to demonstrate a latitudinal gradient on a regional scale (i.e., with samples pooled into 10° latitudinal bins); that peak values occur at 10°–30°N; and that there is a steady decline to 80°N (Clarke and Lidgard 2000, figure 4). However, this relationship is not statistically significant because of a large outlier for the Mediterranean at 43°N. If it is accepted that the Mediterranean has had a distinctive recent geological history (with periods of lowered salinity and high endemism), and thus that this locality can be removed from the analysis, then the latitudinal gradient in beta diversity becomes highly significant.

The Prevalence of Latitudinal Gradients in the Marine Realm

It is important to bear in mind how little we may still know about the true nature of taxonomic diversity levels on coral reefs. For example, in a recent study of just one large lagoon-barrier reef system in northwestern New Caledonia (42 sites; ~295 km^2), Bouchet et al. (2002) found the truly staggering total of 2738 species of molluscs. Admittedly, this was the result of a massive collecting and sorting effort followed by a comprehensive taxonomic study involving a team of specialists. Many of these species are tiny (33.5% measure less than 4.1 mm) and many are rare (32% occur at only a single site), reflecting the fact that much reef diversity could be hidden in cryptic and interstitial environments. Reaka-Kudla (1997) has estimated that the 93,000 described species of all types of coral reef taxa may represent only about 10% of the true total. Apart from the many undescribed tiny forms, there are undoubtedly numerous undetected sibling species and some major reef tracts that remain virtually unexplored.

Thus, the well-studied latitudinal gradients in molluscan diversity may be even steeper than was previously imagined. The potentially enormous number of taxa in the Indo-West Pacific province (IWP) can be contrasted with a total of just 800 species (all groups) for the whole of Antarctica (Clarke and Johnston 2003). Although there are parts of the Antarctic province that are still poorly known, our collections go back over 100 years and contain many tiny, rare taxa (Ponder 1983; Dell 1990; Griffiths et al. 2003).

It would seem only logical to conclude that very steep latitudinal gradients, such as those seen in the molluscs, are in turn underpinned by corresponding gradients in both alpha and beta diversity. There is some theoretical evidence to suggest that both these components of taxonomic diversity should increase in tandem, for the exploitation of novel resources or ecospace is bound to be accompanied, at least to some extent, by the differentiation of habitats (Whittaker 1977; Sepkoski 1988; Ricklefs 1995). Alpha and beta diversity covary widely in nature and it is often difficult to separate their precise effects in the shallow marine realm (Sepkoski 1988; Clarke and Lidgard 2000; Ellingsen and Gray 2000).

We might also expect significant evolutionary radiations through geological time to be underpinned by concurrent increases in both alpha and beta diversity (Sepkoski 1988). Levels of within-habitat diversity in particular seem to have been elevated in the two great radiations of the Phanerozoic: the Ordovician, and the latest Mesozoic-Cenozoic eras (Owen and Crame 2002, and references therein; but see also Alroy et al. 2001). In a recent study using a weighted abundance measure, Powell and Kowaleski (2002) indicated that Cenozoic alpha diversity levels in the shallow marine realm might have increased to a level 2.5 times greater than that in the Early Palaeozoic. Rather less is known about changes in beta diversity through time; however, in a classic early study of Paleozoic marine faunas, Sepkoski (1988) demonstrated a significant increase in beta diversity during the Ordovician period. Such a finding has now been challenged and it may be that either all the increase was concentrated at the very beginning of the Ordovician (Miller 1997b), or there was a genuine fluctuation in levels through the period (Miller and Mao 1995; Patzkowsky 1995). The Neogene proliferation of coral reefs would be expected to lead to a significant increase in tropical beta diversity, but this has not so far been substantiated.

It is possible to summarize some of the foregoing discussion by examining the theoretical relationship between local and regional species richness (Figure 14.1). For example, if there is an essentially straight-line relationship between these two variables (Figure 14.1A), then we can suggest that the same size area will contain more species in the tropics simply as a result of an increase in within-habitat (alpha) diversity; indeed, local and regional richness increase in proportion (Srivastava 1999). Such a straight line is usually referred to as a Type I (or proportional sampling) relationship and the communities falling along it can be regarded as unsaturated (Cornell and Lawton 1999). In contrast to this situation, Type II (or saturated communi-

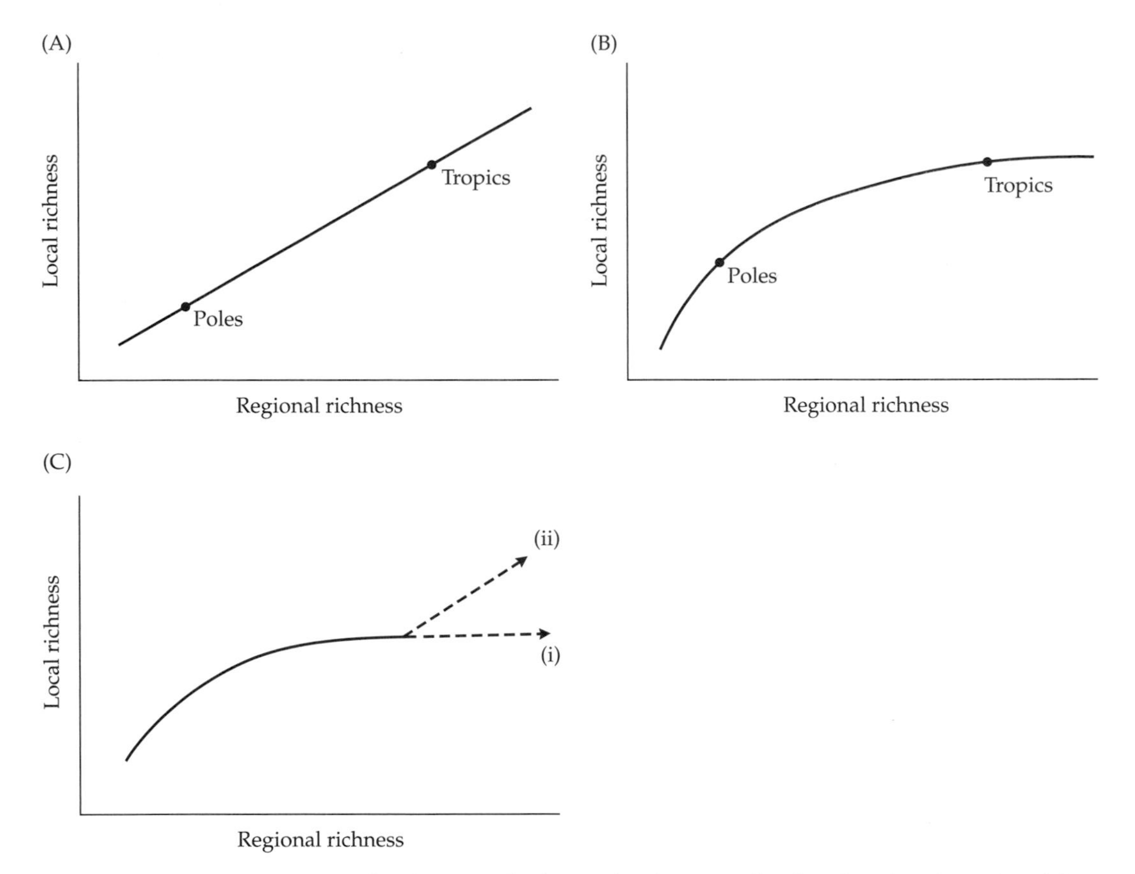

FIGURE 14.1 The theoretical relationship between local and regional species richness. (A) In this instance, there is a straight line relationship between the two variables and we can suggest that the same size area contains more species in the tropics simply as a result of increased within-habitat (or alpha) diversity; local and regional richness increase in proportion and we refer to these communities as 'unsaturated.' (B) In a saturated community, the ratio between local and regional richness becomes increasingly smaller; in this instance the same size area will contain more species in the tropics due to an increase in beta (or between-habitat) diversity. (C) In saturated communities there will be a trend for regional richness to increase but local richness to stay much the same (i); however, on an evolutionary time scale, local diversity may increase through the development of a new trait or the immigration of new taxa into the regional pool (ii). Further details contained in the text. (Based on diagrams in MacArthur 1965; Cornell and Lawton 1992; and Srivastava 1999.)

ties) may show an initial relationship between local and regional diversity but then quickly reach an upper asymptotic value (Figure 14.1B). That is, the ratio between local and regional diversity becomes progressively smaller, and the same size area will contain more species in the tropics due to an increase in between-habitat (beta) diversity (Srivastava 1999).

It is somewhat surprising to find that Type I communities are far commoner in nature than are Type II communities (Cornell and Lawton 1992; Ricklefs 1995; Caley and Schluter 1997; Gaston 2000). Even in the tropical marine realm there is strong evidence for the regional control of individual reef assemblages throughout the IWP (Bellwood and Hughes 2001). Does this mean that beta diversity does not, after all, exert a significant control on the formation of latitudinal diversity gradients? Cornell and Lawton (1992) addressed this key question and pointed out that saturation over ecological time for a given species pool is not necessarily the same as there being hard limits to the richness of local assemblages over evolutionary time. In certain circumstances the asymptotic values characteristic of Type II communities could drift upwards, perhaps with the immigration of new taxa into the regional pool, or with the evolution of some new trait that could allow more species to coexist locally (Figure 14.1C). Increased levels of beta diversity probably play a crucial role in the evolution of species-rich assemblages (Cornell and Lawton 1992), but we still have much to learn about this important interface between theoretical and empirical studies.

Taxonomic Diversity Gradients: Time-Invariant Features on the Earth's Surface?

Taking all the foregoing evidence into consideration, it is possible to conclude that taxonomic diversity gradients at the regional (i.e., gamma diversity) scale occur in many common marine taxa, and that they are in turn underpinned by gradients in both alpha and beta diversity. Perhaps there is some inherent property of tropical marine habitats—such as their greater size, three-dimensional complexity, or higher rates of productivity—that automatically promotes faster rates of alpha and beta diversification (see, e.g., Brown and Lomolino 1998 and Gaston 2000 for a fuller discussion of the causes of high diversity in the tropics). Could the patterns that we see at the present day simply be the cumulative product of such processes operating over tens or even hundreds of millions of years?

Although this is indeed almost certainly the case (see below), there are also some strong indications that other factors were involved in generating the precise patterns that we see at the present day. For example, we now know that the form of the latitudinal gradient in the marine realm is much less regular than was once imagined (Figure 14.2A). In the Northern Hemisphere, diversity values tend to peak at 10°–20°N before falling precipitately at 20°–25°N. Thereafter, values decrease much more slowly and give the whole gradient a distinct stepped or bench-like profile (data taken from a review of latitudinal gradients presented in Crame 2000b; see also Roy et al., this volume). The form of the diversity gradient very definitely does not match that of the latitudinal temperature gradient, and Rosen (1981) suggested that it may in fact represent an abrupt northern edge to the

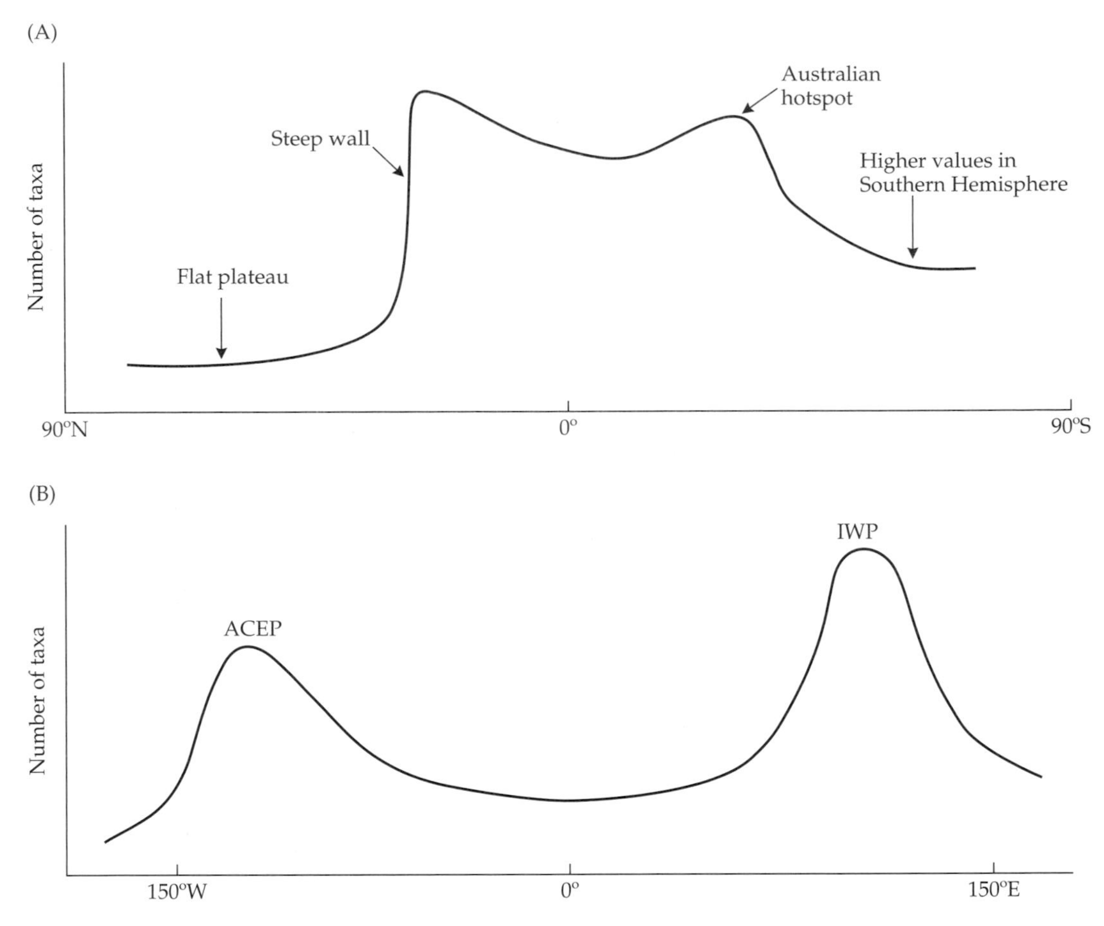

FIGURE 14.2 (A) Hypothetical global latitudinal diversity gradient in the marine realm. (B) Hypothetical global longitudinal diversity gradient in the marine realm. ACEP, Atlantic-Caribbean-Eastern Pacific high-diversity focus; IWP, Indo-West Pacific high-diversity focus. (Based in part on Crame 2000b, Figures 1 and 2.)

coral reef belt that has been accentuated by a series of glacioeustatic sea level fluctuations. Zooxanthellate corals are particularly susceptible to sea level change and frequent exposure during late Neogene climate cycles, which may have served to sharply define the limits of reef growth in both the western Atlantic and western Pacific. It is also possible that the low diversity values of the northern edge of the "plateau" are a response to extensive Quaternary glaciations.

In addition, the Southern Hemisphere gradient does not appear to be a mirror image of that in the north (Figure 14.2A). There are again some low diversity values from the lowest latitude sites (Flessa and Jablonski 1995), but also evidence of a distinct bulge, or hotspot, around Australia (Crame

2000b). It is no coincidence that the richest development of coral reefs in the Southern Hemisphere occurs along the northeastern and northwestern coasts of Australia, and carbonate muds are still widely distributed along much of the south Australian coast (Veron 1995). The Australian continent moved a vast distance northwards during the Cenozoic Era (see below) and seems to be characterized by a genuine admixture of tropical and temperate forms at the present day (Daragh 1985; Crame 2000b). In the highest southern latitudes the form of the gradient is uncertain, but values obtained for almost all major invertebrate taxa are higher than those in the north (Figure 14.2B; Arntz et al. 1997). The longer period of isolation of the Southern Ocean (see below) has led to the adaptive radiation of a number of groups.

It is also important to remember that there are distinct longitudinal gradients in taxonomic diversity in the marine realm. These are most apparent in the tropics where they are delineated by two high diversity foci: an Indo-West Pacific (IWP) focus (running from 110°–150°E), and an Atlantic-Caribbean-East Pacific (ACEP) focus (running from 65°–110°W) (Figure 14.2B; Ellison et al. 1999; Crame 2000b; Briggs, this volume). The IWP focus covers a much larger area and, for almost all common groups of marine taxa, it is richer than the ACEP. Briggs (1996, this volume) has estimated that the total species richness of the IWP is approximately 2.5 times that of the western Atlantic, 3.5 times that of the eastern Pacific, and 7.3 times that of the eastern Atlantic. Similarly, the species richness of trees, shrubs, and ferns in mangrove forests is an order of magnitude greater in the IWP than the ACEP (Ellison et al. 1999). The richness gradient between the two tropical high diversity foci (i.e., ~50°W to 100°E, which encompasses much of the Atlantic and Indian Oceans) is remarkably flat (Figure 14.2B).

There is now a volume of evidence to indicate that levels of global biodiversity increased dramatically through the Cenozoic era (i.e., the last 65 million years of geological time). This was especially so for lower taxonomic levels, which may have increased an order of magnitude in number of species (Signor 1990) (Figure 14.3). Throughout the Cenozoic there were significant radiations of taxa such as amphibians, reptiles, birds, mammals, flowering plants, and insects in the terrestrial realm; similarly, neogastropods, heteroconch bivalves, cheilostome bryozoans, decapod crustaceans, and teleost fish radiated in the marine realm (Hallam 1994; Benton 1999). As a number of these comparatively young groups show strong taxonomic diversity gradients at the present day, it can be inferred that much of this Cenozoic radiation event was centered in tropical and low-latitude regions (Crame 2000a, 2001, 2002).

If this is indeed the case, then it is intriguing to ask why such a pronounced radiation event should have taken place against a background of global cooling. Cenozoic climatic trends are now well established and we know that, from an Early Eocene climatic optimum at 55 my BP, mean oceanic temperatures declined in a series of abrupt steps to those of the present day. Temperature falls of approximately 2°C occurred in both the late Middle

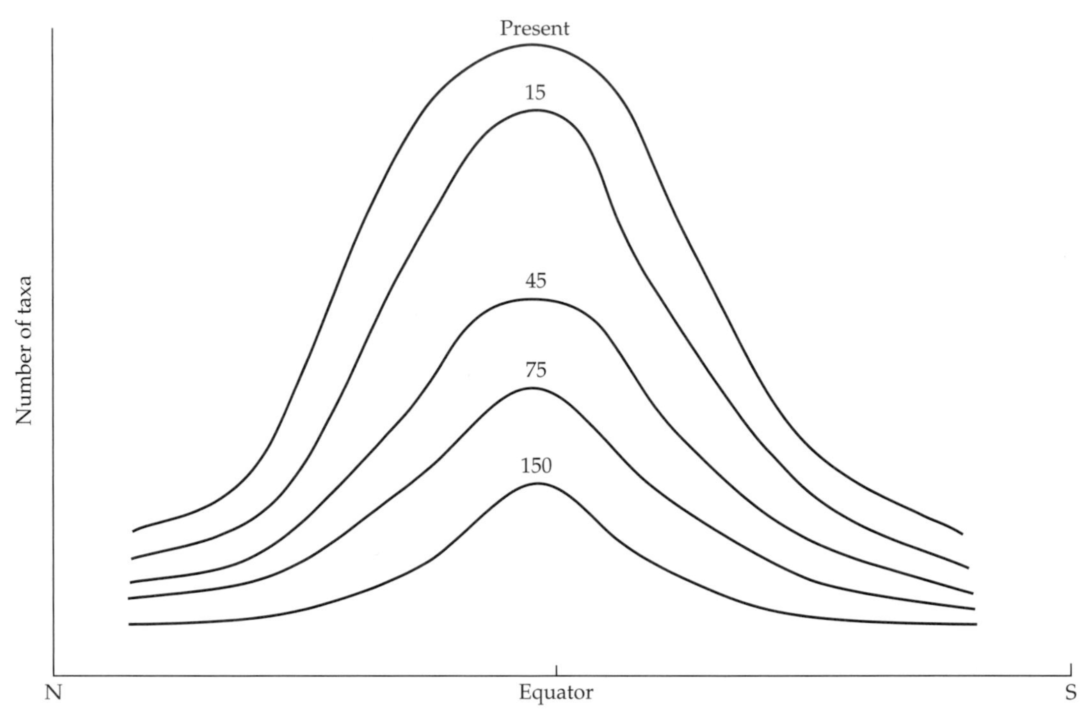

FIGURE 14.3 Theoretical development of the global latitudinal gradient in taxonomic diversity in the marine realm. Five time slices are shown (in million-year increments); number of taxa on the vertical axis. There may have been a particularly steep increase in latitudinal gradients over the last 10–15 million years.

Eocene and at the Middle-Late Eocene boundary, followed by a very rapid 5°C drop at the Eocene-Oligocene boundary (37 my BP) (TABLE 14.1). Further abrupt cooling—possibly by as much as 4° or 5°C—occurred in the Middle-Late Miocene (15–13 my BP), followed by somewhat smaller falls in the latest Miocene-earliest Pliocene (6.2–4.8 my BP), Middle-Late Pliocene (3.6–2.4 my BP), and Late Pliocene-Recent. Even during the Neogene climatic optimum at 17–15 my BP, temperatures only briefly returned to those of the Late Eocene (Clarke and Crame 1992; Frakes et al. 1992; Zachos et al. 2001). It is also apparent that since the Early-Middle Eocene (i.e., approximately 45 my BP), the tropical biome has contracted in area by some 40%–50% (Adams et al. 1990).

It would appear that global cooling may have been promoted by four key Cenozoic tectonic events: (1) physical isolation of Antarctica, (2) closure of the Tethyan Ocean in the Middle East, (3) collision of Australia/New Guinea with Southeast Asia, and (4) uplift of the Central American Isthmus (full details of each of these events are given in Table 14.1). The net effect of all these changes was to effectively switch oceanic circulation from predominantly equatorial to strongly meridional (i.e., north-south) or gyral. This in

TABLE 14.1 *Key tectonic and climatic events of the Cenozoic*

1. Physical Isolation of Antarctica	
Event	**Climate and biological implications**
(i) Early Eocene (50 my BP) Full deep-water separation of South Tasman Rise	(i) Almost complete deep-water isolation of Antarctica. First indications of global cooling at 50–40 my BP. Significant (2°C) temperature falls in both the late Middle Eocene and Middle-Late Eocene boundary. Further isolation of Antarctic marine biota.
(ii) Eocene-Oligocene boundary (37 my BP)	(ii) Major cooling of both surface and bottom waters by 5°C in 75–100 ky BP. Onset of widespread Antarctic glaciation.
(iii) Opening of Drake Passage (~36–23 my BP)	(iii) Precise time of opening uncertain, but deep-water connections probably not established until 23–28 my BP. Almost complete isolation of Antarctic marine biota.
(iv) Middle Miocene (15 my BP) Full establishment of Antarctic Circumpolar Current	(iv) Development of Polar Frontal Zone. Latitudinal temperature gradient similar to that of today.

turn introduced warmer waters into the high latitudes and led to the increases in precipitation necessary for the development of polar ice caps (Crame and Rosen 2002, and references therein).

These tectonic changes are particularly important because they led to the progressive development of thermally defined provinces on essentially north-south trending shorelines. In addition, events 2–4 led to the imposition of a series of east-west trending barriers in tropical and warm temperate regions, so that overall the level of division among biotic provinces during the Cenozoic was the highest seen for the last 500 million years. It has been estimated that, in the shallow marine realm alone, the number of provinces has risen from approximately six in the Late Cretaceous to at least 31 at the present day (Clarke and Crame 2003, and references therein). This is, in effect, differentiation diversity on a global scale where, instead of packing more communities within an individual province (essentially beta diversification), tectonic events resulted in the packing of more provinces within the total biosphere. This high level, or delta (between-region), diversification (*sensu* Magurran 1988) was undoubtedly a major contributor to the process of global diversification during the Cenozoic.

The fossil record is beginning to indicate that a number of coral and mangrove genera common in the IWP today were formerly more widespread in the Early Cenozoic Tethyan Ocean (Rosen 1988; Wilson and Rosen 1988; Ellison et al. 1999). For example, the very common branching coral genus *Acropora* is recorded from the Paleocene of Somalia and Eocene of Europe, but it did not reach either the Caribbean or the IWP until the Oligocene.

TABLE 14.1 *Key tectonic and climatic events of the Cenozoic (continued)*

2. Closure of the Tethyan Ocean in the Middle East	
Event	**Climate and biological implications**
(i) End of Cretaceous (~75–65 my BP) Vast circumpolar-equatorial tropical ocean	(i) Major westerly flowing equatorial current system. Some faunal differentiation in the vast ocean, but no indication of clear high-diversity foci.
(ii) Paleogene (65–23 my BP) Continuity of the tropical Tethyan ocean	(ii) Largely homogeneous tropical fauna. Major pulse in coral reef development at the end of the Oligocene (~23 my BP), but even then there were marked similarities between western Tethys (Mediterranean) and the Caribbean/Gulf of Mexico.
(iii) Early Miocene (20 my BP) Closure of Tethys by the northward movement of the Africa/Arabia landmass	(iii) Westerly flowing tropical current drastically curtailed. Mediterranean Sea excluded from reef belt. Caribbean and eastern Pacific regions become progressively isolated. This marks the beginning of the distinction between the IWP and ACEP foci. Further development through the Neogene (< 20 my BP) sees the relative impoverishment of the ACEP and enrichment of the IWP.

Indeed the genus *Acropora*, with approximately 150 living species, is not at all common on IWP reefs until the late Pliocene (2–3 my BP) (Crame and Rosen 2002, and references therein). It was only after the Early Miocene closure of Tethys in the Middle East and the collision of Australia/New Guinea with Southeast Asia (see Table 14.1) that the IWP high diversity focus began to develop. There is some paleontological evidence to suggest that particularly intense radiations of zooxanthellate corals and reef-associated invertebrates occurred in both the IWP and ACEP in the mid-late Neogene (i.e., last 10–15 million years) (Budd et al. 1996; Collins et al. 1996; Jackson and Johnson 2000; Crame 2001; Crame and Rosen 2002). In essence, marine biotas experienced a Paleogene (65–23 my BP) phase of range retractions and disjunctions, followed by Neogene (23–1.6 my BP) radiations from the two high diversity foci. Much of the diversity we see today may be of very recent origin.

Synopsis

Taxonomic diversity gradients are developed at a regional scale in a wide variety of marine organisms. Latitudinal gradients in particular can be traced from tropical to polar regions and it would seem reasonable to conclude that they are in turn underpinned by corresponding gradients in both within- and between-habitat diversity. As they can be traced back over considerable intervals of geological time, we can also consider them to be ancient, albeit intensifying features on the Earth's surface.

TABLE 14.1 *Key tectonic and climatic events of the Cenozoic (continued)*

3. Collision of Australia/New Guinea with Southeast Asia	
Event	**Climate and biological implications**
(i) Beginning of the Cenozoic era (~65 my BP) Australia/New Guinea separated from Southeast Asia by deep-water gateway (~3000 km wide)	(i) Single tropical (Tethyan) Ocean; no differentiation into Indian and Pacific Oceans
(ii) Paleogene (65–23 my BP) Progressive closure of the Indo-Pacific gateway; northward subduction of Indian-Australian lithosphere beneath the Sunda-Java-Sulawesi arcs	
(iii) Mid-Oligocene (30 my BP) Gap narrowed substantially, but still a clear deep-water passage formed by oceanic crust	
(iv) Latest Oligocene (25 my BP) Major changes in plate boundaries when New Guinea passive margin collided with the leading edge of the eastern Philippines-Halmahera-New Guinea arc system	
(v) Early Miocene (20 my BP) The continent-arc collision closed the deep-water passage between the Indian and Pacific Oceans	(v) Major reorganization of tropical current systems; new shallow-water habitats appear in Indonesian region
(vi) Early-Late Miocene (20–10 my BP) Continued northward movement of Australia/New Guinea; rotation of several plate boundaries and formation of tectonic provinces recognizable today	(vi) Widespread growth of coral reefs in IWP region; huge rise in the numbers of reef and reef-associated taxa; many modern genera and species evolved

Whatever the ultimate cause of taxonomic diversity gradients, they would seem to be reflecting some sort of large-scale variation in evolutionary rates and extinctions. For example, there is a reasonable volume of evidence to suggest that in groups such as the bivalves and planktonic foraminifera, living tropical taxa are significantly younger than their non-tropical counterparts (Flessa and Jablonski 1996). In their comprehensive biogeographic survey of 539 bivalve genera from 117 different localities, Flessa and Jablonski (1996) found the median age of all taxa to be 20.2 million years (i.e., Early Miocene). However, genera confined to only the tropics have a median age of 4.3 million years and those confined to the non-tropics one of 12.3 million years. It would seem reasonable to conclude from these findings that origination rates were significantly higher in the tropics,

TABLE 14.1 *Key tectonic and climatic events of the Cenozoic (continued)*

4. Uplift of Central American Isthmus (CAI)	
Event	**Climate and biological implications**
(i) Mid-Miocene (~15–13 my BP) Sedimentary evidence of earliest phases of shallowing	(i) Still a deep-water connection through the CAI.
(ii) Latest Miocene-Early Pliocene (6–4 my BP) Continued shallowing (to < 100 m depth)	(ii) Major effect on oceanic circulation; Gulf Stream begins to deflect warm, shallow waters northward U.S. seaboard. Initially this led to some warming in northern mid- to high latitudes; however, Gulf Stream water is slightly hypersaline and sinks when it reaches the highest latitudes to form the North Atlantic Deep Water, which then spreads into both the South Atlantic and Pacific Oceans to initiate a major "conveyor belt" of deep-ocean circulation.
(iii) Mid-Pliocene (3.6 my BP) Further closure of the CAI	(iii) North Atlantic thermohaline circulation system intensified; Arctic Ocean effectively isolated from the warm Atlantic, eventually leading to the onset of Northern Hemisphere glaciation at 2.5 my BP.
(iv) Late Pliocene (~3 my BP) Complete closure of the CAI	(iv) Final separation of shallow water Atlantic/Caribbean and eastern Pacific provinces. A further important effect of the closure of the CAI was to reverse the flow of water through the Bering Strait, which in turn influenced Pliocene-Pleistocene patterns of thermohaline circulation in the Arctic and North Atlantic Oceans.

but of course it has to be borne in mind that the quality of the pre-Miocene tropical fossil record is not particularly good.

In my recent investigation of latitudinal gradients in both living and fossil bivalves, I detected a pattern of the steepest gradients occurring in the geologically youngest clades (Crame 2000a, 2002). This is particularly so for the dominant taxon at the present day, the heteroconchs, which underwent a massive Cenozoic radiation. The same phenomenon occurs in the gastropods, where the large neogastropod clade also underwent an extensive Cenozoic tropical radiation (Crame 2001). Such steepening of latitudinal gradients through the Cenozoic could be attributed in part to a retraction mechanism (i.e., as the area of the tropics retracted through the Cenozoic), but a study of the various gradients exhibited by individual clades suggests strongly that they very largely represent evolutionary radiations from tropical and low-latitude regions (Crame 2000a). Support for this contention comes from a recent study of both fossil and living benthic foraminifera by Buzas et al. (2002), who were able to demonstrate a significantly greater range of increase in the alpha diversity of tropical localities (Panama and Costa Rica) than in temperate ones (North American coastal plain) over the last 10 million years. Finally, a center of origin hypothesis within the IWP high diversity focus has been supported by a recent study of the large-scale patterns of species richness within reef fish. Mora et al. (2003) found that both the majority of range midpoints and the highest concentration of endemics occur within the core Indonesia-Philippines region, and it would appear that new taxa have consistently radiated outwards from the core towards the margins of the province.

It is important to realize that tropical high diversity patterns could equally as well be generated by reduced rates of extinction (Stanley 1986, 1990; Rosenzweig 1995). However, we know even less about latitudinal clines in rates of extinction than we do about those in origination, and no clear pattern has yet been established. We do know that, at least during parts of the late Neogene, comparatively high rates of tropical radiation must have been matched by high rates of extinction (Jackson and Johnson 2000), but the pattern farther back in time is much less clear. It is also apparent that we cannot simply attribute taxonomic diversity gradients in the marine realm to higher rates of extinction at the poles. Many polar marine invertebrate taxa seem to be extremely well adapted to life in cold water (Clarke 1993; Arntz et al. 1994), and it may well be that other factors, such as the availability of food or substrate type, govern the composition of polar marine assemblages (Crame 2000a, 2002).

It would also appear that taxonomic diversity gradients steepened considerably during the later part of the Cenozoic era, and perhaps as recently as just 10–15 my BP. We might account for a significant part of this increase through the phenomenon of differentiation (i.e., delta) diversification following the development of a record number of both thermally and tectonically defined marine provinces. Nevertheless, a number of authors have argued that something else must have been involved too, and suggested a direct link between global diversification and a Neogene intensification of glacioeustatic climate cycles (Chown and Gaston 2000; Dynesius and Jansson 2000; Jansson and Dynesius 2002). The mechanisms involved here are complex, but in essence they involve some form of genetic isolation (and thus allopatric speciation) through repeated temperature or sea level fluctuations (for fuller explanations see Crame and Rosen, 2002; Clarke and Crame, 2003). Of course, such mechanisms were not necessarily confined to the tropics and some exciting new molecular evidence is indicating that there may well have been significant late Neogene radiations in the polar regions too (e.g., Held 2000; Page and Linse 2002); however, these were probably never on the same scale or intensity as those observed in the tropical regions. Improving our knowledge of latitudinal clines in the rates of both speciation and extinction must remain at the forefront of biogeographical investigations (Roy et al., this volume).

Frontiers of Marine Biogeography

As with any active research program, the science of marine biogeography is one rich in insights as well as intriguing questions. Below I list some of the most significant findings and important questions that characterize the frontiers of this field.

1. For many groups of marine organisms, both now and in the past, there are simply far more of them in tropical than in polar regions. This is one of the fundamental patterns of life on Earth, but what underlying processes account for this pattern? And, do those causal processes vary among faunal groups, geographic regions, or geological periods?
2. We are beginning to appreciate that, within the marine realm, latitudinal gradients in taxonomic diversity are not nearly so regular as was once imagined. In particular, there may be significant inequalities between the Northern and Southern Hemispheres and these must be further elucidated. We still have much to learn about biodiversity patterns in the deep sea too.
3. Large scale inequalities in both latitudinal and longitudinal gradients point to a substantial historical legacy that can be measured over a geological timescale of millions to tens of millions of years. To what degree are the patterns we study today influenced by the past?
4. There was a substantial increase in global biodiversity through the Cenozoic Era; this may have been by as much as an order of magnitude at the species level. Some exciting evidence has come to light recently to suggest that much of this increase may have been concentrated within the last 10–15 million years. Is this a trend that can be substantiated in a wide range of marine groups?
5. The causes of this abrupt Cenozoic increase in biodiversity are still uncertain. A marked increase in marine provincialism may have been important, as could serial range expansions and contractions induced by intense Neogene glacio-eustatic cycles. It is important to emphasize that this was not just a tropical phenomenon, as some polar marine taxa (especially in the Antarctic) went through intense late Cenozoic radiations too.
6. The global patterns of biodiversity that we see at the present day are not just the product of differential rates of speciation; geographical variation in rates of extinction could have been important too. Within the marine realm we still do not know if there have been significant latitudinal variations in the rates of both background and mass extinctions.

Current and future generations of scientists advancing the frontiers of marine biogeography will no doubt capitalize on major new databases of both living and fossil taxa, coupled with significant advances in molecular phylogenetics. These and related advances offer biogeographers key new insights into the origin and maintenance of global patterns in biodiversity in the marine realm.

Acknowledgments

I would like to thank Professor Jack Briggs and the organizers of the *International Biogeography Society* meeting in Mesquite, Nevada (January 2003) for the invitation to present this paper. Reviews by Andy Clarke, Alex Rogers, and Geerat Vermeij helped to clarify some of the ideas presented herein, although they should in no way be held responsible for any shortcomings or omissions. I would particularly like to thank Mark V. Lomolino for his very thoughtful and comprehensive review. Contribution to British Antarctic Survey core project, "Antarctic Marine Biodiversity: A Historical Perspective."

PART V

CONSERVATION BIOGEOGRAPHY

Mark V. Lomolino

At the foundation of all biogeographic patterns is the *geographic template*: the spatial variation in environmental conditions, which is substantial but highly non-random for all regions of the terrestrial and aquatic realms. Species distributions, gradients in species richness, and all other biogeographic patterns result from differential responses of organisms to the geographic template—adaptation, evolution, dispersal to more suitable environments, or (if all this fails) extinction. All of these processes, and in turn nearly all biogeographic patterns, are also influenced by interactions among organisms (i.e., competition, predation, herbivory, mutualism, commensalisms, and amensalism).

To add one final layer of complexity, we now know that the Earth's geographic template is highly dynamic not only across spatial scales, but across seasons, millennia, and geological periods as well (see Scotese, this volume). Most recently, we have begun to stir the mix and transform the geographic template by degrading, reducing, and fragmenting native ecosystems while at the same time "connecting" long-isolated biotas through countless episodes of species introductions (see Lockwood, this volume).

Our emergence as global-scale ecosystem engineers poses great challenges, but it also presents some important opportunities to advance basic and applied research. The transformation of landscapes and seascapes and the shuffling of the world's biotas constitute manipulative experiments—unwittingly performed, granted, but at a scale seldom if ever rivaled by any planned experiments, and consequently providing unparalleled opportunities to test and hopefully apply theory on the geography and conservation of nature. Emerging out of these great challenges and opportunities is a new discipline, *conservation biogeography*. The term itself is relatively new, perhaps not appearing in the literature until just a few years ago. Yet, as most students of the historical development of science know, many of the modern and truly significant advances in science have their roots deep in the early development of that field.

Biogeographers of the eighteenth and nineteenth centuries were keenly aware of the fact that, not only was the natural world rich in diversity, but what we now call biological diversity was waning under the pressures of expanding industrialized societies. As early as 1839, Charles Darwin remarked on declining diversity and the "downsizing of nature" (*sensu* Lomolino et al. 2001):

> It is impossible to reflect on the state of the American continent without astonishment. Formerly it must have swarmed with great monsters: now we find mere pygmies compared with the antecedent, allied races

Later, in 1876, Alfred Russel Wallace remarked that

> It is clear, therefore, that we are now in an altogether exceptional period of earth's history. We live in a zoologically impoverished world, from which all the hugest, and fiercest, and strangest forms have recently disappeared … .

The so-called "naturalization campaigns," systematic programs by Europeans to establish European plants and animals in their colonies and other distant lands, were clearly underway when Joseph Banks began his voyage with Captain James Cook on the *Endeavor* (1768–1771). These "campaigns" threatened many native and ecologically naïve species of oceanic islands. Upon studying the native fauna of the Galápagos, Darwin (1839) warned that

> A gun is here almost superfluous; for with the muzzle I pushed a hawk off the branch of a tree. … We may infer from these facts, what havoc the introduction of any new beast of prey must cause in a country, before the instincts of the indigenous inhabitants have become adapted to the stranger's craft or power.

Much later, in his discussion of "a wilderness in retreat" in *The Ecology of Invasions by Animals and Plants* (1958), Charles Elton remarked that

> We must make no mistake; we are seeing one of the greatest convolutions of the world's flora and fauna.

These prescient remarks span much of the historical development of biogeography, and all are antecedents to current themes in conservation biogeography. The link between studying patterns in geographic variation of diversity and contributing to its conservation is a common theme indeed. In his foreword to a recent book on Wallace, E. O. Wilson (1999) observed that

> The vastness of the tropical archipelago also provided the knowledge Wallace needed to conceive the biological discipline of biogeography, which has expanded during the late twentieth century into a cornerstone of ecology and conservation biology.

The newly articulated discipline of conservation biogeography, however, has a two-fold emphasis. First, as Wilson and many of his colleagues have demonstrated, biogeographers have many valuable insights for conserving biological diversity. Second, and perhaps the newest emphasis, is that in order to conserve nature, including "the hugest, and fiercest, and strangest forms," we need to conserve their distributions—that is, the geographic, ecological, and evolutionary context of nature.

Each of the contributions in this Part has addressed both of these themes and, at the same time, provides some exemplary models of how we can advance basic research on the geography of nature while simultaneously contributing to conservation. These chapters clearly demonstrate how we can apply lessons from both theoretical and empirical studies for predicting the effects of anthropogenic threats to biological diversity, in particular invasions by exotic species and fragmentation of native ecosystems (see chapters by Lockwood, Rosenzweig, and Heaney). As Lawrence Heaney cogently argues, studies that advance biogeographic theory, coupled with biogeographically informed field research, are essential for successful protection of biological diversity (see also Sánchez-Cordero, this volume).

One theme that deserves much more attention than we were able to provide here is the biogeography of our own species. Many of the patterns we study, whether in basic or applied research, are strongly shaped by human societies who have advanced across the globe in a highly non-random manner for many millennia. We certainly can benefit from new collaborations between anthropologists, landscape ecologists, environmental biologists, and others who view humans not as an annoying complication, but as an integral component of the natural world.

Moreover, we need to develop a much better understanding of not only static patterns in the geography of nature, but biogeographic *dynamics* of imperiled species and those agents that threaten biological diversity, in general. Our abilities to understand the geography of extinction in historic and prehistoric records, and to predict extinctions and map biological diversity in the future rely heavily on our understanding of the geographic dynamics of our own species and on that of our commensals, introduced species, and diseases, some of which (e.g., sylvatic plague, *Yersinia pestis*) affect both humans and imperiled species (see Udvardy 1969; Hengeveld 1989 1990; Flannery 1995; MacPhee and Marx 1997; Diamond 1998; Lomolino and Channell 1995; Channell and Lomolino 2000; Cully and Williams 2001). A more comprehensive knowledge of the factors influencing the spread of future populations of our own species and those of other, problematic species seems essential if we are to preserve the diversity and geography of nature.

Our biogeographers' tool kit now contains an impressive assortment of advanced techniques for identifying and analyzing patterns of biological diversity, including geographic information systems, spatial statistics, powerful computers and data storage capacities, remote sensing, and rapidly advancing abilities to interpolate and predict distributions. The chapters by Heaney and by Sánchez-Cordero and his colleagues provide exemplary case studies and overviews of these new frontiers in predictive biogeography. We are, unfortunately, data-deprived; or, more to the point, we suffer from a Catch-22 of endangered species work: we need to know the most about the species that are the most difficult to study. While Peter Raven and E. O. Wilson (1992) called on us to address the "Linnaean shortfall" (the great discrepancy between the number of described versus existing species), conservation biogeographers deal with an even more challenging shortfall: the paucity of information on the *distributions* of imperiled species. Latin binomials are just not sufficient.

Some of the most important advances in biogeography in general, and conservation biogeography in particular, will be gained by intensifying our efforts to explore and understand the biogeography of less accessible (but often largely endemic and imperiled) biotas and ecosystems—parasites, parasitoids, microbes, and bryophytes; biotas of remote recesses of the biosphere (e.g., those of the abyssal zones); genetically engineered and other anthropogenic "species"; invasive and now "naturalized" species (see Sax 2001); and species inhabiting—and in many ways surviving only in—anthropogenic ecosystems (especially zoos, botanical gardens, and nature reserves).

What we may call the "Wallacean shortfall" (the paucity of information on the geography of nature) is perhaps the most important challenge facing conservation biologists. Although many agencies have devoted their preciously limited energies and funds toward addressing this shortfall (e.g., Conservation International's Rapid Assessment Program), their efforts alone will not suffice. Without a substantially increased campaign to map distributions of imperiled species, both historic and current, we will be hobbled with an implicit and perhaps very tenuous assumption: that the patterns we see for the common and broadly distributed species apply equally to the rarest and most imperiled. I am not sure that we have good reason to believe this—or that we can afford to, only to find out in a few decades that we were wrong and lost many species in the process. Moreover, we do not *need* to assume this. Efficient biological surveys guided by relevant theory are at once highly informative and cost-effective. Such surveys will no doubt provide the raw material for new discoveries in the geography of nature as well as new strategies for conserving the diversity and geography of nature.

CHAPTER 15

How Do Biological Invasions Alter Diversity Patterns?

Julie L. Lockwood

Introduction

As society becomes ever more socially and economically connected through commerce and travel, species increasingly tend to be transported and released well outside of their native ranges. Such human-mediated movement of non-native species represents a substantial departure from natural immigration processes. Human actions tend to move species over larger spatial scales than species were commonly capable of moving on their own (e.g., Carlton and Gellar 1991; Leppakoski and Olenin 2001). Human actions also serve to increase the rate at which species attempt to colonize geographic locations far above natural rates (e.g., Cohen and Carlton 1998). These patterns, plus extinctions at the hands of non-natives, serve to re-shape the distribution of the Earth's biota. Only recently, however, have researchers begun to document the effects of these large-scale (biogeographic) changes in diversity (e.g., Lockwood and McKinney 2001; Scott and Helfman 2001). This chapter reviews this work, and includes a discussion of existing theories that may help us predict how diversity will change in a future that includes even more biotic mixing.

A Brief History of Invasion Research

Humans have been intentionally moving organisms out of their native range and into exotic (non-native) locales since ancient times. For example, as early as the fourth century B.C., the Shu-Yuan du Trade Route linked India and China, and written records from that time relate several exchanges of plants and animals between these two regions (Yan et al. 2001). Indeed, plants have been moved and cultivated since the dawn of civilization, perhaps going back as long ago as 8000 BC (Reichard and White 2001). The establishment of species outside of their native range was seen solely as a utilitarian process by these early societies, and relatively little attention has been paid to patterns in these early invasions, their impact, or their subsequent evolution (although see Hanfling and Kollmann 2002).

Charles Darwin was one of the first to consider what non-native species could tell us about basic ecological and evolutionary principles. Darwin used examples of exotic species established in North America, Australia, and western Europe as fodder for his arguments supporting evolution through natural selection (Darwin 1859). However, it was not until the publication of *The Ecology of Invasions by Animals and Plants* by Charles Elton in 1958, and in 1966 the publication of *The Alien Animals* by George Laycock, that the greater ecological community began to view species invasions as a potential conservation and economic problem. Together these two authors outlined how biological invaders could destroy natural and managed ecosystems by changing food web structure (e.g., predation, competition, and mutualism), or by altering ecosystem processes (e.g., fire regimes). Laycock (1966) accomplished this through detailed exploration of case studies. Elton (1958), echoing Darwin, viewed the establishment and spread of invaders as events subject to known ecological and evolutionary processes. Thus, biological invasions could tell us something fundamental about these processes, and we could use our ecological understanding to combat any ill effects these invaders may have on native species.

This change in how biological invasions were viewed amounted to almost nothing until the mid-1980s. In 1982 the Scientific Committee on Problems of the Environment (SCOPE) decided to launch a project on the ecology of biological invasions. Several regional meetings were organized, and a boom in books and articles about invasions followed (e.g., Mooney and Drake 1986; di Castri et al. 1990; Groves and di Castri 1991). The ecologists that participated in these meetings had built their careers studying basic ecological principles such as competitive exclusion, community succession, ecophysiology, population biology, and biogeography. When they turned their collective attention to the issue of biological invasions, they built the framework for all subsequent work on the issue.

Since the publication of the SCOPE volumes, the study of biological invaders has become *de rigueur*. This has occurred for two reasons. First, there are a growing number of invaders doing spectacular harm to native species and ecosystems, and this has focused research around prevention, detection, and eradication of invaders (Mack et al. 2000). For example, Wilcove et al. (1998) found that the negative impact of invasive species was second only to habitat destruction in importance among factors threatening endangered species in the United States. Recent reviews of the impact of invaders suggest that most losses of native diversity occur when non-native species alter trophic structure of ecosystems (Davis 2003), or when non-native species hybridize with natives (Mack et al. 2000). As the number of non-native species released around the world increases, the challenges to conserve biological diversity will no doubt increase as well.

Second—following the vision of Darwin and Elton—invasive species are seen as providing powerful probes into how ecological communities are assembled and maintained. In particular, our understanding of biogeography and large-scale ecology is inherently limited by the fact that the patterns we observe were formed via mechanisms that we can rarely observe and have little hope of manipulating to fit an experimental design. Thus, we are often forced to infer process from pattern, or to conduct significantly smaller scale experiments that may not capture the full complexity of real systems. Biological invaders represent, in many ways, uncontrolled but potentially insightful experiments on the consequences of species dispersal and colonization. For this reason, they are important components of our investigative arsenal. Thus far, the study of invasions has added to our understanding of the role of interspecific interactions in determining community membership, the rates and modes of speciation, and patterns in range expansion, to name only a few.

This chapter will focus on how biological invaders are re-shaping diversity patterns as measured at comparatively broad scales (e.g., ecoregions and watersheds). In this sense, this chapter is "upsizing" the elements of interest in invasive species detailed in the above paragraphs. Most concern over the loss of diversity has focused on the ill effects of invasive species on one (or a few) co-occurring species in a local community. Rarely have researchers tallied the cumulative effects that species invasions and associated extinctions have on large-scale diversity. Here, I document such changes in diversity as measured within both a given region and across regions. However, mostly as a matter of convenience, I will limit the discussion to changes observed within a given taxonomic group or trophic level (e.g., birds, vascular plants, or freshwater mussels). Finally, I will relate these observed changes in diversity to what we might expect based on our understanding of how increasing dispersal rates should influence diversity at biogeographic scales.

Invasion Process Model

It is always useful to begin discussions of biological invasions by detailing the four-stage process required for successful colonization (Figure 15.1; Richardson et al. 2000; Kolar and Lodge 2002). First, individuals must be transported outside of their natural geographic range. Second, they must then be released or must escape from this transportation vector; species at this stage are termed "introduced." Third, these introduced individuals must establish a self-sustaining population within this new environment. And

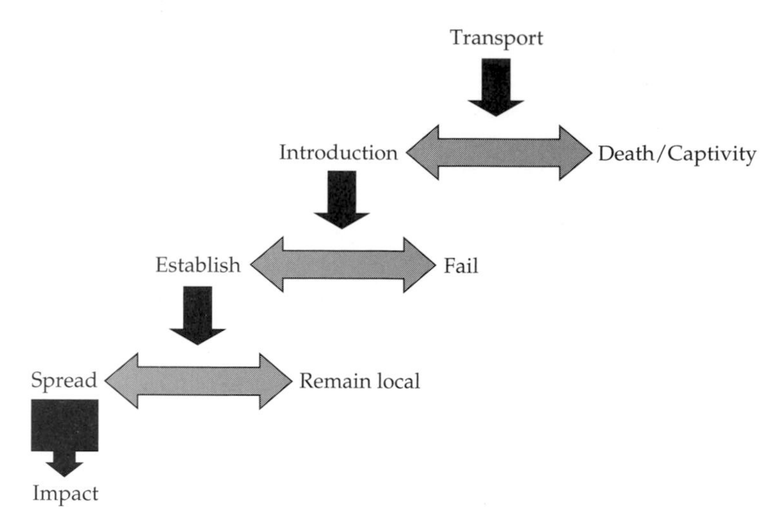

FIGURE 15.1 The four-stage invasion process (black arrows). Each transition acts as a filter that allows only some portion of species from the prior stage to successfully pass through.

fourth, established populations of introduced species may increase in abundance and subsequently spread beyond their point of initial establishment (Richardson et al. 2000). Typically, only the species that make it to the fourth stage cause some ecological or economic harm, and thus the term "invasive" is usually restricted to species at this stage (Davis and Thompson 2000).

Measuring Diversity Changes

For changes in diversity patterns to be detected and measured, individuals of non-native species must establish self-sustaining populations within a non-native range. These established species may be filling the niche left by an extinct native, defining and occupying a new niche, or driving native species toward extinction through interspecific interactions such as competition.

In this section, I utilize several case studies to describe changes in diversity patterns that occurred after several invasions and extinctions. In each case study, the authors assembled a list of species that currently co-exist in a defined area (e.g., ecoregions, watersheds, or islands), including any non-native species that are established there. The number of species on this list represents a snapshot of current diversity. To derive a past, or historical, snapshot of species diversity, authors deleted all non-natives from the current list and added any native species that are known to have become locally extirpated or globally extinct. To describe changes in local diversity, the authors simply compared these two snapshots and calculated net change in species richness (i.e., changes in alpha diversity; see Huston 1994).

In two of these case studies, the authors also calculated changes in species diversity across the various regional localities. Making such a calculation has a long history that goes back at least to Whittaker (1970), who described changes in species assemblages as one moves across an environmental gradient as beta diversity, or spatial turnover. Intuitively this is measured by calculating an inverse measure of beta diversity (i.e., similarity) based on the number or proportion of species shared between two locations (see Shimda and Wilson 1985). In the works that I review here, the authors calculated similarity using either Jaccard's coefficient or an analogous measure.

Here we are interested in whether similarity (or beta diversity) among locations changes after invasion; thus we must compare species lists over time as well as across space. The methods used to derive species lists for two time slices are identical to those used when documenting net changes in alpha diversity. The tricky part comes when trying to decide which locations to compare, since locations near one another will naturally share many more species than those far away. The procedure with the fewest assumptions is to simply compare all locations to each other in a series of pairwise calculations at one time slice (historical), and then again at another time slice (current). The distribution of these differences is then the statistical product that allows you to infer something about the degree of change in spatial diversity. Homogenization occurs when beta diversity decreases (and spatial similarity increases) through time (Rahel 2002).

Alpha Diversity

Figure 15.2 is a histogram showing the results of efforts to document changes in alpha diversity (local species richness) for 103 locations after invasions and extinctions. Most of these localities did not substantially change in species richness (i.e., the most frequently recorded percentage change was either the category that included zero or the category that included a relatively small increase; see Figure 15.2). The second noteworthy result from this exercise is the production of a long right-distributional tail in several of the panels. Thus, in cases where alpha richness did change between historical and current snapshots, it tended to increase. This indicates that the establishment of non-native species tends to either maintain historical values of local species richness, or elevate it beyond historical levels (see also Sax et al. 2002). Only rarely did local species richness decline after invasion.

Since all localities included in Figure 15.2 have experienced several extinctions and non-native species invasions, this latter result is remarkable. It appears to run contrary to our overall conception of how human actions affect biodiversity (e.g., Kennedy et al. 2002; but see also Sax and Gaines 2004; and chapters by Heaney and Rosenzweig, this volume). Decreases in species richness, at least in the near term and when considering one trophic level, will not be apparent at the local scale (see Figure 15.2).

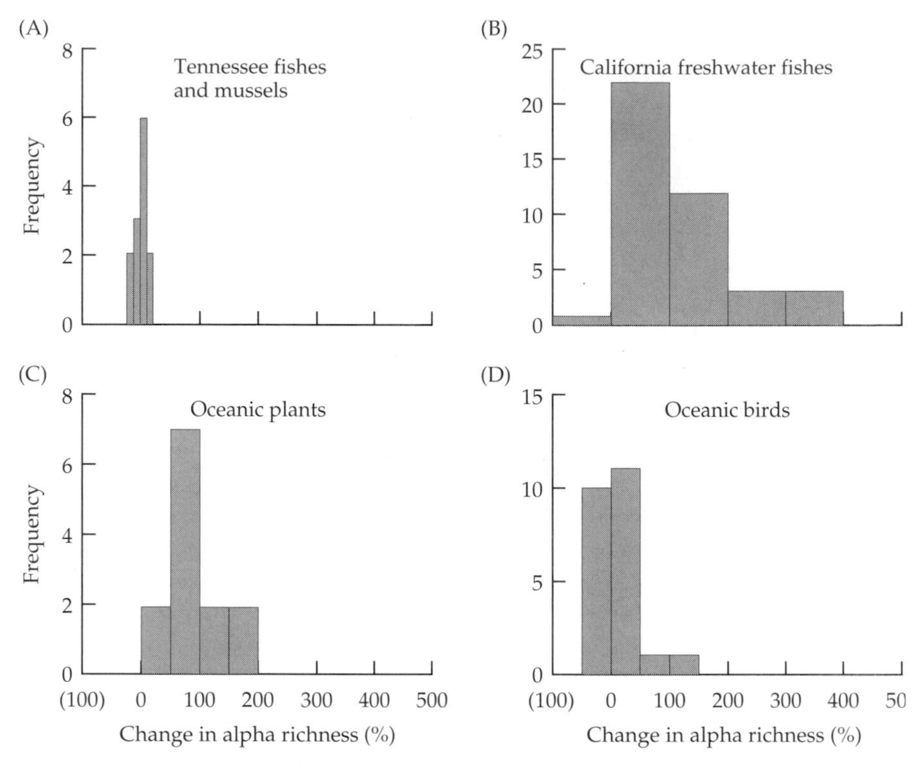

FIGURE 15.2 Percentage change in local (alpha) species richness following non-native species establishment and native species extinction (or extirpation) for (A) freshwater fishes and mussels of Tennessee ecoregions (Duncan and Lockwood 2001), (B) freshwater fishes within California watersheds (Marchetti et al. 2001), (C) oceanic plants, and (D) birds of oceanic islands of the Pacific (Sax et al. 2002). Percentage changes in alpha species richness were calculated by dividing historical (i.e., native) species richness by the net change in richness (i.e., current richness minus historical richness). Thus, a value of 25 on the x-axis indicates that there are now 25% more species at this location than were there historically.

Underlying this apparently anomalous result is a series of secondary questions related to understanding when alpha diversity will tend to increase, when it will remain the same, and when it will decrease slightly. For example, the freshwater fishes and mussels within Tennessee watersheds show the smallest percentage change in local richness, with those changes ranging between a 13% loss and an 18% gain. In contrast, the freshwater fishes of California watersheds show only one drop in local richness (by 18%) and several gains of 200% or more.

A partial explanation for such changes comes from considering barriers to natural dispersal that have left some locations relatively depauperate. For example, the large percentage increases in California fish diversity occur within watersheds that were historically species poor, but that have recently received high numbers of non-native fish introductions. The watersheds of California are fairly isolated from each other and from regional sources of potential natural immigrants (Marchetti et al. 2001). Thus, the introduction (often purposeful) of non-natives has served to overcome this dispersal limitation while most native fishes in these watersheds did not go extinct (although many of these native fish species are currently threatened with extinction).

We see somewhat different patterns for oceanic plants and birds. Birds show little or no increase in local species richness, whereas plants show uniform increases in richness at this scale (see Figure 15.2). Sax et al. (2002) attribute this disparity to differences in dispersal ability between the two groups. The high vagility of birds allows even remote oceanic islands to be regularly visited, and perhaps naturally colonized, by birds native to the region (see also Whittaker, this volume). Sax et al. (2002) suggest that access to regular colonists ensures that oceanic islands are naturally at, or near, their species saturation point. Thus, when human-mediated increases in "dispersal" rates occur for birds, richness will not increase. Any gains in species due to invasions will be met with equivalent numbers of species extinctions. In contrast, vascular plants have very low rates of transoceanic dispersal, and this has apparently limited their natural (or historical) richness on oceanic islands. Thus, when dispersal of vascular plants increases due to human-mediated introduction of non-natives, species richness climbs relatively high.

One explanation for differences in the percentage increase in alpha richness between locations is the timing and magnitude of extinction events. For example, like the watersheds of California, the ecoregions of Tennessee harbor many endemic species as a result of natural limits to dispersal between relatively large river basins (Duncan and Lockwood 2001). However, unlike California, many of the native endemic species (especially the mussels) had already gone extinct well before non-native species arrived, largely as a result of severe habitat alterations such as channeling, over-harvesting, and pollution (Duncan and Lockwood 2001). Thus, despite the substantial number of non-natives introduced into the region, this influx of new species has not been sufficient to swamp the effect of the prior extinction event.

Finally, alpha diversity will respond directly to human biases in terms of which species they transport, and indirectly to the effect of human actions (e.g., urbanization) on dispersal rates of established non-natives (McKinney 2002). Dispersal rates will tend to increase in modern times only for groups of species that attract human attention and are transported frequently as non-natives, or for those who are commensal with humans and therefore "follow" human alterations of environments. In this sense, it is no accident that the taxa depicted in Figure 15.2 are birds, fishes, and plants and not fungi and slime molds. Humans do not purposefully transport and release slime molds, but they do intentionally and frequently transport birds. Even within "attractive" groups like birds and plants, some taxa are much more likely to be transported as non-natives than others (see Lockwood 1999; Mack and Lonsdale 2001). Thus, it is possible that the dramatic increases in numbers of flowering plants on islands is attributable to the relatively high rates at which humans transport and "release" (i.e., plant) flowering plants as ornamentals. Sax et al. (2002) recognized this effect, but to date no studies have been able to tease apart the influence of natural versus human-mediated dispersal rates on observed changes in alpha species richness; it certainly would be a very important subject for future research.

Beta Diversity

Even though alpha diversity may not change or will increase following invasion, species that successfully invade one geographic location may do so in others, thus reducing diversity at broader scales. Rahel (2002) provides a graphical summary of how species invasions and extinctions can combine to change the similarity of two sites (Figure 15.3). Similarity can increase if historically distinct biotas are invaded by a common species, or if historically distinct biotas each lose their unique species but retain the species they

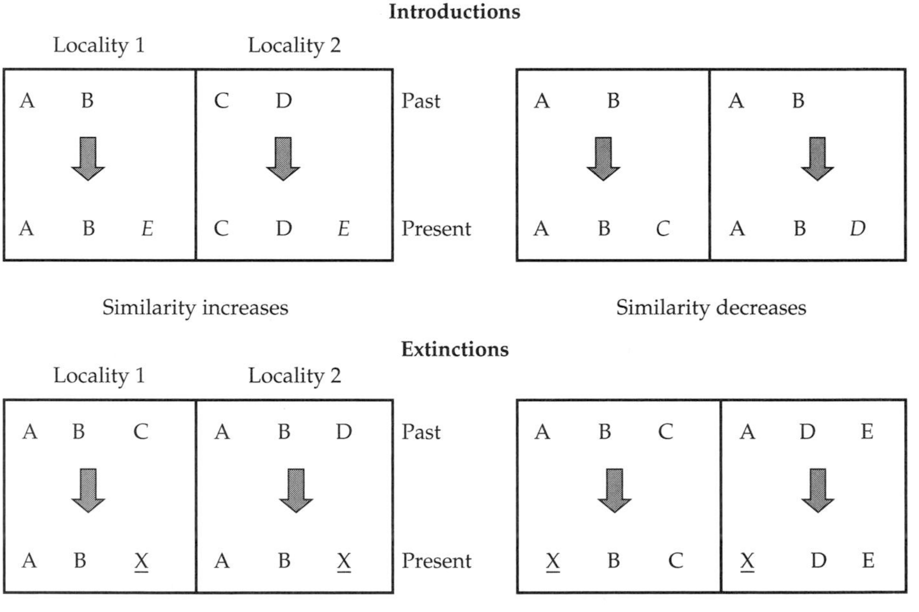

FIGURE 15.3 Possible changes in spatial species turnover (beta diversity) due to establishment of non-native species (shown in italics) and extinction of native species (underlined). Whether two localities increase or decrease in similarity through time depends on the taxonomic identity of the species themselves, and on the geographic distribution of both the natives and non-natives. (Redrawn from Rahel 2002, Figure 2.)

have in common. Similarity can decrease if one of the historically similar biotas gains a unique invader, or if historically distinct biotas lose historically shared species but retain those that are unique to each site. Quite often, species that go extinct are often those with historically narrow geographic distributions. On the other hand, successful invasions tend to occur in many locations, not just one. The combined effect of these increases in dispersal rates is an increase in spatial similarity between formerly distinct biotas (i.e., a reduction in beta diversity).

Empirical research supports these predictions. For example, whereas species richness of the seven Tennessee ecoregions decreased, similarity among these regions increased, especially for the historically most distinct regions (Duncan and Lockwood 2001). Similarity of freshwater fish assemblages among six ecoregions of California also appeared to increase, but similarity among watersheds (within regions) either decreased or remained unchanged after invasions. That is, although local diversity increased within California watersheds, similarity across watersheds largely remained the same or decreased. By comparing similarity scores between rivers before and after they were dammed, Marchetti et al. (2001) also found that dams increase the similarity among freshwater faunas.

Through a series of simulations, Olden and Poff (2003) found that historical richness of native species may have little effect on changes in similarity, although at very low levels of richness, assemblages tend to become more similar through time. The simulated addition of widespread invaders with no subsequent extinctions, localized extinction with no invasion, localized extinction and the addition of widespread invaders, and historically low levels of similarity all lead to increased similarities through time. Differentiation (i.e., decreases in similarity) of these simulated assemblages occurred when non-native species established at local sites but caused no extinctions, when widespread species became extinct, and in situations where there was a historically high level of similarity between sites.

Predicting the Future

Given recent moves toward increased international trade and tourism, the flow of non-native species is likely to continue increasing. What might we expect to happen to biological diversity if barriers to natural dispersal continue to break down? There are currently two non-mutually exclusive theories that can help predict a future with increased biotic mixing. These are island biogeography theory and the unified neutral theory (see chapters by Whittaker and by Turner and Hawkins, this volume).

MacArthur and Wilson's (1967) equilibrium theory of island biogeography theory envisions local species richness, or alpha diversity, to be an equilibrium function of location-specific immigration and extinction rates (MacArthur and Wilson 1967). We can model increases in transport rates of

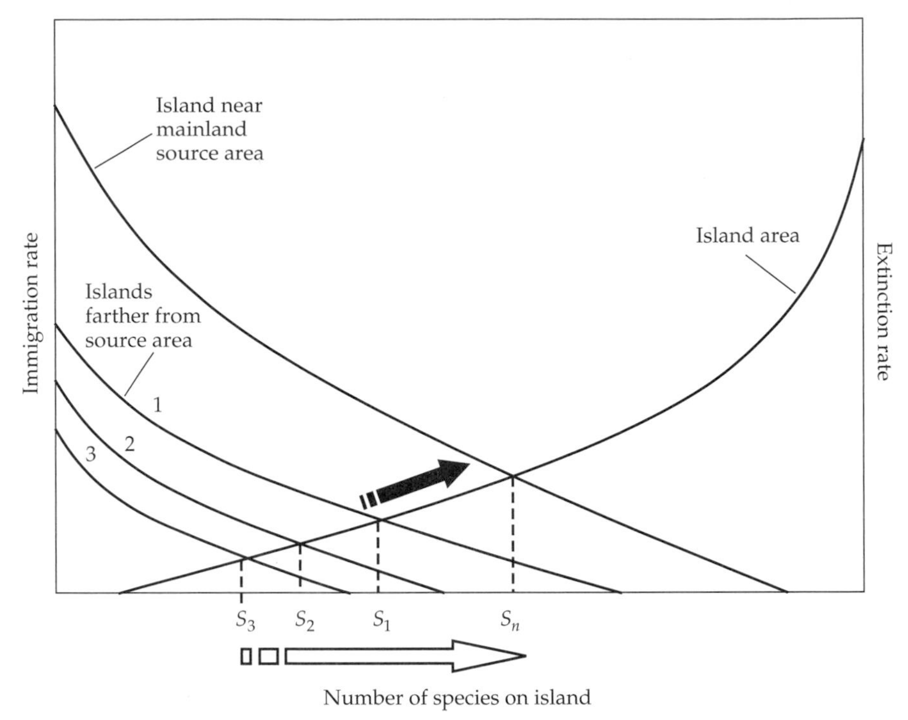

FIGURE 15.4 Standard depiction of the theory of island biogeography, in which immigration rates are considered a function of island isolation from a mainland source species pool and island area. Human-derived increases in dispersal to islands (or other local communities) can be envisioned as a decrease in the island's isolation from the mainland. When isolation is reduced, island species richness (*S*) increases (black arrow). Furthermore, the more isolated the island formerly was, the higher the percentage gain in species it can realize (open arrow).

non-native species as equivalent to increasing immigration rates, and thus the theory predicts that—all else being equal, and if extinctions do not ensue—species richness will increase (Figure 15.4). In addition, those local communities that were formerly very isolated from sources of immigrants will stand to gain the most species (also see chapters by Rosenzweig, Vermeij, and Whittaker, this volume).

What is not clear from using MacArthur and Wilson's equilibrium theory is how we should envision the loss of species from islands. We know that native species often become extinct either through the direct negative effect of invaders, or as a result of habitat alterations that preceded the establishment of invaders. It is also apparent that locations that either did not change or declined in species richness suffered several native species extinctions.

Island biogeography theory shows promise in predicting changes in alpha diversity in a future with increased biotic mixing, but it does not handle changes in beta diversity well; however, Hubbell's (2001) unified neutral

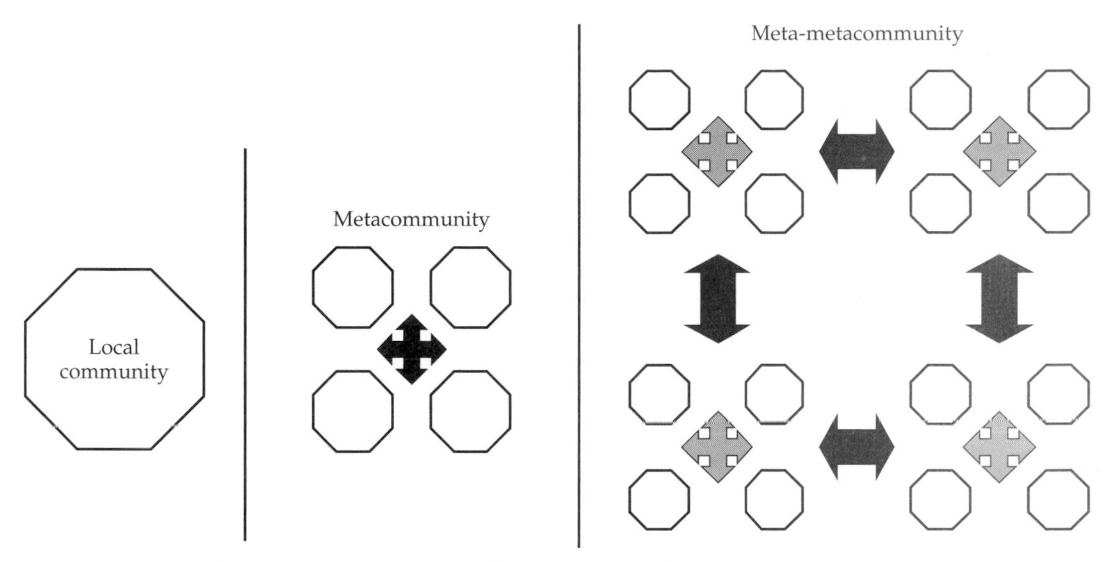

FIGURE 15.5 A schematic of unified neutral theory with a third tier, the meta-metacommunity, added to represent global dispersal events (i.e., introductions of non-native species) as suggested in Davis (2003). This extension of the neutral theory allows for changes in dispersal dynamics between communities (gray arrows) and metacommunities (black arrows) to be incorporated, and for subsequent changes in alpha and beta diversity to be predicted simultaneously.

theory of biodiversity and biogeography does (see Turner and Hawkins, this volume). In this two-tiered model of communities embedded within metacommunities, changes in individual dispersal rates among communities can be shown to affect biodiversity patterns within and among islands (Figure 15.5). In general, we should expect high rates of dispersal among local communities to increase diversity at the local level while reducing diversity at the metacommunity level. This happens because widespread and abundant species drive some local and rare species to extinction. On the other hand, we should expect infrequent dispersal of new species into local communities to allow many rare and endemic species to persist, thus driving metacommunity diversity up (Davis 2003).

Davis (2003) added a third tier to this system—the "meta-metacommunity"—in order to account for global dispersal events, which in today's world is represented by the transport and release of non-native species among regions. In this extension of the unified neutral theory, the metacommunities can be treated as local communities within the meta-metacommunity, and invasions are treated as dispersal events between metacommunities (see Figure 15.5). This extension allows for dispersal to be explicitly modeled at two spatial scales: between communities, and between metacommunities.

Davis (2003) goes on to describe the four diversity patterns expected after divvying up dispersal events across the two scales. When dispersal is limited at both scales, local and metacommunities exhibit high endemism and, as a result, richness of local communities is low, but that of metacommunities is high. If dispersal is unlimited between local communities and kept limited between metacommunities, local richness goes up. However, endemism remains relatively high at the metacommunity level, thus keeping richness high. In the reverse scenario where dispersal is high between metacommunites and low between communities, endemism is low but richness is high at both spatial scales. Finally, in the situation where there are no limits to dispersal at any scale, we see the lowest levels of endemism and the highest levels of richness at both scales.

The advantages of using the extended unified theory to predict changes in diversity due to increased biotic mixing are two-fold. First, dispersal dynamics can be accurately placed within the two spatial scales at which human actions tend to alter natural dispersal patterns of the species. Second, the effect of these changes can be understood at all spatial scales simultaneously. Taking a more detailed look at Figure 15.1, it becomes clear that there are two spatial scales represented. The broadest scale is that of transporting and releasing species outside of their native range, and the smaller scale is that of non-native species establishing and spreading within their new range. This matches well with the idea of dispersal between metacommunities (transport and release) and between communities (establishment and spread). In this sense, the work reviewed above represents partial empirical confirmation of the patterns predicted by Davis (2003). As dispersal becomes less limited at either scale, local richness tends to increase but endemism—the biogeographic trait that produces high beta diversity values—tends to decrease. Further, this theory predicts intermediate scenarios whereby dispersal dynamics can produce differentiation between localities. This flexibility allows us to place empirical examples within a range of possible diversity changes, and then predict future changes by assuming a further erosion of dispersal barriers at both spatial scales.

The potential of the extended unified neutral theory to help explain the variety of alterations in diversity patterns seems large. However, there is much room for both modeling and empirical investigation. In reality, dispersal dynamics are more complex than simply describing them as either limited or not. In the context of understanding changes to diversity due to biological invasions, models of dispersal dynamics must reflect the complexity of both the global transport of species and the local influence of human actions on range expansion following establishment. This requires that we develop a more comprehensive understanding of global transport of non-native species (e.g., Carlton and Gellar 1991), and that we link human actions to patterns in range expansions of non-native species (e.g., Johnson et al. 2001). As the case studies presented above suggest, this is bound to be a complex task, since dispersal dynamics of different taxonomic groups (e.g., plants and birds) are likely to be altered by human actions in distinct ways.

Frontiers of Invasions Research

Biological invasions have implications that range from local ecological and economic harm to the biosecurity of nations (Simberloff 1996). In terms of biological conservation, most studies have concentrated on the occasionally profound impact non-native species have on ecosystem processes, species richness, and genetic integrity of native species (Mack et al. 2000). However, we have thus far largely ignored what the presence of non-native species means to the biogeographical distribution of the world's diversity.

The results described above indicate that non-native species tend to maintain or increase local diversity, but often decrease regional diversity through homogenization. The ongoing and anticipated changes in diversity alter our view of the global biodiversity crises. For example, it suggests we redirect our efforts away from understanding the ecosystem effects of local declines in richness towards documenting relationships between species' identity and distribution of ecosystem services (Sax and Gaines 2004). The increase in similarity across formerly distinct biotas is also of concern because it may alter the ability of regional floras and faunas to respond to future environmental changes, such as global warming. Finally, we can view biological invasions as an unprecedented opportunity to understand how dispersal patterns and interspecific interactions affect biogeographic patterns in species distribution, and to test theories such as the unified neutral theory or the equilibrium theory of island biogeography.

Acknowledgments

I would like to thank Michael Marchetti, Jeffrey Duncan, and Dov Sax for use of their data sets. Dov Sax, Mark V. Lomolino, and an anonymous reviewer provided very helpful comments on previous versions of this chapter.

CHAPTER 16

GIS-Based Predictive Biogeography in the Context of Conservation

Víctor Sánchez-Cordero, Mariana Munguía, and A. Townsend Peterson

Introduction

High rates of habitat destruction—especially deforestation—threaten biodiversity at a global scale. Virtually no ecosystem, marine or terrestrial, exists that is not jeopardized by habitat degradation, which itself has accelerated in recent decades. Recent statistics estimate annual rates of deforestation in several megadiverse countries at 5%, particularly in tropical regions where biodiversity hotspots have been located (Mittermeier et al. 1999). Despite controversy regarding whether loss of natural habitats will result in massive species extinctions in the coming decades, it is widely accepted that biodiversity is being lost as a consequence of significant reductions in natural habitats. Unfortunately, loss of species in deforested habitats can occur rapidly where narrowly distributed endemic species are particularly prone to extinction. Game species, which are generally more widely distributed, are also threatened as a result of habitat destruction, overhunting, and/or overexploitation (Reaka-Kudla et al. 1997). As a consequence, numbers of threatened species have increased dramatically worldwide in recent decades (IUCN 2002).

Biologists and conservationists have raised these concerns in several forums; one conclusion derived from this worrying scenario is that an urgent need exists to reverse trends of biodiversity loss. Hence, it is essential to document biodiversity to know what and where to conserve (Reaka-Kudla et al. 1997). Here we address these issues by (1) stressing the relevance of museum collections as the primary source of information for inferring the distribution of biodiversity in a region; (2) reviewing traditional and new approaches for modeling species' distributions (both potential and projected) using data from museum specimens, electronic maps of ecological parameters, and GIS; (3) using these distributional hypotheses as bases for identifying priority areas for conservation; and (4) posing current challenges and needs for future research on these topics.

Modeling Species' Distributions Using Museum Specimen Data

Museum bio-collections store massive amounts of information on biodiversity worldwide (Soberón 1999). Museum voucher specimens contain primary information on species and their geographic occurrences for documenting biodiversity. This information can be compiled into databases containing species records and georeferenced collecting localities for rapid access. Ideally, we may integrate databases from many different institutions to complete the documentation of biodiversity of particular groups in a given region. Fortunately, this step is being facilitated by the increasing number of institutions providing access to information from their collections, which have been integrated into distributed information facilities over the Internet (Soberón 1999; e.g., see http://elib.cs.berkeley.edu/manis/). Moreover, these efforts can be coordinated with governmental efforts to document national biodiversity; notable examples are the National Commission for the Study and Use of Biodiversity (CONABIO) in Mexico, and the Institute for Biodiversity (INBio) in Costa Rica (both governmental agencies in megadiverse countries). Additional and continuing biological inventories in specific regions are indispensable to complement existing information in museum bio-collections (Peterson et al. 1998).

Nonetheless, biodiversity documentation from museum collections is in many ways incomplete and biased. Despite the fact that naturalists and biologists have conducted inventories and developed collections for more than two centuries, several types of biases are near-universal, and certain biological groups have been collected more extensively and intensively than others (Peterson et al. 1998; Sánchez-Cordero et al. 2001). How useful is museum specimen information for generating accurate understandings of species' distributions? Traditional methods depict species' distributions simply by connecting peripheral localities, tacitly assuming that species are uniformly distributed within this area (e.g., Smith and Smith 1966; Berra 1981; Hall 1981; Howell and Webb 1995). This method clearly defines limits of species' distributions, although it likely overestimates the interior area occupied by a species and may underestimate areas inhabited outside known points. This approach has been used widely as a source of baseline distributional information for identifying biogeographic patterns of species richness and priority areas for conservation in diverse taxa (Brown and Gibson 1993)

The need to generate more accurate distributions to overcome these methodological problems is clear. Recent efforts have addressed these issues via novel techniques for modeling species' ecological niches (*sensu* Grinnell 1917); the results are then projected as potential distributional maps. Quantitative approaches to modeling species' distributions have become increasingly used in biogeography. These models employ uni- and multivariate statistics to correlate biotic and abiotic factors with known occurrences of species using a GIS-based approach, projecting these correlations into distribution maps. Below, we review briefly some of the most common approaches (detailed descriptions are beyond the scope of this contribution, and can be consulted directly in the specialized literature). We emphasize one modeling approach that we have used extensively, however, and we describe its advantages in modeling species' distributions for application to conservation issues.

GAP analysis is a widely used method that involves analyses of the relation of habitat types with known extents of occurrence of species, extrapolating to indicate presence of species in areas where the species has not yet been detected (Scott et al. 1996, 2002). However, a shortcoming is that GAP methodology assumes a direct correlation between species' distributions and habitat types, ignoring that species may respond to broader-scale climatic factors rather than just matching habitat types. Peterson and Kluza (2003) compared GAP with other statistical methodologies for testing predictive efficiency of models (Fielding and Bell 1997), but concluded that GAP models resulted in significant over-reductions in species distributions when compared to other more quantitative methodologies (Fielding and Bell 1997; Peterson et al. 2002a; Peterson and Kluza 2003). Clearly, it is important to consider potential inaccuracies when selecting modeling approaches to predict species' distributions (Peterson et al. 2002a; Peterson and Kluza 2003).

More robust approaches include biotic and abiotic factors in modeling of species' distributions. For example, Austin et al. (1990) used general linear modeling of abiotic factors such as mean annual precipitation, mean annual temperature, solar radiation, and substrate to model *Eucalyptus* distributions in Australia. Fielding and Haworth (1995) used a multivariate statistical approach for modeling distributions of three bird species, relating nesting and specimen records with environmental variables associated to behavioral preferences of these birds. Potential shortcomings to such statistical approaches in modeling species' distributions include (1) the need for large samples of occurrences for species of interest; (2) museum collections data often have resulted from inventories conducted for other purposes, which introduces biases to which statistical approaches are not robust; and (3) assumptions by such methods of a monotonic function relating presence or absence of species to environmental conditions, which are rarely satisfied (Stockwell and Peters 1999).

One of the most prominent ecological niche modeling approaches is BIOCLIM (Nix 1986). This method relates occurrence localities to climatic conditions, and produces a "climatic envelope," which can be projected onto landscapes to identify appropriate conditions for the species. This method has been applied widely to diverse groups, particularly among the Australian biota (Busby 1986; Nix 1986). Similar but more complicated and robust methods such as DOMAIN (Carpenter et al. 1993) and FloraMap (www.floramap-ciat.org) avoid some of the assumptions of the strictly statistical approaches. DOMAIN is a heuristic tool that operates using only presence records and a limited number of biophysical attributes to generate potential species distributions, and it has proven robustness compared with strict statistical models (Carpenter et al. 1993). FloraMap is a correlative modeling tool using climate databases and interpolated climate grids (CIAT) from over 20,000 meteorological stations—mainly in the tropics—for generating probable distribution maps inferred from collecting localities. However, potential problems of these approaches are that they too-frequently overestimate species' ecological niches, as they have some difficulties in dealing with non-climatic variables (e.g., categorical information) (Brown and Gibson 1993).

Ideally, a method for modeling species' distributions would take advantage of the extensive information in museum collections documenting biodiversity, and overcome the shortcomings of the methods described above. A particularly promising approach is the use of genetic algorithms (machine-learning approaches) programmed for combining occurrence data sets with electronic maps of environmental variables. The *Genetic Algorithm for Rule-set Prediction* (GARP, Stockwell and Peters 1999) uses an evolutionary computing approach to develop niche models that can be projected as a potential geographic distribution of a species (Peterson et al. 1999; Stockwell and Peters 1999). Briefly, GARP divides available occurrence points into training and test data sets, then models them in an iterative process of rule selection, evaluation, testing, and incorporation or rejection. Choosing a method from a set of tools (e.g., logistic regression, bioclimatic rules), that method is then applied to the training data, and a rule is developed. Rules may then evolve by a series of perturbations mimicking chromosomal evolution, including point mutation, translocation, and crossing over. Predictive accuracy is evaluated based on the test data (1250 points resampled from the test data, and 1250 points sampled randomly from the study region as a whole). Change in predictive accuracy from one iteration to the next is used to evaluate whether a particular rule is incorporated into or rejected from the model, seeking to select rules that best summarize factors associated with the species' presence; the algorithm runs 1000 iterations or until convergence (Stockwell and Peters 1999; www.lifemapper.org/desktopgarp).

Numerous tests have now indicated excellent prediction success of species' distributions using GARP. These tests have invariably used one data set to build models and another data set to test them. In the best of cases, the test data are drawn from independent field collections designed specifically to test model predictions (e.g., Feria and Peterson 2002); in other cases, tests are achieved by subsetting existing data *a priori* (Peterson and Cohoon 1999; Peterson 2001; Peterson et al. 2002a,b,c; Anderson et al. 2003). In general, GARP appears to be a robust method for predicting species' dis-

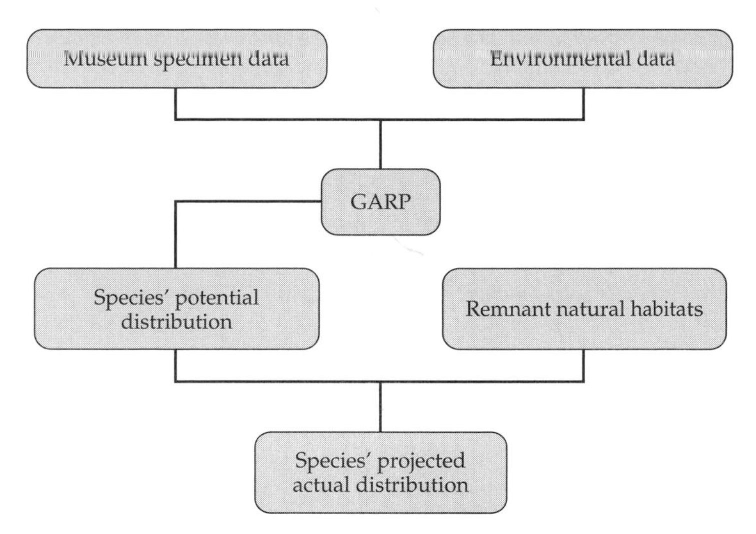

FIGURE 16.1 Flow chart describing the methodology for generating species' potential and projected distributions, based on an ecological niche modeling approach. Primary information is compiled from museum specimen data, including georeferenced species collection localities and electronic maps of environmental variables. The *Genetic Algorithm for Rule-Set Prediction* (GARP) generates an ecological niche of its potential distribution. The species' projected distribution includes all remnant natural habitats within the potential distribution (discerned from current land use/land cover maps). Potential and projected species' distributions can serve as baseline information for biogeographic and conservation studies.

tributions, providing a powerful tool in ecological biogeography (Peterson et al. 1999, 2002b,c; Peterson and Kluza 2003; Illoldi-Rangel et al. in press) (Figure 16.1). Nonetheless, rigorous statistical comparisons between these modeling approaches are still needed to test adequacy and efficiency of species distributional predictions (Peterson and Kluza 2003).

Potential versus Projected Distributions

Ecological niche modeling techniques are used to project predicted distributions onto landscapes to create a potential distribution map. The potential distribution reflects the geographic projection of the species' ecological niche, but not all of this area is necessarily occupied, given effects of other factors external to the model (e.g., land use patterns). As discussed above, high rates of habitat destruction have resulted in significant reduction and fragmentation of natural habitats worldwide, which in turn has reduced species distributions to subsets of their potential areas. How can we evaluate the potential impact of habitat loss on species distributions?

Many scientists hypothesize that reduction of natural habitats negatively affects distributions of native species. We contend that niche modeling of species' distributions can quantify potential impacts of habitat loss on a species-by-species basis. We can quantify reductions of species' distributions by building potential niche models based on climate, topography, and potential (original) vegetation maps, and by projecting model rule sets onto actual land use/land cover maps. The actual distribution of a species is predicted as the areas holding appropriate remnant natural habitats *within* the potential distribution, assuming that deforested areas constitute non-viable ecological conditions (Figure 16.1). These hypotheses can be readily tested by field-validating presence and absence of species in selected fragments of remnant natural habitat versus deforested habitats, respectively (Figure 16.2). Deforested

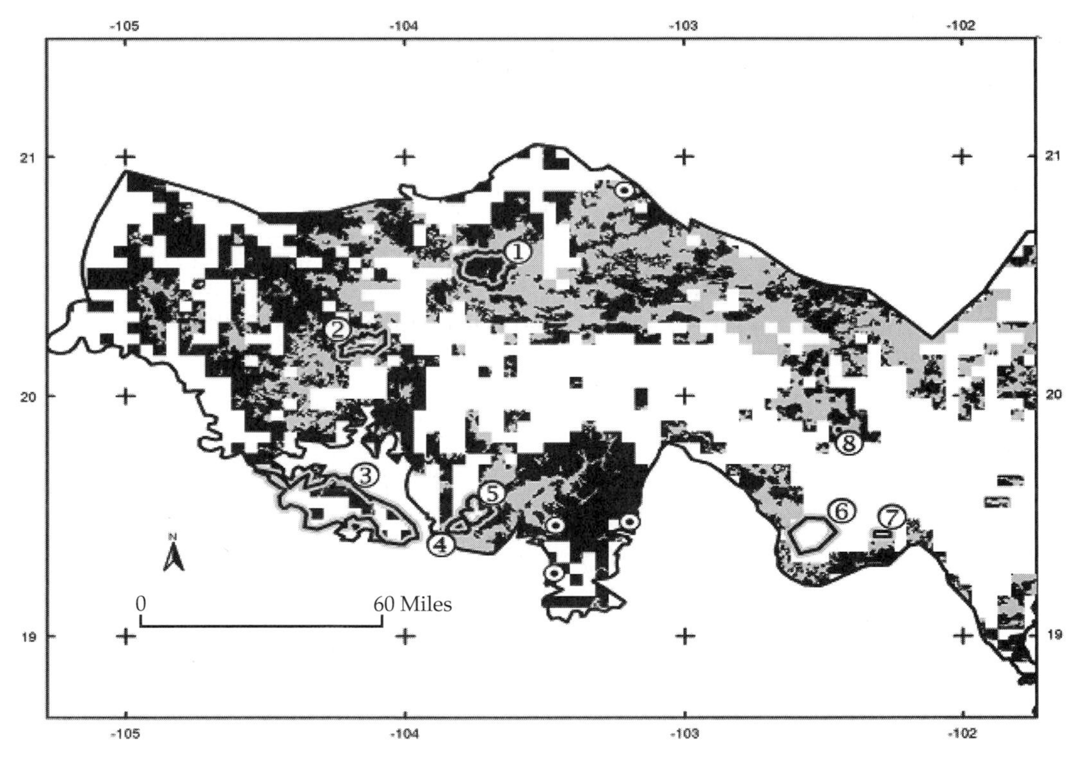

FIGURE 16.2 Ecological niche-based models of potential and projected distributions (black and gray shading, respectively) of *Liomys spectabilis*, an endemic rodent of Central Mexico's Transvolcanic Belt. The potential distribution includes the areas within the species' ecological niche, and the projected distribution includes natural habitats falling within the species' ecological niche, assuming that deforested habitats are non-viable for this species. Points refer to the four georeferenced *L. spectabilis* collecting localities based on museum specimen data. Numbered polygons indicate the decreed Natural Protected Areas (NPAs) in this region. Note that incidence of this species varies among NPAs: *L. spectabilis* is predicted to be present in La Primavera (1), Sierra de Quila (2), Reserva de la Biosfera Sierra de Manantlán (3), and El Jabalí (4); the species is predicted to be absent from Volcán Nevado de Colima (5), Pico de Tancítaro (6), Barranca de Cupatitzio (7), and Lago de Camécuaro (8).

areas are generally devoted to agriculture or urban areas. However, field biologists have noted that many species frequently "use" agricultural areas as feeding sites or refuges, or as routes for dispersal to adjacent habitats (Singleton et al. 2003); yet most native species do not establish reproductive and stable populations in agricultural contexts (Singleton et al. 2003). Prominent exceptions are invasive and pest species, but little is known regarding which species will become pests or invasive once natural habitats are transformed into agricultural fields (Sánchez-Cordero and Martínez-Meyer 2000). Undoubtedly, from a biological and conservation standpoint, natural habitats are the most suitable niches for native species.

GIS-Based Biogeography and Conservation

Our approach of modeling species' distributions based on current land-use patterns provides a testable framework for predicting where species are present or absent, as well as regions in which they have been extirpated due to loss of natural habitat. Potential and projected predictions for species' distributions still need to be field-validated and tested (Peterson et al. 2002a,b). Nonetheless, our models can serve to indicate present-day patterns of distribution under scenarios of habitat degradation, and therefore should provide important insights conserving biological diversity.

Recent approaches to broad-scale biodiversity conservation planning involve efficient selection of *in situ* areas for protection or conservation management (Williams et al. 1996; Sarkar and Margules 2002; Sarkar et al. 2002). Quantitative methods based on robust distributional hypotheses are essential for identification of these priority areas (Egbert et al. 1999; Peterson et al. 2000). We suggest that approaches for modeling species' distributions based on remnant natural habitats can serve as excellent baseline information for addressing such conservation challenges. Such approaches use quantitative ecologically relevant approaches to fill in the gaps in known distributions of species, thus greatly improving the information base. Below, we mention briefly some common methods for prioritizing conservation areas, and we describe some potential applications of our approach to such conservation activities (see also chapters by Rosenzweig, Briggs, and Crame, this volume).

An increasing number of studies have focused on defining priority areas for biodiversity conservation through choosing sites of exceptional species richness, endemism, or complementarity of species representation (Williams et al. 1996; Egbert et al. 1999; Peterson et al. 2000). The most convincing studies for selecting priority areas involve summaries of occurrences of large numbers of species of conservation interest. For example, Williams et al. (1996) selected priority areas for bird conservation in Britain using one of the most comprehensive species distribution database in the world, including more than 170,000 breeding records. They identified areas holding highest species richness (richness hotspots) and endemism (rarity hotspots) as well as sets of areas showing maximal species representation, concluding that all criteria for area selection are required for maximizing bird conservation in Britain (Williams et al. 1996). Further, selection of priority areas for biodiversity conservation require inclusion of as many floristic and faunistic groups as possible, particularly since richness and rarity hotspots do not necessarily coincide geographically between biological groups (Saetersdal et al. 1993; Williams 1998). For example, multi-taxa studies modeling species distributions of endemic birds and mammals from northeast Mexico for selecting priority areas for conservation hold particular promise (Egbert et al. 1999; Peterson et al. 2000). However, as discussed above, information on presence and absence of species for most biological groups is almost always biased and incomplete. Here ecological niche modeling can play a key role, filling gaps in species representation (Egbert et al. 1999; Peterson et al. 2000).

Since many natural protected areas (NPAs) were established based on scenic or political criteria rather than on rigorous analysis for biodiversity protection, we need to determine how much of biodiversity is actually being protected. Once priority areas are identified, we can contrast "optimal" areas with the geographic distribution of NPAs, although rampant habitat destruction can leave little opportunity for expansion of such natural areas. A promising alternative approach is to establish networks of NPAs connected by remnants of natural or secondary habitats to increase species inclusion and permanency (Williams et al. 1996; Egbert et al. 1999; Peterson et al. 2000).

Studies conducted in diverse regions of the world have proposed networks of NPAs for improving biodiversity conservation. Examples include plants and birds in Norway (Saetersdal et al. 1993), plants (Willis et al. 1996) and birds (Fairbanks et al. 2001) in South Africa and China (Godown and Peterson 2003), and terrestrial vertebrates in the United States (Csuti et al. 1997), among other studies. More complex implementations include probabilistic methods to identify reserve networks representing greatest expected numbers of species in Oregon (Polasky et al. 2000). Incorporation of optimization procedures such as flexibility (ability to incorporate crucial information on real conservation problems), efficiency (ability to maximize species richness at the minimum cost), and accountability (solution transparency) have been identified as key attributes for prioritizing natural protected reserve networks (Nicholls and Margules 1993; Pressey et al. 1996; Williams 1998; Rodrigues et al. 2000; Sarkar et al. 2002).

Models of projected distributions of species based on remnant natural habitats within species' ecological niche conditions can be of particular help in proposing networks connecting NPAs. Using models of projected species' distributions, along with the diverse array of methods proposing natural reserve networks, offers an exceptional opportunity to launch conservation strategies in regions and countries with high biodiversity. A major challenge is to overcome existing gaps of knowledge in biodiversity distribution; indeed, new tools based on the vast amounts of information contained in museum bio-collections can generate robust species distribution models as baseline information for biodiversity conservation.

Case Study: Mammal Diversity and Natural Protected Areas in Mexico

To illustrate this suite of ideas, we describe an ongoing study for selecting priority areas and developing an NPA network for mammals of Mexico. We compiled a database of point localities based on museum specimens (see Acknowledgements) as well as recent inventories of mammal species occurring in the Transvolcanic Belt in central Mexico. This region holds an exceptionally species-rich and highly endemic mammalian fauna and includes a high number of decreed natural protected areas (Alcérreca et al. 1989; www.conabio.gob.mx). Basic information on the distribution of this species-rich mammal fauna nevertheless is limited to polygons connecting marginal collecting localities (Hall 1981).

We used GARP to model ecological niches and predict distributions for mammal species in this region. Each locality was georeferenced to the nearest 0.01° of latitude and longitude using 1:250,000 topographic maps (www.inegi.gob.mx). We used digitized maps of environmental variables, including 10 variables that summarize potential vegetation types (www.coanbio.gob.mx): elevation, slope, aspect (from the U.S. Geological Survey's Hydro-1K data set; www.usgs.gov), aspects of climate (from the Intergovermental Panel on Climate Change, http://www.ipcc.ch), mean annual precipitation, mean daily precipitation, maximum daily precipitation, minimum and maximum daily temperature, and mean annual temperature during the period 1960–1990. Ecological niche models were developed on a desktop implementation of GARP (http://www.lifemapper.org/desktopgarp).

Figure 16.2 depicts the potential distribution of *Liomys spectabilis*, a mammal endemic to the Transvolcanic Belt (Hall 1981). This heteromyid is a granivorous rodent known to inhabit only four localities and restricted to tropical deciduous and tropical semideciduous forest at low- to mid-elevations in the state of Jalisco, Mexico (Sánchez-Cordero and Fleming 1993). Its potential geographic distribution as predicted by GARP extends to surrounding areas holding similar regional ecological and environmental conditions (see Figure 16.2). We modeled *L. spectabilis* distribution by projecting the potential distribution model to remnant natural habitats as reflected in the 2000 Agricultural and Land Use Map (Secretario del Medio Ambiente Recursos Naturales y Pesca et al. 2001; www.igeograf.unam.mx) within its potential distribution (see Figure 16.2). Regional deforestation has converted natural habitats into agricultural fields (mostly sugarcane and maize plantations), affecting vast areas of tropical dry and tropical semideciduous forest.

The projected distribution of *L. spectabilis* (gray shading in Figure 16.2) indicates its presence only in scattered patches within the potential distribution and its absence in areas converted to agriculture (areas that are assumed to be non-viable niches for this species). We hypothesized a significant reduction of the *L. spectabilis*' distributional area, with more than a 40% decrease in

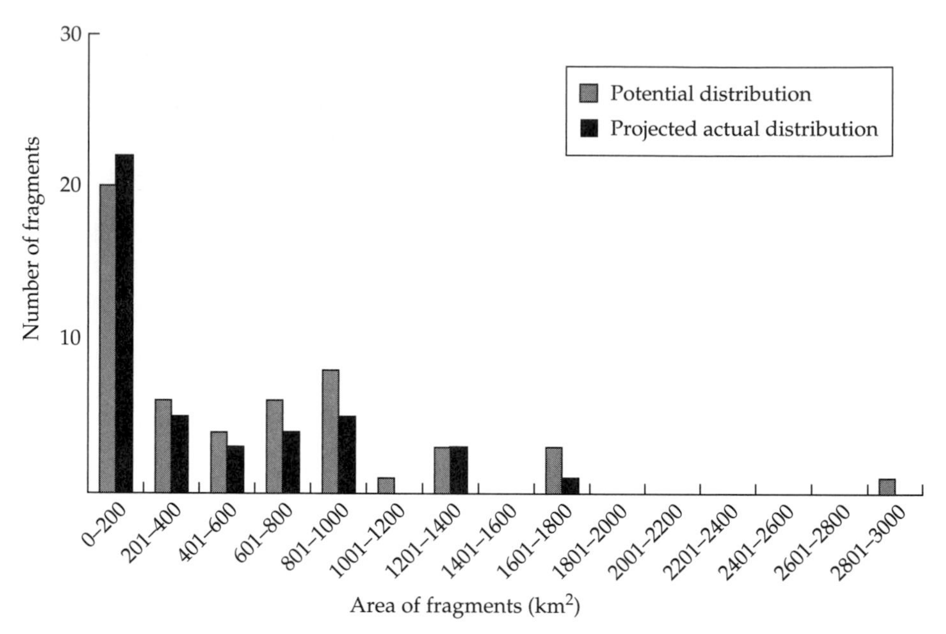

FIGURE 16.3 *L. spectabilis* potential (gray bars) and projected (black bars) distributions in terms of fragmented size, showing a shift in occupancy patterns from relatively large fragments (potential distributions) to smaller fragments (projected distributions) as a consequence of high regional deforestation.

area between potential and projected distributions. Moreover, the projected distribution showed a decreased incidence of this species in larger fragments and an increased incidence in smaller fragments, suggesting increased fragmentation of populations (Figure 16.3). Presumably, deforestation has caused a major impact on the distribution of *L. spectabilis,* and consequently on its conservation status. Our model of the distribution of *L. spectabilis* can be validated by sampling fragments of remnant natural habitats as well as fragments converted to agricultural fields, then comparing the model's predictions and projections to actual presence versus absence, respectively.

Further, we overlapped the projected distribution of *L. spectabilis* with the geographic distribution of NPAs to predict which NPAs might hold populations of this species. *L. spectabilis* was present in La Primavera, Sierra de Quila, Reserva de la Biosfera Sierra de Manantlán, El Jabalí, but was absent from Volcán Nevado de Colima, Pico de Tancítaro, Barranca de Cupatitzio, and Lago de Camécuaro (see Figure 16.2). NPAs potentially holding *L. spectabilis* populations could be interconnected with remnant natural habitats to increase long-term conservation possibilities for the species. Analyses generating potential and projected distributional models for other mammals can be similarly applied to quantify predicted species diversity in the NPAs throughout the region to predict areas of high richness, rarity, and complementarity (Williams et al. 1996; Egbert et al. 1999; Peterson et al. 2000). Finally, by incorporating other faunistic and floristic groups we can conclude with a comprehensive proposal for a natural protected area network for the region. Similar studies can be expanded to other regions and countries since biological and geographic information can be readily obtained from museum collections and digitized environmental variables worldwide.

Frontiers of Predictive Biogeography

High rates of deforestation pose a major challenge for biodiversity conservation worldwide. Museum collections contain massive and essential amounts of information documenting biodiversity, but this resource has been underutilized. However, researchers from universities and government agencies have become increasingly interested in accessing information from scientific collections over the Internet (see for example Manis, http://elib.cs.berkeley.edu/manis/; Red Mundial de Información sobre Biodiversidad, http://www.conabio.gob.mx/remib/; The Speciesanalyst, http://tsadev.speciesanalyst.net/documentation/; and Lifemapper, http://www.lifemapper.org/). It is essential that more academic and governmental institutions become involved in these efforts in order to provide a broader taxonomic and geographic coverage for modeling biodiversity distribution and for conservation of biodiversity.

The rapid development of new techniques integrating museum specimen data and digitized environmental maps in a GIS environment can generate robust niche models and predictions of species' distributions, with great potential for biogeographic and conservation studies. For example, ecological niche models projected as potential distributions can be used to predict species' distributions by projecting remnant natural habitats within the potential distribution, assuming that deforested habitats are not suitable. Such hypothesized distribution can serve as baseline information for underlying patterns of species' distributions and can provide a basis for targeted field inventories. Consequently, field surveys validating predicted distributions should be a research priority in the coming years.

Models of distributions can be incorporated into powerful methods for selecting priority areas for conservation. In addition, analyses of geographic concordance between priority areas for biodiversity conservation and protected areas can lead to development of networks connecting protected areas into more viable systems. One prominent example using a mutitaxa approach for more effective management of natural resources nationwide is the US GAP Analysis Program consisting of federal, state, and regional efforts for mapping species' distributions in the context of habitat associations (Scott et al. 1996, 2002). Multi-national and multi-taxa programs hold excellent promise for developing conservation plans using these approaches even in the most remote and least explored sectors of the Earth's surface.

Further, recent research has raised concerns on the consequences of global climate change on species distributions (Parmesan 1996; Gottfried et al. 1999; Pearson and Dawson 2003). There is increased evidence that climate change is partially responsible for observed range shifts in species' distributions (see Walther et al. 2002 for a review). Theoretically, species respond to environmental changes either by tracking appropriate conditions and moving to adequate ecological niches, or by remaining static and adapting *in situ* to new ecological conditions; if unable to respond to these scenarios, species extinction occurs (Holt 1990). Predicting distributions using ecological niche modeling can thus be particularly useful in exploring potential impacts of climate change on species at local and regional scales. For example, one can predict species' geographic shifts under climate change scenarios assuming a niche-track response, or extinctions due to inability to adequately track environmental conditions or adopt *in situ* (Gottfried et al. 1999; Peterson et al. 1999, 2001, 2002c, and in press; Porter et al. 2000).

Given that species respond preferentially to niche conservatism (Holt 1990; Peterson et al. 1999), distributional predictions based on ecological niche modeling can have useful applications for biodiversity conservation. Recent research modeling mutitaxa distributions under climate change scenarios led to predictions of significant species turnovers in specific regions (Johnston and Schmitz 1997; Gottfried et al. 1999; Porter et al. 2000; Price 2000; Peterson et al. 2002c).

We can also identify some fertile areas for future research. In general, tests of some of the fundamental assumptions of predictive distribution models may provide some important insights for both basic research in biogeography and for its application to conserving biological diversity. We know that during the last glacial recession when habitats shifted and novel ones were created while others disappeared, many species actually shifted their niches and habitat associations (see summary in Brown and Lomolino 1998: 189–200). In addition, ecologists often note that, even within ecological time, habitat associations and realized niches may vary among geographic populations of that species; indeed they often shift, depending on the presence of competitors and predators. To what degree can we assume that the realized niches we have measured are fixed characteristics of a species? Future research should also explore the degree to which distribution patterns are influenced by processes occurring over longer time scales (e.g., dispersal, invasions of exotics, and extinction deficits) and larger spatial scales (see chapters by Lockwood, Heaney, and Whittaker, this volume).

It is clear that current hotspots of endemicity and species richness can suffer geographic shifts in the coming decades. By advancing the frontiers of this research, we may predict new regions of high endemicity and richness under particular climate change scenarios. If so, geographic adjustments in decreed natural protected areas and proposals for new protected areas in selected regions will be critical for biodiversity conservation.

Acknowledgments

We thank Lawrence Heaney for his kind invitation to participate in the inaugural meeting of the *International Biogeography Society*. P. Illoldi-Rangel, M. Linaje, and R. González collaborated in various stages of this work. L. Heaney, P. Illoldi-Rangel, V. Cirelli, and E. Martínez-Meyer critically read drafts of this manuscript and provided valuable comments. We acknowledge the curators of the following mammal collections for providing specimen data: Colección Nacional de Mamíferos (CNMA-IBUNAM); Natural History Museum, University of Kansas (KUNHM); American Museum of Natural History (AMNH); National Museum of Natural History (NMNH); Field Museum of Natural History (FMNH); Museum of Zoology, University of Michigan (UMMZ); The Museum, Michigan State University (MSU); Museum of Vertebrate Zoology, University of California (MVZ); The Museum, Texas Tech University (TTU); and Texas Cooperative Wildlife Collections, Texas A&M University (TAMU-TCWC). This work has been partially funded by CONACyT (Project 35472-V to V. Sánchez-Cordero; Cátedra Patrimonial to A.T. Peterson) and CONABIO (Projects U032 and W036 to V. Sánchez-Cordero).

CHAPTER 17

Applying Species-Area Relationships to the Conservation of Species Diversity

Michael L. Rosenzweig

Introduction

The time has come to ride our horse. Ecology has studied species-area relationships (SPARs) for two centuries. It has collected data at all sorts of spatial scales (e.g., see chapters by Whittaker, by Vermeij, and by Heaney, this volume). It has hammered at them, analyzed them, and debated them. The result is a pretty useful set of biogeographical tools. We need them and should use them wherever it makes sense to do so.

I believe this horse will carry us a long way. It will be especially useful on the road to preserving species from extinction. In this chapter, I will show several examples of such use. One, the SLOSS controversy, has already been brought well in hand (although you may not have thought so before reading this chapter). Another, the impact of exotic species on diversity, is moving along nicely. The other three examples, which I will soon describe, will open the doors to a large amount of research and application.

There are four types of species-area curves (Rosenzweig 1995). The four types correspond to different scales, partly of space, largely of time (Rosenzweig 1999). The scales differ because they involve different combinations of six scale-dependent processes: sampling, dispersal, habitat allocation, immigration, extinction, and speciation. To use these four SPAR types effectively, we must understand each of them.

1. *The point scale*. At this scale, time and space vanish. SPARs result not from biology but from sample-size differences. The larger the space-time sampled, the greater our accumulation of knowledge about an area and a time of fixed heterogeneity.

Each of the other three SPAR types reflects different temporal processes. But because the processes correlate with spatial scale, their SPARs also show up as manifestations of different spatial scales. For ease of presentation, I will invert the scales and speak about the largest scale first.

2. *Interprovincial SPARs*. We see these when we compare the diversities of different evolutionary periods or different biogeographical provinces. These units of space and time generate their species almost entirely from within (by speciation itself). Thus their diversities result from steady states of speciation and global extinction. An interprovincial SPAR represents the macroscale.
3. *Archipelagic SPARs*. We see these when we compare the diversities of different islands in an archipelago. They result from steady states of immigration and extinction on islands. Immigration is faster than speciation (if, as in most archipelagoes, the island is not too far away from a source pool), so it dominates origination rates on archipelagoes. An archipelagic SPAR represents the mesoscale (Holt 1993).
4. *Intraprovincial SPARs*. We see these when we compare the diversities of different-sized pieces of a province. They result from the accumulation of habitat heterogeneity within a region/province as we sample larger and larger fractions of it. Because some of the species in such a sample are represented only by sink populations, maintenance of its diversity requires rapid dispersal from other locations in the province. Intraprovincial SPARs are therefore found at the most rapid scales of time and the smaller scales of space.

To apply them successfully, we must know how these three scales of SPAR fit together. Fortunately, because we know the reasons they differ, we also know a lot about the way they compare to each other.

Imagine an area A at the three scales:

- If A is a piece of a province, it will have the species whose habitats it contains *plus* those species found within it, despite their inability to replace themselves therein. We find such species in A because they disperse often—and regularly—from other areas of the province. In other words, A will have both source species and sink species.

- If A is too remote for regular immigration, it will have *only* source species (i.e., those species with habitat adequate for their intergenerational replacement). Now, if in addition all of its species evolved somewhere else then A, by definition, is a biogeographical island. (Many real islands meet this standard, and many more come close.) However, island A's sample of habitats ought to be similar to provincial piece A. So, without the sink species, island A will have fewer species than provincial piece A.
- Finally, if A is a separate province its species will have originated from within. A moment's pause will convince you that province A must have fewer species than island A because, at any specific diversity, the rate of immigration of species to the island will be much higher than the rate of speciation within the province. Hence the province will not be able to maintain as high a steady state as the island. Notice that province A may meet the geographer's definition of an island and still be a biogeographical province. For example, Heaney (this volume) explains the evidence that the Philippines has more than one such province as regards its mammals (at least).

The rest of the story comes from data conventionally and usefully displayed and analyzed in logarithmic space. At times, other axis types are useful (e.g., when working to reduce the biases of sample size, as I mention below). Intraprovincial SPARs usually have log-transformed slopes in the range 0.1 to 0.2; archipelagic SPARs usually have slopes in the range 0.25 to 0.35 (extending as high as 0.58); and interprovincial SPARs have slopes near 1.0 (often about 0.9 but sometimes as low as 0.6). These slopes are the storied z-values of the biogeographer.

Known z-values are empirical; nevertheless theory is now emerging that makes sense of them (Leitner and Rosenzweig 1997; Allen and White 2003; McGill and Collins 2003). The theory suggests SPARs of different scales should not be straight lines; indeed, data on archipelagic SPARs strongly support that theory (Lomolino 2000c; Lomolino and Weiser 2001), generally exhibiting sigmoidal patterns over large ranges of island area. The theory predicts sigmoidality for intraprovincial SPARs, too, although here the data seem not to agree, suggesting instead that such SPARs are convex-upward on their left sides. But elsewhere I have maintained, and I continue to believe, that the convex-upwardness is a sampling artifact (Rosenzweig 1995). Intraprovincial SPARs free of such artifacts should begin at a left-hand asymptote of point diversity (representing the set of species maintained by resource heterogeneity) and rise—at first only very slowly—as enough area is included to incorporate some habitat heterogeneity as well (Rosenzweig 1999).

Straight or not, as we have seen above archipelagic SPARs must be bounded by the intraprovincial SPAR of their province and the interprovincial SPAR on which their province lies. Indulging in the simplification of lin-

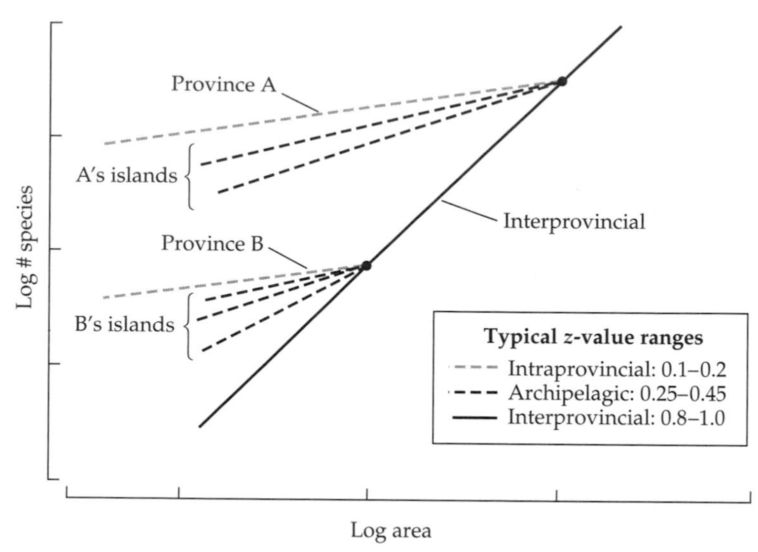

FIGURE 17.1 An idealized set of species-area curves.

earity for the time being results in an idealized picture of a set of SPARs that resembles a set of fans (see Figure 17.1). The simplification to linearity does not weaken the five applications that I now discuss.

SPARs and the SLOSS Controversy

Wilson and Willis (1975) invented the SLOSS controversy: SLOSS asks whether it is better to invest in a *S*ingle *L*arge reserve *O*r *S*everal *S*mall ones (when faced with the choice). Lomolino (1994) provides perhaps the best key to the literature of SLOSS.

SLOSS arose from considering archipelagic SPARs, but has been extended to all scales in a somewhat careless way. For example, in the section following this one, we will see that SLOSS can make no difference at the interprovincial scale. One huge continent will have approximately the same diversity as the sum of several independent smaller ones.

Theoreticians realized that the key to answering SLOSS at smaller scales is being able to predict the overlap of species as one traverses space (Simberloff and Abele 1976; Higgs and Usher 1980). Unfortunately, no theoretician has solved this problem, and I have never been entirely comfortable with the literature's best empirical answers to SLOSS.

Rather ingeniously, Quinn and Harrison (1988) try to solve SLOSS empirically using the species lists of 30 real archipelagoes. Within each archipelago they add islands in two directions—smallest-to-largest and largest-to-smallest—resulting in two accumulation curves per archipelago. (Because of the accumulation feature, these SPARs are unlike traditional archipelagic SPARs, each point of which is a single island.) Quinn and Harrison plot the SPARs in arithmetic space and compare them based on their slopes. Over the left-hand side of the SPAR graph, the rate at which small islands accumulate diversity as their areas and species lists are accumulated exceeds the rate at which large islands do (in 29 of 30 cases). Hence Quinn and Harrison suggest that the SLOSS controversy is settled in favor of several small islands.

My unwillingness to be satisfied with their answer comes from two sources. First, Lomolino (1994) performed a similar test on ten archipelagoes (overlapping those of Quinn and Harrison), and reached the conclusion that the two tactics (small-to-large and large-to-small) achieved the same benefits. Second, the Quinn and Harrison method of analysis has a fatal bias, as follows.

Each Quinn and Harrison SPAR begins at the origin (0,0). It then diverges to allow for the two tactics. For the small-to-large tactic, it next plots the area and diversity of the smallest island. The slope of the line connecting this point to the origin is the archipelago's initial small-to-large slope. But this line comes from the extreme left-hand side of the usual archipelagic SPAR, so it is very steep because there is great convex-upward curvature in the arithmetic plot of such a SPAR (z is about 0.3). In contrast, the large-to-small tactic plots the area and diversity of the largest island, and connects it to the origin. The slope of the connecting line is the archipelago's initial large-to-small slope. But this line comes from connecting the extreme right-hand side of the usual archipelagic SPAR to the origin. Hence it is very shallow (again, because of the convex-upward curvature in the arithmetic plot of an archipelagic SPAR).

Starting at the origin is all the bias required; it explains the Quinn-Harrison result. The only way to get a result in favor of the large-to-small tactic would be to analyze a set of archipelagoes whose z-values exceed unity! Thus the Quinn-Harrison result merely re-states the observation that archipelagic SPARs have fractional z-values.

Lomolino's (1994) analysis does not fall prey to this bias. Like Quinn and Harrison, Lomolino accumulates islands with the two alternative tactics. But his criterion for judging the optimal strategy is to include species per unit area at each step. Using this method, Lomolino found no consistent difference between the two tactics. However, he also found that neither tactic was the "best"; *both* tactics were outperformed merely by adding islands in random order. Improved though it may be, the Lomolino analysis does not deal with the reality that islands in an archipelago are portions of a system. Taking them as individuals assumes implicitly that what is present on each island does not depend on what is present on the others. Does that make a difference? In this section, I will answer that question by introducing and applying a method that treats the archipelagoes as whole systems. My method remains an empirical solution but I have tailored it to the archipelagic scale, which was the original scale of the controversy.

Framing the Question

We assume that enough money exists to set aside a reserve of area A km^2. We can choose to spend this money on a number of small reserves, or a single large one. To reduce the problem to its essentials, we further assume no major habitat differences between the several small areas and the single large area. That is to say, we assume that the alternatives encompass the same sets of major biomes.

Now we predict how data would appear if it did not matter whether we save a single large area or several small areas. If SLOSS made no difference, then the number of species in the combined list of species of all islands would equal the number of species in a large virtual island whose area equaled that of the small islands combined.

How can we determine the diversity of this large virtual island? After all, it is a fictional entity. We can extrapolate the diversity of the fictional island from the archipelagic SPAR of its components. To make this prediction graphical, we construct a SPAR for the archipelago; then we extrapolate it to a point over the area of all of the islands combined. The number of species predicted will be the number of species that exactly reflects the effect of area within the set of small islands—this is the system property of the set of islands.

Now suppose that the actual number of species present in the archipelago exceeds the number predicted for the large island. Then several small islands would have done better than one large island. In contrast, if the actual number of species present in the archipelago is less than the number predicted for the large island, then several small islands will have done worse than one large island.

Results

As an example, consider these real data from the plants of twelve Adirondack mountaintops (Riebesell 1982). If there were a single mountaintop equal in area to the sum of the areas of the twelve—and if SLOSS made no difference—the island would have 19.8 species (Figure 17.2). If the actual number of species of plants on the archipelago exceeds 19.8, the test would favor several small; if the actual number is less, the test would favor single large.

The Adirondack mountaintops actually hold 18 species, so subdividing the land into several islands reduced the overall diversity by 1.8 species. A single large mountaintop would probably have done 10% better than the several small ones. But that is not always the case.

My students and I compiled the data of 37 taxon-archipelago combinations. (As pointed out by Quinn and Harrison [1988], these turn out to be harder to find than one might expect; the list is archived at http://www.-evolutionary-ecology.com/data/islandrefs.html). To qualify for our analysis, data had to meet the following criteria:

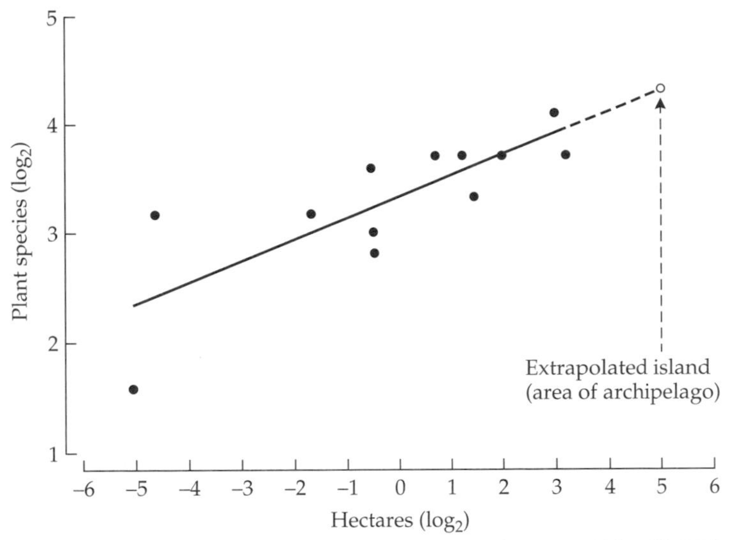

FIGURE 17.2 Creating the prediction of a large area that would be SLOSS-neutral.

- The list of species for each and every island in the test had to be available.
- The *z*-value of the archipelago had to be significant ($p < 0.05$).
- The islands had to be biogeographical islands (i.e., to have obtained most of their species from immigration—not speciation).
- The biota had to have had time to relax to equilibrium. (Thus we avoided the cases of mammals of the southwestern United States sky islands and reptiles of islands in the Sea of Cortez, but used the birds and plants of the latter, as well as the birds of Andean sky islands and the plants of Adirondack mountaintops.)

The outcome of our tests was as close to random as one could imagine. Of the 37 lists that met our criteria, 18 showed a reduced overall diversity; the other 19 showed an increased overall diversity (Figure 17.3). Clearly this result agrees with that of Lomolino (1994) that the effect of fragmenting a reserve is undetectable at this scale.

In a recent magisterial review Harrison and Bruna (1999) have shown that we really lack good evidence to decide the issue of fragmentation. I would agree entirely, and I would add that part of the confusion comes from the likelihood that SLOSS—basically a question of fragmentation—has led us astray. Nevertheless, whether reserves are "single large" or "several small" may well matter at very small scales because of edge effects and population dynamics. And it surely matters at large scales within a continent because a single large reserve would reduce the variety of habitats we set aside. Imagine, for example, that only a single chunk of 5% of the contiguous United States remained as a reserve. No matter where we put it, it will miss most of the nation's habi-

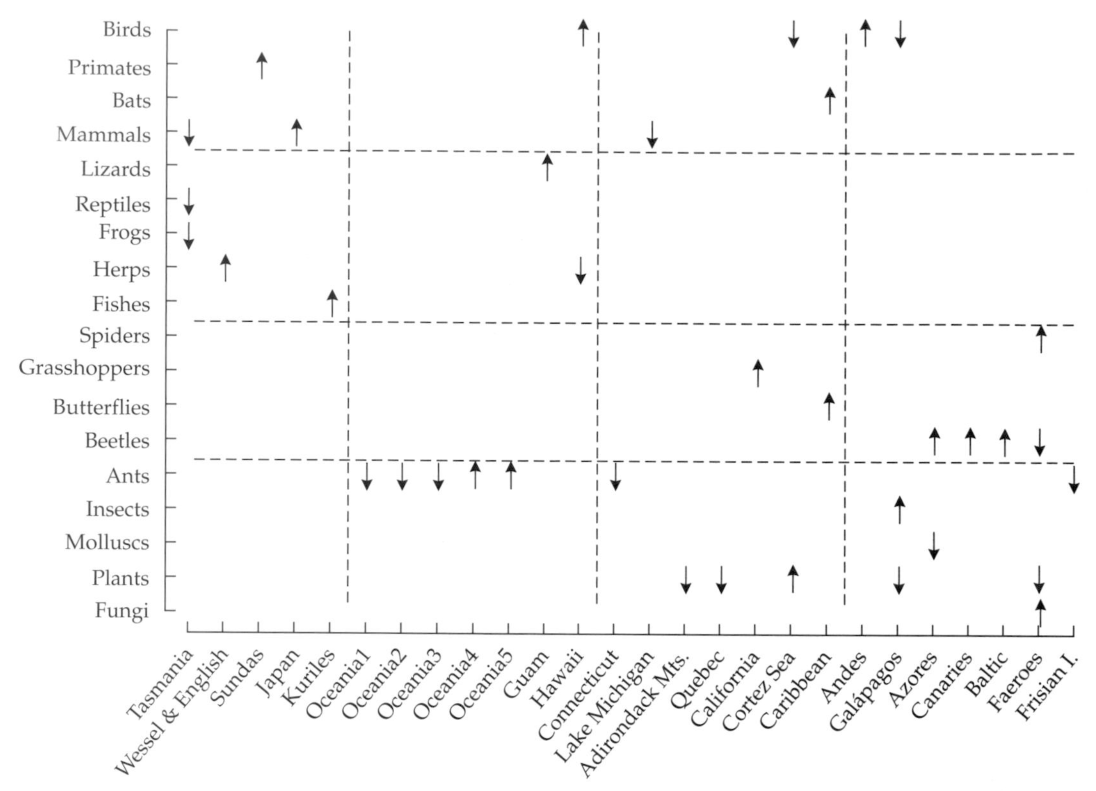

FIGURE 17.3 Thirty-seven tests of SLOSS. Arrows that point up indicate archipelagoes with more species than a single large island would have had. Arrows pointing down represent archipelagoes that had fewer species. Nineteen of 37 arrows point up.

tats. A 5% area is about the size of the state of Montana, and only a very small amount larger than the state of California. What a biotic disaster it would be if, out of the entire lower 48 states, only one of those two states remained.

However, the evidence shows that, at the medium spatial scales for which it was conceived, SLOSS does not seem to matter at all. And it also should not matter at the largest scale—interprovincial—as we shall now see.

SPARs and the Impact of Exotic Species

Some exotic species do considerable ecological damage (Vitousek et al. 1997; Mooney and Cleland 2001). This unassailable fact has led to the reification that exotic species are evil, and that they always and everywhere pose a threat to diversity (Slobodkin 2001). But that is a leap unsupported by science (Sagoff 1999). What can SPARs teach us about the true nature of the threat *vis-à-vis* overall diversity?

The ultimate condition imagined by biogeographers is termed the "New Pangaea" (thanks to Hal Mooney), or the "Homogocene" (thanks to Gordon Orians). Dispersal rates of species among provinces are so high that their identity as separate provinces is disappearing (see chapter by Lockwood, this volume). Restricting ourselves to the Earth's land for purposes of illustration, imagine a single province, composed of all continents, whose area would be the sum of theirs.

Because they are close to linear in arithmetic space, interprovincial SPARs predict that the New Pangaea will experience little or no decline in global steady-state diversities (Rosenzweig 2001). For instance, if they are precisely linear then the number of species in New Pangaea will be the sum of the numbers of species in the old provinces—no gain, no loss. Therefore any reductions to global steady-state diversities caused strictly by exotic species will not last. That is not to say they will not happen—they already have, but the reductions will be temporary.

SPARs also predict that exotics will actually increase average local diversities at least in the long term (Rosenzweig 2001). Our idealized portrait of SPARs at various scales shows this quite clearly. The SPAR that characterizes the New Pangaea will lie above, and be parallel to, even the richest of the old provinces. So any piece of it will also be higher (Figure 17.4). Indeed, Sax and his colleagues (2002) have shown that such increases are occurring already.

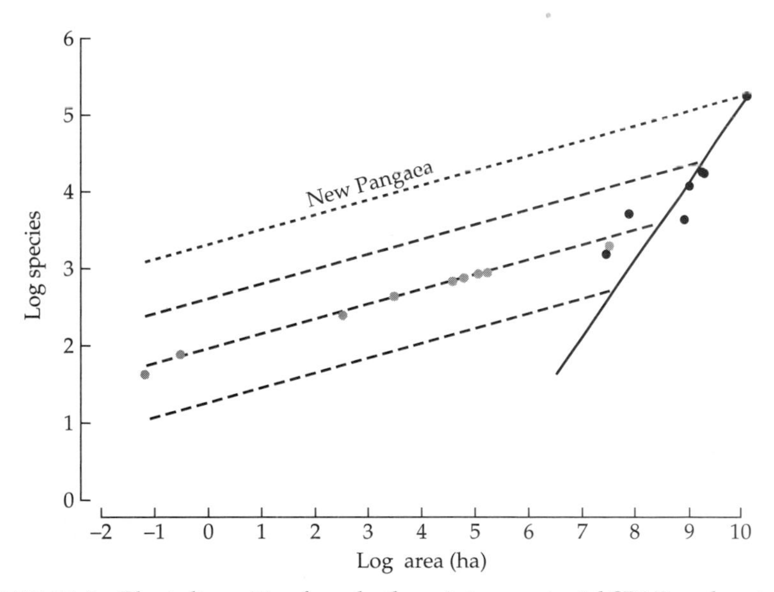

FIGURE 17.4 Plant diversities from both an intraprovincial SPAR and an interprovincial SPAR help to anchor the framework of SPARs that predict diversity in various-sized provinces and portions of them. Because SPARs within different provinces are close to parallel, and because New Pangaea would have the greatest overall area, its pieces would have higher diversity than those of smaller provinces.

SPARs and Indices of Environmental Status

Perhaps because concerns about our environment have grown, environmental indicators have bred like rabbits. In the mid-1990s, the U.S. Environmental Protection Agency (EPA) commissioned the National Research Council to see whether it was possible to develop a very short list of indicators that could be applied and compared broadly across the nation. The idea was not to say that no other indicators might be useful in restricted circumstances. Instead it was to call for a meaningful, practical, uniform set to allow the country to assess its overall environmental state—to give itself a sort of environmental report card.

Chaired by Gordon Orians and staffed by David Policansky, the NRC panel worked for several years and arrived at a short list of indicators, in which SPARs figure prominently (National Research Council 2000). The NRC proposed that one powerful summary of the land's environmental status ought to be the diversity of wild species it supports. Hence the NRC wished to compare the species' densities of the various land uses in the United States to standardized, natural species' densities.

To say that two areas have the same species densities should surely mean that the two areas have diversities that would be equal except for the fact that their areas differ. However, the number of species rises curvilinearly with area. Thus, the density of species cannot be computed merely by simple division (S/A, where S is the number of species and A the area). To arrive at meaningful densities, one needs to use the power laws that fit the curvilinear properties of SPARs. This is readily done. Let density be C in the power equation:

$$S = CA^z \qquad \text{Equation 1}$$

Thus,

$$C = S/A^z \qquad \text{Equation 2}$$

Recall that the curvature—or z-value—of the power equation is roughly constant. So if two different places (i = 1 or 2) in the same ecoregion have equal values of the ratio S_i/A_i^z, then their diversities fall on the same species-area curve. Any difference between them would be due entirely to their different areas. So C equals the density of species.

Of course, this will be strictly true only if the SPAR on which the Si-values lie is perfectly straight (so that z is a constant). Fortunately, experience with SPARs shows that, in logarithmic space, they are usually difficult to distinguish from straight lines. Moreover, even if the SPAR is not perfectly straight, and even if we can only obtain a crude estimate of it, an approximate value of z will improve comparisons immensely compared to the rather meaningless densities that arise if we simply use S/A.

Because it varies very little from taxon to taxon and from habitat to habitat, the curvature (z) does not offer much help in evaluating a place. But densities

(C-values) vary greatly and can be helpful. The larger C, the greater the underlying diversity of a taxon or place. Comparing C's to a standard C will give us a basis for the indicator we need. In this way, species-area relationships can provide a yardstick against which we can evaluate the condition of diversity at scales ranging from small parcels, to entire states, to the nation at large.

I propose measuring the SPARs, and thus the standardized C-values, of a few popular taxa in natural remnants of each of North America's ecoregions as mapped by the World Wildlife Fund (Ricketts et al. 1999). (Popular taxa will help generate the volunteer labor that will be required.) Next, we would measure C in an area of farm (or forest or suburbia) in one of the ecoregions. If C of the farm is less than that in nature, the farm is artificially depauperate. If C of the farm is more than that in nature, the farm has an unsustainably high species density. Thus we take the absolute value of the difference between the C's as our measure of quality. We divide this value by the natural C to get a proportion. (It will be a proportion except in the most unlikely event that a land use has more than twice the species density of its natural ecoregion.) Low proportions signify a close match to nature, so we subtract the proportion from unity to make an indicator that grows with environmental quality.

In sum, the total diversity indicator for land use i of ecoregion j is

$$M_i = 1 - \left\{ \frac{|C_{i,j} - C_{n,j}|}{C_{n,j}} \right\} \qquad \text{Equation 3}$$

where $C_{n,j}$ is the natural species density of the ecoregion.

One easily produces the national indicator from the set of M_i-values. First each is multiplied by the proportion of land it covers. Then the resulting products are summed. Again, the indicator will be a proportion, varying from 0 to 1. The national indicator can be interpreted as the national environmental "footprint" and can be monitored for signs of change.

Finally, anthropogenic habitats may bring about the loss of many native species and the burgeoning of commensals, especially exotics. Thus some people will wish to measure species densities without including any exotic species. We can use the same machinery to achieve an indicator without the exotic species. We simply recalculate $C_{i,j}$ after excluding the exotics. The result is a measure of the proportion of native species expected at a site (of area A) to those actually found there.

SPARs and Reducing the Bias of Sample Size

Before environmental indicators will do much good in practice, we must reduce the amount of time and money it now takes to estimate diversity. In fact, several statistical methods now allow fairly accurate estimation of diversity with an order-of-magnitude reduction in the time and money it would otherwise take.

Two of the three most promising methods depend on extrapolating a certain sort of SPAR to its asymptote. We generate these SPARs by accumulating individuals in areas of fixed heterogeneity. "Fixed heterogeneity" does not mean a single uniform habitat, but rather a set of habitats that does not change as we increase our sample size. Thus we can postulate the presence of a fixed number of species whose list will be complete after we spend an infinite amount of time collecting all the individuals present. That number is the asymptote. I call these "point-scale SPARs" because their increasing diversity comes entirely from their increasing sample sizes and not at all from an increase in habitat heterogeneity; thus it is as if one were accumulating the individuals at a single point entirely lacking spatio-temporal heterogeneity.

To use extrapolation methods, we must work with plots of data that are arithmetic rather than logarithmic. The y-axis is S_{obs} (number of species in the sample); the x-axis is N (number of individuals in the sample). As we add to N, we keep track of S_{obs} on the plot. Then, using non-linear regression, we fit the coefficients of some general formula to the species accumulation curve. The formula should rise with a negative second derivative toward an asymptote. The asymptote is the estimate of how many species are really in the system. The negative second derivative is required because all real data sets rise that way. These formulae are not theories but merely templates like the general linear equation, $y = mx + b$.

The Michaelis-Menten equation was the first extrapolation formula tried by an ecologist (Holdridge et al. 1971):

$$S_{obs} = S\frac{N}{N+a} \qquad \text{Equation 4}$$

where a is the coefficient of curvature and S is the asymptote (i.e., the true number of species in the system).

But the Michaelis-Menten equation fails to go through the point (1,1); instead it traverses the point $(1,\{S/[1 + a]\})$. This is a defect because all samples of one individual will contain one species. One family of formulae that does go through the point (1,1), that may rise with a declining slope, and that does converge on a positive asymptote is

$$S_{obs} = S^{1-N^{-f(N)}} \qquad \text{Equation 5}$$

where $f(N)$ is any positive, unbounded, monotonically increasing function of N. As N rises toward infinity, Equation 5 converges on S—the true diversity of the system.

We have worked with many such functions of N but two seem most promising. Substituting them into Equation 5 produces the extrapolation estimators, Formula 3 and Formula 5. Formula 3 uses

$$f(N) = q \ln N$$

while Formula 5 uses

$$f(N) = qN^q$$

In each case, q is a coefficient of curvature. (Formula numbers have no significance; they are numbered arbitrarily according to the order in which we first used them in an attempt to improve Michaelis-Menten.) Both formulae (and many others) are available in a PC application (Turner et al. 2000).

Here's how all the extrapolation formulae get used. You gradually build up knowledge by repeated sampling of a place. As you add each new sample, you perform a non-linear regression using one of the templates to extrapolate the curve to its asymptote. Figure 17.5 shows one example that employs Formula 5 at three stages: 3, 6, and 15 samples. The raw data were assembled by Gary Poore from the muddy benthos of Port Philip Bay, Australia. The data set has 303 species and 93,173 individuals. We have set the options of the program so that its job is to estimate the number of species in the total collection—303—rather than the number in nature. That way, we know the correct answer and can evaluate the method.

We have tried the formulae on a variety of large data sets and found that extrapolation methods work well. For consistency, I will again show a par-

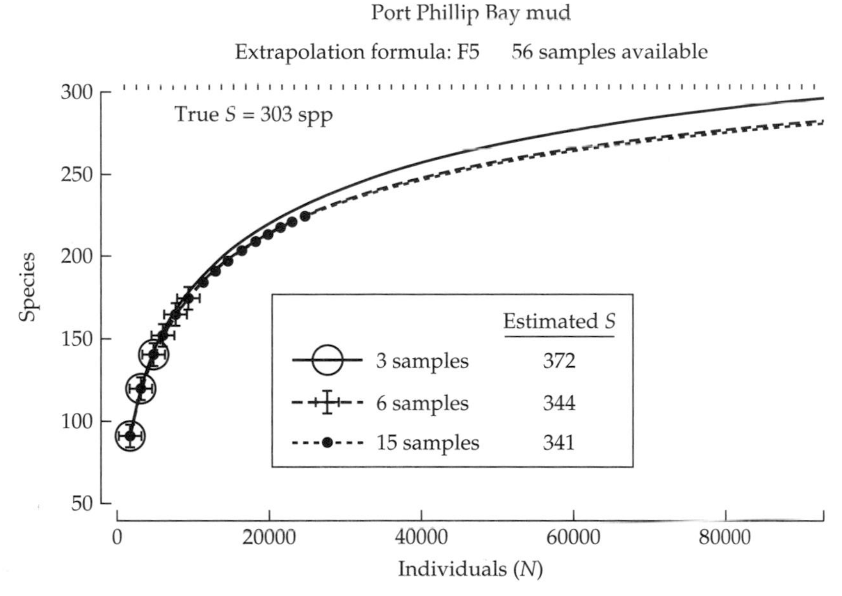

FIGURE 17.5 Non-linear regression using a template to extrapolate the curve to its asymptote. The solid line is fit to only the first three sample points and extrapolated to obtain the estimate. The other two sample sizes (6 and 15) also have individual fits, but are too close for the eye to distinguish at this scale.

ticular example from the muddy benthos data set. I sampled it repeatedly and produced average numbers of individuals and of species for each sample number. Diversity-in-hand rises monotonically with a decelerating slope. In other words, the results become less and less biased as sample size increases (Figure 17.6). That is precisely the point-scale SPAR.

Michaelis-Menten fits this SPAR very well (R^2 = 0.9706) and Formula 5 does even better (R^2 = 0.9997). But Formula 5 does well even when sample sizes are far below 93,000. It makes a reasonable estimate of diversity with only a few samples (see Figure 17.6B). And, as each new sample is added, its estimate of diversity changes only a little.

You can use these methods even if you have only presence/absence data from each sample, and know nothing about abundances. You can even use them when you have yet to sample all habitats of an area (in which case the increases in diversity are coming from a combination of increasing sample size and real increase in diversity). To try this, Rosenzweig et al. (2003) imagined a lepidopterist newly arrived in North America who wanted to know how many species of butterfly live in the United States and Canada. By listing the species in statistically dispersed sets of locales, the explorer could make a very good estimate (< 10% error) with as few as 11 of the continent's 110 ecoregions. The median error was 2.4%.

No estimation method is perfect, of course. But that includes the traditional one of amassing species lists over long periods. The extra accuracy we get from such traditional estimates must be viewed against their shortcomings. For example, we may well wish to know when diversities are in a state of flux and how they are changing. But while we are painstakingly building a traditional estimate, we will not be able to detect the changes. More rapid estimation methods minimize this extra error. Rapid estimation methods might also help to reveal patterns of diversity that have eluded us so far because they exist at very fine spatial scales.

SPARs and Reconciliation Ecology

Biogeographers and conservation biologists often use species-area relationships to evaluate the future of species diversity. Biotic reserves are viewed as islands in a sea of artificial, sterile habitats, and the patterns and theories of island biogeography are applied. The idea is that island biogeography should tell us what proportion of diversity these reserves ought to save (e.g., Brooks et al. 1997). A creative idea, but it is far too optimistic.

Today's reserves are not islands because they are usually all that remain to support so much of life's diversity. Remember, islands by definition get most of their species as immigrants from a mainland source pool. If a species now dependent upon our reserves were to become extinct, it could not return from a source pool—only speciation will be able to add species.

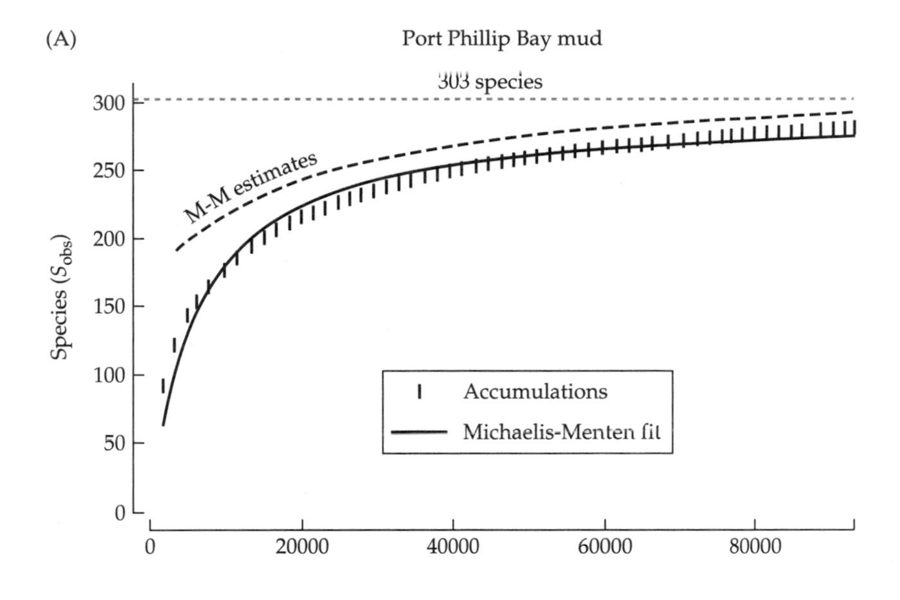

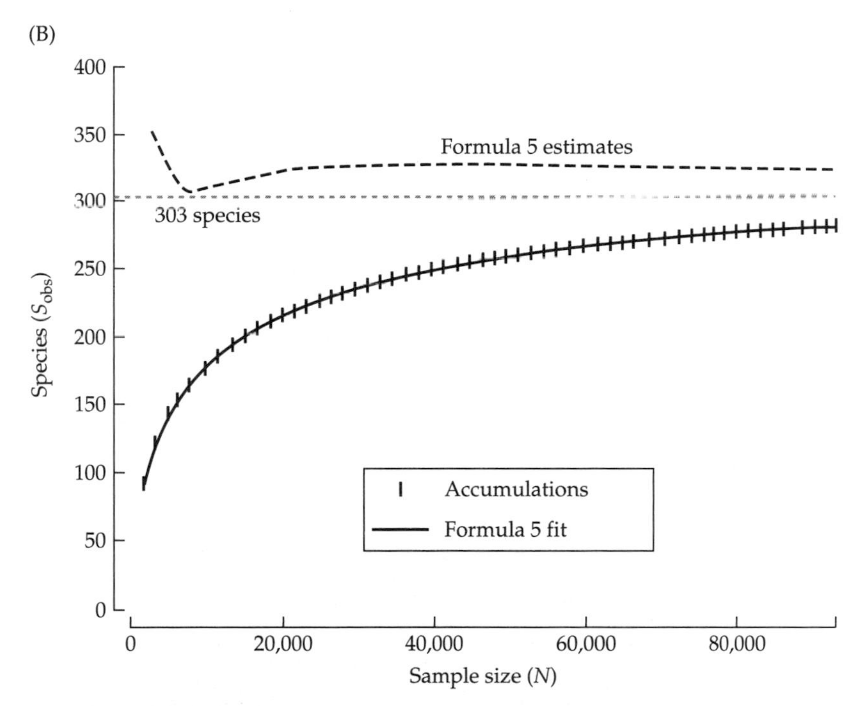

FIGURE 17.6 Estimating diversity with two extrapolation formulae. (A) The Michaelis-Menten formula. This shows the typical negative bias of most estimators when they are given small sample sizes to work with. (B) Formula 5. When given small sample sizes, this method has a much smaller error and often lies close to the horizontal, as in this case.

Please note that I am speaking of *biogeographical islands* here. Island-like provinces such as those revealed by Heaney's work in the Philippines (this volume) have not gotten their diversity from a flow of immigration events. They must support their diversities on their own, but their recognition as provinces means they should fool no one.

So today's system of reserves is more like a set of provinces—albeit small and fragmented—than a set of archipelagoes. That is a crucial distinction. Because of the pronounced curvature of archipelagic SPARs (recall z-values are typically about 0.3), islands do not have to be very large to contain a high percentage of a source pool's species. For example, a single island with only 1% of the area of a source pool can rather reliably harbor about 32% of its species. Given this power of small islands, one can easily imagine that a set of island reserves could conserve most of the species of the pool. However, they cannot conserve as many species if reserves are actually provinces.

Interprovincial SPARs have little or no curvature. Hence a province reduced to 1% of its former area can maintain only 1% of the old diversity; similarly, a province reduced to 5%, can maintain only 5% of the old diversity, etc. Our province-like system of reserves has lost the economies of island biogeography. These economies emerge out of all those things that make an island an island: Islands do not regulate the size of their own species pool; the fineness with which an island's species differentiate among habitats is controlled by co-evolution on their mainland; and islands are too isolated for frequent immigration, yet not isolated enough to have to depend on internal speciation for additions to their species lists. For all these reasons taken together, island SPARs simply decelerate very sharply.

Biogeography's response to this quandary is crystal clear. If saving a small proportion of surface area for diversity will not help very much, then a larger proportion should be acquired. But that would seem to clash with another biological reality (i.e., the geographical dominance of *Homo sapiens*); I won't belabor the point.

The previous paragraph conceals a tacit and false assumption—that land use is a zero-sum game, that the only way to make land available for wild species is to exclude human beings and their works. Reconciliation ecology comes to say—**and to show**—that this assumption is patently false. We can indeed get more land for other species without putting it into reserves. Reconciliation ecology is the science of inventing, establishing, and maintaining new habitats to conserve species diversity in places where people live, work, or play (Rosenzweig 2003a).

We already know that reconciliation ecology works—there are many, many examples. McNeely and Scherr (2002) describe dozens of examples dealing with tropical agriculture, and Rosenzweig (2003b) describes a diverse collection of examples from around the world. Here are a few from the United States.

The pocket parks of Mayor Richard Daley's Chicago convert abandoned gas stations into islands of nature for inner city residents. Prairie Dunes

Country Club's golf course in Hutchinson, Kansas hosts some 35,000 rounds of golf a year, but also manages its roughs to support a large number of wild species (Terman 1997). Florida Power and Light's Turkey Point nuclear- and coal-powered electricity generating station near Miami has 80 miles of cooling canals—canals that the utility company has made into crucial managed habitat for the rare American crocodile (Gaby et al. 1985). And 20,000+ homeowners cooperate with the National Wildlife Federation to build and manage Backyard Habitats™ on their own property.

So reconciliation ecology is far from new. In fact, some examples are centuries old. What is new is the understanding that if civilization wishes to preserve a significant fraction of its diversity, then the science of interprovincial SPARs demands that we practice reconciliation on a grand scale. Thus biogeography through SPARs gives new urgency to reconciliation ecology.

The science required to make reconciliation ecology commonplace may not often be easy to do. No scientific history convinces one of the difficulties better than the reconciliation ecology of natterjack toads (*Bufo calamita*) in England (Denton et al. 1997). It took some 50 scientists about a quarter of a century to bring back this single species in this single place. This amount of effort probably represents an extreme, but we do not yet know.

The Frontiers of Applied SPARs

As a group, the five SPAR applications hint at considerable opportunity for the ecological biogeographer. Some of the applications are incomplete, some suggest further basic research, and some call for further applied research.

Being able to reduce the biases of sampling, promises value in several ways. We need to do so to get more reliable data for our indicators. As the biases disappear from our analyses, we will be able to acquire data at very fine scales. We can use these data to search for rapid changes in diversity and patterns of standing diversity at fine scales of space and time. I for one began worrying about these biases when I tried to estimate point diversities—as space and time vanish, the inaccuracies introduced by sampling biases become intolerable. They overwhelm the signal and lead to such nonsense as negative diversity estimates.

Although we have a plethora of statistical tools for bias reduction, we are just learning how to use them. The valuable methods of Lee and Chao (1994) seem to work well within a fixed set of habitats, but they are not designed to handle data sets in which habitat diversity and sample size increase together (Rosenzweig et al. 2003). Indeed, they break down under those conditions.

But why do we see variability in the accuracy of extrapolation methods? How good are they? Can we develop biologically meaningful estimates of confidence for them? (Those that now exist are confidence limits about the mean estimate, which includes the error of bias—not confidence limits about the true diversity.) Can we develop rules to teach us when to stop sampling?

Why does Formula 5, with its single parameter of curvature, seem to outperform other formulae with two? Does the design flaw in the Michaelis-Menten extrapolator become unimportant when we use it on presence/absence data?

Those would seem to be a daunting list of questions, but there are more. Fortunately, it should not take more than a few years to answer them all. The software for experimenting with the methods exists and makes the research practical (Turner et al. 2000). Meanwhile, the methods already surpass the use of raw diversity data. We should use them (albeit cautiously)—right away!

Biogeographers should also get involved in the work to collect and analyze data for the indicators. In addition to those having to do with SPAR, others on the NRC short list require the biogeographer's touch. Of considerable importance is a vector of land use practices. Actually, it is more like a matrix because each type of land use (e.g., farming) has an orthogonal vector of treatments. All of these land use vectors need to be measured using aerial photography and satellite imagery; then they need to be evaluated by looking at their effects on life. This work—like painting the Golden Gate Bridge—will never end so we might as well get started.

In contrast, I believe the biogeographer's work on exotic species is well in hand. SPARs give us the theory and tell us what to expect of steady states (Rosenzweig 2001), and data are showing us the course of the transitions (e.g., Sax 2002; Sax et al. 2002). What remains is to discover why exotics have sometimes enhanced local diversities and sometimes not. Are such differences likely to be permanent or merely transitional? (Theory predicts the latter.) I do find it ironic that the very concept of an exotic species is rooted in biogeography (see Lockwood, this volume), whereas work on the problems associated with exotics (and they certainly exist) virtually ignored the field of biogeography for a long time.

As far as I can tell, the SLOSS controversy is over. SLOSS does not matter. Practically speaking, I am not sure it ever mattered much. Its value was conceptual. It planted the idea that biogeography might be used for conservation. Wilson and Willis (1975: 525) focused on loss of diversity as they set up the rest of their argument; they would apply "the quantitative theory of island biogeography" to "the problems of diversity maintenance." An inestimable breakthrough, really—as we see now.

That said, why weren't the methods of biogeography quickly used to resolve the question? In this chapter, I have done nothing about SLOSS that could not have been accomplished in 1977 (albeit with fewer data sets). I believe the answer is simple. Because theory produced the question, we resorted to theory for its answer (e.g., Diamond and May 1976; Simberloff and Abele 1976; Higgs and Usher 1980; Simberloff and Abele 1982; Goodman 1987). Instead, we ought to have focused on the data, as did Quinn and Harrison (1988) and Lomolino (1994).

In contrast to SLOSS, reconciliation ecology has hardly begun, and it overflows with challenge and opportunity. Reconciliation ecology is nothing less than the deliberate manipulation of species ranges. One begins by finding out what they are and what makes them so. An immense amount of research lies ahead as we accumulate a library of the habitat requirements of myriad species, and learn how to combine them (National Research Council 2001). It will not be a trivial task to design landscapes of novel, diverse habitats that support as many species as possible while also doing their jobs for people. Will the bioinformatics of habitats addle the brains of those who are meeting the challenges of genetic informatics? And if that does not do it, what happens to them when we admit that natural selection will alter those habitat requirements simply because the organisms will face new environments? I do not know that there has ever been a more rigorous or exciting problem. Certainly, there has not been a more consequential one.

The path that leads to reconciled human habitats also calls for new attitudes and new public and private institutions. But spawned and spurred on by biogeography, reconciliation ecology addresses the new, sterile habitats in which most species cannot function at all. If this new emphasis of conservation biology spreads and influences a substantial proportion of the Earth's area, it can halt the current mass extinction.

Who would have believed that a biogeographical pattern would lead to a vision of the Earth's diversity in the far future? Indeed, who would have thought that it would offer choices to civilization while allowing people to decide how much species diversity to protect? Yet, these promises are contingent upon the next generation of biogeographers. Will they pass the time debating the existence of the toolbox? Or will they advance and perfect the set of tools it contains? They must not shrug their shoulders in hopelessness at the magnitude of the job or the callousness of the human psyche—giving up would be a needless, terrible tragedy. The horse may well need more training, but it can get us there.

Acknowledgments

Gary Poore kindly allowed me to analyze his hard-won data. My graduate class in species diversity of 2000 participated in the hunt for and analysis of SLOSS-relevant data. Particular mention goes to Jessica Brownson, Zachary R. Buchan, John Cox, Daniel L. Ginter, Martin Lingnau, Kevin Russell, and Will Turner for their efforts. Mark V. Lomolino and Larry Heaney provided valuable feedback on the first draft of this chapter.

CHAPTER 18

Conservation Biogeography in Oceanic Archipelagoes

Lawrence R. Heaney

Introduction

The successful conservation of biological diversity will require a wide variety of efforts under highly varied circumstances around the globe. Inspection of conservation-related journals and textbooks will show that most attention has been focused on continental areas. This is not surprising, given that most human populations live in such places, especially the wealthy nations of the temperate regions; nor is the effort unwarranted, given the very real problems faced on continents. Islands, however, which taken together cover just 3% of the Earth's land surface, harbor a disproportionately high diversity of species relative to their area, largely because they often support endemic species (Whittaker 1998 and this volume; see also Mittermeier et al. 1997, 1999). More crucially, a highly disproportionate number of plant and animal extinctions during the last two centuries were those of island endemics (e.g., more than 80% of the birds [Reid and Miller 1989] and more than 70% of the mammals [McPhee and Fleming 2001]). Clearly, island biotas have been more severely impacted by human activities than have continental biotas, and current listings of threatened species strongly indicate that this trend is continuing (IUCN 2002). Thus, while there is no reason to believe that conservation efforts on continents should be lessened, it is clear that efforts on islands should be increased.

A second aspect of the need for conservation on islands has to do with differences in patterns of endemism on two different types of islands. Continental land-bridge islands typically share the majority of their mammalian biota with the adjacent continent; endemism is generally low (typically 0%–10%), and the numbers of species on such islands threatened with extinction is not notably high (Heaney and Patterson 1986; Whittaker 1998; Mittermeier et al. 1999). In contrast, oceanic islands—those that have had no dry-land connection to a continent during their existence—typically have very high levels of endemism, often in the range of 40%–95% (Wagner and Funk 1995; Whittaker 1998). Thus, the disproportionate concentration of biological diversity on islands is due mostly to endemic species on oceanic islands, not on land-bridge islands. Further, most of the extinction of insular species has taken place on oceanic islands, not on continental land-bridge islands, which points to oceanic islands as the group most in need of conservation research and action.

It is the primary thesis of this chapter that biogeographic analysis that develops predictive conceptual models, coupled with investigative, biogeographically informed field research, is essential for successful protection of biological diversity in the large, geologically complex and biologically diverse (but usually poorly known) oceanic archipelagoes of the world. I demonstrate this approach with a synthesis of our current understanding of the biogeography and biological diversity of terrestrial vertebrates using the mammals of the Philippines as a model system.

Study Area: The Philippines

To people who live on continents, islands may seem exotic and rare. However, although they occupy a relatively small proportion of the Earth, islands are in fact quite numerous. In the Indo-Australian region, Indonesia claims over 11,000 islands, the Philippines over 7,000, and other countries in the area several thousand more, totaling well over 20,000 islands in this region (Figure 18.1).

The Philippine Islands range in size from Luzon, at a bit over 100,000 km^2, to tiny, uninhabited islets of less than a hectare (Figure 18.2). The total land area of the Philippines is a little over 300,000 km^2. Luzon and Mindanao are slightly smaller than Cuba. The other islands together total about 100,000 km^2, but none exceeds 15,000 km^2. The climate is generally wet, with different parts of the country averaging between 2 and 10 m of rain per year, and rain forest of several types once covered nearly the entire archipelago.

In the Philippines, as in most countries, the best-known taxa are vertebrates, and they are usually the only groups for which meaningful comparisons can be made. There are about 174 known species of native land mammals (i.e., species that were not introduced by humans, and not including marine species; Table 18.1). Of these 174 species, an extraordinary 111 (64%) are endemic to the Philippine archipelago. Similar figures exist for breeding land birds, reptiles, and amphibians (Table 18.1); of a total 952 native vertebrate species, 542 (57%) are endemic.

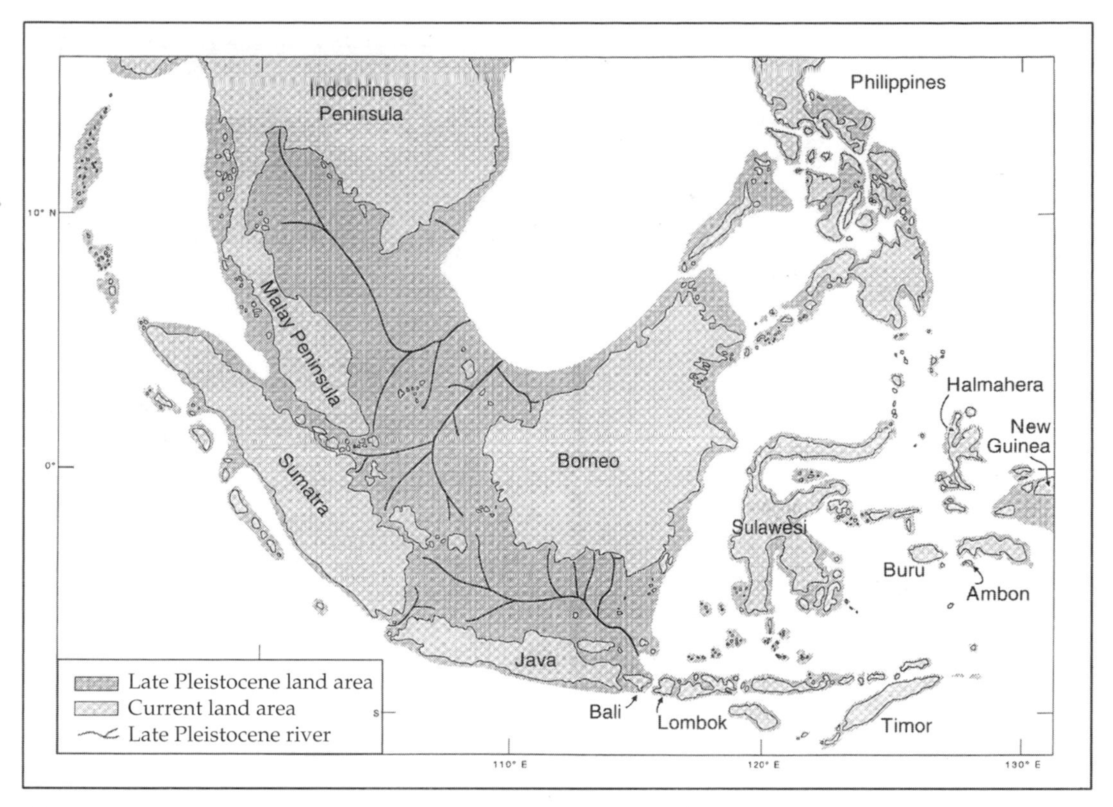

FIGURE 18.1 Indo-Australian region showing continents, and continental and oceanic islands. (Redrawn from Heaney 1991.)

To put these figures into perspective, we can compare them to similar data for Spain, the European country with the highest level of biodiversity for most taxa. Spain has 435 species of native land vertebrates, only 25 of which (6%) are endemic, with most of these endemics living on small oceanic islands offshore of the mainland (Groombridge 1992). This lower biodiversity in Spain occurs on a land area that is about 65% greater than that of the Philippines; clearly, both total biodiversity and (especially) endemism are much greater in the Philippines.

However, Spain is a rather dry, temperate-zone country, whereas the Philippines are positioned in the wet tropics. A second, similar comparison to Brazil, one of the great global centers of biological diversity, shows Brazil to have about 3131 land vertebrate species (more than 3 times the Philippine total) and 788 endemic species (about 45% more than the Philippines). But Brazil is *28 times larger* than the Philippines. Clearly, the concentration of both total diversity and endemic species is much higher in the Philippines than in Brazil.

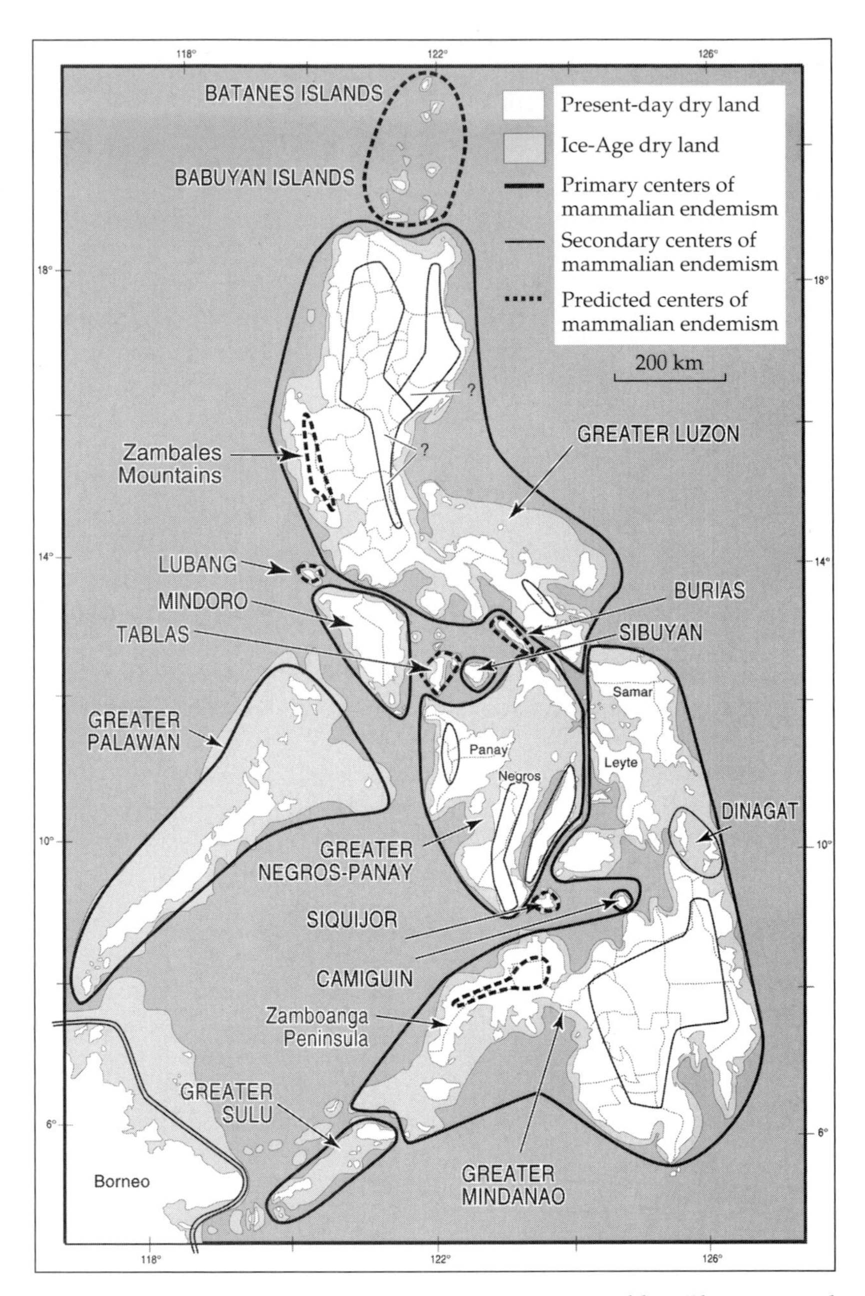

FIGURE 18.2 Map of the Philippines showing current and late Pleistocene islands. Centers of mammalian endemism correspond to deep-water islands that were exposed during Pleistocene periods of low sea level and to isolated mountain ranges on the larger islands. Known centers are shown in solid lines; currently unknown but predicted centers are shown with dotted lines. (Modified from Heaney et al. 1998, Heaney and Mallari 2002, and Ong et al. 2002.)

TABLE 18.1 *Land vertebrate species of the Philippines compared with Spain and Brazil*

	No. species	No. endemic	% Endemic	Land area
Philippines				
Land mammals	174	111	64	
Breeding land birds	395	172	44	
Reptiles	258	168	65	
Amphibians	125	91	73	
Total	952	542	57	300,780 km^2
Spain (total)	435	25	6	451,171 km^2
Brazil (total)	3131	788	25	8,511,965 km^2

Source: From Heaney and Regalado 1998 and Heaney 2002. Current data on Philippine amphibians from R. Brown and A. Diesmos (pers. comm.).

Conservation Status of the Philippines

The Philippine Islands are notable in another, but negative, way. The 1996 Red Book listing of threatened and endangered mammals issued by the IUCN included 49 endemic Philippine species that are in danger of extinction (Baille and Groombridge 1996). This placed the Philippines eighth among nations worldwide for total threatened species. However, all of the countries with greater numbers of threatened mammals are much larger; when area is considered, the Philippines has the greatest concentration of threatened mammals of any country in the world.

The reasons for this concentration of threatened species are complex, but the basic issues are clear. The Philippine Islands have, for the most part, rich volcanic soil. During the Spanish colonial period, most of the fertile lowlands were cleared for development of large plantations that produced rice, hemp, tobacco, sugar, and other commodities for world markets. This clearing of large areas also produced a dramatic upsurge in human population due to various socio-economic factors that promoted large families among impoverished plantation workers. During the American colonial period, clearing for agricultural purposes intensified, and commercial logging of forests for dipterocarps, known as "Philippine mahogany," took place on a large scale. These trends continued after Philippine independence, with the result that mature forest cover declined from about 90% in 1600 to about 65% in 1898, to about 8% today (another 12%–15% is now covered in second-growth forest; Kummer 1992; Environmental Science for Social Change 1999). The geographic progression of deforestation remains highly non-random, clearing first the relatively broad and flat lowlands and then pushing to more isolated, higher, and steeper areas (Rickart et al. 1993; Heaney et al. 1999; see also Lomolino and Perault 2004).

Given this dismal state of affairs, the need to quickly and effectively protect the biodiversity of the Philippines has been recognized by the national government, international aid agencies, national and international conservation organizations, and by the Philippine public at large (e.g., Vitug 1993; Wildlife Conservation Society of the Philippines 1997; Ong et al. 2002). Some of the effort has focused on charismatic large species such as the Philippine eagle and Mindoro dwarf water buffalo, but it has also been widely recognized that there is a great need to protect the biodiversity of the nation as a whole. Thus, much of the focus on conservation has now centered on the question of where to designate and manage national parks that will collectively protect all of the biota of the nation (Ong et al. 2002). This raises a central issue: How is the biodiversity of the Philippines distributed within the archipelago, and how many protected areas (and across what spatial configuration) are needed to conserve that biodiversity? To answer that question, we turn to the mammals of the archipelago, which have been studied intensively for the last two decades.

Mammalian Diversity in the Philippines: An Overview

The mammalian fauna of the Philippines contains a wide range of species, some representatives of which are shown in Figure 18.3 (Heaney et al. 1998). The fauna includes shrews, gymnures, a few species of tree shrews, a variety of squirrels, one of the two members of the mammalian order Dermoptera ("flying lemurs"), a pangolin, a few small carnivores, wild pigs and deer, a dwarf water buffalo, and a few primates. The most diverse groups are the bats, with about 75 species in 6 families (including giant fruit bats with a more than 5-foot wingspan, and tiny flat-headed bats that roost inside hollow bamboo stems). Murid rodents represent at least 55 species.

The Philippine murids are technically considered to be "rats and mice," but many look and behave very little like these familiar rodents. Several species are highly specialized to feed on earthworms; others are giants that feed on tender young leaves in the treetops. Collectively, the 55 species are probably the result of about five separate colonizations from the Asian mainland, followed by extensive diversification within the archipelago; the two largest endemic radiations include at least 26 and 14 species, respectively (Figure 18.4). These species have diversified to an extent that exceeds that of Darwin's finches of the Galápagos Islands, with extensive morphological, ecological, and behavioral differences having arisen among species derived from a single ancestral population that reached the Philippine Islands from the Asian mainland about 10–20 my BP (Heaney and Rickart 1990; S. A. Jansa and L. R. Heaney, unpubl. data).

FIGURE 18.3 Representative bats and small mammals from the Philippines. (From Heaney et al. 1998; drawings by Jodi Sedlock.) ▶

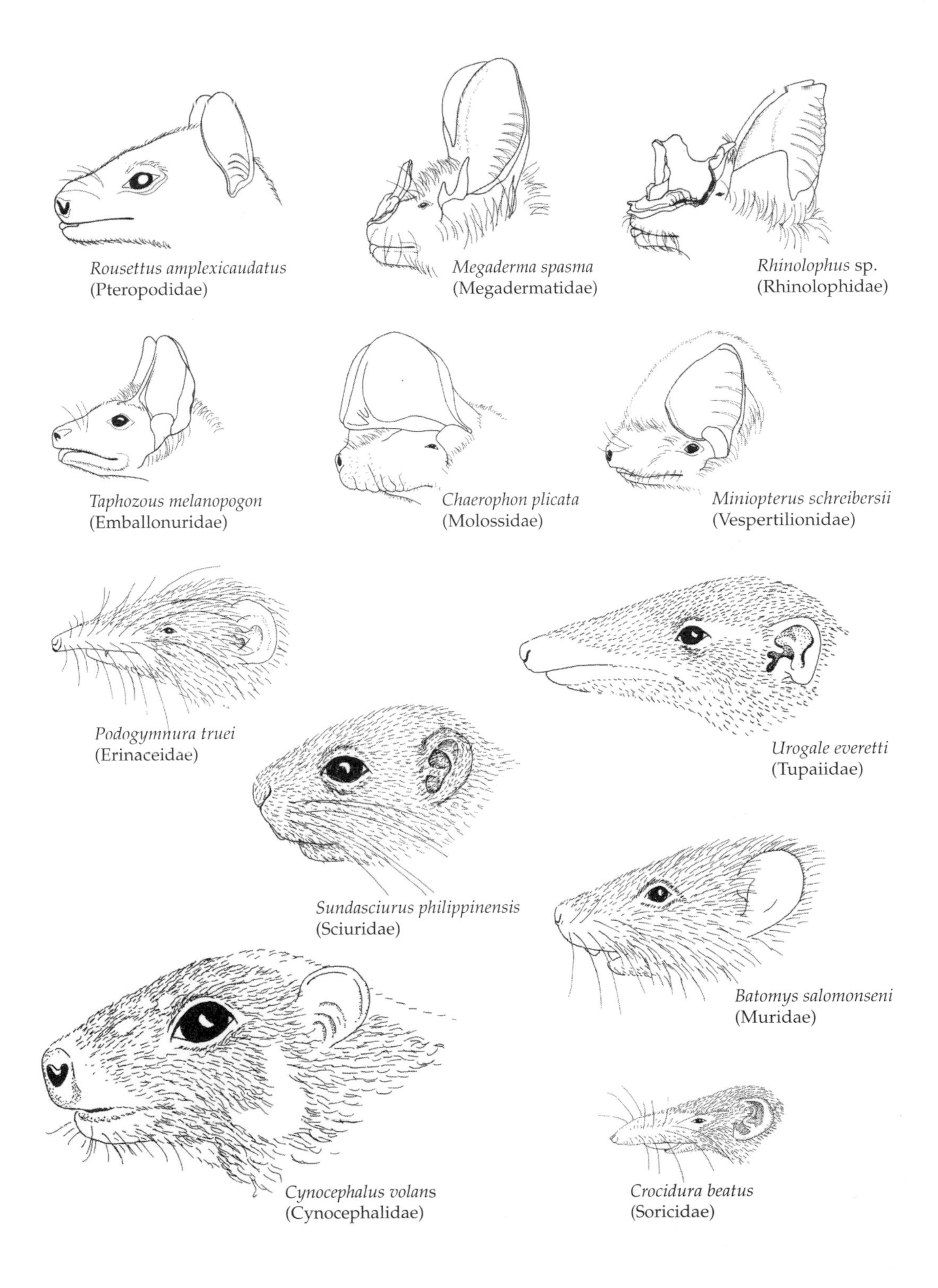
Rousettus amplexicaudatus
(Pteropodidae)
Megaderma spasma
(Megadermatidae)
Rhinolophus sp.
(Rhinolophidae)
Taphozous melanopogon
(Emballonuridae)
Chaerophon plicata
(Molossidae)
Miniopterus schreibersii
(Vespertilionidae)
Podogymnura truei
(Erinaceidae)
Urogale everetti
(Tupaiidae)
Sundasciurus philippinensis
(Sciuridae)
Batomys salomonseni
(Muridae)
Cynocephalus volans
(Cynocephalidae)
Crocidura beatus
(Soricidae)

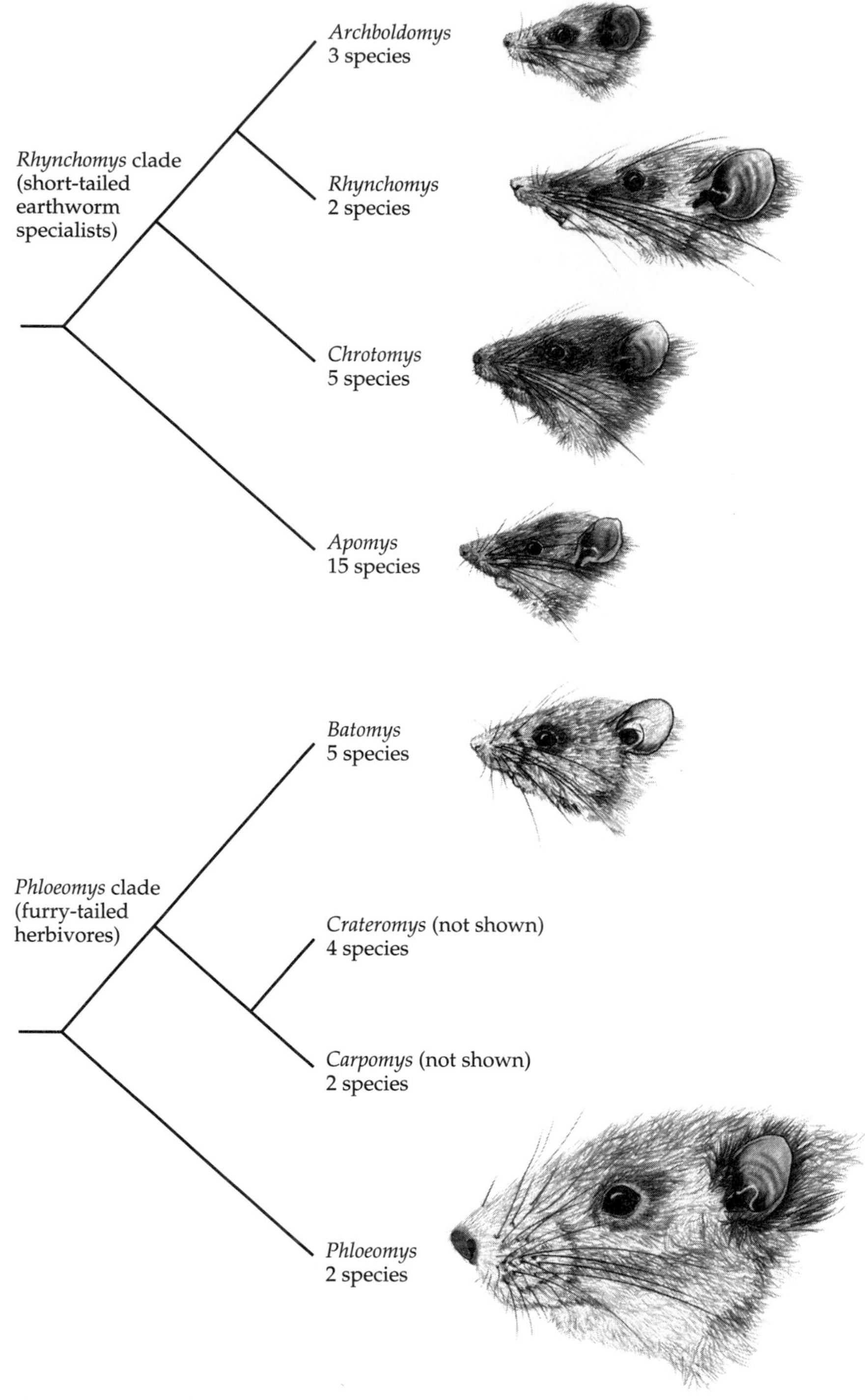

FIGURE 18.4 Phylogeny of two endemic Philippine radiations, the *Rhynchomys* and the *Phloeomys* clades. (Phylogeny from S. A. Jansa and L. R. Heaney, in prep.; drawings by Maggie Robertson.)

Knowledge of the distribution and diversity of Philippine mammals was very limited up until the 1980s. Most of the available distribution maps dated from 1934 (Taylor 1934) and showed only two or three localities for most species. No attempt had been made to show centers of endemism or diversity for mammals, though such maps had been made for birds, reptiles, land snails, and some groups of flowering plants in the 1920s (Dickerson 1928). Compilation of specimen records from museums in the Philippines, the United States, and Europe allowed documentation of patterns of distribution and apparent patterns of natural colonization, extinction, and speciation (Heaney 1986; Heaney and Rickart 1990). Field inventories in previously poorly known areas were undertaken to determine the accuracy of existing information, to gather what was typically the first ecological information on the endemic species, and to investigate conceptual issues in equilibrium island biogeography (e.g., Heideman and Heaney 1989; Heaney 1991; Rickart et al. 1991, 1993). These data also allowed the first applications to conservation of the mammalian fauna (e.g., Hauge et al. 1986; Heaney and Heideman 1987; Heaney and Utzurrum 1991; Heaney 1993; Rickart 1993).

Remarkably, during the last 20 years at least 26 previously unknown species of mammals—over 15% of the fauna—have been discovered (Heaney et al. 1998, and unpubl. data). While such a high rate of discovery is exciting, it could be taken as an indication of a substantial problem for conservation planning: Does this mean that many centers of biodiversity are currently unknown, and may therefore go unprotected? The answer to this crucial question appears to be "no": most of the new species were discovered not by chance, but by development of two complementary conceptual models of biogeographic patterns in the Philippines as well as field studies that tested these models, which may be as valuable to conservation as they are to a conceptual understanding of biogeographical processes. They have allowed the documentation of quite a few centers of endemism that were unknown or unrecognized until recently; they also predict the location and richness of other possible centers that have not yet been investigated.

Modeling Mammalian Biodiversity in the Philippines

Why does the Philippine archipelago have such a high level of mammalian diversity and, especially, such a high level of endemism? The general association of high levels of endemism with oceanic islands suggests that many of the answers lie in the geography (size, isolation, and topographic relief) and geological history of the islands.

The geological and tectonic history of Indo-Australia is one of the most complex of any part of the world, but a number of recent publications have greatly clarified the development of the region, including the locations of not only geological "rock units," but also of dry-land islands (papers in Hall and Holloway 1998; Hall 2002; see also Scotese, this volume). It is now known that the modern Philippine archipelago originated in three different areas, and that these three primary rock units have been gradually coalescing since about 35–40 my BP.

The first of these units originated as a portion of the continental margin next to modern southeastern China; this unit rafted away from the edge of the continent due to a small spreading zone that developed about 35 my BP. At that time, this rock unit (which later gave rise to modern Palawan and Mindoro as well as parts of some nearby islands) was submerged deep underwater and capped by marine sediments. This rock unit did not rise above sea level until it came into proximity with the second major unit (which gave rise to modern Luzon) beginning about 10 my BP. The origin of modern Luzon took place as a series of small, *de novo* islands about 35 my BP, far to the southeast of its current location and far from any larger land masses. This "proto-Luzon" group gradually moved northwest, along with at least one fairly large island that uplifted by about 15 my BP as proto-Luzon approached proto-Mindoro.

The third primary geological unit, proto-Mindanao, originated at about the same time as proto-Luzon, but farther to the southeast (near the location of modern New Guinea). No substantial islands developed from this unit before about 12 my BP, when it began to approach proto-Luzon. A phase of fairly widespread uplift occurred from 8–6 my BP and another uplift occurred from 5–3 my BP as coalescence of the archipelago proceeded. This uplift and associated volcanic activity enlarged the large islands, and also produced additional small islands that are currently isolated from other islands (such as Camiguin and Sibuyan; see Figure 18.2). The current heights of the islands are generally close to their maximum.

Changes in sea level during the Pliocene and Pleistocene were a second major geological/historical factor influencing the extent of islands (see also Betancourt, this volume; and Jackson, this volume). During each of the estimated 21 periods of Pleistocene glacial development, continental glaciers developed when more snow fell during winter than could melt during summer. The water that formed the massive continental glaciers originated in the seas through evaporation, and thus as the glaciers thickened sea level gradually declined.

During the most recent glacial episode, which peaked about 18,000 years ago, sea level was about 120 m lower than its current level (Fairbanks 1989). This change in sea level, probably not the most extreme of the Pleistocene, had a massive effect on the extent of dry land in Southeast Asia (see Figure 18.1). The broad, shallow seas covering the Sunda Shelf (which now extends from Thailand, Viet Nam, and adjacent areas to Borneo, Sumatra, and Java) disappeared, leaving vast areas of dry land. This area, which now internally shares a fairly homogeneous mammalian fauna, is bounded by a series of sharp declines to very deep water that cut it off from islands to the west, south, and east. The islands beyond the edge of the continental shelf are oceanic islands, some of which are parts of large archipelagoes, and some of which are quite distant from any other land.

Within the Philippines, the current islands expanded somewhat during the periods of low sea level, some of them joining with other islands to form much larger islands than we see today (see Figure 18.2). However, with the exception of the Palawan region, which was connected to Borneo as part of the Asian continent during earlier glacial epsiodes when sea level was even lower, all of the islands remained unconnected to the Asian mainland. Simultaneously, deep (though usually narrow) channels cut between the various islands, so that the Philippines remained as a set of isolated islands of varying sizes even during periods of low sea level.

This distinctive history of sets of islands that have emerged *de novo* above the surface of the ocean and have retained their individual histories suggested a solution to the puzzle of distribution patterns in the Philippines. My collaborators and I hypothesized that each island (or set of islands) surrounded by deep water would have retained isolated faunas, and therefore might have endemic species. Initial investigations based on existing data strongly supported this hypothesis since the fauna of each deep-water island (Greater Luzon, Greater Mindanao, Greater Palawan, etc.) is fairly homogeneous internally (except as noted below), but each had 45% to 80% endemism among the non-volant mammals relative to adjacent deep-water islands (Heaney 1986). Moreover, both the total number of species and the number of endemic non-volant mammals was correlated with the area of the island. We formalized these observations as a predictive model: each deep-water island is a unique center of endemism, and the number of endemic species of mammals is tightly correlated with that island's area. This led us to predict that Sibuyan, which had an unknown fauna at the time, would be a small but distinctive center of endemic biodiversity, with probably four endemic species of mammals, based on its area. When this prediction proved true (Goodman and Ingle 1993; Heaney 2000), it led us to question whether prior information about Camiguin was correct since it was not known to have any endemic mammals (Heaney 1984); indeed, the model predicted two endemics on the island of Camiguin. Subsequent fieldwork on Camiguin revealed two previously unknown endemic mammals (Heaney and Tabaranza 1997; Rickart et al. 2002). The discovery of new species on Panay (Gonzales and Kennedy 1996) and Mindoro (Ruedas 1995) convinced us that the model was robust (Heaney 2002). We now predict that Tablas and Lubang Islands (see Figure 18.2), currently virtually unknown, will be revealed as additional centers of endemism, with one endemic mammal species (and many endemic plants and invertebrates) on each island.

During the course of our efforts to conduct comprehensive inventories of certain islands, we encountered a second pattern that is also crucial to understanding endemism on this set of tropical oceanic islands. Early publications on some endemic mammals from the Philippines mentioned specimens of some species from high elevations, and others from low elevation. We naturally looked for them in such places, by conducting systematic surveys along elevational transects of mountains on several islands. We found that diversity of bats conformed to expected patterns along the transects: species richness among both insectivorous and frugivorous species was highest in the lowlands and declined progressively with increasing elevation (e.g., Heaney et al. 1989). However, the non-volant mammals (predominantly insectivorans and rodents) showed a pattern that contradicted all expectations: many species did not occur in the lowland rain forest, but rather in montane or mossy rain forest at elevations above 1000 m. Overall species richness increased from sea level to mountaintops, reaching maximum diversity (and abundance) at or near the tops of the ca. 2000 m mountains we initially studied (Heaney and Rickart 1990; Rickart et al. 1991). Further studies on higher mountains showed that diversity probably peaks at about the area of transition from montane forest to mossy forest (often at 1600 m to 2000 m, depending on local conditions; Heaney 2001). This pattern of maximum species richness at mid-elevations was subsequently documented at several sites worldwide (papers in Heaney and Lomolino 2001), and now appears to be the most common elevational pattern for small mammals and a range of other organisms.

As a corollary to this, some species were confined within an island to only a single mountain range or single isolated mountain; indeed, the higher the elevation, the greater the number of species with restricted distributions (Heaney and Rickart 1990; Rickart 1993), as has often been noted in continental areas (e.g., Merriam, 1890; Peterson et al. 1993). We hypothesized a general pattern: mammalian species confined to high elevations will be restricted to single mountain ranges, and each isolated mountain range will have endemic species occurring at a high elevation. We tested this hypothesis with additional surveys that resulted in the discovery of one new species in southern Luzon (Rickart and Heaney 1991), two on Mindanao (Rickart et al. 1998 and in press), and two more in northern Luzon (Rickart et al. 1998, and unpubl. data); other simultaneous discoveries elsewhere in the archipelago confirmed the pattern. We now hypothesize that each distinct mountain range (or large isolated mountain) on a given large island (or set of islands as shown in Figure 18.2) will not only be a unique center of mammalian biodiversity, but also that the number of species will be correlated with the area of the mountain range that is above the transition zone from lowland to montane forest. For example, we predict that the currently poorly known Zambales Mountains of Luzon and the Zamboanga Peninsula of Mindanao will prove to be just such montane centers of endemism, with one or two endemic mammal species on each.

Setting Priorities for Conservation

By combining our two predictive models—that each deep-water island is a unique center of endemism, and each isolated mountain range on the larger islands is a subcenter of endemism—we can map the locations of centers of mammalian endemism (see Figure 18.2). If a protected area of mature rain forest (of suitable size, and including all habitat types) were present in each of these areas, virtually all of the endemic mammals of the Philippines should have at least one place to maintain a stable population.

Two questions immediately arise. First, do distribution patterns of other organisms conform to those of mammals? And second, were there no biogeographic models available in the past? The first question was addressed at the Philippine Biodiversity Conservation Priority-Setting Program organized by the Philippine Department of Environment and Natural Resources (DENR) and Conservation International's Philippine program (Ong et al. 2002). More than 40 biologists who specialize in vertebrates, insects, and plants compared their information on distribution and endemism patterns. They found that their information, though incomplete in differing ways, was highly compatible with the patterns shown in Figure 18.2. Moreover, on many occasions as the workshop progressed, participants realized that their prior observations of, for example, endemic species on Camiguin or Siquijor that previously seemed inexplicable and unique were part of a much broader pattern of overall biotic endemism.

Only three classes of exceptions were apparent at the time. First, as an example, no endemic mammals are known from Siquijor, although the model predicts one or two. Field studies (Lepiten 1997) provided a stark explanation: the island has been nearly entirely deforested, with only a few small patches of badly degraded second growth remaining. If endemic mammals were once present, they are now extinct. While there is too little forest to maintain any endemic mammals, Siquijor still has remnant populations of endemic insects and plants (e.g., Ong et al. 2002; Danielsen and Treadaway, in press). From such examples we learned that, sadly, conservation efforts are too late for some groups of organisms, though not for others.

Second, it became clear that there are special cases that involve highly distinctive habitats. For example, amphibians and reptiles show subcenters of endemism in areas where there is extensive exposure of limestone, and some plant groups show centers of endemism in areas of ultrabasic soil (Ong et al. 2002). Such special cases are clearly missed by our biogeographic model, and they must be investigated and included in conservation planning for biotic protection to be effective.

Third, colonizing ability has a substantial impact on how many centers and subcenters of endemism a given group shows. For example, because all birds in the Philippines can fly, not surprisingly they show lower levels of endemism compared to non-volant mammals (see Table 18.1); indeed, birds often have no endemic species on deep-water islands that are small and near to species-rich larger islands. For example, Sibuyan has no endemic species of birds, though it has four endemic mammals. In contrast, frogs disperse more poorly than mammals, and so they have more subcenters of endemism on the larger islands than do mammals. Nevertheless, the overall pattern of frog endemism is very similar to that of mammals and birds (Ong et al. 2002).

The second question identified at the start of this section—whether there were no prior biogeographic models of endemism—can be answered simply: there were several. A classic compilation of biodiversity information from the 1920s (Dickerson 1928) showed patterns of endemism for many groups (though not including mammals), and many subsequent studies compiled existing data on certain groups (e.g., birds: Hauge et al. 1986, and Peterson et al. 2000; mammals: Heaney and Mallari 2002) and sought for evidence in gaps in the current system of national parks. While these analyses have been highly useful, they all suffer from a common weakness: they necessarily assume, usually implicitly, that their data were comprehensive (i.e., that their data included all species that existed). None were based on models that could be used predictively, and so they missed several centers of endemism that are now known to exist. Further, it seems likely that some of the centers would have remained unknown for a long time, perhaps long enough for them, too, to become entirely deforested.

But surely the national government and international agencies must have had some comprehensive biogeographic description in mind when they planned conservation action in such a geologically and biotically complex country? Again, the answer is yes. Documents that set national policy (e.g., Philippine Department of Environment and Natural Resources 1997) included a map that showed the "terrestrial biotic regions" of the Philippines. This map has many elements that are consistent with current data, and it certainly made positive contributions to conservation by encouraging protection efforts in many different parts of the archipelago. However, parts of this map are contradictory to current and past information: many of the smaller centers of endemism are not recognized, and there is no discernable biogeographic model inherent in it. When I tried to determine the source of the decisions reflected on the map, I was told simply that it had been drawn by one person, that he had retired, and that no one knew his sources of information. This map had guided national conservation policy for at least 20 years. Our new predictive model was adopted for national conservation planning in 2002 (Ong et al. 2002) because of the congruity of patterns of different groups of organisms and its utility in clarifying the reasons for congruity.

Can Biodiversity Be Preserved without Biogeography?

If there were infinite time and funding for field studies, field biologists could conduct inventories of the biota of each of the 7000 islands in the Philippines. It would then be possible to compile a comprehensive listing of the distribution of every species in the country. Such an empirical database would allow precise and accurate identification of every center of endemism, and it would provide perfect data for developing a geographically explicit national conservation plan.

Unfortunately, funding remains highly limited, and the number of sufficiently knowledgeable field biologists, though growing, is still small. Moreover, the rate of destruction of both mature and second-growth forest is still very high. By 2020 the amount of forest remaining is projected to reach critically low levels even in those few parts of the country that supported good forest in the late 1990s (Environmental Science for Social Change 1999).

Data presented here demonstrate that predictive biogeographic models, which are then used to design field investigations, are essential to protect biodiversity. With such models we can predict areas of endemism that have otherwise escaped detection (e.g., Camiguin and Sibuyan in the Philippines), and some areas that are still virtually unknown (e.g., Lubang and Tablas Islands and the Zambales Mountains). When these models fare well under testing, their predictions give us both the means to understand biogeographic processes and the means to begin timely conservation action in places that might otherwise be ignored. This is essential in regions such as the Philippines and Indonesia where islands number in the thousands and corresponding data are few.

The Philippine model leads me to three important predictions. First, each of the tens of thousands of oceanic islands (or sets of islands that were united during times of low sea level) in Indonesia, the Caribbean Sea, and elsewhere around the world will be found to be a center of endemism (the smaller and less isolated islands especially for invertebrates and low-dispersal plants; the larger islands for mammals and birds as well). Second, I predict that isolated, high-mountain areas on any large island or continent in the tropics (and perhaps elsewhere as well) will be subcenters of endemism. Third, I predict that shallow waters of all isolated oceanic archipelagoes will be centers of endemism (see also Vermeij, this volume). Predictive conceptual models of this type fill a gap left by GIS-based predictive modeling, which can predict distributions of known species (Scott et al. 2002; Sánchez-Cordero et al., this volume) but have no means of predicting the existence of unknown species. Indeed, without inclusion of predictive conceptual biogeographic models such as those discussed here, conservation planning in the richly biodiverse and highly threatened archipelagoes of the world, and in equivalent marine and continental areas, will not succeed.

Acknowledgments

In many senses, this paper is the product of efforts by dozens of individuals with whom I have been privileged to work. I cannot thank them all individually, but extend my thanks just the same; Nonito Antoque, Danilo Balete, Carlo Custodio, Renato Fernandez, Paul Heideman, Nina Ingle, Guy Musser, Eric Rickart, Blas Tabaranza, and Ruth Utzurrum deserve special recognition. The Barbara Brown and Ellen Thorne Smith Funds of the Field Museum, the National Science Foundation, and the MacArthur Foundation have all provided essential financial support. Clara Simpson produced the map, Maggie Robertson and Jodi Sedlock made the drawings, and Danilo Balete, Julie Lockwood, Mark V. Lomolino, and Eric Rickart provided welcome suggestions that improved various drafts of the manuscript. The Department of Environment and Natural Resources, through the Protected Areas and Wildlife Bureau, provided permits and encouragement. I thank the Geological Sciences Department, Northwestern University, for use of office space during a sabbatical leave.

Concluding Remarks

James H. Brown

As indicated in the Preface, the present volume is a direct result of the inaugural meeting of the *International Biogeography Society* in January of 2003. The chapters in this volume are based on plenary talks at that meeting. I gave the capstone address at the meeting, and the following remarks are based on that talk. These remarks encompass a very personal perspective on the past history, the present state, and the future prospects of biogeography.

Being invited to give capstone talks and write concluding chapters is a sign of increasing seniority and advancing age. Dating my career as a biogeographer from my 1971 paper on mammals on mountaintops, I have been working in the field for more than 30 years. Given the rapid pace of modern science, that is long enough to have witnessed major changes. But my primary focus here will be on the present and the future of biogeography because, after all, this book's title and contents emphasize "frontiers."

Past History

Biogeography, as a recognized discipline, did not exist when I began working in the 1970s. Of course, the intellectual roots go back much farther. In the Introduction to this volume, Lomolino and Heaney recount how the seminal contributions extend from the naturalists of the eighteenth century through the ecologists, systematists and other evolutionary biologists, and paleontologists of the mid-twentieth century. Many of papers that embody these historic advances are reproduced, with commentary, in the recent book *Foundations of Biogeography* edited by Lomolino, Sax, and Brown (2004).

Biogeography began to coalesce into a vigorous unified discipline in the 1980s. What caused this? I see three themes. First, exciting advances both within and outside biogeography that were made in the 1960s and 1970s began to gain wide attention. Examples include the equilibrium theory of island biogeography, cladistic or vicariance biogeography, allopatric speciation, paleobiology and macroevolution, plate tectonics, and Earth system science. Second, some senior scientists and many of the younger ones initiated research on these topics. More importantly, they began to recognize the unifying empirical and theoretical themes that gave coherence and integration to the emerging discipline. Similar biogeographic patterns in plants and animals, and on land and in the oceans and fresh waters, led investigators to erect and evaluate hypotheses that the congruent patterns were caused by the same ecological processes or historic events. Third, these advances began to be incorporated into the training of young scientists. There developed a mutually stimulatory feedback between new textbooks, symposium volumes, and journals, and new courses and discussion groups in colleges and universities. The title "biogeography" began to replace zoogeography, plant ecology, geographical ecology, and other narrower descriptors in these books and curricula.

Since the 1980s there has been an explosion of activity. New datasets and analytical tools raised new research questions. New investigators with diverse backgrounds and skills were attracted to address them, and these investigators increasingly called themselves biogeographers. They worked across the old boundaries that had divided the field and developed connections to other disciplines. Personally, I have found this explosion both exciting and intimidating. Perhaps it was just the foolhardy arrogance of youth, but when Arthur Gibson and I wrote the first edition of *Biogeography* in the early 1980s, I felt that I had a reasonable command of the breadth and depth of the field. When Mark Lomolino and I wrote the second edition in the late 1990s, I felt woefully inadequate, realizing that I had been unable to keep up with many of the important advances.

Present State

Today biogeography is thriving. Only a few other emerging disciplines, such as conservation biology, global change science, and complexity science, are experiencing comparable success. New technological and analytical tools have provided grist for the empiricists mills—and they have been spinning gold out of bytes. Technology allows us to see, measure, and think about the world on geographic and global spatial scales and to work backward through Earth history on geological time scales. Now there are data on the distributions of organisms (thousands of maps and georeferenced data files compiled from the literature, museum databases, and new biological surveys); on the environment (from ground, ocean, and satellite-based sensors); on Earth history (from new fossil finds and dating technologies); and on the gene sequences and phylogenetic relationships of individuals and species (from molecular techniques). Advances in computer technology have greatly enhanced our abilities to store, process, analyze, and visualize these data. Advances in informatics, geographic information systems, geostatistics, time-series analysis; algorithms for gene sequencing, phylogenetic reconstruction, predicting weather, climate, and oceanographic changes; and simulation models of complicated spatial and temporal phenomena all allow biogeographers to explore interrelationships among these data.

This rapidly expanding empirical base and analytical power is forcing a reevaluation of the conceptual foundations of biogeography. As recently as five or ten years ago, I was telling students that the accumulating data were supporting existing theories; although empirical evaluations sometimes revealed isolated exceptions, and required slight modifications, I thought then that these exceptions and modifications only served to strengthen the existing framework. Was I ever wrong! Now I am convinced that most of this framework is either fatally flawed or unrealistically simplistic, or requires major modification. To illustrate this perspective, I present seven assertions that challenge several cherished concepts and theories of twentieth-century biogeography. At least some eminent biogeographers agree with some of these, because my comments repeat some of the themes presented in the preceding chapters. I present these assertions here very briefly, without citing references, primarily for heuristic purposes: to stimulate debate and additional theoretical and empirical research.

1. *Species diversity of most taxa inhabiting islands and insular habitats is far from a dynamic equilibrium between opposing rates of colonization and extinction.* On oceanic islands, the force of colonization is weak and diversity increases gradually over time, often due to *in situ* speciation rather than independent colonization events. The rate of accumulation of species has been accelerated recently by the naturalization of exotic species. On continental islands and recently fragmented habitats, the force of extinction is weak and diversity declines slowly over time. Only when the area is very small is rapid, coordinated extinction likely.

2. *Much—perhaps most—speciation is not allopatric.* In birds, mammals, and some other groups, most speciation probably does occur as a result of geographic isolation caused by founder or vicariance events. But in fish, insects, plants, microbes, and many other groups, speciation more frequently occurs *in situ*, due primarily to selection for divergent phenotypes to exploit multiple environmental opportunities.

3. *Different taxonomic groups living together in the same place usually do not have congruent biogeographic histories.* The component taxa originated at different times, responded differently to environmental conditions, and had different probabilities of dispersal, colonization, speciation, and extinction. Consequently, biotas of both islands and continents have been assembled by multiple processes and the different taxa exhibit distinctive patterns of biogeographic and phylogenetic history.
4. *Range shifts in response to major geological and environmental changes have not been unidirectional and coincident across different species, lineages, and functional groups.* Ecological communities are not so integrated and coevolved that they respond as units to local or global environmental change. With the notable exception of rare obligate interdependencies, each kind of organism tends to shift its abundance and distribution individualistically as environmental changes affect its unique tolerances and requirements.
5. *Diversity of most lineages cannot be explained simply by a difference between speciation and extinction rates.* By definition, the diversity of a lineage at any time reflects the cumulative history of speciation and extinction events. Every monophyletic lineage begins and ends with a single species, regardless of whether or not the lineage attains high diversity in the interim. In order for lineages to long persist, speciation rates must approximately equal extinction rates; any sustained imbalance would lead to exponential rates of diversification or extinction.
6. *Latitudinal and elevational patterns of diversity are often different.* Whereas in most groups species richness declines continuously from the Equator to the Poles, richness on mountains frequently shows a peak at intermediate elevations. Whatever the processes that cause these patterns—environmental conditions, historical processes of dispersal, speciation, and extinction, or random placement of ranges—elevational gradients should not be viewed as smaller-scale models for latitudinal diversity gradients.
7. *The current rate and magnitude of human-caused global extinctions are not the greatest in Earth history.* As serious as these extinctions are, they do not yet approach the magnitudes of the Cretaceous-Tertiary, Permo-Triassic, and perhaps other mass extinctions in the fossil record. Furthermore, human activities have not caused reduced diversity in most kinds of organisms at all spatial scales. Global diversity of nearly all taxa has indeed decreased. In many groups, however, local diversity has actually increased, because colonizations of exotic species have exceeded extinctions of native species.

I see the present state of biogeography as one of enormous intellectual ferment. New data from new technologies and methodologies are challenging the assumptions and predictions of current theories and are threatening to overturn many conceptual foundations of the field.

Future Promise

It is hard to imagine a more exciting time to be a biogeographer. There are major challenges, and how these challenges are addressed will determine the future. Most seriously, it will determine whether biogeography continues to coalesce, integrate, and advance as a unified discipline, or whether it will break apart once again into specialized research programs that seldom communicate with one another.

One thing seems clear. The empirical advances will continue. There will be no stopping the technological breakthroughs and the resulting accumulation and analysis of data. The knowledge base for understanding the distribution, diversity, and conservation of living things will continue to increase explosively. It is not clear, however, that this increase in factual knowledge will continue to advance and unify the field. I see four big challenges, and each comes with a warning bell.

The first challenge is to find ways to rigorously and creatively use all of the massive quantities of data that will become available. Just because information exists—somewhere—does not mean it is accessible to those who could be using it. Major issues concern the quality of data and the need for accompanying metadata to make clear *how* the data were obtained (which constrains how they can appropriately be used). I will gloss over issues of data storage and security, which are important issues but are problems more for engineers than for biogeographers to solve.

There are important issues of how the data are to be processed, analyzed, integrated, and synthesized. This is where both rigor and creativity will be essential. The more information becomes available, the greater the challenge of sorting through the complexity to find pervasive patterns and test hypotheses about mechanisms.

The warning bells that I hear also concern the training and mindsets of younger scientists. At the risk of sounding like an old fogy, I am concerned that most prospective scientists spend more of their formative years sitting in front of computers than outside messing around in nature. This is not all bad, but it can lead to a certain arrogance and carelessness. The vast quantities of information which are becoming available on the Web should be and will be used. Computers can do much of the work of compiling, processing, and analyzing these data. But I am worried when I see data being downloaded from the Web and run through statistical software packages with little concern for the source and quality of the data, the assumptions underlying the statistical procedures, or first-hand knowledge of the organisms and environments.

Most importantly, the information will remain largely valueless until it is used to address meaningful problems in basic and/or applied science. This is where field training and naturalists' intuition will remain essential. For example, when a museum record is downloaded from the Web, it is important to realize what it refers to—it is a specimen in a museum drawer that was collected in the field at a specific place and date and subsequently cataloged and identified—and that there is some possibility of error in entering and transcribing the data and identifying the specimen. Furthermore, the biological surveys that collected the specimens usually were not designed to address the questions currently being asked. So, for example, lack of records from a locality may mean inadequate collecting rather than absence of a species, and historical records of a species do not necessarily mean that that species is still present. Most importantly, to use the data to do good science requires not only quality control, analytical thinking, and creative insight, but also the experience and taste to tackle interesting and important questions. For these things, electronic data and computers to manipulate them can be an enormous boon, but they cannot replace human minds and experiences.

The second big challenge for biogeography will be to continue its development as an interdisciplinary science. By this I mean that it must continue to expand its realm of inquiry and to develop and expand interfaces with other disciplines. Vibrant new disciplines often emerge at the interfaces of preexisting specialties. Much of the current excitement in biogeography stems from its incorporation and synthesis of data and concepts from other disciplines: molecular genetics for phylogenetic reconstruction and phylogeography; plate tectonics, fossil records, and paleoecology for historical biogeography; and remotely sensed changes in climate, vegetation, and human activities for biodiversity conservation. To continue these trends requires not only taking advantage of the technologies and data, but also incorporating the perspectives, ideas, and theories, of other disciplines. I do not see how this will be accomplished without greatly increased interdisciplinary collaboration.

As my colleague Stephanie Forrest says, "The most effective way to do interdisciplinary science is through collaborations among disciplinary scientists." Traditionally trained systematists or ecologists can still make important contributions to biogeography, especially if they bring their special expertise to collaborations with, for example, molecular geneticists, paleontologists, Earth scientists (geologists, climatologists, and oceanographers), GIS and remote sensing experts, or computer scientists. The trend in biogeography, as in most sciences, is for increased collaboration, meaning more multi-investigator research programs and more co-authored publications. But there is a world of difference between collaboration within a narrowly focused, hierarchically organized research group composed of a professor, postdocs, and graduate students, and interactions that bring together as equals scientists from two or more disciplines. Biogeography needs more of the latter if it is to continue to thrive at the interfaces among biology, the Earth sciences, and—increasingly—the social sciences.

My concern here is that, for all the lip service given to the need for interdisciplinary science, it has yet to realize its promise. I can attest from experience that interdisciplinary collaboration is more easily talked about than achieved. It is hard to find collaborators with appropriate interests and expertise; to communicate across the barriers of concepts, vocabulary, and institutional organizations; and to develop working relationships that produce worthwhile science. In our research on metabolic scaling, we have hit upon one successful model. We work intensively in extended 4- to 6-hour sessions at weekly intervals in small, egalitarian groups (In our case composed of 3–6 senior scientists, postdocs, and grad students, and including ecologists, molecular biologists, and physicists). This is probably not the only model, but I suspect that to do real interdisciplinary science it will be necessary for individuals with very different backgrounds and expertise to find ways to work closely together. It will, of course, require more support from institutions and funding agencies.

The third challenge will be to develop new biogeographic theory. The potential advances from incorporating new technology, data, and interdisciplinary perspectives need to be synthesized, unified, and presented to the world in a conceptual context. No discipline can long survive without a conceptual foundation. Without principles to give them weight and show their significance, all facts are equally important. If much current biogeographic theory is either wrong, too simplistic, or relevant to only specific instances (and, as indicated above, I believe this is the case), then it must be jettisoned and replaced or at least modified. I am not advocating a single grand unified theory of biogeography. I foresee something much more like the conceptual basis of evolutionary biology, where the universal molecular structure and function of the genome, the laws of Mendelian and quantitative genetics, the theory of natural selection, and still-to-be-perfected theories of speciation and diversification all play important and complementary roles.

So, what form will the theoretical foundation of biogeography take? I only wish I knew. I do have a few thoughts. The task need not be all that daunting. Many of the ingredients are already in place. Some components are well established first principles or theoretical foundations of existing disciplines that need to be imported and integrated into biogeography. Examples include principles of molecular genetics, evolutionary theory, plate tectonic theory, concepts of oceanic and atmospheric circulation, and principles of population growth. It will be a challenge to put together these and other ingredients to create a conceptual foundation for biogeography—one that can account for geographic patterns of species diversity, understand the taxonomic specificity of the Cretaceous-Tertiary and end-Pleistocene extinctions, predict the shifts in geographic ranges in response to global change, and interpret the variation in the success and impacts of exotic species.

My concern here is that too few biogeographers are inclined to seek conceptual unification and to try their hand at theory. By its very nature, biogeography emphasizes the enormous variation in nature—across all forms of life and over the entire surface of the Earth. Most biogeographers seem more interested in explaining this variation than in identifying the common themes and explaining them in terms of first principles. There is an increasing amount of "modeling," but most of it uses multiple variables and specific parameter values to design computer simulation models to address very narrow problems. There is a need for empirical studies and models that focus on details, specifics, and variation. There is also a need to balance such approaches with theoretical approaches that focus on the big picture, general patterns, and fundamental processes. The reticulation, diversity, and pluralism that Lomolino and Heaney emphasize in this book's Introduction should not be construed to preclude the search for general theories. I deliberately used the plural: *theories*. I do not envision, at least for a very long time, a single grand unifying theory for all of biogeography. I do see the potential to develop multiple theories, which over time will give the discipline an increasingly interconnected conceptual foundation.

The fourth and final challenge is to make biogeography more useful and relevant. Society increasingly demands that biogeography apply its unique expertise to practical problems. Modern human beings are transforming the world on geographic and global scales. *Homo sapiens* has become the dominant species on Earth. It has spread to all corners of the globe, fundamentally altered the land, sea, and atmosphere, drastically changed the abundance, distribution, and diversity of other organisms, and created an unprecedented social and economic systems based on fossil fuel energy and agricultural, industrial, and information technology. Biogeography, in collaboration with other relevant disciplines, has the responsibility to address these human-caused changes: to understand the causes, interpret the consequences, and make predictions about the future.

The biogeography of the future must be expanded to include humans and all of their activities. There is precedence for this in some countries and institutions, where biogeography has strong historical ties to (human) geography and the social sciences. As biogeography becomes increasingly applied and humanized, however, it cannot abandon its roots and principles in basic science. Humans must obey the same scientific laws that govern the physical universe and life on Earth. The immense problems that confront us—energy, water, food, disease, and pollution—can all be understood in terms of existing scientific principles. This does not mean that the solutions will be easy or straightforward. Indeed, these problems intersect with each other, and with social and political issues that make solutions extremely difficult. But the basic sciences, with biogeography in a prominent place among them, must play a major role.

My chief concern here is that biogeography could fail to recognize and respond to the full extent of the challenge. The discipline could continue to become more applied, but limit its applications to investigating and mitigating the impacts of human activities on other organisms: changes in biodiversity in different groups of animals and plants and at varying spatial scales, design of nature preserves and other conservation measures, and shifts in abundances and ranges of species in response to environmental change. But the big issue is the future of humanity itself, and the next few decades will be critical. Unless humans find practical, lasting solutions to the problems of energy, water, food, disease, and pollution, human civilization cannot survive. The challenge is no less than to find a way for Earth to support both a diverse biota and some advanced form of advanced human civilization with its unique society, culture, economy, information, and technology. Biogeography must play a major role in responding to this challenge.

Literature Cited

Numbers in brackets at the end of each reference indicate the chapter or chapters in which the work is cited.

Aassine, S. and M. C. El Jaï. 2002. Vegetation dynamics modeling: A method for coupling local and space dynamics. *Ecological Modeling* 154: 237–249. [2]

Aava, B. 2001. Can resource use be the link between productivity and species richness in mammals? *Biodiversity and Conservation* 10: 2011–2022. [11]

Ackerly, D. D. 2003. Community assembly, niche conservatism and adaptive evolution in changing environments. *International J. Plant Sciences* 164: S165–S184. [3; Part 1]

Adams, C. G., D. E. Lee and B. R. Rosen. 1990. Conflicting isotope and biotic evidence for tropical sea-surface temperatures during the Tertiary. *Palaeogeography, Palaeoclimatology, Palaeoecology* 77: 289–313. [14]

Adams, J. M. and F. I. Woodward. 1989. Patterns in tree species richness as a test of the glacial extinction hypothesis. *Nature* 339: 699–701. [9, 11]

Alcérreca, C., O. Flores, D. Gutiérrez, M. Herzig, R. Pérez and V. Sánchez-Cordero. 1988. *Fauna Silvestre y Areas Naturales Protegidas*. Fundación Siglo Veintiuno A. C. [16]

Allen, A. P. and E. P. White. 2003. A unified theory for macroecology based on spatial patterns of abundance. *Evolutionary Ecology Research* 5: 493–499. [17]

Allen, A. P., J. H. Brown and J. F. Gillooly. 2002. Global biodiversity, biochemical kinetics and the energetic-equivalence rule. *Science* 297: 1545–1548. [8, 11]

Allen, G. R. 1975. *Damselfishes of the South Seas*. T.F.H. Publications, Neptune City, NJ. [13]

Allen, J. R. M., U. Brandt, A. Brauer, H.-W. Hubberten, B. Huntley, J. Keller, M. Kraml, A. Mackensen, J. Mingram, J. F. W. Negendank, N. R. Nowaczyk, H. Oberhänsli, W. A. Watts, S. Wulf and B. Zolitschka. 1999. Rapid environmental changes in southern Europe during the last glacial period. *Nature* 400: 740–743. [Part 1]

Alley, R. B. 2000. *The Two-Mile Time Machine: Ice Cores, Abrupt Climate Change and Our Future*. Princeton University Press, Princeton, NJ. [3]

Alley, R. B. and P. U. Clark. 1999. The deglaciation of the Northern Hemisphere: A global perspective. *Annual Review of Earth and Planetary Sciences* 27: 149–182. [3]

Alroy, J., C. R. Marshall, R. K. Bambach, K. Bezusko, M. Foote, F. T. Fursich, T. A. Hansen, S. M. Holland, L. C. Ivany, D. Jablonski, D. K. Jacobs, D. C. Jones, M. A. Kosnik, S. Lidgard, S. Low, A. I. Miller, P. M. Novack-Gottshall, T. D. Olszewski, M. E. Patzkowsky, D. M. Raup, K. Roy, J. J. Sepkoski, Jr., M. G. Sommers, P. J. Wagner and A. Webber. 2001. Effects of sampling standardization on estimates of Phanerozoic marine diversification. *Proc. National Academy of Sciences USA* 98: 6261–6266. [14]

Althoff, D. M. and O. Pellmyr. 2002. Examining genetic structure in a bogus yucca moth: A sequential approach to phylogeography. *Evolution* 56: 1632–1643. [5, 7]

Anderson, R. M. and R. M. May. 1991. *Infectious Diseases of Humans: Dynamics and Control*. Oxford University Press, Oxford. [10]

Anderson, R. P., D. Lew and A. T. Peterson. 2003. Evaluating predictive models of species' distributions: Criteria for selecting optimal models. *Ecological Modelling* 162: 211–232. [16]

Andersson, L. 1996. An ontological dilemma: Epistemology and methodology of historical biogeography. *J. Biogeography* 23: 269–277. [5, 7]

Andrews, P. and E. M. O'Brien. 2000. Climate, vegetation, and predictable gradients in mammal species richness in southern Africa. *Journal of Zoology, London* 251: 205–231. [9, 11; Part 3]

Angel, M. V. 1997. Pelagic biodiversity. In R. F. G. Ormond, J. D. Gage and M. V. Angel (eds.), *Marine Biodiversity: Patterns and Processes*. Cambridge University Press, Cambridge, pp. 35–68. [14]

Anstey, R. L. 1978. Taxonomic survivorship and morphologic complexity in Paleozoic bryozoan genera. *Paleobiology* 4: 407–418. [8]

Anstey, R. L. 1986. Bryozoan provinces and patterns of generic evolution and extinction in the Late Ordovician or North America. *Lethaia* 19: 33–51. [8]

Arbogast, B. S. and G. J. Kenagy. 2001. Comparative phylogeography as an integrative approach to historical biogeography. *J. Biogeography* 28: 819–825. [5]

Arnaud, P. M. 1974. Contribution à la bionomie marine benthique des régions antarctiques et subantarctiques. *Téthys* 6: 567–653. [12]

Arntz, W. E., J. Gutt and M. Klages. 1997. Antarctic marine biodiversity. In B. Battaglia, J. Valencia and D. W. H. Walton (eds.), *Antarctic Communi-ties: Species, Structure, and Survival*. Cambridge University Press, Cambridge, pp. 3–14. [14]

Aronson, R. B. and R. E. Plotnick. 1998. Scale-independent interpretations of macroevolutionary dynamics. In M. L. McKinney and J. A. Drake (eds.), *Biodiversity Dynamics: Turnover of Populations, Taxa, and Communities*. Columbia University Press, New York, pp. 430–450. [14]

Arroyo, M. T. K., M. Riveros, A. Peñaloza, L. Cavieres and A. M. Faggi. 1996. Phytogeographic relationships and regional richness patterns of the cool temperate rainforest flora of southern South America. In R. G. Lawford, P. B. Alaback and E. Fuentes (eds.), *High-Latitude Rainforests and Associated Ecosystems of the West Coast of the Americas*. Springer-Verlag, New York, pp. 134–172. [10]

Arundel, S. 2002. Modeling climate limits of plants found in Sonoran Desert packrat middens. *Quaternary Research* 58: 112–121. [2]

Asher, J., M. Warren, R. Fox, P. Harding, G. Jeffcoate and S. Jeffcoate. 2001. *Millennium Atlas of the Butterflies in Britain and Ireland*. Oxford University Press, Oxford. [9]

Astorga, A., M. Fernandez, E. E. Boschi and N. Lagos. 2003. Two oceans, two taxa, and one mode of development: Latitudinal diversity patterns of South American crabs and test for possible causal processes. *Ecology Letters* 6: 420–427. [10]

Austin, M. P., A. O. Nichols and C. R. Margules. 1990. Measurements of the realized qualitative niche: Environmental niches for five *Eucalyptus* species. *Ecological Monographs* 60: 161–177. [16]

Avise, J. C. 2000. *Phylogeography: The History and Formation of Species*. Harvard University Press, Cambridge, MA. [5, 7; Parts 1, 2, and 4]

Avise, J. C., J. Arnold, R. M. Ball, E. Bermingham, T. Lamb, J. E. Neigel, C. A. Reeb and N. C. Saunders. 1987. Intraspecific phylogeography: The mitochondrial-DNA bridge between population genetics and systematics. *Annual Review of Ecology and Systematics* 18: 489–522. [5]

Bacon, F. 1620. *Novum Organum Scientiarum*, Ludguni Batav. [4]

Badgley, C. and D. L. Fox. 2000. Ecological biogeography of North American mammals: Species density and ecological structure in relation to environmental gradients. *J. Biogeography* 27: 1437–1467. [9]

Baille, J. and B. Groombridge. 1996. *1996 IUCN Red List of Threatened Animals*. Species Survival Commission, IUCN, Gland, Switzerland, www.redlist.org. [18]

Baker, P. A., C. A. Rigsby, G. O. Seltzer, S. C. Fritz, T. K. Lowenstein, N. P. Bacher and C. Veliz. 2001a. Tropical climate changes at millennial and orbital timescales on the Bolivian Altiplano. *Nature* 409: 698–701. [2]

Baker, R. G., E. A. Bettis III, D. P. Schwert, D. G. Horton, C. A. Chumbley, L. A. Gonzalez and M. K. Reagan. 1996. Holocene paleoenvironments of northeast Iowa. *Ecological Monographs* 66: 203–234. [3]

Bakker, R. T. 1980. Dinosaur heresy—dinosaur renaissance: Why we need endothermic archosaurs for a comprehensive theory of bioenergetic evolution. In R. D. K. Thomas and E. C. Olson (eds.), *A Cold Look at the Warm-Blooded Dinosaurs*, pp. 351–462. Westview Press, Boulder, CO. [12]

Baldwin, J. D., A. L. Bass, B. W. Bowen and W. H. Clark, Jr. 1998. Molecular phylogeny and biogeography of the marine shrimp *Penaeus*. *Molecular Phylogenetics and Evolution* 10: 399–407. [13]

Ball, I. R. 1975. Nature and formulation of biogeographic hypotheses. *Systematic Zoology* 24: 407–430. [Part 4]

Balmford, A., J. L. Moore, T. Brooks, N. Burgess, L. A. Hansen, P. Williams and C. Rahbek. 2001. Conservation conflicts across Africa. *Science* 291: 2616–2619. [11]

Banford, H. M., E. Bermingham, B. B. Collette and S. S. McCafferty. 1999. Phylogenetic systematics of the *Scomberomorus regalis* (Teleostei: Scom-bridae) species group: Molecules, morphology, and biogeography of Spanish mackerels. *Copeia* 1999: 596–613. [Part 4]

Barlow, N. D. 1994. Size distributions of butterfly species and the effect of latitude on species size. *Oikos* 71: 326–332. [8]

Barrett, P. H., P. J. Gautrey, S. Herbert, D. Kohn and S. Smith (eds.). 1987. *Charles Darwin's Notebooks, 1836–1844*. Cornell University Press, Ithaca, NY [6]

Bartlein, P. J., K. H. Anderson, P. M. Anderson, M. E. Edwards, C. J. Mock, R. S. Thompson, R. S. Webb, T. Webb III and C. Whitlock. 1998. Paleoclimate simulations for North America over the past 21,000 years: Features of the simulated climate and comparisons with paleoenvironmental data. *Quaternary Science Reviews* 17: 549–585. [2, 3]

Be, A. W. H. 1977. An ecological, zoogeographic, and taxonomic review of recent planktonic foraminifera. In A. T. S. Ramsay (ed.), *Oceanic Micropalaeontology*, Vol. 1. Academic Press, San Diego, pp. 1–88. [14]

Beerli, P. and J. Felsenstein. 1999. Maximum-likelihood estimation of migration rates and effective population numbers in two populations using a coalescent approach. *Genetics* 152: 763–773. [5]

Beever, E. A., P. E. Brussard and J. Berger. 2003. Patterns of apparent extirpation among isolated populations of pikas Ochotona princeps. in the Great Basin. *J. Mammalogy* 84: 37–54. [11]

Bellwood, D. R. 1996. The Eocene fishes of Monte Bolca: The earliest coral reef fish assemblage. *Coral Reefs* 15: 11–19. [13]

Bellwood, D. R. and P. C. Wainwright. 2002. The history and biogeography of fishes on coral reefs. In P. F. Sale (ed.), *Coral Reef Fishes*, pp. 5–32. Academic Press, San Diego. [13]

Bellwood, D. R. and T. P. Hughes. 2001. Regional-scale assembly rules and biodiversity of coral reefs. *Science* 292: 1532–1534. [11]

Bennett, K. D. 1997. *Evolution and Ecology: The Pace of Life*. Cambridge University Press, Cambridge. [3]

Bennett, P. M. and I. P. F. Owens. 1997. Variation in extinction risk among birds: Chance or evolutionary predisposition? *Proc. Royal Society of London Series B* 264: 401–408. [8]

Benton, M. J. 1999. The history of life: Large databases in palaeontology. In D. A. T. Harper (ed.), *Numerical Palaeobiology*. Wiley, Chichester, UK, pp. 249–283. [14]

Benzie, J. A. H. 1998. Genetic structure of marine organisms and Southeast Asian biogeography. In R. Hall and J. D. Halloway (eds.), *Biogeography and Geological Evolution of Southeast Asia*, pp. 197–209. Backhuys Publishers, Leiden, NL. [13]

Benzie, J. A. H. and S. T. Williams. 1997. Genetic structure of the giant clam (*Tridacna maxima*) population in the West Pacific is not consistent with dispersal by present-day ocean currents. *Evolution* 51: 768–783. [13]

Bermingham, E. and C. Moritz. 1998. Comparative phylogeography: Concepts and applications. *Molecular Ecology* 7: 367–369. [5]

Bernardi, G., D. R. Robertson, K .E. Clifton and E. Azzurro. 2000 Molecular systematics, zoogeography, and evolutionary ecology of the Atlantic parrotfish *Sparisoma*. *Molecular Phylogenetics and Evolution* 15: 292–300. [Part 4]

Berra, T. M. 1981. *An Atlas of Distribution of the Fresh Water Fish Families of the World*. University of Nebraska Press. Lincoln, NE. [16]

Betancourt, J. L. and B. Saavedra. 2002. Nuevo método paleoecológico para el estudio de zonas áridas en Sudamérica: Paleomadrigueras de roedores [New paleoecological method for Quaternary studies in arid lands of South America: Rodent middens]. *Revista Chilena de Historia Natural* 75: 527–546. [2]

Betancourt, J. L., C. Latorre, J. A. Rech, J. Quade and K. A. Rylander. 2000. A 22,000-yr record of monsoonal precipitation from northern Chile's Atacama Desert. *Science* 289: 1542–1546. [2]

Betancourt, J. L., K. A. Rylander, C. Peñalba and J. L. McVickar. 2001. Late Quaternary vegetation history of Rough Canyon, south-central New Mexico, USA. *Palaeogeography, Palaeoclimatology, Palaeoecology* 165: 71–95. [2]

Betancourt, J. L., K. A. Rylander, L. Scafati, D. Melendi, S. Monge, S. Lopez, F. A. Roig, Sr., F. A. Roig, Jr. and W. Volkheimer. 2003. Long-term stability of Monte Desert uplands inferred from chinchilla rat middens in the Sierra de Cacheuta near Mendoza, Argentina. *J. Biogeography,* submitted. [2]

Betancourt, J. L., T. R. Van Devender and P. S. Martin (eds.). 1990. *Packrat Middens: The Last 40,000 Years of Biotic Change*. University of Arizona Press, Tucson. [2, 3]

Betancourt, J. L., W. S. Schuster, J. B. Mitton and R. S. Anderson. 1991. Fossil and genetic history of a pinyon pine *Pinus edulis.* isolate. *Ecology* 72: 1685–1697. [2]

Beurois, J. 1975. Étude Ecologique et Halieutique des Fondes de Péche et des Espèces d'Intéret Commerciel (Langoustes et Poissons) des Iles Saint-Paul et Amsterdam (Océan Indien). Comité National Français de Recherches Antarctiques 37: 1–91. [12]

Bhattarai, K. R. and O. R. Vetaas. 2003. Variation in plant species richness of different life forms along a subtropical elevation gradient in the Himalayas, east Nepal. *Global Ecology and Biogeography* 12: 327–340. [10, 11; Part 3]

Birks, H. J. B. 1989. Holocene isochrone maps and patterns of tree spreading in the British Isles. *J. Biogeography* 16: 503–540. [11]

Black, M. B., K. M. Halanych, P. A. Y. Maas, W. R. Hoeh, J. Hashimoto, D. Desbruyères, R. A. Lutz and R. C. Vrijenhoek. 1997. Molecular systematics of vestimentiferan tubeworms from hydrothermal vents and cold-water seeps. *Marine Biology* 130: 141–149. [Part 4]

Blackburn, T. M. and K. J. Gaston. 1996. Spatial patterns in the body sizes of bird species in the New World. *Oikos* 77: 436–446. [8]

Blackburn, T. M. and K. J. Gaston. 1997. The relationship between the geographic area and the latitudinal gradient in species richness in New World birds. *Evolutionary Ecology* 11: 195–204. [9]

Blackburn, T. M. and K. J. Gaston (eds.). 2003. *Macroecology: Concepts and Consequences*. Blackwell Science, Oxford. [10]

Blakey, R. C. 2003. *Carboniferous-Permian Global Paleogeography of Assembly of Pangaea*. Symposium on Global Correlations and Their Implications for the Assembly of Pangea, August 10-16, 2003, Utrecht, Netherlands. International Congress on Carboniferous and Permian Stratigraphy (abstract). [1]

Bocharova N. Yu. and C. R. Scotese. 1993. *Revised Global Apparent Polar Wander Paths and Global Mean Poles*. PALEOMAP Project Progress Report 56-1293. Department of Geology, University of Texas at Arlington. [1]

Bond, G. C., H. Heinrich, W. S. Broecker, L. D. Labeyrie, J. McManus, J. Andrews, S. Huon, R. Jantschik, S. Clasen, C. Simet, K. Tedesco, M. Klas, G. Bonani and S. Ivy. 1992. Evidence for massive discharges of icebergs into the North Atlantic Ocean during the last glacial period. *Nature* 360: 245–249. [3]

Boone, R. B. and W. B. Krohn. 2000. Partitioning sources of variation in vertebrate species richness. *J. Biogeography* 27: 457–470. [9, 11]

Booth, R. K., S. T. Jackson and C. E. D. Gray. 2004. Paleoecology and high-resolution paleohydrology of a kettle peatland in upper Michigan. *Quaternary Research* 61: 1–13. [3]

Bouchard, P., D. R. Brooks and D. K. Yeates. In press. Mosaic macroevolution in Australian wet tropics arthropods: Community assemblage by taxon pulses. In C. Moritz and E. Bermingham (eds.), *Rainforest: Past, Present, Future*. University of Chicago Press, Chicago. [7]

Bouchet, P., P. Lozouet, P. Maestrati and V. Heiros. 2002. Assessing the magnitude of species richness in tropical marine environments: Exceptionally high numbers of molluscs at a New Caledonia site. *Biological J. Linnean Society* 75: 421–436. [12, 14]

Boucot A. J., C. Xu and C. R. Scotese. 2004. *Atlas of Lithological Indicators of Climate*. Society of Economic and Petroleum Geologists, in review. [1]

Bowen, B. W. and W. S. Grant. 1997. Phylogeogra-phy of the sardines (*Sardinops* spp.): Assessing biogeographic models and population histories in temperate upwelling zones. *Evolution* 51: 1601–1610. [Part 4]

Bowen, B. W., A. M. Clark, F. A. Abreu-Grobois and A. Cheves. 1998. Global phylogeography of the Ridley Sea Turtles (*Lepidochelys* spp.) as inferred from mitochondrial sequences. *Genetica* 101: 179–189. [13]

Bowler, P. J. 1996. *Life's Splendid Drama*. University of Chicago Press, Chicago. [6]

Bradley, R. S. 1999. *Paleoclimatology: Reconstructing Climates of the Quaternary*, 2nd Ed. Academic Press, San Diego. [3]

Bremer, K. 1992. Ancestral areas: A cladistic reinterpretation of the center of origin concept. *Systematic Biology* 41: 436–445. [6]

Brey, T., M. Klages, C. Dahm, M. Gorny, J. Gutt, S. Hain, M. Stiller, W. E. Arntz, J.-W. Wagele and A. Zimmerman. 1994. Antarctic benthic diversity. *Nature* 368: 297. [14]

Briggs, J. C. 1955. A monograph of the clingfishes (Order Xenopterygii). *Stanford Ichthyological Bulletin* 6: 1–224. [13]

Briggs, J. C. 1966. Zoogeography and evolution. *Evolution* 20: 282–289. [13]

Briggs, J. C. 1974. *Marine Zoogeography*. McGraw Hill, New York. [8, 12, 13; Part 4]

Briggs, J. C. 1987. *Biogeography and Plate Tectonics*. Elsevier, Amsterdam. [13]

Briggs, J. C. 1995. *Global Biogeography*. Elsevier, Amsterdam [Part 4]

Briggs, J. C. 1996. Tropical diversity and conservation. *Conservation Biology* 10: 713–718. [14]

Briggs, J. C. 1999. Modes of speciation: Marine Indo-West Pacific. *Bulletin of Marine Science* 65: 645–656. [Part 4]

Briggs, J. C. 1999a. Coincident biogeographic patterns: Indo-West Pacific. *Evolution* 53: 326–335. [13]

Briggs, J. C. 1999b. Extinction and replacement in the Indo-West Pacific Ocean. *J. Biogeography* 26: 777–783. [13]

Briggs, J. C. 1999c. Modes of speciation: Marine Indo-West Pacific. *Bulletin of Marine Science* 65: 645–656. [13; Part 4]

Briggs, J. C. 2000. Centrifugal speciation and centers of origin. *J. Biogeography* 27: 1183–1188. [13]

Briggs, J. C. 2002. Letter to the editor re: Roberts et al. 2002. *Science* 296: 1026. [13]

Briggs, J. C. 2003a. The biogeographic and tectonic history of India. *J. Biogeography* 30: 381–388. [1]

Briggs, J. C. 2003b. Centers of origin as evolutionary engines. *J. Biogeography* 30: 1–18. [13]

Briggs, J. C. and C. J. Humphries. 2004. Early classics. In M. V. Lomolino, D. F. Sax and J. H. Brown (eds.) *Foundations of Biogeography: Classic Works with Commentaries*. The University of Chicago Press, Chicago, pp. 5–13. [Part 3]

Brooks, D. R. 1981. Hennig's parasitological method: A proposed solution. *Systematic Zoology* 30: 229–249. [7]

Brooks, D. R. 1985. Historical ecology: A new approach to studying the evolution of ecological associations. *Annals of the Missouri Botanical Gardens* 72: 660–680. [6]

Brooks, D. R. 1988. Scaling effects in historical biogeography: A new view of space, time, and form. *Systematic Zoology* 38: 237–244. [7]

Brooks, D. R. 1990. Parsimony analysis in historical biogeography and coevolution: Methodological and theoretical update. *Systematic Zoology* 39: 14–30. [6, 7]

Brooks, D. R. and D. A. McLennan. 1991. *Phylogeny, Ecology, and Behavior*. University of Chicago Press, Chicago. [6]

Brooks, D. R. and D. A. McLennan. 2001. A comparison of a discovery-based and an event-based method of historical biogeography. *J. Biogeography* 28: 757–767. [5]

Brooks, D. R. and D. A. McLennan. 2002. *The Nature of Diversity: An Evolutionary Voyage of Discovery*. University of Chicago Press, Chicago. [5, 6, 7; Part 2]

Brooks, D. R. and D. A. McLennan. 2003. Extending phylogenetic studies of coevolution: Secondary Brooks parsimony analysis, parasites, and the Great Apes. *Cladistics* 19: 104–119. [7]

Brooks, D. R. and E. O. Wiley. 1988. *Evolution as Entropy: Toward a Unified Theory of Biology*. University of Chicago Press, Chicago. [7]

Brooks, D. R., M. G. P. van Veller and D. A. McLennan. 2001. How to do BPA, really. *J. Biogeography* 28: 343–358. [4, 5, 7; Part 2]

Brooks, T. M., S. L. Pimm and N. J. Collar. 1997. Deforestation predicts the number of threatened birds in insular Southeast Asia. *Conservation Biology* 11: 382–394. [17]

Brown, J. H. 1971. Mammals on mountaintops: Non-equilibrium insular biogeography. *American Naturalist* 105: 467–478. [11]

Brown, J. H. 1981. Two decades of homage to Santa Rosalia: Toward a general theory of diversity. *American Zoologist* 21: 877–888. [9, 11]

Brown, J. H. 1989. Species diversity. In A. A. Myers and R. S. Giller (eds.), *Analytic Biogeography*. Chapman and Hall, New York, pp. 59–89. [10]

Brown, J. H. 1995. *Macroecology*. University of Chicago Press, Chicago. [2, 7, 10]

Brown, J. H. 1999. Macroecology: Progress and prospect. *Oikos* 87: 3–14. [10]

Brown, J. H. and A. C. Gibson. 1993. *Biogeography*. C.V. Mosby, St. Louis, MO. [16]

Brown, J. H. and A. Kodric-Brown. 1977. Turnover rates in insular biogeography: Effect of immigration on extinction. *Ecology* 58: 445–449. [11, 12]

Brown, J. H. and M. V. Lomolino. 1998. *Biogeography*, 2nd Ed. Sinauer Associates, Sunderland, MA. [1, 3, 9, 10, 11, 12, 14; Parts 1, 3, 4]

Brown, J. H. and B. A. Maurer. 1987. Evolution of species assemblages: Effects of energetic constraints and species dynamics on the diversification of the North American avifauna. *American Naturalist* 130: 1–17. [7]

Brown, J. H. and B. A. Maurer. 1989. Macroecology: The division of food and space among species on continents. *Science* 243: 1145–1150. [7]

Brown, J. H. and D. F. Sax. 2004. Gradients of species diversity: Why are there so many species in the Tropics? In M. V. Lomolino, D. F. Sax and J. H. Brown (eds.) *Foundations of Biogeography: Classic Works with Commentaries*. The University of Chicago Press, Chicago, pp. 1145–1154. [Part 3]

Brown, J. H. and M. V. Lomolino. 2000. Concluding remarks: Historical perspective and the future of island biogeography theory. *Global Ecology and Biogeography* 9: 87–92. [7, 11]

Brown, J. H., J. F. Gillooly, G. B. West and V. M. Savage. 2003. The next step in macroecology: From general empirical patterns to universal ecological laws. In T. M. Blackburn and K. J. Gaston (eds.), *Macroecology: Concepts and Consequences*. Blackwell Science, Oxford, pp. 408–423. [10]

Brown, W. L., Jr. 1957. Centrifugal speciation. *Quarterly Review of Biology* 32: 247–277. [13]

Brown, W. L., Jr. 1987. Punctuated equilibrium excused: The original examples fail to support it. *Biological J. Linnean Society* 31: 383–404. [13]

Browne, J. 1983. *The Secular Ark: Studies in the History of Biogeography*. Yale University Press, New Haven. [Part 1]

Brundin, L. 1966. Transantarctic relationships and their significance as evidenced by midges. *Kungliga Svenska Vetenskapsakademiens Handlinger* (Series 4): 11: 1–472. [4; Part 2]

Brundin, L. Z. 1988. Phylogenetic biogeography. In A. A. Myers and P. S. Giller (eds.), *Analytical Biogeography*. Chapman and Hall, New York, pp. 343–369. [6]

Brunsfeld, S. J., J. Sullivan, D. E. Soltis and P. S. Soltis. 2002. Comparative phylogeography of northwestern North America: A synthesis. In *Integrating Ecology and Evolution in a Spatial Context*. The 14th Special Symposium of the British Ecological Society. Royal Holloway College, University of London, pp. 319–339. [Part 2]

Buckley, R. C. and S. B. Knedlhans. 1986. Beachcomber biogeography: Interception of dispersing propagules by islands. *J. Biogeography* 13: 68–70. [11]

Bucklin, A. and P. H. Wiebe. 1998. Low mitochondrial diversity and small effective population sizes of the copepods *Calanus finmarchicus* and *Nannocalanus minor*: Possible impact of climate variation during recent glaciation. *Journal of Heredity* 89: 383–392. [Part 4]

Budd, A. F., K. G. Johnson and T. A. Stenmann. 1996. Plio-Pleistocene turnover and extinctions in the Caribbean reef-coral fauna. In J. B. C. Jackson, A. F. Budd and A. G. Coates (eds.), *Evolution and Environment in Tropical America*. University of Chicago Press, Chicago, pp. 168–204. [14]

Buffon, G. L. Leclerc, Compte de. 1761. *Histoire Naturelle Generale*. Imprimerie Royale, Paris. [Part 2]

Buffon, G. L. LeClerc, Compte de. 1776. Histoire naturelle generale et particuliere. *Servant de Suite à l'histoire des animaux Quadrupedes* (Suppl. III). Imprimerie Royale, Paris. [4]

Burness, G. P., J. Diamond and T. Flannery. 2001. Dinosaurs, dragons, and dwarfs: The evolution of maximal body size. *Proc. National Academy of Sciences USA* 98: 14,518–14,523. [12]

Burney, D. A., H. F. James, L. P. Burney, S. L. Olson, W. Kikuchi, W. L. Wagner, M. Burney, D. McCloskey, D. Kikuchi, F. G. Grady, R. Gage III and R. Nishek. 2001. Fossil evidence for a diverse biota from Kaua'i and its transformation since human arrival. *Ecological Monographs* 71: 615–641. [Part 1]

Burnham, R. J. and A. Graham. 1999. The history of neotropical vegetation: New developments and status. *Annals of the Missouri Botanical Garden* 86: 546–589. [Part 1]

Busby, J. R. 1986. A biogeoclimatic analysis of *Nothofagus cunninghamii* (Hook) Oerst in southern Australia. *Australian J. Ecology* 11: 1–7. [16]

Bush, M. B. 2002. Distributional change and conservation on the Andean flank: A paleoecological perspective. *Global Ecology and Biogeography* 11: 463–473. [Part 1]

Bush, M. B. and R. J. Whittaker. 1991. Krakatau: Colonization patterns and hierarchies. *J. Biogeography* 18: 341–356. [11]

Bush, M. B. and R. J. Whittaker. 1993. Non-equilibration in island theory of Krakatau. *J. Biogeography* 20: 453–458. [11]

Buzas, M. A. and S. J. Culver. 1999. Understanding regional species diversity through the log series distribution of occurrences. *Diversity and Distributions* 8: 187–195. [14]

Buzas, M. A., L. S. Collins and S. J. Culver. 2002. Latitudinal difference in biodiversity caused by higher tropical rate of increase. *Proc. National Academy of Sciences USA* 99: 7841–7843. [14]

Caley, M. J. and D. Schluter. 1997. The relationship between local and regional diversity. *Ecology* 78: 70–80. [14]

Calsbeek, R., J. N. Thompson and J. E. Richardson. 2003. Patterns of molecular evolution and diversification in a biodiversity hotspot: The Califor-nia Floristic Province. *Molecular Ecology* 12: 1021–1029. [Part 2]

Cantera, J. R. and P. M. Arnaud. 1984. Les Gastéropodes Prosobranches des Iles Kerguelen et Crozet (Sud de l'Océan Indien): Comparaison Ecologique et Particularités Biologiques. Comité National Français des Recherches Antarctiques 56: 1–169. [12]

CAPE Project Members. 2001. Holocene paleoclimate data from the Arctic: Testing models of

global climate change. *Quaternary Science Reviews* 20: 1275–1287. [3]
Cardillo, M. 1999. Latitude and rates of diversification in birds and butterflies. *Proc. Royal Society of London Series B* 266: 1221–1225. [9, 10]
Cardillo, M. 2002. The life-history basis of latitudinal diversity gradients: How do species traits vary from the poles to the equator. *J. Animal Ecology* 71: 79–87. [10]
Cardillo, M. and L. Bromham. 2001. Body size and risk of extinction in Australian mammals. *Conservation Biology* 15: 1435–1440. [8]
Cardillo, M., J. S. Huxtable and L. Bromham. 2003. Geographic range size, life history, and rates of diversification in Australian mammals. *J. Evolutionary Biology* 16: 282–288. [10]
Carlquist, S. 1995. Introduction. W. L. Wagner and V. A. Funk (eds.), *Hawaiian Biogeography: Evolution on a Hotspot Archipelago*. Smithsonian Series in Comparative Evolutionary Biology. Smithsonian Institution Press, Washington, D.C., pp. 1–13. [4]
Carlton, J. T. and J. B. Gellar. 1991. 1000 points of invasion: Rapid oceanic dispersal of coastal organisms and implications for evolutionary biology, ecology, and biogeography. *American Zoologist* 31: A127. [15]
Carpenter, G., A. N. Gillison and J. Winter. 1993. DOMAIN: A flexible modeling procedure for mapping potential distributions of animals and plants. *Biodiversity and Conservation* 2: 667–680. [16]
Carson, H. L. 2003. Mate choice theory and the mode of selection in sexual populations. *Proc. National Academy of Sciences of the USA* 100: 6584–6587. [12]
Carson, H. L. and D. A. Clague. 1995. Geology and biogeography of the Hawaiian Islands. In W. L. Wagner and V. A. Funk (eds.), *Hawaiian Biogeography: Evolution on a Hotspot Archipelago*. Smithsonian Series in Comparative Evolution-ary Biology. Smithsonian Institution Press, Washington, D.C., pp. 14–29. [4]
Carson, H. L., D. E. Hardy, H. T. Spieth and W. S. Stone. 1970. The evolutionary biology of the Hawaiian Drosophilidae. In M. K. Hecht and W. C. Steere (eds.), *Essays in Evolution and Genetics in Honor of Theodosius Dobzhansky*, pp. 437–543. Appleton-Century Crofts, New York. [12]
Castilla, J. C., S. A. Navarrete and J. Lubchenco. 1993. Southeastern Pacific coastal environments: Main features, large-scale perturbations, and global climate change. In H. A. Mooney, E. R. Fuentes and B. I. Kronberg (eds.), *Earth System Responses to Global Change: Contrasts between North and South America*. Academic Press, New York, pp. 167–188. [10]
Chace, F. A. 1966. Decapod crustaceans from St. Helena Island, South Atlantic. *Proc. United States National Museum* 118: 622–662. [12]
Channell, R., and M. V. Lomolino. 2000. Dynamic biogeography and conservation of endangered species. *Nature* 403: 84–86. [Part 5]
Chapman, V. J. 1976. *Mangrove Vegetation*. Strauss and Cramer GmBH, Germany. [13]
Chase, J. M. and M. A. Leibold. 2002. Spatial scale dictates the productivity-biodiversity relationship. *Nature* 416: 427–430. [11]
Chatterjee, S. and C. R. Scotese. 1999. The breakup of Gondwana and the evolution and biogeography of the Indian plate. *Proc. Indian National Science Academy* 65(3): 397–425. [1]
Chown, S. L. and K. J. Gaston. 2000. Areas, cradles, and museums: The latitudinal gradient in species richness. *Trends in Ecology and Evolution* 15: 311–315. [10, 14]
Ciampaglio, C. N., M. Kemp and D. W. McShea. 2001. Detecting changes in morphospace occupation patterns in the fossil record: Characterization and analysis of measures of disparity. *Paleobiology* 27: 695–715. [8]
Clark, J. S. and J. S. McLachlan. 2003. Stability of forest biodiversity. *Nature* 423: 635–638. [9]
Clarke, A. 1992. Is there a latitudinal diversity cline in the sea? *Trends in Ecology and Evolution* 7: 286–287. [10, 14]
Clarke, A. 1993. Temperature and extinction in the sea: A physiologist's view. *Paleobiology* 19: 499–518. [14]
Clarke, A. and J. A. Crame. 1992. The Southern Ocean benthic fauna and climate change: A historical perspective. *Philosophical Transactions of the Royal Society of London Series B* 338: 299–309. [14]
Clarke, A. and J. A. Crame. 1997. Diversity, latitude, and time: Patterns in the shallow sea. In R. F. G. Ormond, J. D. Gage and M. V. Angel (eds.), *Marine Biodiversity: Patterns and Processes*. Cambridge University Press, Cambridge, pp. 122–147. [10]
Clarke, A. and J. A. Crame. 2003. The importance of historical processes in global patterns of biodiversity. In T. M. Blackburn and K. J. Gaston (eds.), *Macroecology: Concepts and Consequences*. Blackwell Science, Oxford, pp. 130–154. [9, 14]
Clarke, A. and N. M. Johnston. 2003. Antarctic marine benthic diversity. *Oceanography and Marine Biology: An Annual Review* 41: 47–114. [14]
Clarke, A. and S. Lidgard. 2000. Spatial patterns of diversity in the sea: Bryozoan species rich-

ness in the North Atlantic. *J. Animal Ecology* 69: 799–814. [14]

Clement, M., D. Posada and K. A. Crandall. 2000. TCS: A computer program to estimate gene genealogies. *Molecular Ecology* 9: 1657–1660. [5]

Clinebell, H. R. R., O. L. Phillips, A. H. Gentry, N. Stark and H. Zuuring. 1995. Prediction of Neotropical tree and liana richness from soil and climatic data. *Biodiversity and Conservation* 4: 56–90. [11]

Cocks, L. R. M. and C. R. Scotese. 1991. The global biogeography of the Silurian period. In M. G. Bassett (ed.), *The Murchison Symposium.* Special Papers in Palaeontology 44, Geological Society of London, pp. 109–122. [1]

Cody, M. L. and E. Velarde. 2002. Land birds. In T. J. Case, M. L. Cody and E. Ezcurra (eds.), *A New Island Biogeography of the Sea of Cortés.* Oxford University Press, New York, pp. 271–312. [10]

Coelho, P. A. and R. Ramos de Araùjo. 1972. A contribuição e a distribuição da fauna de decapodos do litoral leste da America do sul entre as latitudes de 5°N e 39°S. *Trabalhos Oceanograficos da Universidade Federal de Pernambuco* 13: 133–236. [12]

Cohen, A. N. and J. T. Carlton. 1998. Accelerating invasion rate in a highly invaded estuary. *Science* 279: 555–558. [15]

Collins, L. S., A. F. Budd and A. G. Coates. 1996. Earliest evolution associated with closure of the Tropical American Seaway. *Proc. National Academy of Sciences USA* 93: 6069–6072. [14]

Colwell, R. K. 1997. *Estimates: Statistical Estimation of Species Richness and Shared Species from Samples. Version 5. User's Guide and Application.* Depart-ment of Ecology and Evolutionary Biology, University of Connecticut, Storrs, CT. http:/ / viceroy.eeb.uconn.edu/estimates. [14]

Colwell, R. K. and D. C. Lees. 2000. The mid-domain effect: Geometric constraints on the geography of species richness. *Trends in Ecology and Evolution* 15: 70–76. [11]

Condit, R., P. S. Ashton, P. Baker, S. Bunyavejchewin, S. Gunatilleke, N. Gunatilleke, S. P. Hubbell, R. B. Foster, A. Itoh, J. V. LaFrankie, H. S. Lee, E. Losos, N. Manokaran, R. Sukumar and T. Yamakura. 2000. Spatial patterns in the distribution of tropical tree species. *Science* 288: 1414–1418. [11]

Connell, J. H. 1978. Diversity in tropical rain forests and coral reefs. *Science* 199: 1302–1310. [11]

Connell, J. H. and E. Orias. 1964. The ecological regulation of species diversity. *American Naturalist* 98: 399–414. [9]

Connin, S. L., J. L. Betancourt and J. Quade. 1998. Late Pleistocene C_4 plant dominance and summer rainfall in the southwestern U.S.A. from isotopic study of herbivore teeth. *Quaternary Research* 2: 170–193. [2]

Cook, E. R., D. M. Meko, D. W. Stahle and M. K. Cleaveland. 1999. Drought reconstruction for the continental United States. *J. Climate* 12: 1145–1162. [3]

Cook, P. J. 1990. *Australia: Evolution of a Continent.* BMR Palaeogeographic Group. Australian Government Publishing Service, Canberra. [1]

Cook, T. D. and A. W. Bally. 1975. *Stratigraphic Atlas of North and Central America.* Princeton University Press, Princeton, NJ. [1]

Cope, J. C. W., J. K. Ingham and P. F. Rawson. 1992. *Atlas of Paleogeography and Lithofacies.* Geological Society of London, Memoir 13. [1]

Cordani, U. G., E. J. Milani, A. T. Filho and D. A. Campos (eds.). 2000. *Tectonic Evolution of South America.* Brazil 2000, 31st International Geological Congress, August 6–17, 2000, Rio de Janeiro. [1]

Cornell, H. V. and J. H. Lawton. 1992. Species interactions, local and regional processes, and limits to the richness of ecological communities: A theoretical perspective. *J. Animal Ecology* 61: 1–12. [14]

Cousins, S. H. 1989. Species richness and energy theory. *Nature* 340: 350–351. [8]

Cowling, R. M., D. M. Richardson, R. J. Schulze, M. T. Hoffman, J. J. Midgley and C. Hilton-Taylor. 1997. Species diversity at the regional scale. In R. M. Cowling, D. M. Richardson and S. M. Pierce (eds.), *Vegetation of Southern Africa.* Cambridge University Press, Cambridge, pp 447–473. [11]

Cox, B. and P. D. Moore. 2000. *Biogeography: An Ecological and Evolutionary Approach,* 6th Ed. Blackwell Science, Oxford. [14]

Cracraft, J. 1988. Deep-history biogeography: Retrieving the historical pattern of evolving continental biotas. *Systematic Zoology* 37: 221–236. [5, 6]

Craig, R., L. Shiozawa, J. L. Betancourt, T. L. Burgess and E. Pierson. 2003. Climatic envelopes for species distributions using a statistical downscaling model: Saguaro, *Carnegiea giganteae,* in the Sonoran Desert, USA and Mexico. In preparation. [2]

Crame, J. A. 1996. Antarctica and the evolution of taxonomic diversity gradients in the marine realm. *Terra Antarctica* 3: 121–134. [8]

Crame, J. A. 1997. An evolutionary framework for the Polar regions. *J. Biogeography* 24: 1–9. [10, 14]

Crame, J. A. 2000a. Evolution of taxonomic diversity gradients in the marine realm: Evidence from the composition of Recent bivalve faunas. *Paleobiology* 26: 188–214. [13, 14]

Crame, J. A. 2000b. The nature and origin of taxonomic diversity gradients in marine bivalves. In E. M. Harper, J. D. Taylor and J. A. Crame (eds.), *The Evolutionary Biology of the Bivalvia*. Special Publications 177, Geological Society of London, pp. 347–360. [14]

Crame, J. A. 2001. Taxonomic diversity gradients through geological time. *Diversity and Distributions* 7: 175–189. [14]

Crame, J. A. 2002. Evolution of taxonomic diversity gradients in the marine realm: A comparison of Late Jurassic and Recent bivalve faunas. *Paleobiology* 28: 184–207. [8, 10, 14]

Crame, J. A. and B. R. Rosen. 2002. Cenozoic palaeogeography and the rise of modern biodiversity patterns. In J. A. Crame and A. W. Owen (eds.), *Palaeobiogeography and Biodiversity Change: The Ordovician and Mesozoic-Cenozoic Radiations*. Special Publications 194, Geological Society of London, pp. 153–168. [14]

Craw, R. C. 1983. Panbiogeography and vicariance cladistics: Are they truly different? *Systematic Zoology* 32: 431–438. [4]

Craw, R. C. and R. D. M. Page. 1988. Panbiogeography: Method and metaphor in the new biogeography. In M.-W. Ho and S. W. Fox (eds.), *Panbiogeography: Method and Metaphor in the New Biogeography*. John Wiley and Sons, New York, pp. 163–189. [4]

Craw, R. C., J. R. Grehan, J. R. and M. J. Heads. 1999. *Panbiogeography: Tracking the History of Life*. Oxford Biogeography Series No. 11. Oxford University Press, New York. [Part 2]

Cressey, R. F., B. Collette and J. Russo. 1983. Copepods and scombrid fishes: A study in host-parasite relationships. *Fishery Bulletin* 81: 227–265. [7]

Crisci, J. V. 2001. The voice of historical biogeography. *J. Biogeography* 28: 157–168. [5]

Crisci, J. V., M. M. Cigliano, J. J. Morrone and S. Roig-Junent. 1991. Historical biogeography of southern South America. *Systematic Zoology* 40: 152–171. [7]

Crisci, J. V., L. Katinas and P. Posadas. 2003. *Historical Biogeography: An Introduction*. Harvard University Press, Cambridge, MA. [4; Part 2]

Critchfield, W. B. 1984. Impact of the Pleistocene on the genetic structure of North American conifers. In R. M. Lanner (ed.), *Proceedings of the Eighth North American Forest Biology Workshop, July 30–August 1, 1984, Utah State University, Logan, UT*. Pages 70–118. [3]

Croizat, L. 1952. *Manual of Phytogeography*. Dr. W. Junk B.V., The Hague. [4]

Croizat, L. 1958. *Panbiogeography* (3 volumes). Published by the author, Caracas. [4, 5; Part 2]

Croizat, L. 1964. *Space, Time, Form: The Biological Synthesis*. Published by the author, Caracas. [4]

Croizat, L., G. Nelson and D. E. Rosen. 1974. Centers of origin and related concepts. *Systematic Zoology* 23: 265–287. [4, 6]

Csuti, B., S. Polasky, P.-H. Williams, R. L. Pressey, L. D. Camm and M. Kershaw. 1997. A comparison of reserve selection algorithms using data on terrestrial vertebrates in Oregon. *Biological Conservation* 80: 83–97. [16]

Cuffey, K. M. and G. D. Clow. 1997. Temperature, accumulation and ice sheet elevation in central Greenland through the last deglacial transition. *J. Geophysical Research* 102(C12): 26383–26396. [3]

Cully, J. F., Jr. and E. S. Williams. 2001. Interspecific comparisons of sylvatic plague in prairie dogs. *J. Mammalogy*, 82, 894–905. [Part 5]

Culver, S. J. and M. A. Buzas. 2000. Global latitudinal species diversity gradient in deep-sea benthic foraminifera. *Deep-Sea Research I* 47: 259–275. [14]

Cunningham, C. W. and T. M. Collins. 1994. Developing model systems for molecular biogeography: Vicariance and interchange in marine invertebrates. In B. Schierwater, B. Streit, G. P. Wagner and R. DeSalle (eds.), *Molecular Ecology and Evolution: Approaches and Applications*. Birkhauser Verlag, Berlin, pp. 405–433. [5; Part 4]

Currie, D. J. 1991. Energy and large-scale patterns of animal—and plant—species richness. *Ameri-can Naturalist* 137: 27–49. [9, 10, 11]

Currie, D. J. 2001. Projected effects of climate change on patterns of vertebrate and tree species richness in the conterminous United States. *Ecosystems* 4: 216–225. [9]

Currie, D. J. and V. Paquin. 1987. Large-scale biogeographical patterns of species richness of trees. *Nature* 329: 326–327. [9, 11]

Cushman, J. H., J. H. Lawton and B. F. J. Manly. 1993. Latitudinal patterns in European ant assemblages: Variation in species richness and body size. *Oecologia* 95: 30–37. [8]

Daly, C., R. P. Neilson and D. L. Phillips. 1994. A statistical-topographic model for mapping climatological precipitation over mountainous terrain. *J. Applied Meteorology* 33: 140–158. [2]

Dalziel, I. W. D. 1991. Pacific margins of Laurentia and East Antarctica-Australia as a conjugate rift pair: Evidence and implications for an

Eocambrian supercontinent. *Geology* 19: 598–601. [1]

Dalziel, I. W. D. 1997. Neoproterozoic-Paleozoic geography and tectonics: Review, hypothesis, and environmental speculation. *Geological Society of America Bulletin* 109: 16–42. [1]

Dammerman, K. W. 1948. *The Fauna of Krakatau (1883–1933)*. Verhandelingen der Koninklijke Nederlandse Akademie Van Wetenschappen Afdeling Natuurkunde Tweede Sectie, N. V. Noord-Hollandsche Uitgevers Maatschappij, Amsterdam, 44: 1–594. [11]

Dana, J. D. 1853. On an isothermal oceanic chart: Illustrating the geographical distribution of marine animals. *American J. Science* Series 2, 16: 153–167, 314–327. [8; Part 4]

Danielsen, F. and C. G. Treadaway. In press. Priority conservation areas for Lepidoptera (Rhopalocera) in the Philippine Islands. *Animal Conservation*. [18]

Darlington, P. J. 1943. Carabidae of mountains and islands: Data on the evolution of isolated faunas, and on atrophy of wings. *Ecological Monographs* 13: 38–61. [7]

Darlington, P. J. Jr. 1957. *Zoogeography: The Geographical Distribution of Animals*. John Wiley and Sons, New York. [Part 2]

Darragh, T. A. 1985. Molluscan biogeography and biostratigraphy of the Tertiary of southeastern Australia. *Alcheringa* 9: 83–116. [14]

Darwin, C. 1839. *The Voyage of the Beagle* (Chapter 17). Harper, New York. [Part 5]

Darwin, C. 1859. *On the Origin of Species by Means of Natural Selection, or the Preservation of Favoured Races in the Struggle for Life*. John Murray, London. Facsimile of First Edition reprinted by Harvard University Press, Cambridge, MA. [4, 6, 12, 13, 15; Part 4]

Davis, A. J., L. J. Jenkinson, J. H. Lawton, B. Shorrocks and S. Woods. 1998. Making mistakes when predicting shifts in species range in response to global warming. *Nature* 391: 783–786. [2]

Davis, M. A. 2003. Biotic globalization: Does competition from introduced species threaten biodiversity? *BioScience* 53: 481–489. [15]

Davis, M. A. and K. Thompson. 2000. Invasion terminology: Should ecologists define their terms differently than others? No, not if we want to be of any help. *Bulletin of the Ecological Society of America* 82: 206. [15]

Davis, M. B. 1969. Climatic changes in southern Connecticut recorded by pollen deposition at Rogers Lake. *Ecology* 50: 409–422. [Part 3]

Davis, M. B. 1976. Pleistocene biogeography of temperate deciduous forests. *Geoscience and Man* 13: 13–26. [3]

Davis, M. B. 1981. Quaternary history and the stability of forest communities. In D.C. West, H. H. Shugart and D. B. Botkin (eds.), *Forest Succession: Concepts and Applications*. Springer-Verlag, New York, pp. 132–154. [3]

Davis, M. B. 1983. Quaternary history of deciduous forests of eastern North America and Europe. *Annals of the Missouri Botanical Garden* 70: 550–563. [3]

Davis, M. B. 2000. Palynology after Y2K: Understanding the source area of pollen in sediments. *Annual Review of Earth and Planetary Sciences* 28: 1–18. [3]

Davis, M. B. and R. G. Shaw. 2001. Range shifts and adaptive responses to Quaternary climate change. *Science* 292: 673–679. [3; Part 1]

Davis, M. B., R. R. Calcote, S. Sugita and H. Takahara. 1998. Patchy invasion and the origin of a hemlock-hardwoods forest mosaic. *Ecology* 79: 2641–2659. [3]

De Broyer, C. and K. Jazdzewski. 1996. Biodiversity of the Southern Ocean: Towards a new synthesis for the Amphipoda (Crustacea). *Bolletin del Museo Civico di Storia Naturale di Verona* 20: 547–568. [14]

de Candolle, A. P. 1820. *Essai Elementaire de Geographie Botanique*. Flevrault, Strasbourg and Paris. [4]

de Jong, H. 1998. In search of historical biogeographic patterns in the western Mediterranean terrestrial fauna. *Biological J. Linnaean Society* 65: 99–164. [7]

de Queiroz, A. 2002. Contingent predictability in evolution: Key traits and diversification. *Systematic Biology* 51: 917–929. [10]

Deevey, E. S. 1949. Biogeography of the Pleistocene. Part 1. Europe and North America. *Geological Society of America Bulletin* 60: 1315–1416. [3]

Dell, R. K. 1990. Antarctic Mollusca: With special reference to the fauna of the Ross Sea. *The Royal Society of New Zealand* 27: 1–297. [10, 14]

den Hartog, C. 1970. *The Sea Grasses of the World*. Nederland Academy, Tweede Reeks. [13]

Denton, J. S., S. P. Hitchings, T. J. C. Beebee and A. Gent. 1997. A recovery program for the natterjack toad (*Bufo calamita*) in Britain. *Conservation Biology* 11: 1329–1338. [17]

Dercourt, J., L. E. Ricou and B. Vrielynck (eds.). 1993. *Atlas Tethys Palaeoenvironmental Maps*. Gauthier-Villars, Paris. [1]

DeSalle, R. 1995. Molecular approaches to biogeographic analyses of Hawaiian Drosophilidae. In W. L. Wagner and V. A. Funk (eds.), *Hawaiian Biogeography: Evolution on a Hotspot Archipelago*. Smithsonian Series in Comparative Evolutionary Biology. Smithsonian Institution Press, Washington, D.C., pp. 72–89. [4]

di Castri, F., A. J. Hansen and M. Debussche. 1990. *Biological Invasions in Europe and the Mediterranean Basin*. Kluwer Academic Publishers, New York. [15]

Dial, K. P. and N. Czaplewski. 1990. Do woodrat middens accurately reflect the animals' environment or diets? In J. L. Betancourt, T. R. Van Devender and P. S. Martin (eds.), *Packrat Middens: The Last 40,000 Years of Biotic Change*. The Woodhouse Mesa Study, University of Arizona Press, Tucson, pp. 43–58. [2]

Diamond, J. 1974. Colonization of exploded volcanic islands by birds: The supertramp strategy. *Science* 184: 803–806. [12]

Diamond, J. 1988. Factors controlling species diversity: Overview and synthesis. *Annals of the Missouri Botanical Garden* 75: 117–129. [10]

Diamond, J. 1991. A new species of rail from the Solomon Islands and convergent evolution of insular flightlessness. *Auk* 108: 461–470. [12]

Diamond, J. M. 1998. *Guns, Germs and Steel*. W.W. Norton, New York. [Part 5]

Diamond, J. M. and R. M. May. 1976. Island biogeography and the design of natural reserves. In R. M. May (ed.), *Theoretical Ecology: Principles and Applications*. W. B. Saunders, Philadelphia, pp. 163–186. [17]

Dickerson, R. E. 1928. *Distribution of Life in the Philippines*, 2: 1–322. Monograph, Bureau of Science, Manila. [18]

Dietz, R. S. and J. C. Holden. 1970. The breakup of Pangaea. *Scientific American* 223: 30–41. [7]

Dilcher, D. L. 2000. Geological history of the vegetation in southeast United States. SIDA, *Botanical Miscellany*, 181–211. [Part 1]

Diniz-Filho, J. A. F., L. M. Bini and B. A. Hawkins. 2003. Spatial autocorrelation and red herrings in geographical ecology. *Global Ecology and Biogeography* 12: 53–64. [9, 11]

Dixon, A. F. G., P. Kindlmann, J. Leps and J. Holman. 1987. Why are there so few species of aphids, especially in the tropics? *American Naturalist* 129: 580–592. [9]

Dommergues, J.-L., S. Montuire and P. Neige. 2002. Size patterns through time: The case of the Early Jurassic ammonite radiation. *Paleobiology* 28: 423–434. [8]

Donaldson, T. J. 1986. Distribution and species richness patterns of Indo-West Pacific Cirrhitidae: Support for Woodland's hypothesis. In T. Uyeno, R. Arai, T. Taniuchi and K. Matsura (eds.), *Indo-Pacific Fish Biology*, pp. 623–628. Ichthyological Society of Japan, Tokyo. [13]

Donoghue, M. J. and B. R. Moore. 2003. Toward an integrative historical biogeography. *Integrative and Comparative Biology* 43: 261–270. [4]

Donoghue, M. J., J. A. Doyle, J. Gauthier, A. G. Kluge and T. Rowe. 1989. The importance of fossils in phylogeny reconstruction. *Annual Reviews of Ecology and Systematics* 20: 431–460. [6]

Dudley, R. 2000. *The Biomechanics of Insect Flight: Form, Function, Evolution*. Princeton University Press, Princeton, NJ. [12]

Duncan, J. R. and J. L. Lockwood. 2001. Spatial homogenization of the aquatic fauna of Tennessee: Extinction and invasion following land use change and alteration. In J. L. Lockwood and M. L. McKinney (eds.), *Biotic Homogenization*. Kluwer Academic/Plenum Press, New York. [15]

Duncan, R. A. 1981. Hot spots in the southern oceans: An absolute frame of reference for motion of the Gondwana continents. *Tectonophysics* 74: 29–42. [1]

Duncan, R. A. and M. A. Richards. 1991. Hot spots, mantle plumes, flood basalts, and true polar wander. *Reviews of Geophysics* 29: 31–50. [1]

Dynesius, M. and R. Jansson. 2000. Evolutionary consequences of changes in species' geographical distributions driven by Milankovitch climate oscillations. *Proc. National Academy of Sciences USA* 97: 9115–9120. [9, 14]

Ebach, M. C. 1999. Paralogy and the center of origin concept. *Cladistics* 15: 387–391. [4]

Ebach, M. C. 2003. Area cladistics. *Biologist* 50: 169–172. [4]

Ebach, M. C. and C. J. Humphries. 2002. Cladistic biogeography and the art of discovery. *J. Biogeography* 20: 427–444. [4, 5; Part 2]

Ebach, M. C. and C. J. Humphries. 2003. Ontology of biogeography. *J. Biogeography* 30: 959–962. [5]

Ebach, M. C. and D. M. Williams. 2004. Congruence and language. *Taxon* 53: 113–118. [4]

Ebach, M. C., C. J. Humphries and D. M. Williams. 2003a. Phylogenetic biogeography deconstructed. *J. Biogeography* 30: 1285–1296. [4]

Ebach, M. C., R. A. Newman, C. J. Humphries and D. M. Williams. 2003b. *3item: Three-item analysis for area cladistics*. Published by the authors, Natural History Museum, London. [4]

Eble, G. 1998. The role of development in evolutionary radiations. In M. L. McKinney and J. A. Drake (eds.), *Biodiversity Dynamics*. Columbia University Press, New York, pp. 132–161. [8]

Edmunds, G. F. 1975. Phylogenetic biogeography of mayflies. *Annals of the Missouri Botanical Gardens* 62: 251–263. [4]

Edwards, C. W., C. F. Fulhorst and R. D. Bradley. 2001. Molecular phylogenetics of the *Neotoma albigula* species group: Further evidence of a paraphyletic assemblage. *J. Mammalogy* 82: 267–279. [2]

Edwards, S. V. and P. Beerli. 2000. Perspective: Gene divergence, population divergence and the variance in coalescence time in phylogeographic studies. *Evolution* 54: 1839–1854. [5]

Egbert, S. L., A. T. Peterson, V. Sánchez-Cordero and K. Price. 1999. Modeling conservation priorities in Veracruz, Mexico. In S. Morain (ed.), *GIS Solutions in Natural Resource Management*. OnWord Press, Santa Fe, NM, pp. 141–150. [16]

Ekman, S. 1935. *Tiergeographie des Meeres*. Akademische Verlagsgesellschaft, Leipzig. [Part 4]

Ekman, S. 1953. *Zoogeography of the Sea*. Sidgwick and Jackson, London. [13; Part 4]

Eldholm, O. and M. F. Coffin. 2000. Large igneous provinces and plate tectonics. In M. A. Richards, R. G. Gordon and R. D. van der Hilst (eds.), *The History and Dynamics of Global Plate Motions*. Geophysical Monograph Series 121: 309–326. [1]

Eldredge, N. 1979. Alternative approaches to evolutionary theory. *Bulletin of the Carnegie Museum of Natural History* 13: 7–19. [10]

Eldredge, N. and J. Cracraft. 1980. Phylogenetic Patterns and the Evolutionary Process: Method and Theory in Comparative Biology. Columbia University Press, New York. [6]

Eldredge, N. and S. J. Gould. 1972. Punctuated equilibria: An alternative to phyletic gradualism. In J. M. Schopf (ed.), *Models in Paleobiology*, pp. 82–115. Freeman, Cooper and Co., San Francisco. [13]

Elias, S. A. 1994. *Quaternary Insects and Their Environments*. Smithsonian Institution Press, Washington, D.C. [3]

Elias, S. A., T. R. Van Devender and R. De Baca. 1995. Insect fossil evidence of Late Glacial and Holocene environments in the Bolson de Mapimi, Chihuahuan Desert, Mexico: Compar-isons with the paleobotanical record. *Palaios* 10: 454–464. [2]

Ellingsen, K. E. 2001. Biodiversity of a continental shelf soft-sediment macrobenthos community. *Marine Ecology Progress Series* 218: 1–15. [14]

Ellingsen, K. E. and J. S. Gray. 2002. Spatial patterns of benthic diversity: Is there a latitudinal gradient along the Norwegian continental shelf? *J. Animal Ecology* 71: 373–389. [14]

Ellison, A. M., E. J. Farnsworth and R. E. Merkt. 1999. Origins of mangrove ecosystems and the mangrove biodiversity anomaly. *Global Ecology and Biogeography* 8: 95–115. [14]

Elton, C. 1958. *The Ecology of Invasions by Animals and Plants*. Methuen, London. [15; Part 5]

Endler, J. A. 1982. Problems in distinguishing historical from ecological factors in biogeography. *American Zoologist* 22: 441–452. [7]

Enghoff, H. 1995. Historical biogeography of the Holarctic: Area relationships, ancestral areas and dispersal of non-marine animals. *Cladistics* 11: 223–263. [6]

Enghoff, H. 1996. Widespread taxa, sympatry, dispersal and an algorithm for resolved area cladograms. *Cladistics* 12: 349–364. [6]

Enquist, B. J., G. B. West, E. L. Charnov and J. H. Brown. 1999. Allometric scaling of production and life-history variation in vascular plants. *Nature* 401: 907–911. [11]

Environmental Science for Social Change. 1999. *Decline of Philippine Forests*. ESSC, Inc. and Bookmark, Makati, Philippines. [18]

Erwin, D. H. 2001. Lessons from the past: Biotic recoveries from mass extinctions. *Proc. National Academy of Sciences USA* 98: 5399–5403. [8]

Erwin, T. L. 1979. Thoughts on the evolutionary history of ground beetles: Hypotheses generated from comparative faunal analyses of lowland forest sites in temperate and tropical regions. In T. L. Erwin, G. E. Ball and D. R. Whitehead (eds.), *Carabid Beetles: Their Evolution, Natural History, and Classification*. Dr. W. Junk B.V., The Hague, pp. 539–592. [7]

Erwin, T. L. 1981. Taxon pulses, vicariance, and dispersal: An evolutionary synthesis illustrated by carabid beetles. In G. Nelson and D. E. Rosen (eds.), *Vicariance Biogeography: A Critique*. Columbia University Press, New York, pp. 159–196. [7]

Erwin, T. L. 1985. The taxon pulse: A general pattern of lineage radiation and extinction among carabid beetles. In G. E. Ball (ed.), *Taxonomy, Phylogeny and Zoogeography of Beetles and Ants*. Dr. W. Junk, Dordrecht, NL, pp. 437–472. [7]

Erwin, T. L. 1991. An evolutionary basis for conservation strategies. *Science* 253: 750–752. [7]

Erwin, T. L. and J. Adis. 1982. Amazonian inundation forests: Their role as short-term refuges and generators of species richness and taxon pulses. In G. Prance (ed.), *Biological Diversification in the Tropics*. Columbia University Press, New York, pp. 358–371. [7]

Espizua, L. E. 1999. Chronology of Late Pleistocene advances in the Río Mendoza Valley, Argentina. *Global and Planetary Change* 22: 193–200. [2]

Evans, B. J., J. Supriatna, N. Andayani, M. I. Setiadi, D. C. Cannatella and D. J. Melnick. 2003. Monkeys and toads define areas of endemism on Sulawesi. *Evolution* 57: 1436–1443. [Part 2]

Fairbanks, D. H., B. Reyers and A. S. van Jaarsveld. 2001. Species and environment representation: Selecting reserves for the retention of avian diversity in KwaZulu-Natal, South Africa. *Biological Conservation* 98: 365–379. [16]

Fairbanks, R. G. 1989. A 17,000-year glacio-eustatic sea level record: Influence of glacial melting on the Younger Dryas event and deep-sea circulation. *Nature* 342: 637–642. [18]

Faith, D. P. 1992. Conservation evaluation and phylogenetic diversity. *Biological Conservation* 61: 1–10. [Part 2]

FAUNMAP Working Group. 1994. FAUNMAP: A database documenting Late Quaternary distributions of mammal species in the United States. *Illinois State Museum Scientific Papers 25.* [3]

FAUNMAP Working Group. 1996. Spatial responses of mammals to Late Quaternary environmental fluctuations. *Science* 272: 1601–1606. [3; Part 1]

Feria, T. P. and A. T. Peterson. 2002. Prediction of bird community composition based on point-occurrence data and inferential algorithms: A valuable tool in biodiversity assessments. *Diversity and Distributions* 8: 49–56. [16]

Fielding, A. H. and J. F. Bell. 1997. A review of methods for the assessment of predictions errors in conservation presence/absence models. *Environmental Conservation* 24: 38–49. [16]

Fielding, A. H. and P. F. Haworth. 1995. Testing the generality of bird-habitat models. *Conservation Biology* 9: 1466–1481. [16]

Fieldsa, J. and J. C. Lovett. 1997. Geographical patterns of old and young species in African forest biota: The significance of specific montane areas as evolutionary centres. *Biodiversity and Conservation* 6: 325–346. [9]

Fitch, W. M. 1971. Toward defining the course of evolution: Minimum change for a specific tree topology. *Systematic Zoology* 20: 406–416. [6]

Flannery, T. F. 1995. *The Future Eaters: An Ecological History of the Australasian Lands and People.* G. Braziller, New York. [Part 5]

Flessa, K. W. and D. Jablonski. 1995. Biogeography of Recent marine bivalve molluscs and its implications for paleobiogeography and the geography of extinction: A progress report. *Historical Biology* 10: 25–47. [14]

Flessa, K. W. and D. Jablonski. 1996. Geography of evolutionary turnover. In D. Jablonski, D. H. Erwin and J. Lipps (eds.), *Evolutionary Paleobiol-ogy*. Chicago University Press, Chicago, pp. 376–397. [13, 14]

Foin, T. C. 1976. Plate tectonics and the biogeography of the Cypraeidae (Mollusca: Gastropoda). *J. Biogeography* 3: 19–34. [13]

Foote, M. 1996. Models of morphological diversification. In D. Jablonski, D. H. Erwin and J. H. Lipps (eds.), *Evolutionary Paleobiology*. University of Chicago Press, Chicago, pp. 62–86. [8]

Foote, M. 1997. The evolution of morphological diversity. *Annual Review of Ecology and Systematics* 28: 129–152. [8]

Forbes, E. 1843. Report on the Mollusca and Radiata of the Aegean Sea. *Reports of the British Association of Science* 1843 (1844): 130–193. [Part 4]

Forbes, E. 1856. Map of the distribution of marine life. In A. K. Johnston (ed.), *The Physical Atlas of Natural Phenomena*. Lea and Blanchard, Philadelphia. [Part 4]

Forbes, E. 1859. *The Natural History of European Seas*. John Van Voorst, London. [Part 4]

Ford, E. B. 1945. *Butterflies*. Collins, London. [9]

Forest, J. and D. Guinot. 1962. Remarques biogéographiques sur les crabes des archipels de la Société et des Tuamotu. *Cahier du Pacifique* 4: 41–75. [12]

Forey, P. L. 1981. Biogeography. In P. L. Forey (ed.), *Chance, Change and Challenge: The Evolving Biosphere*, Vol. 2. British Museum Natural History and Cambridge University Press, London and Cambridge, pp. 241–245. [4]

Fornari, M., F. Risacher and G. Féraud. 2001. Dating of paleolakes in the central Altiplano of Bolivia. *Palaeogeography, Palaeoclimatology, Palaeoecology* 172: 269–282. [2]

Forster, J. R. 1778. Observations Made During a Voyage Round the World, on Physical Geography, Natural History, and Ethic Philosophy. G. Robinson, London. [9]

Foster, J. B. 1964. Evolution of mammals on islands. *Nature* 202: 234–235. [12]

Frakes, L. A., J. E. Francis and J. I. Syktus. 1992. *Climate Modes of the Phanerozoic*. Cambridge University Press, Cambridge. [14]

Francis, A. P. and D. J. Currie. 1998. Global patterns of tree species richness in moist forests: Another look. *Oikos* 81: 598–602. [9]

Francis, A. P. and D. J. Currie. 2003. A globally consistent richness-climate relationship for angiosperms. *American Naturalist* 161: 523–536. [8, 11]

Franklin, J. 1995. Predictive vegetation mapping: Geographical modeling of biospatial patterns in relation to environmental gradients. *Progress in Physical Geography* 19: 474–499. [2]

Fraser, R. H. 1998. Vertebrate species richness at the mesoscale: Relative roles of energy and heterogeneity. *Global Ecology and Biogeography Letters* 7: 215–220. [11]

Fraser, R. H. and D. J. Currie. 1996. The species richness-energy hypothesis in a system where historical factors are thought to prevail: Coral reefs. *American Naturalist* 148: 138–159. [8, 9]

Freckleton, R. P. and A. R. Watkinson. 2003. Are all plant populations metapopulations? *J. Ecology* 91: 321–324. [11]

Frey, J. K. 1993. Modes of peripheral isolate formation and speciation. *Systematic Biology* 42: 373–381. [13]

Fricke, R. 1988. *Systematik und Historische Zoogeography der Callionymidae (Teleostei): Des Indischen Ozeans*, 2 Vol. Ph.D. Dissertation. Albert-Ludwigs-Universitat, Freiberg im Breisgau, Germany. [13]

Fryer, G. and M. J. Lucas. 2001. A century and a half of change in the butterfly fauna of the Huddersfield area of Yorkshire. *The Naturalist* 126: 49–112. [9]

Funk, V. A. 2004. Revolutions in historical biogeography. In M. Lomolino, D. Sax and J. Brown (eds.), *Foundations in Biogeography*. University of Chicago Press, Chicago. pp. 647–657. [Part 2]

Funk, V. A. and D. R. Brooks. 1990. Phylogenetic systematics as the basis of comparative biology. *Smithsonian Contributions to Botany* 73: 1–45. [6, 7]

Funk, V. A. and W. L. Wagner. 1995. Biogeographic patterns in the Hawaiian Islands. In W. L. Wagner and V. A. Funk (eds.), *Hawaiian Biogeography: Evolution on a Hotspot Archipelago*. Smithsonian Series in Comparative Evolutionary Biology. Smithsonian Institution Press, Washington, D.C., pp. 1–13. [4]

Gaby, R., M. P. McMahon, F. J. Mazzoti, W. N. Gillies and J. R. Wilcox. 1985. Ecology of a population of *Crocodylus acutus* at a power plant site in Florida. *J. Herpetology* 19: 189–198. [17]

Gage, J. D., and P. A. Tyler. 1991. Deep-sea biology: A natural history of organisms at the deep-sea floor. Cambridge University Press, Cambridge. [Part 4]

Gahagan, L. M., C. R. Scotese, J.-Y. Royer, D. T. Sandwell, J. K. Winn, R. L. Tomlins, M. I. Ross, J. S. Newman, R. D. Mueller, C. L. Mayes, L. A. Lawver and C. E. Heubeck. 1988. Tectonic fabric of the ocean basins from satellite altimetry data. In C. R. Scotese and W. W. Sager (eds.), Mesozoic and Cenozoic plate reconstructions. *Tectono-physics* 155: 1–26. [1]

Gaines, S. D. and J. Lubchenco. 1982. A unified approach to marine plant-herbivore interactions. II. Biogeography. *Annual Review of Ecology and Systematics* 13: 111–138. [10]

Gajewski, K., R. Vance, M. Sawada, I. Fung, L. D. Gignac, L. Halsey, J. John, P. Maisongrande, P. Mandell, P. J. Mudie, P. J. H. Richard, A. G. Sherin, J. Soroko and D. H. Vitt. 2000. The climate of North America and adjacent ocean waters ca. 6 ka. *Canadian J. Earth Sciences* 37: 661–681. [3]

Gallardo, C. S. and P. E. Penchaszadeh. 2001. Hatching mode and latitude in marine gastropods: Revisiting Thorson's paradigm in the Southern Hemisphere. *Marine Biology* 138: 547–552. [10]

Garcia, A. and H. Lagiglia. 1999. A 30,000-year old megafauna dung layer from Gruta del Indio Mendoza, Argentina. *Current Research in the Pleistocene* 16: 116–118. [2]

Garth, J. S. 1965. Brachyuran decapod Crustacea of Clipperton Island. *Proc. California Academy of Sciences* 33: 1–46. [12]

Garth, J. S. 1973. The brachyuran crabs of Easter Island. *Proc. California Academy of Sciences* 39: 311–336. [12]

Gaston, K. J. 1996. What is biodiversity? In K. J. Gaston (ed.), *Biodiversity*. Blackwell Science, Oxford, pp. 1–9. [8]

Gaston, K. J. 2000. Global patterns in biodiversity. *Nature* 405: 220–227. [8, 10, 14]

Gaston, K. J. 2003. *The Structure and Dynamics of Geographic Ranges*. Oxford University Press, Oxford. [10]

Gaston, K. J. and T. M. Blackburn. 1995. Birds, body size and the threat of extinction. *Philosophical Transactions of the Royal Society of London Series B* 347: 205–212. [8]

Gaston, K. J. and P. H. Williams. 1996. Spatial patterns in taxonomic diversity. In K. J. Gaston (ed.), *Biodiversity: A Biology of Numbers and Difference*. Blackwell Science, Oxford, pp. 202–229. [14]

Gaston, K. J., S. L. Chown and R. D. Mercer. 2001. The animal species-body size distribution of Marion Island. *Proc. National Academy of Sciences USA* 98: 14493–14496. [10]

Gaston, K. J., P. H. Warren and P. M. Hammond. 1992. Predator/non-predator ratios in beetle assemblages. *Oecologia* 90: 417–421. [8]

Gauthier, J., A. G. Kluge and T. Rowe. 1988. Amniote phylogeny and the importance of fossils. *Cladistics* 4: 105–209. [6]

Gaylord, B. and S. D. Gaines. 2000. Temperature or transport? Range limits in marine species mediated solely by flow. *American Naturalist* 155: 769–789. [8]

Gilbert, C. R. 1972. Characteristics of the Western Atlantic reef-fish fauna. *Quarterly Journal of the Florida Academy of Sciences* 35: 130–144. [Part 4]

Gilbert, F. S. 1980. The equilibrium theory of island biogeography, fact or fiction? *J. Biogeography* 7: 209–235. [11]

Givens, C. R. and F. M. Givens. 1987. Age and significance of fossil white spruce *Picea glauca*, Tunica Hills, Louisiana-Mississippi. *Quaternary Research* 27: 283–296. [3]

Givnish, T. J., K. J. Sytsma, J. F. Smith and W. S. Hahn. 1995. Molecular evolution, adaptive radiation and geographic speciation in *Cyanea* (Campanulaceae, Lobelioideae). In W. L. Wagner and V. Funk (eds.), *Hawaiian Biogeography: Evolution on a Hot Spot Archipelago.* Smithsonian Institution Press, Washington, D.C., pp. 288–337. [4]

Glover, A., G. Paterson, B. Bett, J. Gage, M. Sibuet, M. Sheader and L. Hawkins. 2001. Patterns in polychaete abundance and diversity from the Madeira Abyssal Plain, northeast Atlantic. *Deep Sea Research I* 48: 217–236. [14]

Godown, M. E. and A. T. Peterson. 2000. Preliminary distributional analysis of U.S. endangered bird species. *Biodiversity and Conservation* 9: 1313–1322. [16]

Godwin, H. 1956. *History of the British Flora.* Cambridge University Press, Cambridge. [Part 3]

Godwin, H. 1975. *History of the British Flora*, 2nd Ed. Cambridge University Press, Cambridge. [3]

Goloboff, P. 1998. *NONA,* Version 1.9. Distributed by American Museum of Natural History, New York. [4]

Golonka, J. 2000. Cambrian-Neogene plate tectonic maps. *Rozprawy Habilitacyjne* Nr 350. Wydawnictwo Uniwersytetu Jagiellonskiego, Krakow. [1]

Golonka, J., M. I. Ross and C. R. Scotese. 1996. Phanerozoic paleogeographic and paleoclimatic modeling maps. In A. F. Embry, B. Beauchamp and D. J. Glass (eds.), *PANGEA: Global Environments and Resources.* Canadian Society of Petroleum Geologists Memoir 17, pp. 1–48. [1]

Gomez, E. D., P. M. Alino, H. T. Yap and W. Y. Licuanan. 1994. A review of the status of Philippine reefs. *Marine Pollution Bulletin* 29: 62–68. [13]

Gonzales, P. C. and R. S. Kennedy. 1996. A new species of *Crateromys* (Rodentia: Muridae) from Panay, Philippines. *J. Mammalogy* 77: 25–40. [18]

Good, R. O'D. 1931. A theory of plant geography. *New Phytologist* 30: 149–171. [3]

Goodman, D. 1987. The demography of chance extinction. In M. E. Soulé (ed.), *Viable Populations for Conservation.* Cambridge University Press, Cambridge, pp. 11–34. [17]

Goodman, S. M. and N. R. Ingle. 1993. Sibuyan Island in the Philippines: Threatened and in need of conservation. *Oryx* 27: 174–180. [18]

Gosliner, T. M. and R. Draheim. 1996. Indo-Pacific gastropod biogeography: How we know what we don't know. *American Malacological Bulletin* 12: 37–43. [13]

Gottfried, M., H. Pauli, K. Reiter and G. Grabherr. 1999. A fine-scaled predictive model for changes in species distribution patterns of high mountain plants induced by climate warming. *Diversity and Distributions* 5: 241–251. [16]

Gould, S. J. 1994. Ernst Mayr and the centrality of species. *Evolution* 48: 31–35. [13]

Gould, S. J. 2002a. *I Have Landed: The End of a Beginning in Natural History.* Harmony Books, New York. [9]

Gould, S. J. 2002b. *The Structure of Evolutionary Theory.* Belknap Press of Harvard University Press, Cambridge, MA. [9]

Gould, S. J. and N. Eldredge. 1993. Punctuated equilibrium comes of age. *Nature* 366: 223–227. [13]

Grant, P. R. 1986. *Ecology and Evolution of Darwin's Finches.* Princeton University Press, Princeton, NJ. [12]

Grant, P. R. (ed.). 1998. *Evolution on Islands.* Oxford University Press, Oxford. [11]

Grant, W. S. 1987. Genetic divergence between congeneric Atlantic and Pacific Ocean fishes. In R. Ryman and F.M. Utter (eds.), *Population Genetics and Fishery Management.* University of Washington Press, Seattle, pp. 225–246. [Part 4]

Graves, J. E. 1998. Molecular insights into the population structure of cosmopolitan marine fishes. *Journal of Heredity* 89: 427–437. [Part 4]

Gray, J. and A. J. Boucot (eds.). 1979. *Historical Biogeography, Plate Tectonics, and the Changing Environment.* Oregon State University Press, Corvallis. [Part 4]

Gray, J. S. 1997. Gradients in marine biodiversity. In R. F. G. Ormond, J. D. Gage and M. V. Angel (eds.), *Marine Biodiversity: Pattern and Process.* Cambridge University Press, Cambridge, pp. 18–34. [14]

Gray, J. S. 2000. The measurement of marine species diversity, with an application to the benthic fauna of the Norwegian continental shelf. *J. Experimental Marine Biology and Ecology* 250: 23–49. [14]

Gray, J. S. 2001. Antarctic marine benthic biodiversity in a worldwide latitudinal context. *Polar Biology* 24: 633–641. [14]

Gray, S. T., J. L. Betancourt, C. Fastie and S. T. Jackson. 2003. Patterns and sources of multidecadal oscillations in drought-sensitive tree-ring records from the central and southern Rocky Mountains. *Geophysical Research Letters Vol. 30*, No. 6, 1316, doi: 10.1029/2002GL016154. [3]

Grayson, D. K. 2000. Mammalian responses to Middle Holocene climatic change in the Great Basin of the western United States. *J. Biogeography* 27: 181–192. [2]

Grayson, D. K. and D. B. Madsen. 2000. Biogeographic implications of recent low-elevation recolonization by *Neotoma cinerea* in the Great Basin. *J. Mammalogy* 81: 1100–1105. [2, 11]

Greenwood, J. J. D. 2001. Birds, biodiversity of. *Encyclopedia of Biodiversity*, Vol. 1. Academic Press, New York, pp. 489–520. [9]

Griffiths, H. J., K. Linse and J. A. Crame. 2003. SOMBASE – Southern Ocean Molluscan Database: A tool for biogeographic analysis in diversity and ecology. *Organisms, Diversity, and Evolution* 3: 207–213. [14]

Grinnell, J. 1917. Field tests of theories concerning distributional control. *American Naturalist* 51: 115–128. [16]

Grismer, L. L. 1994. The origin and evolution of the peninsular herpetofauna of Baja California, Mexico. *Herpetological Natural History* 2: 51–106. [5]

Grissino-Mayer, H. D. 1996. A 2129-year reconstruction of precipitation for northwestern New Mexico, USA. In J. S. Dean, D. M. Meko and T. W. Swetnam (eds.), *Tree Rings, Environment and Humanity, Proceedings of the International Conference, Tucson, Arizona, 17–21 May 1994*. Pages 191–204. [3]

Groombridge, B. (ed.). 1992. *Global Biodiversity: Status of the Earth's Living Resources*. Chapman and Hall, London. [18]

Groves, R. H. and F. di Castri. 1991. *Biogeography of Mediterranean Invasions*. Cambridge University Press, Cambridge. [15]

Grytnes, J. A. and O. R. Vetaas. 2002. Species richness and altitude: A comparison between null models and interpolated plant species richness along the Himalayan altitudinal gradient, Nepal. *American Naturalist* 159: 294–304. [11]

H-Acevedo, D. and D. J. Currie. 2003. Does climate determine broad-scale patterns of species richness? A test of the causal link by natural experiment. *Global Ecology and Biogeography* 12: 461–473. [11; Part 3]

Hafner, D. J. and B. R. Riddle. 2003. Mammalian phylogeography and evolutionary history of northern Mexico's deserts. In G. Ceballos and J.-L. Catron (eds.), *Biodiversity, Ecosystems, and Conservation in Northern Mexico*. Oxford University Press, Oxford, in press. [2]

Hall, C. A. 1964. Shallow-water marine climates and molluscan provinces. *Ecology* 45: 226–234. [8]

Hall, E. R. 1981. *The Mammals of North America*, Vol. I and II. Ronald Press, New York. [16]

Hall, R. 2002. Cenozoic geological and plate tectonic evolution of Southeast Asia and the Southwest Pacific: Computer-based reconstructions, model, and animations. *J. Asian Earth Sciences* 20: 353–431. [18]

Hall, R. and J. D. Holloway (eds.). 1998. *Biogeogra-phy and Geological Evolution of Southeast Asia*. Backhuys Publishers, Leiden, The Netherlands. [18]

Hallam, A. (ed.) 1973. *Atlas of Paleobiogeography*. Elsevier, Amsterdam. [Part 4]

Hallam, A. 1977. Jurassic bivalve biogeography. *Paleobiology* 3: 58–73. [6]

Hallam, A. 1994. *An Outline of Phanerozoic Biogeography*. Oxford University Press, Oxford. [14; Part 4]

Hanfling, B. and J. Kollmann. 2002. An evolutionary perspective of biological invasions. *Trends in Ecology and Evolution* 17: 545–546. [15]

Haq, B. U., J. Hardenbol and P. R. Vail. 1987. Chronology of fluctuating sea levels since the Triassic (250 mya to present). *Science* 235: 1156–1167. [1]

Harper, D. A. T. and J.-Y. Rong. 2001. Palaeozoic brachiopod extinctions, survival and recovery: Patterns within the rhynchonelliformeans. *Geological J.* 36: 317–328. [8]

Harries, P. J. 1999. Repopulations from Cretaceous Mass Extinctions: Environmental and/or Evolutionary Controls? Geological Society of America Special Paper 332: 345–364. [8]

Harris, A. H. 1984. *Neotoma* in the late Pleistocene of New Mexico and Chihuahua. In H. H. Genoways and M. R. Dawson (eds.), *Contributions in Quaternary Vertebrate Paleontology: A Volume in Memorial to John E. Guilday*. Carnegie Museum of Natural History Special Publication 8, Pittsburgh, pp. 164–178. [2]

Harrison, S. and E. Bruna. 1999. Habitat fragmentation and large-scale conservation: What do we know for sure? *Ecography* 22: 225–232. [17]

Hauge, P., J. Terborgh, B. Winter and J. Parkinson. 1986. Conservation priorities in the Philippine Archipelago. *Forktail* 2: 83–91. [18]

Hausdorf, B. 1988. Weighted ancestral area analysis and a solution of the redundant distribution problem. *Systematic Biology* 47: 445–456. [4]

Hausdorf, B. 2002. Units in biogeography. *Systematic Biology* 51: 648–652. [4]

Hawkins, B. A. 1995. Latitudinal body-size gradients for the bees of the eastern United States. *Ecological Entomology* 20: 195–198. [8]

Hawkins, B. A. 2001. Ecology's oldest pattern? *Trends in Ecology and Evolution* 16: 470. [9]

Hawkins, B. A. and J. A. Diniz-Filho. 2002. The mid-domain effect cannot explain the diversity gradient of Nearctic birds. *Global Ecology and Biogeography* 11: 419–426. [11]

Hawkins, B. A. and J. H. Lawton. 1995. Latitudinal gradients in butterfly body sizes: Is there a general pattern? *Oecologica* 102: 31–36. [8]

Hawkins, B. A. and E. E. Porter. 2001. Area and the latitudinal diversity gradient for terrestrial birds. *Ecology Letters* 4: 595–601. [9]

Hawkins, B. A. and E. E. Porter. 2003a. Does herbivore diversity depend on plant diversity? The case of California butterflies. *American Naturalist* 161: 40–49. [9]

Hawkins, B. A. and E. E. Porter. 2003b. Water-energy balance and the geographic pattern of species richness of western Palearctic butterflies. *Ecological Entomology* 28: 678–686. [9]

Hawkins, B. A. and E. E. Porter. 2003c. Relative influences of current and historical factors on mammal and bird diversity patterns in deglaciated North America. *Global Ecology and Biogeography* 12: 475–481. [9, 11]

Hawkins, B. A., J. A. F. Diniz-Filho and E. E. Porter. 2003a. Productivity and history as predictors of the latitudinal diversity gradient of terrestrial birds. *Ecology* 84: 1608–1623. [9]

Hawkins, B. A., R. Field, H. V. Cornell, D. J. Currie, J.-F. Guégan, D. M. Kaufman, J. T. Kerr, G. G. Mittelbach, T. Oberdorff, E. M. O'Brien, E. E. Porter and J. R. G. Turner. 2003b. Energy, water, and broad-scale geographic patterns of species richness. *Ecology* 84: 3105–3117. [9]

Haxby, W. F. 1987. *Gravity Field of the World's Oceans* (color map). Lamont-Doherty Geological Observatory, Palisades, NY. [1]

Haxeltine, A. and I. C. Prentice. 1996. BIOME3: An equilibrium terrestrial biosphere model based on ecophysiological constraints, resource availability and competition among plant functional types. *Global Biogeochemical Cycles* 10: 693–710. [2]

Hay, W. W., R. M. DeConto, C. N. Wold, K. M. Wilson, S. Voigte, M. Schulz, A. R. Wold, W. -C. Dullo, A. B. Ronov, Balukhovsky and E. Soding. 1999. Alternative global Cretaceous paleogeography. In Barrera, E. and C. C. Johnson (eds.), *Evolution of the Cretaceous Ocean-Climate System*. Geological Society of America Special Paper 332,, Boulder, CO. [1]

Heads M. J.. 1989. Integrating earth and life sciences in New Zealand natural history: The parallel arcs model. *New Zealand J. Zoology* 16: 549–585. [4]

Heads, M. J. 1990. Mesozoic tectonics and the deconstruction of biogeography: A new model of Australasian biology. *J. Biogeography* 17: 223–225. [4]

Heaney, L. R. 1978. Island area and body size of insular mammals: Evidence from the tri-colored squirrel (*Callosciurus prevosti*) of Southeast Asia. *Evolution* 34: 29–44. [12]

Heaney, L. R. 1984. Mammals from Camiguin Island, Philippines. *Proc. Biological Society of Washington* 97: 119–125. [18]

Heaney, L. R. 1986. Biogeography of the mammals of Southeast Asia: Estimates of colonization, extinction, and speciation. *Biological J. Linnean Society* 28: 127–165. [18]

Heaney, L. R. 1991. An analysis of patterns of distribution and species richness among Philippine fruit bats (Pteropodidae). *Bulletin of the American Museum of Natural History* 206: 145–167. [18]

Heaney, L. R. 1993. Biodiversity patterns and the conservation of mammals in the Philippines. *Asia Life Sciences* 2: 261–274. [18]

Heaney, L. R. 2000. Dynamic disequilibrium: A long-term, large-scale perspective on the equilibrium model of island biogeography. *Global Ecology and Biogeography* 9: 59–74. [7, 11, 18; Part 2]

Heaney, L. R. 2001. Small mammal diversity along elevational gradients in the Philippines: An assessment of patterns and hypotheses. *Global Ecology and Biogeography* 10: 15–39. [11, 18]

Heaney, L. R. 2002. Biological diversity in the Philippines: An introduction to megadiversity in a nation of islands. Box 1, pp. 2–8. In P. Ong, L. E. Afuang and R. G. Rosell-Ambal (eds.), *Philippine Biodiversity Conservation Priorities: A Second Iteration of the National Biodiversity Strategy and Action Plan*. Philippine Department of the Environment and Natural Resources, Quezon City. [18]

Heaney, L. R. and P. D. Heideman. 1987. Philippine fruit bats, endangered and extinct. *Bats* 5: 3–5. [18]

Heaney, L. R. and M. V. Lomolino (eds.). 2001. Diversity patterns of small mammals along elevational gradients. Special issue, *Global Ecology and Biogeography* 10: 1–109. [18]

Heaney, L. R. and N. A. D. Mallari. 2002. A preliminary analysis of current gaps in the protection of threatened Philippine terrestrial mammals. *Sylvatrop* (2000) 10: 28–39. [18]

Heaney, L. R. and B. D. Patterson (eds.). 1986. *Island Biogeography of Mammals*. Academic Press, London. [18]

Heaney, L. R. and J. C. Regalado, Jr. 1998. *Vanishing Treasures of the Philippine Rain Forest*. The Field Museum, Chicago. [18]

Heaney, L. R. and E. A. Rickart. 1990. Correlations of clades and clines: Geographic, elevational, and phylogenetic distribution patterns among Philippine mammals. In G. Peters and R. Hutterer (eds.), *Vertebrates in the Tropics*, pp. 321–332. Museum Alexander Koenig, Bonn. [18]

Heaney, L. R. and B. R. Tabaranza, Jr. 1997. A preliminary report on mammalian diversity and conservation status of Camiguin Island, Philippines. *Sylvatrop* (1995) 5: 57–64. [18]

Heaney, L. R. and R. C. B. Utzurrum. 1991. A review of the conservation status of Philippine land mammals. *Association of Systematic Biologists of the Philippines Communications* 3: 1–13. [18]

Heaney, L. R., D. S. Balete, L. Dolar, A. C. Alcala, A. Dans, P. C. Gonzales, N. Ingle, M. Lepiten, W. Oliver, E. A. Rickart, B. R. Tabaranza, Jr. and R. C. B. Utzurrum. 1998. A synopsis of the mammalian fauna of the Philippine Islands. *Fieldiana Zoology* (new series) 88: 1–61. [18]

Heaney, L. R., D. S. Balete, E. A. Rickart, R. C. B. Utzurrum and P. C. Gonzales. 1999. Mammalian diversity on Mt. Isarog, a threatened center of endemism on southern Luzon Island, Philip-pines. *Fieldiana Zoology* (new series) 95: 1–62. [18]

Heaney, L. R., P. D. Heideman, E. A. Rickart, R. C. B. Utzurrum and J. S. H. Klompen. 1989. Elevational zonation of mammals in the central Philippines. *J. Tropical Ecology* 5: 259–280. [18]

Heatwole, H. and R. Levins. 1972. Biogeography of the Puerto Rican Bank: Flotsam transport of terrestrial animals. *Ecology* 53: 112–117. [12]

Heck, K. L. and E. D. McCoy. 1978. Long distance dispersal and the reef building corals of the Eastern Pacific. *Marine Biology* 48: 349–356. [13]

Heck, K. L. and E. D. McCoy. 1979. Biogeography of sea grasses: Evidence from associated organisms. *Proceedings of the First International Symposium on Marine Biogeography and Evolution in the Southern Hemisphere* 1: 109–127. [13]

Heideman, P. D. and L. R. Heaney. 1989. Population biology and estimates of abundance of fruit bats (Pteropodidae) in Philippine submontane rainforest. *J. Zoology* 218: 565–586. [18]

Held, C. 2000. Phylogeny and biogeography of serolid isopods (Crustacea, Isopoda, Serolidae) and the use of ribosomal expansion segments in molecular systematics. *Molecular Phylogenetics and Evolution* 15: 165–178. [14]

Hengeveld, R. 1989. *Dynamics of Biological Invasions*. Chapman and Hall, New York. [Parts 1, 5]

Hengeveld, R. 1990. *Dynamic Biogeography*. Cambridge University Press, Cambridge. [Part 5]

Hennig, W. 1950. *Grundzüge einer Theorie der Phylogenetischen Systematik*. Deutscher Zentralverlag, Berlin. [4]

Hennig, W. 1965. Phylogenetic systematics. *Annual Review of Entomology* 10: 97–116. [4]

Hennig, W. 1966. *Phylogenetic Systematics*. Univer-sity of Illinois Press, Urbana. [4, 7; Parts 2, 4]

Hess, H. H. 1962. History of ocean basins. In A. E. Engel, J. H. L. James and B. F. Leonard (eds.), *Petrologic Studies: A Volume in Honor of A. F. Buddington*. Geological Society of America, CO, pp. 599–620. [4]

Hessler, R. R. and H. L Sanders. 1967. Faunal diversity in the deep sea. *Deep-Sea Research* 14: 65–78. [Part 4]

Hewitt, G. M. 1993. Postglacial distribution and species substructure: Lessons from pollen, insects, and hybrid zones. In D. R. Lees and D. Edwards (eds.), *Evolutionary Patterns and Processes*, Linnaean Society Symposium Series 14. Academic Press, London, pp. 97–123. [9]

Hewitt, G. M. 1996. Some genetic consequences of ice ages, and their role in divergence and speciation. *Biological Journal of the Linnaean Society* 58: 247–276. [9]

Hewitt, G. M. 1999. Post-glacial re-colonization of European biota. *Biological J. Linnean Society* 68: 87–112. [3]

Hewitt, G. M. 2000. The genetic legacy of the Quaternary ice ages. *Nature* 405: 907–913. [3]

Higgs, A. J. and M. B. Usher. 1980. Should nature reserves be large or small? *Nature* 285: 568–569. [17]

Hillebrand, H. 2004. On the generality of the latitudinal diversity gradient. *American Naturalist* 163: 192–211. [9]

Hoernle, K., P. van den Bogaard, R. Werner, B. Lissinna, F. Hauff, G. Alvarado and D. Garbe-Schönberg. 2002. Missing history (16–71 Ma) of the Galápagos hotspot: Implications for the tectonic and biological evolution of the Americas. *Geology* 30: 795–798. [12]

Hoffman, P. F. 1991. Did the breakout of Laurentia turn Gondwanaland inside out? *Science* 252: 1409–1412. [1]

Hofreiter, M., J. L. Betancourt, A. P. Sbriller, V. Markgraf and H. G. McDonald. 2003. Phylogeny, diet and habitat of an extinct ground sloth from Cuchillo Curá, Neuquén Province, southwest Argentina. *Quaternary Research* 59: 364–378. [2]

Hofreiter, M., H. N. Poinar, W. G. Spaulding, K. Bauer, P. S. Martin, G. Possnert and S. Pääbo. 2000. A molecular analysis of ground sloth diet through the last glaciation. *Molecular Ecology* 9: 1975–1984. [2]

Holdridge, L. R. 1967. *Life Zone Ecology*. Tropical Science Center, San Jose, Costa Rica. [2]

Holdridge, L. R., W. C. Grenke, W. H. Hatheway, T. Liang and J. A. Tosi. 1971. *Forest Environments in Tropical Life Zones*. Pergamon Press, Oxford. [17]

Holmes, A. 1931. Radioactivity and Earth movements. *Transactions of the Geological Society of Glasgow* 18: 559–606. [4]

Holmgren, C., J. L. Betancourt, K. A. Rylander, J. Roque, O. Tovar, H. Zeballos, E. Linares and J. Quade. 2001. Holocene vegetation history from fossil rodent middens near Arequipa, Peru. *Quaternary Research* 56: 242–251. [2]

Holmgren, C., M. C. Peñalba, K. A. Rylander and J. L. Betancourt. 2003. A 16,000 ^{14}C yr BP packrat midden series from the U.S.A-Mexico border lands. *Quaternary Research,* in press. [2]

Holt, R. D. 1990. The microevolutionary consequences of climate change. *Trends in Ecology and Evolution* 5: 311–315. [16]

Holt, R. D. 1993. Ecology at the mesoscale: The influence of regional processes on local communities. In R. E. Ricklefs and D. Schluter (eds.), *Species Diversity in Ecological Communities*. University of Chicago Press, Chicago, pp. 77–88. [17]

Holthuis, L. B. and E. Sivertsen. 1967. The Crustacea Decapoda, Mysidacea and Cirripedia of the Tristan da Cunha archipelago, with a review of the "*frontalis*" subgroup of the genus *Jasus*. *Results of the Norwegian Scientific Expedition to Tristan da Cunha (1937–1938)* 52: 1–55. [12]

Holthuis, L. B., A. J. Edwards and H. R. Lubbock. 1980. The decapod and stomatopod Crustacea of St. Paul's Rocks. *Zoologische Mededelingen* 56: 27–51. [12]

Hooghiemstra, H. 1994. Pliocene-Quaternary floral migration, evolution of northern Andean ecosystems, and climate change: Implications from the closure of the Panamanian isthmus. *Profil* 7: 413–425. [Part 1]

Hooghiemstra, H. and A. M. Van Cleef. 1995. Pleistocene climatic change and environmental and generic dynamics in the North Andean montane forest and páramo. In S. P. Churchill et al. (eds.), *Biodiversity and Conservation of Neotropical Montane Forests*. New York Botanical Garden, pp. 35–49. [Part 1]

Hooghiemstra, H. and T. van der Hammen. 1998. Neogene and Quaternary development of the neotropical rain forest: The forest refugia hypothesis, and a literature overview. *Earth Science Reviews* 44: 147–183. [Part 1]

Hooker, J. D. 1853. *The Botany of the Antarctic voyage of H.M.S. Discovery Ships* Erebus *and* Terror *in the Years 1839–1843*. Lovell Reeve, London. [Part 4]

Hooker, J. D. 1867. Lecture on insular floras. Delivered before the British Association for the Advancement of Science at Nottingham, August 27, 1866. [Parts 2, 4]

Hovenkamp, P. 1997. Vicariance events, not areas, should be used in biogeographical analysis. *Cladistics* 13: 67–79. [5, 6]

Howell, S. N. and S. Webb. 1995. *A Guide to the Birds of Mexico and Northern Central America*. Oxford University Press, Oxford. [16]

Hubbell, S. P. 2001. *The Unified Neutral Theory of Biodiversity and Biogeography*. Princeton University Press, Princeton, NJ. [9, 10, 11, 15; Part 3]

Hubbell, S. P. and J. K. Lake. 2003. The neutral theory of biodiversity and biogeography, and beyond. In T. M. Blackburn and K. J. Gaston (eds.), *Macroecology: Concepts and Consequences*. Blackwell Science, Oxford, pp. 45–63. [9]

Hugall, A., C. Moritz, A. Moussalli and. J. Stanisic. 2002. Reconciling paleodistribution models and comparative phylogeography in the wet tropics rainforest land snail *Gnarosophia bellendenkerensis*. *Proc. National Academy of Science USA* 99: 6112–6117. [Part 2]

Hughes, N. F. (ed.). 1973. *Organisms and Continents through Time*. Oxford Palaeontological Association. Distributed by Blackwell, London. [Part 4]

Hughes, T. P., D. R. Bellwood and S. R. Connolly. 2002. Biodiversity hotspots, centers of endemicity, and the conservation of coral reefs. *Ecology Letters* 5: 775–784. [13]

Hulver, M. 1985. *Cretaceous Marine Paleogeography of Africa*. Master's Thesis, University of Chicago. [1]

Humphries, C. J. 2000. Form, space and time: Which comes first? *J. Biogeography* 27: 11–15. [5; Part 2]

Humphries, C. J. 2004. From dispersal to geographic congruence: Comments on cladistic biogeography in the 20th century. In D. M. Williams and P. L. Forey (eds.), *Milestones in*

Systematics. CRC Press, Boca Raton, FL, pp. 225–260. [4]

Humphries, C. J. and L. Parenti. 1986. Cladistic biogeography. *Oxford Monographs in Biogeography* 2: 1–98. [6]

Humphries, C. J. and L. R. Parenti. 1999. *Cladistic Biogeography: Interpreting Patterns of Plant and Animal Distributions*, 2nd Ed. Oxford Biogeogra-phy Series No. 12. Oxford University Press, Oxford. [4. 7; Part 2]

Humphries, C. J. and P. H. Williams. 1994. Cladograms and trees in biodiversity. In D. M. Siebert and D. Williams, *Models in Phylogenetic Reconstruction*. Oxford University Press, Oxford. [Part 2]

Hunn, C. A and P. Upchurch. 2001. The importance of time/space in diagnosing the causality of phylogenetic events: Towards a "chronobiogeographical" paradigm? *Systematic Biology* 50: 391–407. [4]

Hunter, K. L., J. L. Betancourt, B. R. Riddle, T. R. Van Devender, K. L. Cole and W. G. Spaulding. 2001. Ploidy race distributions since the Last Glacial Maximum in the North American desert shrub, *Larrea tridentata*. *Global Ecology and Biogeography* 10: 521–533. [2]

Huntley, B. 1990. Dissimilarity mapping between fossil and contemporary pollen spectra in Europe for the past 13,000 years. *Quaternary Research* 33: 360–376. [3]

Huntley, B. 1991. How plants respond to climate change: Migration rates, individualism and the consequences for plant communities. *Annals of Botany* 67(Suppl. I): 15–22. [3]

Huntley, B. 1999. The dynamic response of plants to environmental change and the resulting risks of extinction. In G. M. Mace, A. Balmford and J. R. Ginsberg (eds.), *Conservation in a Changing World*. Cambridge University Press, Cambridge, pp. 69–85. [3]

Huntley, B. and T. Webb III. 1988. *Vegetation History*. Kluwer Academic, Dordrecht, NL. [Part 1]

Huntley, B. and T. Webb III. 1989. Migration: Species' response to climatic variations caused by changes in the earth's orbit. *J. Biogeography* 16: 5–19. [3]

Huntley, B., P. J. Bartlein and I. C. Prentice. 1989. Climatic control of the distribution and abundance of beech (Fagus L.) in Europe and North America. *J. Biogeography* 16: 551–560. [3]

Hurlbert, A. H. and J. P. Haskell. 2003. The effect of energy and seasonality on avian species richness and community composition. *American Naturalist* 161: 83–97. [9]

Huston, M. A. 1994. *Biological Diversity: The Coexistence of Species on Changing Landscapes*. Cambridge University Press, Cambridge. [9, 10, 15]

Huston, M. A. 1999. Local processes and regional patterns: Appropriate scales for understanding variation in the diversity of plants and animals. *Oikos* 86: 393–401. [9]

Hutchinson, G. E. 1959. Homage to Santa Rosalia, or why are there so many kinds of animals? *American Naturalist* 93: 145–159. [9, 10]

Hutchison, C. S. 1989. *Geological Evolution of South-East Asia*. Oxford University Press, Oxford. [1]

Huxley, T. H. 1870. Anniversary address. In M. Foster and E. R. Lankester (eds.), *The Scientific Memoirs of Thomas Henry Huxley*. Macmillan, London, pp. 510–550. [6]

Illoldi-Rangel, P., V. Sánchez-Cordero and A. T. Peterson. 2004. Predicting distributions of Mexican mammals using ecological niche modeling. *J. Mammalogy* 85: 658–662. [16]

IUCN. 2002. *2002 Red List of Threatened Species*. Species Survival Commission, IUCN, Gland, Switzerland. www.redlist.org. [16, 18]

Izsak, C. and A. R. G. Price. 2001. Measuring β-diversity using a taxonomic similarity index, and its relation to spatial scale. *Marine Ecology Progress Series* 215: 69–77. [14]

Jablonski, D. 1993. The tropics as a source of evolutionary novelty through geological time. *Nature* 364: 142–144. [9]

Jablonski, D. 1995. Extinction in the fossil record. In R. M. May and J. H. Lawton (eds.), *Extinction Rates*. Oxford University Press, Oxford, pp. 25–44. [8]

Jablonski, D. 1996. Body size and macroevolution. In D. Jablonski, D. H. Erwin and J. H. Lipps (eds.), *Evolutionary Paleobiology*. University of Chicago Press, Chicago, pp. 256–289. [8]

Jablonski, D. 1998. Geographic variation in the molluscan recovery from the end-Cretaceous extinction. *Science* 279: 1327–1330. [8]

Jablonski, D. 2000. Micro- and macroevolution: Scale and hierarchy in evolutionary biology and paleobiology. *Paleobiology* 26(Suppl. to No. 4): 15–52. [8]

Jablonski, D. 2001. Lessons from the past: Evolutionary impacts of mass extinctions. *Proc. National Academy of Sciences USA* 98: 5393–5398. [8]

Jablonski, D. 2002. Survival without recovery after mass extinctions. *Proc. National Academy of Sciences USA* 99: 8139–8144. [8]

Jablonski, D. and R. A. Lutz. 1983. Larval ecology of marine benthic invertebrates: Paleobiological implications. *Biological Reviews* 58: 21–89. [10]

Jablonski, D. and D. M. Raup. 1995. Selectivity of end-Cretaceous marine bivalve extinctions. *Science* 268: 389–391. [8]

Jablonski, D. and K. Roy. 2003. Geographical range and speciation in fossil and living molluscs. *Proc. Royal Society of London Series B* 270: 401–406. [10; Part 1]

Jablonski, D. and J. W. Valentine. 1981. Onshore-offshore gradients in Recent Eastern Pacific shelf faunas and their paleobiogeographic significance. In G. G. E. Scudder and J. L. Reveal (eds.), *Evolution Today: Proceedings of the 2nd International Congress of Systematic and Evolutionary Biology*, pp. 443–455. [8]

Jablonski, D., K. Roy and J. W. Valentine. 2000. Analyzing the latitudinal diversity gradient in marine bivalves. In E. M. Harper, J. D. Taylor and J. A. Crame (eds.), *The Evolutionary Biology of the Bivalvia*. Geological Society of London Special Publication 177: 361–365. [8]

Jablonski, D., K. Roy and J. W. Valentine. 2003. Evolutionary macroecology and the fossil record. In T. M. Blackburn and K. J. Gaston (eds.), *Macroecology: Concepts and Consequences*. Blackwell Science, Oxford, pp. 368–390. [8]

Jackson, J. B. C. 1974. Biogeographic consequences of eurytopy and stenotopy among marine bivalves and their biogeographic significance. *American Naturalist* 104: 541–560. [10]

Jackson, J. B. C. and K. G. Johnson. 2000. Life in the last few million years. In D. H. Erwin and S. L. Wing (eds.), *Deep Time: Paleobiology's Perspective*. Suppl. to *Paleobiology* 26(4): 221–235. [14]

Jackson, J. B. C., M. X. Kirby, W. H. Berger, K. A. Bjorndal, L. W. Botsford, B. J. Bourque, R. H. Bradbury, R. Cooke, J. Erlandson, J. A. Estes, T. P. Hughes, S. Kidwell, C. B. Lange, H. S. Lenihan, J. M. Pandolfi, C. H. Peterson, R. S. Steneck, M. J. Tegner and R. R. Warner. 2001. Historical overfishing and the recent collapse of coastal ecosystems. *Science* 293: 629–638. [Part 1]

Jackson, S. T. 1994. Pollen and spores in Quaternary lake sediments as sensors of vegetation composition: Theoretical models and empirical evidence. In A. Traverse (ed.), *Sedimentation of Organic Particles*. Cambridge University Press, Cambridge, pp. 253–286 [3]

Jackson, S. T. 2000. Out of the garden and into the cooler? A Quaternary perspective on deep-time paleoecology. In R.A. Gastaldo and W.A. DiMichele (eds.), *Evolution of Phanerozoic Terrestrial Ecosystems*. The Paleontological Society Papers, Vol. 6. New Haven, CT, pp. 287–308. [3]

Jackson, S. T. and R. K. Booth. 2002. The role of Late Holocene climate variability in the expansion of yellow birch in the western Great Lakes region. *Diversity and Distributions* 8: 275–284. [3]

Jackson, S. T. and C. R. Givens. 1994. Late Wisconsinan vegetation and environment of the Tunica Hills region, Louisiana–Mississippi. *Quaternary Research* 41: 316–325. [3]

Jackson, S. T. and M. E. Lyford. 1999. Pollen dispersal models in Quaternary plant ecology: Assumptions, parameters and prescriptions. *Botanical Review* 65: 39–75. [3]

Jackson, S. T. and J. T. Overpeck. 2000. Responses of plant populations and communities to environmental changes of the Late Quaternary. *Paleobiology* 26(Suppl.): 194–220. [3]

Jackson, S. T. and C. Weng. 1999. Late Quaternary extinction of a tree species in eastern North America. *Proc. National Academy of Sciences USA* 96: 13847–13852. [3]

Jackson, S. T. and D. R. Whitehead. 1991. Holocene vegetation patterns in the Adirondack Mountains. *Ecology* 72: 641–653. [3]

Jackson, S. T. and J. W. Williams. 2004. The modern analog method in Quaternary paleoecology. *Annual Review of Earth and Planetary Sciences*, in review. [3]

Jackson, S. T., J. L. Betancourt, M. E. Lyford and S. T. Gray. In review. A 40,000-year woodrat midden record of vegetational and biogeographic dynamics in northeastern Utah. *J. Biogeography*. [2]

Jackson, S. T., J. T. Overpeck, T. Webb III, S. E. Keattch and K. H. Anderson. 1997. Mapped plant-macrofossil and pollen records of Late Quaternary vegetation change in eastern North America. *Quaternary Science Reviews* 16: 1–70. [3]

Jackson, S. T., R. S. Webb, K. H. Anderson, J. T. Overpeck, T. Webb III, J. Williams and B. C. S. Hansen. 2000. Vegetation and environment in unglaciated eastern North America during the last glacial maximum. *Quaternary Science Reviews* 19: 489–508. [3]

Jansson, R. and M. Dynesius. 2002. The fate of clades in a world of recurrent climatic change: Milankovitch oscillations and evolution. *Annual Review of Ecology and Systematics* 33: 741–777. [14]

Janzen, D. H. 1970. Herbivores and the number of tree species in tropical forests. *American Naturalist* 104: 501–508. [11]

Jernvall, J., P. C. Wright. 1998. Diversity components of impending primate extinctions. *Proc.*

National Academy of Sciences USA 95: 11279–11283. [8]

Johnson, D. L. 1980. Problems in the land vertebrate zoogeography of certain islands and the swimming power of elephants. *J. Biogeography* 7: 383–398. [11]

Johnson, L. E., A. Ricciardi and J. T. Carlton. 2001. Overland dispersal of aquatic invasive species: A risk assessment of transient recreational boating. *Ecological Applications* 11: 1789–1799. [15]

Johnston, A. K. 1856. *The Physical Atlas of Natural Phenomenon*. Blackwood and Son, Edinburgh. [8]

Johnston, K. M. and O. J. Schmitz. 1997. Wildlife and climate change: Assessing the sensitivity of selected species to simulated doubling of atmospheric CO_2. *Global Change Biology* 3: 531–544. [16]

Jokiel, D. L. 1989. Rafting of reef corals and other organisms at Kwajalein Atoll. *Marine Biology* 101: 483–493. [13]

Jokiel, D. L. 1990. Transport of reef corals into the Great Barrier Reef. *Nature* 347: 665–667. [13]

Jokiel, D. L. and F. J. Martinelli. 1992. The vortex model of coral reef biogeography. *J. Biogeography* 19: 449–458. [13]

Jones, C. G., J. H. Lawton and M. Shachack. 1994. Organisms as ecosystem engineers. *Oikos* 69: 373–386. [10]

Jones, K. E., A. Purvis and J. Gittleman. 2003. Biological correlates of extinction risk in bats. *American Naturalist* 161: 601–614. [10]

Jones, P. D., T. J. Osborn and K. R. Briffa. 2001. The evolution of climate over the last millennium. *Science* 292: 662–667. [3]

Jordan, G. J. 1997. Evidence of Pleistocene plant extinction and diversity from Regatta Point, western Tasmania, Australia. *Botanical J. Linnean Society* 123: 45–71. [3]

Jurdy, D. M., M. Stefanick and C. R. Scotese. 1995. Paleozoic plate dynamics. *J. Geophysical Research* 100: 17965–17975. [1]

Kammer, T. W., T. K. Baumiller and W. I. Ausich. 1998. Evolutionary significance of differential species longevity in Osagean-Meramecian (Mississippian) crinoid clades. *Paleobiology* 24: 155–176. [8]

Karl, S. A., S. Schutz, D. Desbruyîres, R. C. Vrijenhoek and R. Lutz. 1996. Molecular analysis of gene flow in the hydrothermal-vent clam, *Calyptogena magnifica*. *Molecular Marine Biology and Biotechnology* 5: 193–202. [Part 4]

Kaspari, M. and E. L. Vargo. 1995. Colony size as a buffer against seasonality: Bergmann's Rule in social insects. *American Naturalist* 145: 610–632. [8]

Kay, E. A. 1967. Position and relationships of marine molluscan fauna of the Hawaiian Islands. *Venus* 25: 94–104. [12]

Kay, E. A. 1979. *Hawaiian Marine Shells*. Bishop Museum Press, Honolulu, HI. [12]

Kay, E. A. 1984. Patterns of speciation in the Indo-West Pacific. In F. J. Radovsky, P. H. Raven and S. H. Sohmer (eds.), *Biogeography of the Tropical Pacific*, pp. 15–31. Bishop Museum Special Publication 72. [12]

Kay, E. A. 1990. Cypraeidae of the Indo-Pacific: Fossil history and biogeography. *Bulletin of Marine Science* 47: 23–34. [13]

Kay, E. A. 1991. The marine mollusks of the Galápagos: Determinants of marine faunas. In M. J. James (ed.), *Galápagos Marine Invertebrates: Taxonomy, Biogeography, and Evolution in Darwin's Islands*, pp. 235–252. Plenum, New York. [12]

Kazmin, V. G. and L. M. Natopov (eds.). 1998. Paleogeographic Atlas of Northern Eurasia, 26 Maps: Devonian, Carboniferous, Permian, Triassic, Jurassic, Cretaceous, Paleogene and Neogene 380. 6–6.7 Ma. Institute of Tectonics of Lithospheric Plates, Moscow. [1]

Kendall, M. A. 1996. Are Arctic soft-sediment macrobenthic communities impoverished? *Polar Biology* 16: 393–399. [14]

Kendall, M. A. and M. Aschan. 1993. Latitudinal gradients in the structure of macrobenthic communities: A comparison of Arctic, temperate and tropical sites. *J. Experimental Marine Biology and Ecology* 172: 157–169. [14]

Kennedy, T. A., S. Naeem, K. M. Howe, J. M. H. Knops, D. Tilman and P. Reich. 2002. Biodiversity as a barrier to ecological invasion. *Nature* 417: 636–638. [15]

Kerr, J. T. 2001. Butterfly species richness patterns in Canada: Energy, heterogeneity, and the potential consequences of climatic change. *Conser-vation Ecology* 5: 10. [online] URL: http://www. consecol.org/vol5.iss1/art10. [9]

Kerr, J. T. and D. J. Currie. 1999. The relative importance of evolutionary and environmental controls on broad-scale patterns of species richness in North America. *Ecoscience* 6: 329–337. [9]

Kerr, J. T. and L. Packer. 1997. Habitat as a determinant of mammal species richness in high-energy regions. *Nature* 385: 252–254. [8]

Kerr, J. T. and L. Packer. 1999. The environmental basis of North American species richness patterns among *Epicauta* (Coleoptera: Meloidae). *Biodiversity and Conservation* 8: 617–628. [9]

Kerr, J. T., T. R. E. Southwood and J. Cihlar. 2001. Remotely sensed habitat diversity predicts butterfly richness and community similarity in Canada. *Proc. National Academy of Sciences USA* 98: 11365–11370. [9]

Kerr, J. T., R. Vincent and D. J. Currie. 1998. Lepidopteran richness patterns in North America. *Ecoscience* 5: 448–453. [9]

Keymer, J. E., P. A. Marquet, J. X. Velasco-Hernández and S. A. Levin. 2000. Extinction thresholds and metapopulation persistence in dynamics landscapes. *American Naturalist* 156: 478–494. [10]

Kitchell, J. A. 1990. The reciprocal interaction of organism and effective environment: Learning more about the "and." In R. M. Ross and W. D. Allmond (eds.), *Causes of Evolution: A Paleontological Perspective*. University of Chicago Press, Chicago, pp. 151–169. [10]

Kitching, I. J., P. L. Forey, C. J. Humphries and D. M. Williams. 1998. *Cladistics: The Theory and Practice of Parsimony Analysis*, 2nd Ed. The Systematics Association Publication No. 10. Oxford University Press, Oxford. [4]

Klicka, J. and R. M. Zink. 1997. The importance of recent Ice Ages in speciation: A failed paradigm. *Science* 277: 1666–1669. [6]

Knowles, L. L. and W. P. Maddison. 2002. Statistical phylogeography. *Molecular Ecology* 11: 2623–2635. [5]

Knowlton, N. 1993. Sibling species in the sea. *Annual Review of Ecology and Systematics* 24: 189–216. [13]

Knowlton, N., L. A. Weigt, L. A. Solórzano, D. K. Mills and E. Bermingham. 1993. Divergence in proteins, mitochondrial DNA, and reproductive compatibility across the Isthmus of Panama. *Science* 260: 1629–1632. [Part 4]

Knox, G. A. 1960. Littoral ecology and biogeography of the southern oceans. *Proc. Royal Society of London Series B* 152: 567–624. [12]

Kohn, A. J. 1980. *Conus kahiko*, a new Pleistocene gastropod from Oahu, Hawaii. *J. Paleontology* 54: 534–541. [12]

Kohn, A. J. 1997. Why are coral reef communities so diverse? In R. F. G. Ormond, J. D. Gage and M. V. Angel (eds.), *Marine Biodiversity: Pattern and Process*. Cambridge University Press, Cambridge, pp. 201–215. [14]

Kolar, C. S. and D. M. Lodge. 2002. Ecological predictions and risk assessment for alien fishes in North America. *Science* 298: 1233–1236. [15]

Kosuge, S. 1969. Fossil mollusks of Oahu, Hawaii Islands. *Bulletin of the National Science Museum of Japan* 12: 753–764. [12]

Kuch, M., N. Rohland, J. L. Betancourt, C. L. Latorre, S. Steppan and H. N. Poinar. 2002. Molecular analysis of an 11,700-year old rodent midden from the Atacama Desert, Chile. *Molecular Ecology* 11: 913–924. [2, 3]

Kullman, L. 1998. Palaeoecological, biogeographical, and palaeoclimatological implications of early Holocene immigration of Larix sibirica Ledeb. into the Scandes Mountains, Sweden. *Global Ecology and Biogeography Letters* 7: 181–188. [11]

Kummer, D. M. 1992. *Deforestation in the Postwar Philippines*. University of Chicago Press, Chicago. [18]

Kussakin, O. G. (ed.). 1982. *Marine Biogeography*. Nauka, Moscow. [Part 4]

Kutzbach, J. E., R. Gallimore, S. Harrison, P. Behling, R. Selin and F. Laarif. 1998. Climate and biome simulations for the past 21,000 years. *Quaternary Science Reviews* 17: 473–506. [3]

Lack, D. 1969. The numbers of bird species on islands. *Bird Study* 16: 193–209. [11]

Lack, D. 1976. *Island Biology Illustrated by the Land Birds of Jamaica*. Blackwell Science, Oxford. [11]

Ladd, H. S. 1960. Origin of the Pacific island molluscan fauna. *American J. Science* 258A: 137–150. [13]

Laland, K. N., F. J. Odling-Smee and M. W. Feldman. 1999. Evolutionary consequences of niche construction and their implications for ecology. *Proc. National Academy of Sciences USA* 96: 10242–10247. [10]

Lambshead, P. J. D., C. J. Brown, T. J. Ferrero, N. J. Mitchell, C. R. Smith, L. E. Hawkins and J. Tietjen. 2002. Latitudinal diversity patterns of deep-sea marine nematodes and organic fluxes: A test from the central equatorial Pacific. *Marine Ecology Progress Series* 236: 129–135. [14]

Lambshead, P. J. D., J. Tietjen, T. Ferrero and P. Jensen. 2000. Latitudinal diversity gradients in the deep sea with special reference to North Atlantic nematodes. *Marine Ecology Progress Series* 194: 159–167. [14]

Lanner, R. M. and T. R. Van Devender. 1998. The recent history of pinyon pines in the American Southwest. In D. M. Richardson (ed.), *Ecology and Biogeography of Pinus*. Cambridge University Press, Cambridge, pp. 171–182. [2, 3]

Latham, R. E. and R. E. Ricklefs. 1993. Global patterns of tree species richness in moist forests: Energy-diversity theory does not account for variation in species richness. *Oikos* 67: 325–333. [9]

Latorre, C., J. L. Betancourt, K. A. Rylander and J. A. Quade. 2002. Vegetation invasions into Absolute Desert: A 45,000-year rodent midden record from the Calama-Salar de Atacama Basins, Chile. *Geological Society of America Bulletin* 114: 349–366. [2]

Latorre, C., J. L. Betancourt, K. A. Rylander, J. Quade and O. Matthei. 2003. A vegetation history from the arid prepuna of northern Chile 22°–23°S over the last 13,500 years. *Palaeogeography, Palaeoclimatology, Palaeoecology*, 194: 223–246. [2]

Lawlor, T. E. 1998. Biogeography of great mammals: Paradigm lost? *J. Mammalogy* 79: 1111–1130. [11]

Laycock, G. 1966. *The Alien Animals*. American Museum of Natural History Press, Garden City, NY. [15]

Lazarus, D., H. Hilbrecht, C. Spencer-Cervato and H. Thierstein. 1995. Sympatric speciation and phyletic change in *Globortalia truncatulinoides*. *Paleobiology* 21: 28–51. [13]

Lee, S.-M. and A. Chao. 1994. Estimating population size via sample coverage for closed capture-recapture models. *Biometrics* 50: 88–97. [17]

Lee, T. E., B. R. Riddle and P. L. Lee. 1996. Speciation in the desert pocket mouse (*Chaetod-ipus penicillatus* Woodhouse). *J. Mammalogy* 77: 58–68. [5]

Leitner, W. A. and M. L. Rosenzweig. 1997. Nested species-area curves and stochastic sampling: A new theory. *Oikos* 79: 503–512. [17]

Lennon, J. J. 2000. Red shifts and red herrings in geographical ecology. *Ecography* 23: 101–113. [11]

Lennon, J. J., J. J. D. Greenwood and J. R. G. Turner. 2000. Bird diversity and environmental gradients in Britain: A test of the species-energy hypothesis. *J. Animal Ecology* 69: 581–598. [9]

Leonard, J. A., R. K. Wayne and A. Cooper. 2000. Population genetics of Ice Age brown bears. *Proc. National Academy of Sciences USA* 97: 1651–1654. [3]

Lepiten, M. V. 1997. The mammals of Siquijor Island, central Philippines. *Sylvatrop* (1995) 5: 1–17. [18]

Leppakoski, E. and S. Olenin. 2001. The meltdown of biogeographic peculiarities of the Baltic Sea: The interaction of natural and man-made processes. *Ambio* 30: 202–209. [15]

Lessios, H. A., B. D. Kessing, D. R. Robertson and G. Paulay. 1999. Phylogeography of the pantropical sea urchin *Eucidaris* in relation to land barriers and ocean currents. *Evolution* 53: 806–817. [13; Part 4]

Levin, D. A. and A. C. Wilson. 1976. Rates of evolution in seeded plants: Net increase in diversity of chromosome number and species number through time. *Proc. National Academy of Sciences USA* 73: 2086–2090. [10]

Levin, L. A., R. J. Etter, M. A. Rex, A. J. Gooday, C. R. Smith, J. Pineda, C. T. Stuart, R. R. Hessler and D. Pawson. 2001. Environmental influences on regional deep-sea species diversity. *Annual Review of Ecology and Systematics* 32: 51–93. [14; Part 4]

Lewontin, R. 2000. *The Triple Helix: Gene, Organism, and Environment*. Harvard University Press, Cambridge, MA. [9]

Lewontin, R. C. 1983. Gene, organism, and environment. In D. S. Bendall (ed.), *Evolution from Molecules to Men*. Cambridge University Press, Cambridge, pp. 151–172. [10]

Li, S. J., Y. L. Song and Z. G. Zeng. 2003. Elevational gradients of small mammal diversity on the northern slopes of Mt. Qilian, China. *Global Ecology and Biogeography* 12(6): 449. [11]

Lieberman, B. S. 1997. Early Cambrian paleogeography and tectonic history: A biogeographic approach. *Geology* 25: 1039–1042. [6]

Lieberman, B. S. 1999. Testing the Darwinian legacy of the Cambrian radiation using trilobite phylogeny and biogeography. *J. Paleontology* 73: 176–181. [6]

Lieberman, B. S. 2000. *Paleobiogeography: Using Fossils to Study Global Changes, Plate Tectonics, and Evolution*. Topics in Geobiology 16. Kluwer Academic/Plenum, New York. [6, 7]

Lieberman, B. S. 2001. Applying molecular phylogeography to test paleoecological hypotheses: A case study involving Amblema plicata (Mollusca: Unionidae). In W. D. Allmon and D. J. Bottjer (eds.), *Evolutionary Paleoecology*. Columbia University Press, New York, pp. 83–103. [6]

Lieberman, B. S. 2002. Phylogenetic biogeography with and without the fossil record: Gauging the effects of extinction and paleontological incompleteness. *Palaeogeography, Palaeoclimatology and Palaeoecology* 178: 39–52. [6; Part 1]

Lieberman, B. S. 2003a. Unifying theory and methodology in biogeography. *Evolutionary Biology* 33: 1–25. [6; Part 1]

Lieberman, B. S. 2003b. Taking the pulse of the Cambrian radiation. *J. Integrative and Comparative Biology* 43: 229–237. [6]

Lieberman, B. S. 2003c. Paleobiogeography: The relevance of fossils to biogeography. *Annual Reviews of Ecology, Evolution and Systematics* 34: 51–69. [6]

Lieberman, B. S. and N. Eldredge. 1996. Trilobite biogeography in the Middle Devonian: Geological processes and analytical methods. *Paleobiology* 22: 66–79. [6, 7]

Liebherr, J. K. and A. E. Hajek. 1990. A cladistic test of the taxon cycle and taxon pulse hypotheses. *Cladistics* 6: 39–59. [7]

Lieth, H. 1975. Modelling the primary productivity of the world. In H. Lieth and R. H. Whittaker (eds.), *Primary Productivity of the Biosphere*. Springer-Verlag, New York, pp. 237–263. [9]

Linder, H. P. 2001. On areas of endemism, with an example from African Restionaceae. *Systematic Biology* 50(6): 892–912. [4]

Lockwood, J. L. 1999. Using taxonomy to predict success among introduced avifauna: The importance of transport and establishment. *Conserva-tion Biology* 13: 560–567. [15]

Lockwood, J. L. and M. L. McKinney. 2001. *Biotic Homogenization*. Kluwer Academic/Plenum Press, New York. [15]

Lockwood, R. 1998. K-T Extinction and Recovery: Taxonomic versus Morphological Patterns in Veneroid Bivalves. Geological Society of America Abstract 30: 286. [8]

Lockwood, R. 2003. Abundance not linked to survival across the end-Cretaceous mass extinction: Patterns in North American bivalves. *Proc. National Academy of Sciences USA* 100: 2478–2482. [8]

Lomolino, M. V. 1985. Body size of mammals on islands: The island rule re-examined. *American Naturalist* 125: 310–316. [12]

Lomolino, M. V. 1990. The target area hypothesis: The influence of island area on immigration rates of non-volant mammals. *Oikos* 57: 297–300. [11]

Lomolino, M. V. 1994. An evaluation of alternative strategies for building networks of nature reserves. *Biological Conservation* 69: 243–249. [17]

Lomolino, M. V. 2000a. A call for a new paradigm of island biogeography. *Global Ecology and Biogeography* 9: 1–6. [7, 10]

Lomolino, M. V. 2000b. A species-based theory of insular zoogeography. *Global Ecology and Biogeography* 9: 39–58. [7, 10]

Lomolino, M. V. 2000c. Ecology's most general, yet protean pattern: The species-area relationship. *J. Biogeography* 27: 17–26. [11, 17]

Lomolino, M. V. 2001. Elevation gradients of species-density: Historical and prospective views. *Global Ecology and Biogeography* 10: 3–13. [11]

Lomolino, M. V. 2002. "... There are areas too small, and areas to large, to show clear diversity patterns." R. H. MacArthur (1972: 191). *J. Biogeogra-phy* 29: 555–557. [11]

Lomolino, M. V., and R. Channell. 1995. Splendid isolation: Patterns of range collapse in endangered mammals. *J. Mammalogy* 76: 335–347. [Part 5]

Lomolino, M. V. and D. R. Perault. 2004. Geographic gradients of deforestation and mammalian communities in a fragmented, temperate rain forest landscape. *Global Ecology and Biogeo-graphy* 13(1): 55. [18]

Lomolino, M. V. and M. D. Weiser. 2001. Towards a more general species-area relationship: Diversity on all islands, great and small. *J. Biogeography* 28: 431–445. [11, 17]

Lomolino, M. V., D. F. Sax and J. H. Brown (eds.). 2004. *Foundations of Biogeography: Classic Papers with Commentaries*. Chicago University Press, Chicago. [3, 11; Part 2]

Lomolino, M. V., R. Channell, D. R. Perault, and G. A. Smith. 2001. Downsizing nature: Anthropogenic dwarfing of species and ecosystems. In M. McKinney and J. Lockwood (eds.), *Biotic Homogenization: The Loss of Diversity through Invasion and Extinction*. [Part 5]

Lopez de Casenave, J., V. R. Cueto and L. Marone. 1998. Granivory in the Monte desert: Is it less intense than in other arid zones of the world? *Global Ecology and Biogeography Letters* 7: 197–204. [2]

Lopez Gappa, J. 2000. Species richness of marine Bryozoa in the continental shelf and slope off Argentina (south-west Atlantic). *Diversity and Distributions* 6: 15–27. [14]

Losos, J. B. and D. Schluter. 2000. Analysis of an evolutionary species-area relationship. *Nature* 408: 847–850. [7]

Lyell, C. 1830–1832. *Principles of Geology*. 3 vols. John Murray, London. [4; Part 4]

Lyell, C. 1832. *Principles of Geology*, Vol. 2. University of Chicago Press, Chicago. [6]

Lyford, M. E., J. L. Betancourt, S. T. Gray and S. T. Jackson. 2003a. Influence of landscape structure and climate variability in a late Holocene natural invasion. *Ecological Monographs*, in press. [2]

Lyford, M. E., R. G. Eddy, S. T. Gray and S. T. Jackson. 2003b. Validating the use of woodrat *Neotoma* middens for documenting natural invasions. *J. Biogeography*, in press. [2]

Lyford, M. E., S. T. Jackson, J. L. Betancourt and S. T. Gray. 2003c. Influence of landscape structure and climate variability on a late Holocene plant migration. *Ecological Monographs*, in press. [3]

Lyons, S. K. and M. R. Willig. 1999. A hemispheric assessment of scale dependence in latitudinal gradients of species richness. *Ecology* 80: 2483–2491. [14]

MacArthur, R. H. 1965. Patterns of species diversity. *Biological Reviews* 40: 510–533. [14]

MacArthur, R. H. and E. O. Wilson. 1963. An equilibrium theory of insular zoogeography. *Evolution* 17: 373–387. [7, 11; Part 3]

MacArthur, R. H. and E. O. Wilson. 1967. *The Theory of Island Biogeography*. Monographs in Population Biology, Princeton University Press, Princeton, NJ. [7, 9, 11, 12, 15; Part 3]

Mack, R. N. and W. M. Lonsdale. 2001. Humans as global plant dispersers: Getting more than we bargained for. *BioScience* 51: 95–102. [15]

Mack, R. N., D. Simberloff, W. M. Lonsdale, H. Evans, M. Clout and F. A. Bazzaz. 2000. Biotic invasions: Causes, epidemiology, global consequences, and control. *Ecological Applications* 10: 689–710. [15]

MacLeod, N. 2002. Geometric morphometrics and geological shape-classification systems. *Earth Science Review* 59: 27–47. [8]

MacPhee, R. D. M. and P.A. Marx. 1997. The 40,000-year plague: Humans, hyperdisease, and first-contact extinctions. In S. M. Goodman and B. D. Patterson (eds.), *Natural Change and Human Impact in Madagascar*. Smithsonian Institution Press, Washington, D.C., pp. 169–217. [Part 5]

Magurran, A. E. 1988. *Ecological Diversity and its Measurement*. Croom Helm, London. [14]

Magurran, A. E. and P. A. Henderson. 2003. Explaining the excess of rare species in natural species abundance distributions. *Nature* 422: 714–716. [10]

Mahoney, J. J. and M. F. Coffin, eds. 1997. Large igneous provinces: Continental, oceanic and planetary flood volcanism. *Geophysical Monograph* 100. [1]

Maldonado, A. 2003. *Cambios Vegetacionales y Climáticos en Chile Durante el Holoceno: Análisis de Polen Fósil en el Extremo Norte del Cinturón de Vientos del Oeste*. Ph.D. Dissertation Thesis, University of Chile, Santiago. [2]

Mallory, W. W. (ed.). 1972. *Geological Atlas of the Rocky Mountain Region*. Rocky Mountain Association of Geologists, Denver. [1]

Manchester, S. R. 1999. Biogeographical relationships of North American Tertiary floras. *Annals of the Missouri Botanical Garden* 86: 472–522. [Part 1]

Manley, G. 1974. Central England temperatures: Monthly means 1659 to 1973. *Quarterly J. Royal Meteorological Society* 100: 389–405. [9]

Manne, L. L., S. L. Pimm, J. M. Diamond and T. M. Reed. 1998. The form of the curves: A direct evaluation of MacArthur and Wilson's classic theory. *J. Animal Ecology* 67: 784–794. [11]

Manning, R. B. and F. A. Chace. 1990. Decapod and stomatopod Crustacea from Ascension Island, South Atlantic Ocean. *Smithsonian Contributions to Zoology* 503: 1–91. [12]

Marchetti, M. P., T. Light, J. Feliciano, T. Armstrong, Z. Hogen, J. Viers and P. B. Moyle. 2001. Homogenization of California's fish fauna through abiotic change. In J. L. Lockwood and M. L. McKinney (eds.), *Biotic Homogenization*. Kluwer Academic/Plenum Press, New York. [15]

Mares, M. A. and M. I. Rosenzweig. 1978. Granivory in North and South American deserts: Rodents, birds and ants. *Ecology* 59: 235–241. [2]

Margules, C. R. and R. L. Pressey. 2000. Systematic conservation planning. *Nature* 405: 243–253. [16]

Markgraf, V. (ed.). 2001. *Interhemispheric Climate Linkages*. Academic Press, San Diego, CA. [3]

Markgraf, V., J. L. Betancourt and K. A. Rylander. 1997. Late Holocene rodent middens from Río Limay, Neuquén Province, Argentina. *The Holocene* 7: 325–329. [2]

Markwick, P. J. and R. Lupia. 2002. Palaeontological databases for palaeobiogeography, palaeoecology, and biodiversity: A question of scale. In J. A. Crame and A. W. Owen (eds.), *Palaeobiogeogra-phy and Biodiversity Change: The Ordovician and Mesozoic-Cenozoic Radiations*. Special Publica-tions 194, Geological Society of London, pp. 169–178. [14]

Marone, L., J. Lopez de Casenave and V. R. Cueto. 2000. Granivory in southern South American deserts: Conceptual issues and current evidence. *Bioscience* 50: 123–132. [2]

Marquet, P. A. 2002. The search for general principles in ecology. *Nature* 418: 723. [10]

Marquet, P. A. and H. Cofre. 1999. Large temporal and spatial scales in the structure of mammalian assemblages in South America: A macroecological approach. *Oikos* 85: 299–309. [10]

Marquet, P. A. and J. X. Velasco-Hernández. 1997. A source-sink patch occupancy metapopulation model. *Revista Chilena de Historia Natural* 70: 371–380. [10]

Marquet, P. A., J. X. Velasco-Hernández and J. E. Keymer. 2003. Patch dynamics, habitat degradation and space in metapopulations. In G. A. Bradshaw and P. A. Marquet (eds.), *How Landscapes Change: Human Disturbance and Ecosystem Fragmentation in the Americas*, pp. 239–254. Springer-Verlag, New York. [10]

Marshall, C. J. and J. K. Liebherr. 2000. Cladistic biogeography of the Mexican transition zone. *J. Biogeography* 27: 203–216. [7]

Marticorena, C., O. Matthei, R. Rodriguez, M. T. K. Arroyo, M. Muñoz, F. Squeo and G. Arancio. 1998. Catalogo de la flora vascular de la Segunda Región Región de Antofagasta, Chile. *Gayana Botánica* 55: 23–83. [2]

Martin, P. S. and R. G. Klein (eds.). 1984. *Quaternary Extinctions: A Prehistoric Revolution*. University of Arizona Press, Tucson. [3]

Martin, P. S. and P. J. Mehringer, Jr. 1965. Pleistocene pollen analysis and biogeography of the Southwest. In H. E. Wright and D. J. Frey (eds.), *Quaternary of the United States*. Princeton University Press, Princeton, NJ, pp. 433–451. [2]

Mathews, J. R. 1955. *Origin and Distribution of the British Flora*. Hutchinson, London. [9]

Matthew, W. D. 1915, 1939. *Climate and Evolution*, 2nd Ed. New York Academy of Sciences, New York. [Part 2]

Matthew, W. D. 1939. *Climate and Evolution*, 2nd Ed. New York Academy of Sciences, New York. [6]

Maurer, B. A. 1999. *Untangling Ecological Complexity: The Macroscopic Perspective*. University of Chicago Press, Chicago. [7]

Maurer, B. A. 2000. Macroecology and consilience. *Global Ecology and Biogeography* 9: 275–280. [7]

Maurer, B. A., J. H. Brown and R. D. Rusler. 1992. The micro and macro in body size evolution. *Evolution* 46: 939–953. [10]

Mayden, R. L. 1988. Vicariance biogeography, parsimony and evolution in North American freshwater fishes. *Systematic Zoology* 37: 329–355. [6]

Maynard Smith, J. and E. Szathmary. 1995. *The Major Transitions in Evolution*. W. H. Freeman, Oxford. [7]

Mayr, E. 1942. *Systematics and the Origin of Species*. Columbia University Press, New York. [Part 2]

Mayr, E. 1954. Change of genetic environment and evolution. In J. Huxley, A. C. Hardy and E. B. Ford (eds.), *Evolution as a Process*, pp. 157–180. Allen and Unwin, London. [13]

McAuliffe, J. R. 1994. Landscape evolution, soil formation, and ecological patterns and soil processes in Sonoran Desert bajadas. *Ecological Monographs* 64: 111–148. [2]

McAuliffe, J. R. and T. R. Van Devender. 1998. A 22,000-year record of vegetation change in the north-central Sonoran Desert. *Palaeogeography, Palaeoclimatology, Palaeoecology* 141: 253–275. [2]

McCoy, E. D. and K. L. Heck. 1976. Biogeography of corals, sea grasses, and mangroves. *Systematic Zoology* 25: 201–210. [10]

McDowall, R. M. 1978. Generalized tracks and dispersal in biogeography. *Systematic Zoology* 27: 88–104. [Part 4]

McElhinny, M. W. and P. McFadden. 1998. *Paleomagnetism: Continents and Oceans*. International Geophysics Series 73. Academic Press, San Diego. [1]

McGhee, G. R., Jr. 1999. *Theoretical Morphology*. Columbia University Press, New York. [8]

McGill, B. and C. Collins. 2003. A unified theory for macroecology based on spatial patterns of abundance. *Evolutionary Ecology Research* 5: 469–492. [17]

McGill, B. J. 2003. A test of the unified neutral theory of biodiversity. *Nature* 422: 381–385. [9]

McGlone, M. S. 1996. When history matters: Scale, time, climate, and tree diversity. *Global Ecology and Biogeography Letters* 5: 309–314. [9]

McKenna, M. C. 1975. Fossil mammals and Early Eocene North Atlantic land continuity. *Annals of the Missouri Botanical Garden* 62: 335–353. [6]

McKerrow, W. S and C. R. Scotese. 1990. *Palaeozoic Biogeography and Paleogeography*. Geological Society of London Memoir 12. [1]

McKinney, M. L. 1997. Extinction vulnerability and selectivity: Combining ecological and paleontological views. *Annual Review of Ecology and Systematics* 28: 495–516. [8]

McKinney, M. L. 2002. Do human activities raise species richness? Contrasting patterns in the United States plants and fishes. *Global Ecology and Biogeography* 11: 343–348. [15]

McKinney, M. L. and J. L. Lockwood. 1999. Biotic homogenization: A few winners replacing many losers in the next mass extinction. *Trends in Ecology and Evolution* 14: 450–453. [Part 3]

McLachlan, J. S. 2003. Population growth of eastern North American trees since the last glacial maximum. Ph.D. Thesis, Duke University, Durham, NC. 106 pp. [3]

McLennan, D. A. and D. R. Brooks. 2002. Complex histories of speciation and dispersal: An example using some Australian birds. *J. Biogeography* 29: 1055–1066. [7]

McNab, B. K. 1971. On the ecological significance of Bergmann's Rule. *Ecology* 52: 845–854. [8]

McNab, B. K. 1994. Energy conservation and the evolution of flightlessness in birds. *American Naturalist* 144: 628–642. [12]

McNeely, J. A. and S. J. Scherr. 2002. *Ecoagriculture: Strategies to Feed the World and Save Wild Biodiversity*. Island Press, Washington, DC. [17]

McPhee, R. D. and C. Fleming. 2001. Brown-eyed, milk-giving, and extinct: Losing mammals since A.D. 1500. In M. J. Novacek (ed.), *The Biodiversity Crisis: Losing What Counts*, pp. 95–99. The New Press, New York. [18]

Medawar, P. B. 1984. *The Limits of Science*. Harper and Row, New York. [3]

Meert, J. G. and R. Van der Voo. 1994. The Neoproterozoic (1000–540 Ma) glacial intervals: No more snowball earth? *Earth and Planetary Science Letters* 123: 1–13. [1]

Menard, H. W. 1986. *The ocean of truth*. Princeton University Press, Princeton. [Part 4]

Menon, A. G. K. 1977. A systematic monograph of the tongue soles of the genus *Cynoglossus* Hamilton-Buchanen (Pisces: Cynoglossidae). *Smithsonian Contributions to Zoology* 238: 1–129. [13]

Mensing, S. A., R. G. Elston, Jr., G. L. Raines, R. J. Tausch and C. L. Nowak. 2000. A GIS model to predict the location of fossil packrat *Neotoma* middens in central Nevada. *Western North American Naturalist* 60: 111–120. [2]

Merriam, C. H. 1890. Results of a biological survey of the San Francisco Mountain region and the desert of the Little Colorado, Arizona. *North American Fauna* 3: 1–136. [18]

Middlemiss, F. A. and P. F. Rawson (eds.). 1971. *Faunal Provinces in Space and Time.* Seel House Press, Liverpool. [Part 4]

Mileikovsky, S. A. 1971. Types of larval development in marine bottom invertebrates, their distribution and ecological significance: A re-evaluation. *Marine Biology* 10: 193–213. [10]

Miller, A. I. 1997a. Dissecting global diversity patterns: Examples from the Ordovician radiation. *Annual Review of Ecology and Systematics* 28: 85–104. [Part 1]

Miller, A. I. 1997b. A new look at age and area: The geographical and environmental expansion of genera during the Ordovician. *Paleobiology* 23: 410–419. [14]

Miller, A. I. 1998. Biotic transitions in global marine diversity. *Science* 281: 1157–1160. [Part 1]

Miller, A. I. and S. Mao. 1995. Association of orogenic activity with the Ordovician radiation of marine life. *Geology* 23: 305–308. [14]

Mitchell, C. E. and A. G. Power. 2003. Release of invasive plants from fungal and viral pathogens. *Nature* 421: 625–626. [12]

Mittelbach, G. G., C. F. Steiner, S. M. Scheiner, K. L. Gross, H. L. Reynolds, R. B. Waide, M. R. Willig, S. I. Dodson and L. Gough. 2001. What is the observed relationship between species richness and productivity? *Ecology* 82: 2381–2396. [11]

Mittelbach, G. G., S. M. Scheiner and C. F. Steiner. 2003. What is the observed relationship between species richness and productivity? Reply. *Ecology* 84: 3390–3395. [9, 11]

Mittermeier, R. A., N. Myers and C. G. Mittermeier. 1999. *Hotspots, Earth's Biologically Richest and Most Endangered Terrestrial Ecoregions.* CEMEX, Mexico City. [16, 18]

Mittermeier, R. A., P. R. Gil and C.G. Mittermeier (eds.). 1997. *Megadiversity: Earth's Biologically Wealthiest Nations.* CEMEX, Monterrey, Mexico. [18]

Mitton, J. B., B. R. Kreiser and R. G. Latta. 2000. Glacial refugia of limber pine *Pinus flexilis* James. inferred from the population structure of mitochondrial DNA. *Molecular Ecology* 9: 91–97. [2]

Mooney, H. A. and E. E. Cleland. 2001. The evolutionary impact of invasive species. *Proc. National Academy of Sciences USA* 98: 5446–5451. [17]

Mooney, H. and J. A. Drake. 1986. *Ecology of Biological Invasions of North America and Hawaii.* Springer-Verlag, New York. [15]

Moore, P. G. 1977. Additions to the littoral fauna of Rockall, with a description of *Araeolaimus penelope* sp. nov. (Nematoda: Axonomlaimidae). *J. Marine Biological Association of the United Kingdom* 57: 191–200. [12]

Moores, E. M. 1991. Southwest U.S.-East Antarctica (SWEAT) connection: A hypothesis. *Geology* 19: 425–428. [1]

Mora, C., P. M. Chittaro, P. F. Sale, J. P. Kritzer and S. A. Ludsin. 2003. Patterns and processes in reef fish diversity. *Nature* 421: 933–936. [13, 14]

Morafka, D. J. 1977. A Biogeographical Analysis of the Chihuahuan Desert through its herpetofauna. Dr. W. Junk B.V., The Hague. [5]

Morgan, W. J. 1981. Hot spot tracks and the opening of the Atlantic and Indian Oceans. In C. Emiliani (ed.), *The Sea*, Vol. 7. Wiley, New York, pp. 443–475. [1]

Moritz, C. 2002. Strategies to protect biological diversity and the evolutionary processes that sustain it. *Systematic Biology* 51: 238–254. [13]

Moritz, C., K. S. Richardson, S. Ferrier, G. B. Monteith, J. Stanisic, S. E. Williams and T. Whiffin. 2001. Biogeographic concordance and efficiency of taxon indicators for establishing conservation priority in a tropical rainforest biota. *Proc. Royal Society of London Series B* 268: 1875–1881. [Part 2]

Morrison, L. W. 2002a. Island biogeography and metapopulation dynamics of Bahamian ants. *J. Biogeography* 29: 387–394. [11]

Morrison, L. W. 2002b. Determinants of plant species richness on small Bahamian islands. *J. Biogeography* 29: 931–941. [11]

Mossop, G. and I. Shetsen. 1994. *Geological Atlas of the Western Canada Sedimentary Basin*. Canadian Society of Petroleum Geologists, Calgary. [1]

Mueller, R. D., J.-Y. Royer and L. A. Lawver. 1993. Revised plate motions relative to hot spots from combined Atlantic and Indian Ocean hotspot tracks. *Geology* 16: 275–278. [1]

Mueller, R. D., W. R. Roest, J.-Y. Royer, L. M. Gahagan and J. G. Sclater. 1996. *Age of the Ocean Floor*. Report MGG-12, Data Announcement 96-MGG-04. National Geophysical Data Center, Boulder, CO. [1]

Muñoz-Duran, J. 2002. Correlates of speciation and extinction rates in the Carnivora. *Evolutionary Ecology Research* 4: 963–991. [10]

Murphy, R. W. and G. Aguirre-Léon. 2002. Nonavian reptiles: Origins and evolution. In T. J. Case, M. L. Cody and E. Ezcurra (eds.), *A New Island Biogeography of the Sea of Cortés*. Oxford University Press, Oxford, pp. 181–220. [5]

Myers, A. A. 1997. Biogeographic barriers and the development of marine biodiversity. *Estuarine, Coastal, and Shelf Sciences* 44: 241–248. [14]

Myers, R. F. 1989. *Macronesian Reef Fishes*. Coral Graphics, Barigada, Guam. [13]

Nason, J. D., J. L. Hamrick and T. H. Fleming. 2002. Historical vicariance and postglacial colonization effects on the evolution of genetic structure in *Lophocereus*, a Sonoran Desert columnar cactus. *Evolution* 56: 2214–2226. [5]

National Research Council. 2000. *Ecological Indicators for the Nation*. National Academy Press, Washington, DC. [17]

National Research Council. 2001. *Grand Challenges in Environmental Sciences*. National Academy Press, Washington, DC. [17]

Nee, S. and R. M. May. 1997. Extinction and the loss of evolutionary history. *Science* 278: 692–694. [8]

Neige, P. 2003. Spatial patterns of disparity and diversity of the Recent cuttlefishes (Cephalopoda) across the Old World. *J. Biogeography* 30: 1125–1137. [8]

Neilson R. P. 1995. A model for predicting continental-scale vegetation distribution and water balance. *Ecological Applications* 5: 362–385. [2]

Nelson, G. J. 1969. The problem of historical biogeography. *Systematic Zoology* 18: 243–246. [4]

Nelson, G. J. 1974. Historical biogeography: An alternative formalization. *Systematic Zoology* 23: 555–558. [4; Part 4]

Nelson, G. J. 1978. From Candolle to Croizat: Comments on the history of biogeography. *J. History of Biology* 11: 269–305. [4]

Nelson, G. J. 1982. Cladistique et biogeographie. *Comptes Rendus de la Société de Biogeographie* 58: 75–94. [4]

Nelson, G. J. 1983. Vicariance and cladistics: Historical perspectives with implications for the future. In R. W. Sims, J. H. Price and P. E. S. Whalley (eds.), *Evolution, Time and Space: The Emergence of the Biosphere*. The Systematics Association Special, Vol. 23. Academic Press, London, pp. 469–492, [4]

Nelson, G. J. 1984. Cladistics and biogeography. In T. Duncan and T. F. Stuessy (eds.), *Cladistics: Perspectives on the Reconstruction of Evolutionary History*. Columbia University Press, New York, pp. 273–293. [4]

Nelson, G. J. 1985. A decade of challenge: The future of biogeography. In A. E. Leviton and M. L. Aldrich (eds.), *Plate Tectonics and Biogeography*. 187–196. Special Book Publication of Earth Sciences History 42: 91–196. [4]

Nelson, G. J. 1986. Models and prospects of biogeography. In A. Pierrot-Bults, C. S. van der Spoel, B. J. Zahuranec and R. K. Johnson (eds.), *Pelagic Biogeography: Proceedings of an International Conference, The Netherlands May 29–June 5 1985*. Pages 214–218. [4]

Nelson, G. J. 1994. Homology and systematics. In B. K. Hall (ed.), *Homology: The Hierarchical Basis of Comparative Biology*. Academic Press, San Diego, CA, pp. 101–149. [4]

Nelson, G. J. and P. Y. Ladiges. 1991a. *TAX: MSDOS Program for Cladistic Systematics*. Published by the authors, New York and Melbourne. [4]

Nelson, G. J. and P. Y. Ladiges. 1991b. Standard assumptions for biogeographic analysis. *Australian Systematic Botany* 4: 41–58. [4]

Nelson, G. J. and P. Y. Ladiges. 1991c. Three-area statements: Standard assumptions for biogeographic analysis. *Systematic Zoology* 40: 470–485. [4]

Nelson, G. J. and P. Y. Ladiges. 1994. *TASS Versions 1.4, 1.5. Three Area Subtrees*. Published by the authors, New York and Melbourne. [4]

Nelson, G. J. and P. Y. Ladiges. 1995. *TASS Version 2.0. Three Area Subtrees*. Published by the authors, New York and Melbourne. [4]

Nelson, G. J. and P. Y. Ladiges. 1996. Paralogy in cladistic biogeography and analysis of paralogy-free subtrees. *American Museum Novitates* 3167: 1–58. [4]

Nelson, G. J. and P. Y. Ladiges. 2001. Gondwana, vicariance biogeography and the New York School revisited. *Australian J. Botany* 49: 389–409. [4]

Nelson, G. J. and N. I. Platnick. 1980. A vicariance approach to historical biogeography. *Bioscience* 30: 339–343. [4]

Nelson, G. J. and N. Platnick. 1981. *Systematics and Biogeography: Cladistics and Vicariance*. Columbia University Press, New York. [4, 6, 7; Parts 2, 4]

Nelson, G. J. and D. E. Rosen (eds.). 1981. *Vicariance Biogeography: A Critique*. Columbia University Press, New York. [4; Part 4]

Newell, N. D. 1971. An outline history of tropical organic reefs. *American Museum Novitates* 2465: 1–37. [14]

Newman, W. A. 1979. On the biogeography of balanomorph barnacles of the southern ocean including new balanid taxa: A subfamily, two genera, and three species. *Proc. International Symposium on Marine Biogeography and Evolution in the Southern Hemisphere* 1: 279–306. [12]

Newton, I. 1969. Winter fattening in the bullfinch. *Physiological Zoology* 42: 96–107. [9]

Nicholls, A. O. and C. R. Margules. 1993. An upgraded reserve selection algorithm. *Biological Conservation* 64: 165–169. [16]

Niklas, K. J. and B. J. Enquist. 2001. Invariant scaling relationships for interspecific plant biomass production rates and body size. *Proc. Natural Academy of Sciences USA* 98: 2922–2927. [11]

Nix, H. A. 1986. A biogeographic analysis of Australian elapid snakes. In H. A. Nix (ed.), *Atlas of Australian Elapid Snakes*. Bureau of Flora and Fauna, Canberra, pp. 4–15. [16]

Nokleberg, W. J., L. M. Parfenov, J. W. H. Monger, I. O. Norton, A. I. Khanchuk, D. B. Stone, C. R. Scotese, D. W. Scholl and K. Fujita. 2001. *Phanerozoic Tectonic Evolution of the Circum-North Pacific*. U.S.G.S. Professional Paper 1626, Denver,. [1]

Noller, J. S., J. M. Sowers and W. R. Lettis (eds.). 2000. *Quaternary Geochronology: Methods and Applications*. American Geophysical Union, Washington, D.C. [3]

Noonan, G. R. 1988. Biogeography of North American and Mexican insects and a critique of vicariance biogeography. *Systematic Zoology* 37: 366–384. [6]

Norris, R. D. 1991. Biased extinction and evolutionary trends. *Paleobiology* 17: 388–399. [8]

Norris, R. D. 2002. Extinction selectivity and ecology in planktonic foraminifera. *Palaeogeography, Palaeoclimatology, Palaeoecology* 95: 1–17. [8]

Norton, I. O. 1995. Plate motion in the North Pacific: The 43 Ma nonevent. *Tectonics* 14(5): 1080–1094. [1]

Norton, I. O. 2000. Global Hotspot reference frames and plate motion. In M. A. Richards, R. G. Gordon and R. D. van der Hilst (eds.), *The History and Dynamics of Global Plate Motions*. Geophysical Monograph Series 121, pp. 339–357. [1]

Nowak, C. L., R. S. Nowak, R. J. Tausch and P. E. Wigand. 1994. Tree and shrub dynamics in northwestern Great Basin woodland and shrub steppe during the Late Pleistocene and Holocene. *American J. Botany* 81: 265–277. [2]

Nowak, R. S., C. L. Nowak and R. J. Tausch. 2000. Probability that a fossil absent from a sample is also absent from the paleolandscape. *Quaternary Research* 54: 144–154. [2]

Nur, A. and A. Ben Avraham. 1981. Lost Pacifica continent: A mobilistic speculation. In G. Nelson and D. E. Rosen (eds.), *Vicariance Biogeography: A Critique*. Columbia University Press, New York, pp. 341–358. [4]

O'Brien, E. M. 1993. Climatic gradients in woody plant species richness: Towards an explanation based on an analysis of Southern Africa's woody flora. *J. Biogeography* 20: 181–198. [11]

O'Brien, E. M. 1998. Water-energy dynamics, climate, and prediction of woody plant species richness: An interim general model. *J. Biogeography* 25: 379–398. [9, 11]

O'Brien, E. M., R. Field and R. J. Whittaker. 2000. Climatic gradients in woody plant (tree and shrub) diversity: Water-energy dynamics, residual variation, and topography. *Oikos* 89: 588–600. [9, 11]

O'Brien, E. M., R. J. Whittaker and R. Field. 1998. Climate and woody plant diversity in southern Africa: Relationships at species, genus, and family levels. *Ecography* 21: 495–509. [11]

Oberdorff, T., B. Hugueny and J.-F. Guégan. 1995. Global scale patterns of fish species richness in rivers. *Ecography* 18: 345–352. [9]

Oberdorff, T., B. Hugueny and J.-F. Guégan. 1997. Is there an influence of historical events on contemporary fish species richness in rivers? Comparisons between Western Europe and North America. *J. Biogeography* 24: 461–467. [9]

Oberdorff, T., S. Lek and J.-F. Guégan. 1999. Patterns of endemism in riverine fish of the northern hemisphere. *Ecology Letters* 2: 75–81. [9]

Odling-Smee, F. J., K. N. Laland and M. W. Feldman. 1996. Niche construction. *American Naturalist* 147: 641–648. [10]

Olden, J. D. and N. L. Poff. 2003. Toward a mechanistic understanding and prediction of biotic homogenization. *American Naturalist* 162(4): 442–460. [15]

Ong, P., L. E. Afuang and R. G. Rosell-Ambal (eds.). 2002. *Philippine Biodiversity Conservation Priorities: A Second Iteration of the National Biodiversity Strategy and Action Plan*. Philippine Department of the Environment and Natural Resources, Quezon City. [18]

Orians, G. H. and O. T. Solbrig (eds.). 1977. *Convergent Evolution in Warm Deserts of Argentina and the United States.* US/IBP Synthesis Series 3. Academic Press, New York. [2]

Ormund, R. F. G. and C. M. Roberts. 1997. The biodiversity if coral reef fishes. In R. F. G. Ormund, J. D. Gage and M. V. Angel (eds.), *Marine Biodiversity*, pp.215–257. Cambridge University Press, Cambridge. [13]

Ortmann, A. 1896. *Grundzuge der marinen Tiergeographie.* G. Fischer, Jena. [Part 4]

Osborn, H. F. 1900. The geological and faunal relations of Europe and America during the Tertiary period and the theory of successive invasions of an African fauna. *Science* 11: 561–574. [6]

Ostergaard, J. M. 1928. Fossil marine mollusks of Oahu. *Bernice Pauahi Bishop Museum Bulletin* 51: 1–32. [12]

Ostergaard, J. M. 1939. Report on fossil Mollusca of Molokai and Maui. *Occasional Papers of the Bernice Pauahi Bishop Museum* 15(6): 67–77. [12]

Otto-Bliesner, B. L., E. Becker and N. Becker. 1994. *Atlas of Phanerozoic Paleoclimate Simulated by a Global Climate Model.* Center for Earth System History, Report #1, November 1994. University of Texas at Arlington. [1]

Overpeck, J. T., R. S. Webb and T. Webb III. 1992. Mapping eastern North American vegetation changes of the past 18 ka: No-analogs and the future. *Geology* 20: 1071–1074. [3]

Overpeck, J. T., T. Webb III and I. C. Prentice. 1985. Quantitative interpretation of fossil pollen spectra: Dissimilarity coefficients and the method of modern analogs. *Quaternary Research* 23: 87–108. [3]

Owen, A. W. and J. A. Crame. 2002. Palaeobiogeography and the Ordovician and Mesozoic-Cenozoic biotic radiations. In J. A. Crame and A. W. Owen (eds.), *Palaeobiogeogra-phy and Biodiversity Change: The Ordovician and Mesozoic-Cenozoic Radiations*. Special Publica-tions 194, Geological Society of London, pp. 1–11. [14]

Owens, I. P. F., P. M. Bennett and P. H. Harvey. 1999. Species richness among birds: Body size, life history, sexual selection or ecology? *Proc. Royal Society of London Series B* 266: 933–939. [10]

Page, R. D. M. 1989a. *COMPONENT Version 1.5.* University of Auckland, Auckland. [4; Part 2]

Page, R. D. M. 1989b. Comments on component-compatibility in historical biogeography. *Cladistics* 5: 167–182. [4; Part 2]

Page, R. D. M. 1990. Component analysis: A valiant failure? *Cladistics* 6: 119–136. [4]

Page, R. D. M. 1993a. *COMPONENT, Version 2.0.* The Natural History Museum, London. [4]

Page, R. D. M. 1993b. Genes, organisms and areas: The problem of multiple lineages. *Systematic Biology* 42: 77–84. [4]

Page, T. J. and K. Linse. 2002. More evidence of speciation and dispersal across the Antarctic Polar Front through molecular systematics of Southern Ocean *Limatula* (Bivalvia: Limidae). *Polar Biology* 25: 818–826. [14]

Palmer, M. W. and P. S. White. 1994. Scale dependence and the species-area relationship. *American Naturalist* 144: 717–740. [14]

Palumbi, S. R. 1994. Genetic divergence, reproductive isolation, and marine speciation. *Annual Review of Ecology and Systematics* 25: 547–572. [13]

Palumbi, S. R. 1997. Molecular biogeography of the Pacific. *Coral Reefs* 16(Suppl.): 547–552. [13]

Palumbi, S. R., G. Grabowski, T. Duda, L. Geyer and N. Tachino. 1997. Speciation and population genetic structure in tropical Pacific sea urchins. *Evolution* 51: 1506–1517. [13]

Pandolfi, J. M. 1992. Successive isolation rather than evolutionary centers for the origination of Indo-Pacific reef corals. *J. Biogeography* 19: 593–609. [13, 14]

Parmesan, C. 1996. Climate and species' range. *Nature* 382: 765–766. [16]

Parmesan, C., N. Ryrholm, C. Stefanescu, J. K. Hill, C. D. Thomas, H. Descimon, B. Huntley, L. Kaila, J. Kullberg, T. Tammaur, W. J. Tennet, J. A. Thomas and M. Warren. 1999. Poleward shifts in geographical ranges of butterfly species associated with global warming. *Nature* 399: 579–583. [9]

Parrish, J. T., A. M. Ziegler and C. R. Scotese. 1982. Rainfall patterns and the distribution of coals and evaporites in the Mesozoic and Cenozoic. *Palaeogeography, Palaeoclimatology, Palaeoecology* 40: 67–101. [1]

Parrish, T. 2002. *Krakatau: Genetic Consequences of Island Colonization*. Ph.D. Thesis, University of Utrecht, Netherlands Institute of Ecology, Heteren. [11]

Paterson, G. L. J., G. D. F. Wilson, N. Cosson and P. A. Lamont. 1998. Hessler and Jumars (1974) revisited: Abyssal polychaete assemblages from the Atlantic and Pacific. *Deep Sea Research II* 45: 225–251. [14]

Patterson, C. 1981a. Methods of paleobiogeography. In G. Nelson and D. E. Rosen (eds.), *Vicariance Biogeography: A Critique*. Columbia University Press, New York, pp. 446–489. [4]

Patterson, C. 1981b. The development of the North American fish fauna: A problem of historical biogeography. In P. L. Forey (ed.), *Chance, Change and Challenge: The Evolving Biosphere*, Vol. 2. British Museum of Natural History and Cambridge University Press, London and Cambridge, pp. 265–281. [4]

Patterson, C. 1982. Morphological characters and homology. In K. A. Joysey and A. E. Friday (eds.), *Problems of Phylogenetic Reconstruction*. Academic Press, London, pp. 21–74. [4]

Patterson, C. 1983. Aims and methods in biogeography. In R.W. Sims, J. H. Price and P. E. S. Whalley (eds.), *Evolution, Time and Space: The Emergence of the Biosphere*. Academic Press, New York, pp. 1–28. [6]

Patton, J. L. and S. T. Alvarez-Castañeda. 2002. Phylogeography of the desert woodrat, *Neotoma lepida*, with comments on systematics and biogeographic history. In V. Sánchez-Cordero and R. A. Medellín (eds.), *Contribuciones Mastozoologicas en Homenaje a Bernardo Villa*. [2]

Patzkowsky, M. E. 1995. Ecological aspects of the Ordovician radiation of articulate brachiopods. In J. D. Cooper, M. L. Droser and S. C. Finney (eds.), *Ordovician Odyssey: Short Papers for the Seventh International Symposium on the Ordovician System*. Pacific Section SEPM, Fullerton, CA, pp. 413–414. [14]

Paulay, G. 1989. Marine invertebrates of the Pitcairn Islands: Species composition and biogeography of corals, molluscs, and echinoderms. *Atoll Research Bulletin* 326: 1–28. [12]

Paulay, G. 1990. Effects of Late Cenozoic sea-level fluctuations on the bivalve faunas of tropical oceanic islands. *Paleobiology* 16: 415–434. [12]

Paulay, G. 1994. Biodiversity on oceanic islands: Its origin and extinction. *American Zoologist* 34: 134–144. [11]

Paulay, G. 1996. Dynamic clams: Changes in the bivalve fauna of Pacific Islands as a result of sea-level fluctuations. *American Malacological Bulletin* 12: 45–57. [12]

Paulay, G. 1997. Diversity and distribution of reef organisms. In C. E. Birkeland (ed.), *Life and Death of Coral Reefs*, pp. 298–353. Chapman and Hall, New York. [13, 14]

Pearson, R. G. and T. P. Dawson. 2003. Predicting the impacts of climate change on the distribution of species: Are bioclimate envelope models useful? *Global Ecology and Biogeography* 12: 361–371. [16]

Pearson, S. and J. L. Betancourt. 2002. Understand-ing arid environments using fossil rodent middens. *J. Arid Environments* 50: 499–511. [2]

Pedicino, L., S. W. Leavitt and J. L. Betancourt. 2002. Historical variations in $\delta^{13}C_{leaf}$ of herbarium specimens in the southwestern U.S. *Western North American Naturalist* 62: 348–359. [2]

Pendall, E., J. L. Betancourt and S. W. Leavitt. 1999. Paleoclimatic significance of ?D and ?^{13}C in pinyon pine needles from packrat middens spanning the last 40,000 years. *Palaeogeography, Palaeoclimatology, and Palaeoecology* 147: 53–72. [2]

Petchey, O. L. and K. J. Gaston. 2002. Extinction and the loss of functional diversity. *Proc. Royal Society of London Series B* 269: 1721–1727. [8]

Peterson, A. T. 2001. Predicting species geographic distributions based on ecological niche modeling. *Condor* 103: 599–605. [16]

Peterson, A. T. and K. P. Cohoon. 1999. Sensitivity of distributional prediction algorithms to geographic data completeness. *Ecological Modeling* 117: 159–164. [16]

Peterson, A. T. and D. A. Kluza. 2003. New distributional modelling approaches for gap analysis. *Animal Conservation* 6: 47–54. [16]

Peterson, A. T., L. G. Ball and K. M. Brady. 2000. Distribution of birds of the Philippines: Biogeography and conservation priorities. *Bird Conservation International* 10: 149–167. [18]

Peterson, A. T., L. G. Ball and K. P. Cohoon. 2002b. Predicting distributions of tropical birds. *Ibis* 144: E27–E32. [16]

Peterson, A. T., S. L. Egbert, V. Sánchez-Cordero and K. V. Price. 2000. Geographic analysis of conservation priorities for biodiversity: A case study of endemic birds and mammals in Veracruz, Mexico. *Biological Conservation* 93: 85–94. [16]

Peterson, A. T., V. O. A. Flores, P. L. S. Leon, B. J. E. Llorente, M. M. A. Luis, S. A. G. Navarro-Siguenza, C. M. G. Torres and I. Vargas. 1993. Conservation priorities in northern Middle America: Moving up in the world. *Biodiversity Letters* 1: 33–38. [18]

Peterson, A. T., A. G. Navarro-Siguenza and H. Benítez-Díaz. 1998. The need for continued scientific collecting: A geographic analysis of Mexican bird specimens. *Ibis* 140: 288–294. [16]

Peterson, A. T., M. A. Ortega-Huerta, J. Bartley, V. Sánchez-Cordero, J. Soberón, R. H. Buddemeier and D. R. B. Stockwell. 2002c. Future projections for Mexican faunas under global climate change scenarios. *Nature* 416: 626–629. [16]

Peterson, A. T., V. Sánchez-Cordero, J. Soberón, J. Bartley, R. W. Buddemeier and A. Navarro-Siguenza. 2001. Effects of global climate change on geographic distributions of Mexican Cracidae. *Ecological Modeling* 144: 21–30. [16]

Peterson, A. T., J. Soberón and V. Sánchez-Cordero. 1999. Conservatism of ecological niches in evolutionary time. *Science* 285: 1265–1267. [16]

Peterson, A. T., D. R. B. Stockwell and D. A. Kluza. 2002a. Distributional prediction based on ecological niche modeling of primary occurrence data. In J. M. Scott, P. J. Heglund and M. I. Morrison (eds.), *Predicting Species Occurrences: Issues of Scales and Accuracy*. Island Press, Washington, DC, pp. 617–623. [16]

Peterson, A. T., H. Tian, E. Martínez-Meyer, J. Soberón and V. Sánchez-Cordero. In press. Modeling ecosystems shifts and individual species distribution shifts. In T. Lovejoy (ed.), *Climate Change and Biodiversity*. Smithsonian Institute Press, Washington, DC. [16]

Petit J. R., J.-M. Barnola, I. Basile, M. Bender, J. Chappellaz, M. Davis, G. Delaygue, M. Delmotte, V. M. Kotlyakov, M. Legrand, V. Y. Lipenkov, C. Lorius, L. Pepin, C. Ritz, E. Saltzman, M. Stievenard, J. Jouzel, D. Raynaud and N. I. Barkov. 1999. Climate and atmospheric history of the past 420,000 years from the Vostok ice core, Antarctica. *Nature* 399: 429–436. [3]

Petit, R. J., S. Brewer, S. Bordács, K. Burg, R. Cheddadi, E. Coart, J. Cottrell, U. M. Csaikl, B. van Dam, J. D. Deans, S. Espinel, S. Fineschi, R. Finkeldey, I. Glaz, P. G. Goicoechea, J. S. Jensen, A. O. König, A. J. Lowe, S. F. Madsen, G. Mátyás, R. C. Munro, F. Popescu, D. Slade, H. Tabbener, S. G. M. de Vries, B. Ziegenhagen, J.-L. de Beaulieu and A. Kremer. 2002a. Identification of refugia and post-glacial colonization routes of European white oaks based on chloroplast DNA and fossil pollen evidence. *Forest Ecology and Management* 156: 49–74. [3; Part 1]

Petit, R. J., U. M. Csaikl, S. Bordács, K. Burg, E. Coart, J. Cottrell, B. van Dam, J. D. Deans, S. Dumolin-Lapègue, S. Fineschi, R. Finkeldey, A. Gillies, I. Glaz, P. G. Goicoechea, J. S. Jensen, A. O. König, A. J. Lowe, S. F. Madsen, G. Mátyás, R. C. Munro, M. Olalde, M.-H. Pemonge, F. Popescu, D. Slade, H. Tabbener, D. Taurchini, S. G. M. de Vries, B. Ziegenhagen and A. Kremer. 2002b. Chloroplast DNA variation in European white oaks: Phylogeography and patterns of diversity based on data from over 2600 populations. *Forest Ecology and Management* 156: 5–26. [3; Part 1]

Petren, K and T. J. Case. 2002. Updated mtDNA phylogeny for *Sauromalus* and implications for the evolution of gigantism. In T. J. Case, M. L. Cody and E. Ezcurra (eds.), *A New Island Biogeography of the Sea of Cortés*. Oxford University Press, New York, pp. 574–579. [5]

Philippine Department of Environment and Natural Resources. 1997. *Philippine Biodiversity: An Assessment and Action Plan*. Bookmark, Makati, Philippines. [18]

Phillips, O. L., P. Hall, A. H. Gentry, S. A. Sawyer and R. Vasquez. 1994. Dynamics and species richness of tropical rain forests. *Proc. National Academy of Sciences USA* 91: 2805–2809. [11]

Pianka, E. R. 1966. Latitudinal gradients in species diversity: A review of concepts. *American Naturalist* 100: 33–46. [9, 10]

Piccoli, G., S. Sartori and A. Franchino. 1987. Benthic molluscs of shallow Tethys and their destiny. In K. G. McKenzie (ed.), *Proceedings of the International Symposium on Shallow Tethys*, Vol 2A, pp. 333–373. A.A. Balkema, Rotterdam. [13]

Pielou, E. C. 1979. *Biogeography*. John Wiley and Sons, Chichester, UK. [14]

Pierrot-Bults, A. C. et al. (eds.). 1986. *Pelagic Biogeography*. UNESCO, Amsterdam, Paris. [Part 4]

Pindell, J. L., R. Higgs and J. F. Dewey. 1998. Cenozoic palinspastic reconstruction, paleogeographic evolution, and hydrocarbon setting of the northern margin of South America. In J. L. Pindell and C. L. Drake (eds.), *Paleogeographic Evolution and Non-Glacial Eustasy*. Northern South America. SEPM (Society for Sedimentary Geology) Special Publication 58:, pp. 45–86. [1]

Planz, J. V. 1992. *Molecular Phylogeny and Evolution of the American Woodrats, Genus Neotoma (Muridae)*. Ph.D. Dissertation, Biological Sciences, University of North Texas, Denton, TX. [5]

Platnick, N. I. 1976. Concepts of dispersal in historical biogeography. *Systematic Zoology* 25: 294–295. [6]

Platnick, N. I. 1979. Philosophy and the transformation of cladistics. *Systematic Zoology* 28: 537–546. [4]

Platnick, N. I. and G. J. Nelson. 1978. A method of analysis for historical biogeography. *Systematic Zoology* 27: 1–16. [4, 5, 6; Part 2]

Poinar, H. N., M. Hofreiter, G. W. Spaulding, P. S. Martin, B. A. Stankiewicz, H. Bland, R. P. Evershed, G. Possnert and S. Paabo. 1998. Molecular coproscopy: Dung and diet of the extinct ground sloth *Nothrotheriops shastensis*. *Science* 281: 402–406. [2, 3]

Poinar, H. N., M. Kuch, G. McDonald, P. S. Martin and S. Pääbo. 2003. Nuclear gene sequences from a late Pleistocene sloth coprolite. *Current Biology* 13: 1150–1152. [2]

Polasky, S., J. D. Camm, A. R. Solow, B. Csuti, D. White and R. Ding. 2000. Choosing reserve networks with incomplete species information. *Biological Conservation* 94: 1–10 [16]

Pollock, D. E. 1990. Palaeoceanography and speciation in the spiny lobster genus *Jasus*. *Bulletin of Marine Science* 46: 387–405. [12]

Ponder, W. F. 1983. Rissoaform gastropods from the Antarctic and sub-Antarctic. *British Antarctic Survey Scientific Reports* 108. [14]

Poore, G. C. B. and G. D. F. Wilson. 1993. Marine species richness. *Nature* 361: 597–598. [14]

Popper, K. R. 1968. *The Logic of Scientific Discovery*. Harper and Row, New York. [7]

Porter, W. P., S. Budaraju and N. Ramankutty. 2000. Calculating climate effects on birds and mammals: Impacts on biodiversity, conservation, population parameters, and global community structure. *American Zoologist* 40: 597. [16]

Posada, D. and K. A. Crandall. 2001. Intraspecific gene genealogies: Trees grafting into networks. *Trends in Ecology & Evolution* 16: 37–45. [5]

Posada, D., K. A. Crandall and A. R. Templeton. 2000. GeoDis: A program for the cladistic nested analysis of the geographical distribution of genetic haplotypes. *Molecular Ecology* 9: 487–488. [5]

Poulin, R. and J.-F. Guégan. 2000. Nestedness, anti-nestedness, and the relationship between prevalence and intensity in ectoparasite assemblages of marine fish: A spatial model of species coexistence. *International J. Parasitology* 30: 1147–1152. [9]

Powell, A. W. 1973. The patellid limpets of the world. Patellidae. *Indo-Pacific Mollusca* 3: 75–206. [10]

Powell, C. McA., Z. X. Li, M. W. McElhinny, J. G. Meert and J. K. Park. 1993. Paleomagnetic constraints on the timing of the Neoproterozoic breakup of Rodinia and the Cambrian formation of Gondwana. *Geology* 21: 889–892. [1]

Powell, M. G. and M. Kowaleski. 2002. Increase in evenness and sampled alpha diversity through the Phanerozoic: Comparison of early Paleozoic and Cenozoic marine fossil assemblages. *Geology* 30: 331–334. [14]

Preece, R. C. 1995. The composition and relationships of the marine molluscan fauna of the Pitcairn Islands. *Biological J. Linnean Society* 56: 339–358. [12]

Preece, R. C. 1997. The spatial response of non-marine Mollusca to past climate changes. In B. Huntley, W. Cramer, A. V. Morgan, H. C. Prentice and J. R. M. Allen (eds.), *Past and Future Rapid Environmental Changes: The Spatial and Evolutionary Responses of Terrestrial Biota*. Springer-Verlag, New York, pp. 163–177. [3]

Prentice, I. C. 1988. Records of vegetation in space and time: The principles of pollen analysis. In B. Huntley and T. Webb III (eds.), *Vegetation History*. Kluwer Academic Publishers, Dordrecht, NL, pp. 17–42. [3]

Prentice, I. C., P. J. Bartlein and T. Webb III. 1991. Vegetation and climate change in eastern North America since the last glacial maximum. *Ecology* 72: 2038–2056. [3]

Pressey, R. L., H. P. Possingham and C. R. Margules. 1996. Optimality in reserve selection algorithms: When does it matters and how much? *Biological Conservation* 76: 259–267. [16]

Price, A. R. G., M. J. Keeling and C. J. O'Callaghan. 1999. Ocean-scale patterns of "biodiversity" of Atlantic asteroids determined from taxonomic distinctness and other measures. *Biological J. Linnean Society* 66: 187–203. [14]

Price, J. 2000. Modeling the potential impacts of climate change on the summer distributions of Massachusetts passerines. *Bird Observer* 28: 224–230. [16]

Price, J. P. and D. A. Clague. 2002. How old is the Hawaiian biota? Geology and phylogeny suggest recent divergence. *Proc. Royal Society of London Series B* 269: 2429–2435. [12]

Purvis, A., P.-M. Agapow, J. L. Gittleman and G.M. Mace. 2000. Non-random extinction and the loss of evolutionary history. *Science* 288: 328–330. [8]

Qian, H. and R E Ricklefs 1999. A comparison of the taxonomic richness of vascular plants in China and the United States. *American Naturalist* 154: 160 181. [9]

Quinn, J. F. and S. P. Harrison. 1988. Effects of habitat fragmentation and isolation on species richness: Evidence from biogeographic patterns. *Oecologia* 75: 132–140. [17]

Radtkey, R. R., S. M. Fallon and T. J. Case. 1997. Character displacement in some *Cnemidophorus* lizards revisited: A phylogenetic analysis. *Proc. National Academy of Sciences USA* 94: 9740–9745. [5]

Rahbek, C. and G. R. Graves. 2001. Multi-scale assessment of patterns of avian species richness. *Proc. National Academy of Sciences USA* 98: 4534–4539. [9]

Rahel, F. J. 2002. Homogenization of freshwater faunas. *Annual Review of Ecology and Systematics* 33: 291–315. [15]

Randall, J. E. 1998. Zoogeography of shore fishes of the Indo-Pacific region. *Zoological Studies* 37: 227–268. [13]

Randall, J. E., G. R. Allen and R. C. Steene. 1990. *Fishes of the Great Barrier Reef and Coral Sea*. University of Hawaii Press, Honolulu, HI. [12]

Raup, D. M. and D. Jablonski. 1993. Geography of end-Cretaceous marine bivalve extinctions. *Science* 260: 971–973. [8]

Raven, P. H. and E. O. Wilson. 1992. A fifty-year plan for biodiversity surveys. *Science* 81: 525–542. [Part 5]

Rawlinson, P. A., R. A. Zann, S. van Balen and I. W. B. Thornton. 1992. Colonization of the Krakatau islands by vertebrates. *Geojournal* 28: 225–231. [11]

Ray, J. 1721. *Three Physico-Theological Discourses, Concerning I. The Primitive Chaos, and Creation of the World. II. The General Deluge, Its Causes and Effects. III. The Dissolution of the World, and Future Conflagration*, 4th Ed. William and John Innys, London. 456 p. [Part 1]

Reaka, M. L. 1980. Geographic range, life history patterns, and body size in a guild of coral-dwelling mantis shrimps. *Evolution* 34: 1019–1030. [12]

Reaka-Kudla, M. L. 1996. The global biodiversity of coral reefs: A comparison with rain forests. In M. L. Reaka-Kudla, D. E. Wilson and E. O. Wilson (eds.), *Biodiversity II*. Joseph Henry Press, Washington, DC, pp. 83–108. [14]

Reaka-Kudla, M., D. E. Wilson and E. O. Wilson. 1997. *Biodiversity II: Understanding and Protecting Our Biological Resources*. Joseph Henry Press. Washington, DC. [16]

Rees, P. M. 2002. Land-plant diversity and the end-Permian mass extinction. *Geology* 30: 827–830. [8]

Reeves, C. and M. de Wit. 2000. Making ends meet in Gondwana: Retracing the transforms of the Indian Ocean and reconnecting continental shear zones. *Terra Nova* 12(6): 272–282. [1]

Rehder, H. A. 1980. The marine mollusks of Easter Island (Isla de Pascua) and Sala y Gomez. *Smithsonian Contributions to Zoology* 289: 1–167. [12]

Rehfeldt, G. E. 1999. Systematics and genetic structure of Ponderosae taxa Pinaceae inhabiting the mountain islands of the Southwest. *American J. Botany* 86: 741–752. [2]

Rehfeldt, G. E., C. C. Ying and W. R. Wyckoff. 2001. Physiologic plasticity, evolution and impacts of a changing climate on *Pinus contorta*. *Climate Change* 50: 355–376. [3]

Rehfeldt, G. E., C. C. Ying, D. L. Spittlehouse and D. A. Hamilton, Jr. 1999. Genetic responses to climate for *Pinus contorta* in British Columbia: Niche breadth, climate change and reforestation. *Ecological Monographs* 69: 375–407. [3]

Reichard, S. H. and P. White. 2001. Horticulture as a pathway of invasive plant introductions in the United States. *Bioscience* 51: 103–113. [15]

Reid, D. G. 1985. Habitat and zonation patterns of *Littoraria* species (Gastropoda: Littorinidae) in Indo-Pacific mangrove forests. *Biological J. Linnean Society* 26: 39–68. [12]

Reid, D. G. 2001. *The genus* Nodilittorina von Martens, *1897 (Gastropoda: Littorinidae) in the Indo-Malayan region*. Phuket Marine Biological Center Special Publication 25: 433–449. [12]

Reid, W. V. and K. R. Miller. 1989. *Keeping Options Alive: The Scientific Basis for Conserving Biodiversity*. World Researches Institute, Washington, DC. [18]

Rensch, B. 1959. *Evolution Above the Species Level*. Methuen, London. [10]

Rex, M. A., C. T. Stuart and G. Coyne. 2000. Latitudinal gradients of species richness in the deep-sea benthos of the North Atlantic. *Proc. National Academy of Sciences USA* 97: 4082–4085. [14; Part 4]

Rex, M. A., C. T. Stuart, R. R. Hessler, J. A. Allen, H. L. Sanders and G. D. F. Wilson. 1993. Global-scale latitudinal patterns of species diversity in the deep-sea benthos. *Nature* 365: 636–639. [14]

Rhode, D. 2001. Packrat middens as a tool for reconstructing historic ecosystems. In D. Egan and E. Howell (eds.), *Historical Ecology Handbook: A Restorationist's Guide to Reference Ecosystems*. Island Press, Covelo, CA, pp. 257–293. [2]

Richardson, D. M., P. Pysek, M. Rejmanek, M. G. Barbour, F. D. Panetta and C. J. West. 2000. Naturalization and invasion of alien plants: Concepts and definitions. *Diversity and Distributions* 6: 93–107. [15]

Rickart, E. A. 1993. Diversity patterns of mammals along elevational and disturbance gradients in the Philippines: Implications for conservation. *Asia Life Sciences* 2: 251–260. [18]

Rickart, E. A. and L. R. Heaney. 1991. A new species of *Chrotomys* (Muridae) from Luzon Island, Philippines. *Proc. Biological Society of Washington* 104: 387–398. [18]

Rickart, E. A., L. R. Heaney, D. S. Balete and B. R. Tabaranza, Jr. 1998. A review of the genera *Crunomys* and *Archboldomys* (Rodentia: Muridae: Murinae) with descriptions of two new species from the Philippines. *Fieldiana Zoology* (new series) 89: 1–24. [18]

Rickart, E. A., L. R. Heaney, P. D. Heideman and R. C. B. Utzurrum. 1993. The distribution and ecology of mammals on Leyte, Biliran, and Maripipi islands, Philippines. *Fieldiana Zoology* (new series) 72: 1–62. [18]

Rickart, E. A., L. R. Heaney and B. R. Tabaranza, Jr. 2002. Review of *Bullimus* (Muridae: Murinae) and description of a new species from Camiguin Island, Philippines. *J. Mammalogy* 83: 421–436. [18]

Rickart, E. A., L. R. Heaney and B. R. Tabaranza, Jr. 2003. A new species of *Limnomys* (Rodentia: Muridae: Murinae) from Mindanao Island, Philippines. *J. Mammalogy* 84: 1443–1455. [18]

Rickart, E. A., L. R. Heaney and R. B. Utzurrum. 1991. Distribution and ecology of small mammals along an elevational transect in southeastern Luzon, Philippines. *J. Mammalogy* 72: 458–469. [18]

Ricklefs, R. E. 1995. 3.2 The distribution of biodiversity. In V. H. Heywood (ed.), *Global Biodiversity Assessment*. Cambridge University Press, Cambridge, pp. 139–173. [14]

Ricklefs, R. E. and E. Bermingham. 2001. Nonequilibrium diversity dynamics of the Lesser Antillean avifauna. *Science* 294: 1522–1524. [5, 6, 11, 12]

Ricklefs, R. E. and R. E. Latham. 1993. Global patterns of diversity in mangrove floras. In R. E. Ricklefs and D. Schluter (eds.), *Species Diversity in Ecological Communities*, pp. 215–229. Univer-sity of Chicago Press, Chicago. [13]

Ricklefs, R. E. and D. B. Miles. 1994. Ecological and evolutionary inferences from morphology: An ecological perspective. In P. C. Wainwright and S. M. Reilly (eds.), *Ecological Morphology*. Uni-versity of Chicago Press, Chicago, pp. 13–41. [8]

Ricklefs, R. E. and K. O'Rourke. 1974. Aspect diversity in moths: A temperate-tropical comparison. *Evolution* 29: 313–324. [8]

Ricklefs, R. E. and D. Schluter. 1993. Species diversity: Regional and historical influences. In R. E. Ricklefs and D. Schluter (eds.), *Species Diversity in Ecological Communities: Historical and Geo-graphical Perspectives*, pp. 350–363. University of Chicago Press, Chicago. [14]

Ricklefs, R. E., R. E. Latham and H. Qian. 1999. Global patterns of tree species richness in moist forests: Distinguishing ecological influences and historical contingency. *Oikos* 86: 369–373. [9]

Riddle, B. R. 1995. Molecular biogeography in the pocket mice (*Perognathus* and *Chaetodipus*) and grasshopper mice (*Onychomys*): The late Cenozoic development of a North American aridlands rodent guild. *J. Mammalogy* 76: 283–301. [5]

Riddle, B. R. 1998. The historical assembly of continental biotas: Late Quaternary range-shifting, areas of endemism, and biogeographic structure in the North American mammal fauna. *Ecography* 21: 437–446. [5]

Riddle, B. R. and D. J. Hafner. In press. A stepwise approach to integrating phylogeographic and phylogenetic biogeographic perspectives on the history of a core North American warm deserts biota. *J. Arid Environments*. [5]

Riddle, B. R. and R. L. Honeycutt. 1990. Historical biogeography in North American arid regions: An approach using mitochondrial-DNA phylogeny in grasshopper mice (genus *Onychomys*). *Evolution* 44: 1–15. [5]

Riddle, B. R., D. J. Hafner, L. F. Alexander and J. R. Jaeger. 2000a. Cryptic vicariance in the historical assembly of a Baja California peninsular desert biota. *Proc. National Academy of Sciences USA* 97: 14438–14443. [5; Part 2]

Riddle, B. R., D. J. Hafner and L. F. Alexander. 2000b. Comparative phylogeography of Baileys' pocket mouse (*Chaetodipus baileyi*) and the *Peromyscus eremicus* species group: Historical vicariance of the Baja California Peninsular desert. *Molecular Phylogenetics and Evolution* 17: 161–172. [5]

Riddle, B. R., D. J. Hafner and L. F. Alexander. 2000c. Phylogeography and systematics of the *Peromyscus eremicus* species group and the historical biogeography of North American warm regional deserts. *Molecular Phylogenetics and Evolution* 17: 145–160. [5]

Riebesell, J. F. 1982. Arctic-alpine plants on mountaintops: Agreement with island biogeography theory. *American Naturalist* 119: 657–674. [17]

Rieppel, O. 2003. Semaphoronts, cladograms and the roots of total evidence. *Biological J. Linnean Society* 80(1): 167–186. [4]

Rimington, E. 1992. *Butterflies of the Doncaster District*. Sorby Record Special Series No. 9. Sorby Natural History Society and Sheffield City Museums, Sheffield, UK. [9]

Roberts, C.M., C. J. McClean, J. E. N. Veron, J. D. Hawkins, G. R. Allen, D. E. McAllister, C. G. Mittermeier, F. W. Schueler, M. Spalding, F. Wells, C. Vynne and T. B. Werner. 2002. Marine biodiversity hotspots and conservation priorities for tropical reefs. *Science* 295: 1280–1284. [13]

Rocha, L. A. 2003a. Ecology, the Amazon barrier, and speciation in western Atlantic *Halichoeres*

(Labridae). Ph.D. dissertation, University of Florida, Gainesville. 88 pp. [Part 4]

Rocha, L. A. 2003b. Patterns of distribution and processes of speciation in Brazilian reef fishes. *Journal of Biogeography* 30: 1161–1171. [Part 4]

Rocha, L. A., A. L. Bass, D. R. Robertson and B. W. Bowen. 2002. Adult habitat preferences, larval dispersal, and the comparative phylogeography of three Atlantic surgeonfishes (Teleostei: Acanthuridae). *Molecular Ecology* 11: 243–252. [Part 4]

Rodrigues, A. S., J. O. Cerdeira and K. J. Gaston. 2000. Flexibility, efficiency, and accountability: Adapting reserve selection algorithms to more complex conservation problems. *Ecography* 23: 565–574. [16]

Rogers, A. R. and H. Harpending. 1992. Population-growth makes waves in the distribution of pairwise genetic-differences. *Molecular Biology and Evolution* 9: 552–569. [5]

Rohde, K. 1992. Latitudinal gradients in species diversity: The search for the primary cause. *Oikos* 65: 514–527. [8, 9, 10, 14]

Rohde, K. 1998. Latitudinal gradients in species diversity: Area matters, but how much? *Oikos* 82: 184–190. [9, 10]

Rohde, K. 1999. Latitudinal gradients in species diversity and Rapoport's rule revisited: A review of recent work, and what can parasites teach us about the causes of the gradients? *Ecography* 22: 593–613. [10]

Romm, J. 1994. A new forerunner to continental drift. *Nature* 367: 407–408. [4]

Ronov, A., V. Khain and Balukhovsky. 1989. *Atlas of Lithological-Paleogeographical Maps of the World: Mesozoic and Cenozoic of Continents and Oceans*. U.S.S.R. Academy of Sciences, Leningrad. [1]

Ronquist, F. 1994. Ancestral areas and parsimony. *Systematic Biology* 43: 267–274. [6]

Ronquist, F. 1997. Dispersal-vicariance analysis: A new approach to the quantification of historical biogeography. *Systematic Biology* 46: 195–203. [7; Part 2]

Ronquist, F. 1998. Phylogenetic approaches in coevolution and biogeography. *Zoologica Scripta* 26: 313–322. [6]

Ronquist, F. 1998. Three-dimensional cost-matrix optimization and maximum co-speciation. *Cladistics* 14: 167–172. [4]

Root, T. 1988a. *Atlas of Wintering North American Birds: An Analysis of Christmas Bird Count Data.* University of Chicago Press, Chicago. [9]

Root, T. 1988b. Energy constraints on avian distributions and abundances. *Ecology* 69: 330–339. [9]

Root, T. 1988c. Environmental factors associated with avian distributional boundaries. *J. Biogeography* 15: 489–505. [9]

Rosa, D. 1918. *Ologenesi: Nuova Teoria dell'Evoluzione e della Distribuzione Geografica dei Viventi*. Bem-porad, Firenze. [4]

Rosen, B. R. 1981. The tropical high diversity enigma—the corals' eye view. In P. L. Forey (ed.), *The Evolving Biosphere*. British Museum (Natural History), London and Cambridge University Press, Cambridge, pp. 103–129. [14]

Rosen, B. R. 1984. Reef coral biogeography and climate through the late Cainozoic: Just islands in the sun or a critical pattern of islands? In R. Brenchley (ed.), *Fossils and Climate*, pp.201–260. John Wiley, Chichester, UK. [13]

Rosen, B. R. 1988. From fossils to earth history: Applied historical biogeography. In A. A. Myers and P. S. Giller (eds.), *Analytical Biogeography: An Integrated Approach to the Study of Animal and Plant Distributions*. Chapman and Hall, London and New York, pp. 437–481. [4]

Rosen, B. R. 1988. Progress, problems, and patterns in the biogeography of reef corals and other tropical marine organisms. *Helgoländer Meeresuntersuchungen* 42: 269–301. [14]

Rosen, D. E. 1975 [1976]. A vicariance model of Caribbean biogeography. *Systematic Zoology* 24: 431–464. [4]

Rosen, D. E. 1978. Vicariant patterns and historical explanation in biogeography. *Systematic Zoology* 27: 159–188. [4, 6]

Rosen, D. E. 1979. Fishes from the uplands and intermontane basins of Guatemala: Revisionary studies and comparative geography. *Bulletin of the American Museum of Natural History* 162: 267–376. [4]

Rosenblatt, R. H. and R. S. Waples. 1986. A genetic comparison of allopatric populations of shore fish species from the eastern and central Pacific Ocean: Dispersal or vicariance? *Copeia* 1986: 275–284. [Part 4]

Rosenzweig, M. L. 1968. Net primary productivity of terrestrial environments: Predictions from climatological data. *American Naturalist* 102: 67–84. [9]

Rosenzweig, M. L. 1992. Species diversity gradients: We know more and less than we thought. *J. Mammalogy* 73: 715–730. [9]

Rosenzweig, M. L. 1995. *Species Diversity in Space and Time*. Cambridge University Press, Cambridge. [8, 9, 10, 11, 13, 14, 17]

Rosenzweig, M. L. 1999. Species diversity. In J. McGlade (ed.), *Advanced Theoretical Ecology: Principles and Applications*. Blackwell Science, Oxford, pp. 249–281. [17]

Rosenzweig, M. L. 2001. The four questions: What does the introduction of exotic species do to diversity? *Evolutionary Ecology Research* 3: 361–367. [17]

Rosenzweig, M. L. 2003a. Reconciliation ecology and the future of species diversity. *Oryx* 37: 194–205. [17]

Rosenzweig, M. L. 2003b. *Win-Win Ecology: How the Earth's Species Can Survive in the Midst of Human Enterprise*. Oxford University Press, New York. [17]

Rosenzweig, M. L. 2003c. How to reject the area hypothesis of latitudinal gradients. In T. M. Blackburn and K. J. Gaston (eds.), *Macroecology: Concepts and Consequences.* Blackwell Science, Oxford, pp. 87–106. [9]

Rosenzweig, M. L. and E. A. Sandlin. 1997. Species diversity and latitude: Listening to area's signal. *Oikos* 80: 172–176. [10, 11]

Rosenzweig, M. L., W. R. Turner, J. G. Cox and T. H. Ricketts. 2003. Estimating diversity in unsampled habitats of a biogeographical province. *Conservation Biology* 17: 864–874. [17]

Roughgarden, J. 1979. *Theory of Population Genetics and Evolutionary Ecology: An Introduction.* Macmillan Publishing Co., New York. [9]

Rowe, K. C., E. J. Heske, P. W. Brown and K. N. Paige. 2004. Surviving the ice: Northern refugia and postglacial colonization. *Proc. National Academy of Sciences USA* 101: 10355–10359. [3]

Rowling, J. K. 1997. *Harry Potter and the Sorcerer's Stone*. Scholastic, New York. [1]

Roy, K. 1996. The roles of mass extinction and biotic interaction in large-scale replacements: A reexamination using the fossil record of stromboidean gastropods. *Paleobiology* 22: 436–452. [8]

Roy, K. and M. Foote. 1997. Morphological approaches to measuring biodiversity. *Trends in Ecology and Evolution* 12: 277–281. [8]

Roy, K. and K. K. Martien. 2001. Latitudinal distribution of body size in north-eastern Pacific marine bivalves. *J. Biogeography* 28: 485–493. [8]

Roy, K., D. P. Balch and M. E. Hellberg. 2001. Spatial patterns of morphological diversity across the Indo-Pacific: Analyses using strombid gastropods. *Proc. Royal Society of London Series B* 268: 2503–2508. [8]

Roy, K., D. Jablonski and K. K. Martien. 2000. Invariant size-frequency distributions along a latitudinal gradient in marine bivalves. *Proc. National Academy of Sciences USA* 97: 13150–13155. [8]

Roy, K., D. Jablonski and J. W. Valentine. 1994. Eastern Pacific molluscan provinces and latitudinal diversity gradient: No evidence for "Rapoport's Rule." *Proc. National Academy of Sciences USA* 91: 8871–8874. [8, 14]

Roy, K., D. Jablonski and J. W. Valentine. 1996. Higher taxa in biodiversity studies: Patterns from eastern Pacific marine mollusks. *Philosophical Transactions of the Royal Society of London Series B* 351: 1605–1613. [8]

Roy, K., D. Jablonski, J. W. Valentine and G. Rosenberg. 1998. Marine latitudinal diversity gradients: Tests of causal hypotheses. *Proc. National Academy of Sciences USA* 95: 3699–3702. [8, 9, 10, 14]

Roy, K., D. Jablonski and J. W. Valentine. 2000. Dissecting latitudinal diversity gradients: Functional groups and clades of marine bivalves. *Proc. Royal Society of London Series B* 267: 293–299. [8, 10]

Roy, K., D. Jablonski and J. W. Valentine. 2001. Climate change, species range limits and body size in marine bivalves. *Ecology Letters* 4: 366–370. [8]

Rudwick, M. S. 1976. *The Meaning of Fossils*, 2nd Ed. Science History Publications, New York. [6]

Ruedas, L. A. 1995. Description of a new large-bodied species of *Apomys* Mearns 1905 (Mammalia: Rodentia: Muridae) from Mindoro Island, Philippines. *Proc. Biological Society of Washington* 108: 302–318. [18]

Ruedi, M. 1996. Phylogenetic evolution and biogeography of Southeast Asian shrews (genus Crocidura: Soricidae). *Biological J. Linnaean Society* 58: 197–219. [7]

Ruggiero, A. 1999. Spatial patterns in the diversity of mammal species: A test of the geographic area hypothesis in South America. *Ecoscience* 6: 338–354. [9]

Rutherford, S., S. D'Hondt and W. Prell. 1999. Environmental controls on the geographic distribution of zooplankton diversity. *Nature* 400: 749–753. [9, 14]

Saba, S. and A. Toyos. 2003. Seed removal by birds, rodents and ants in the Austral portion of the Monte Desert. *J. Arid Environments* 53: 115–124. [2]

Saetersdal, M., J. M. Line and H. J. B. Birks. 1993. How to maximize biological diversity in nature reserve selection: Vascular plants and breeding birds in deciduous woodlands in western Norway. *Biological Conservation* 66: 131–138. [16]

Sagoff, M. 1999. What's wrong with alien species? *Report of the Institute for Philosophy and Public Policy* 19: 16–23. [17]

Salvat, B. 1969. Dominance biologique de quelques mollusques dans les atolls fermés

(Tuamotu, Polynésie). *Malacologia* 9: 187–190. [12]

Salvat, B. 1971. Mollusques lagunaires et récifaux de l'île de Raevavae (Australes, Polynésie). *Malacological Reviews* 4: 1–15. [12]

Sampson, S. D., L. M. Witmer, C. A. Forster, D. W. Krause, M. O'Connor, P. Dodson and F. Ravoavy. 1998. Predatory dinosaur remains from Madagascar: Implications for the Cretaceous biogeography of Gondwana. *Science* 280: 1048–1051. [1]

Sánchez-Cordero, V. and T. H. Fleming. 1993. Ecology of tropical heteromyids. In *Biology of the Family Heteromyidae*. Special Publications American Society of Mammalogists No. 10, pp. 596–617. [16]

Sánchez-Cordero, V. and E. Martínez-Meyer. 2000. Museum specimen data predict crop damage by rodents. *Proc. National Academy Science USA* 97: 7074–7077. [16]

Sánchez-Cordero, V., A. T. Peterson and P. Escalante-Pliego. 2001. El modelado de la distribución de especies y la conservación de la diversidad biológica. In H. M. Hernández, A. N. García-Aldrete, F. Álvarez, M. Ulloa (comps.), *Enfoques Contemporáneos para el Estudio de la Biodiversidad*. Ediciones Científicas Universitarias FCE, pp. 359–379. [16]

Sanderson, M. J. 2002. Estimating absolute rates of molecular evolution and divergence times: A penalized likelihood approach. *Molecular Biology and Evolution* 19: 101–109. [5]

Santelices, B. and P. A. Marquet. 1998. Seaweeds, latitudinal diversity patterns, and Rapoport's rule. *Diversity and Distributions* 4: 71–75. [10, 14]

Santini, F. and R. Winterbottom. 2002. Historical biogeography of Indo-western Pacific coral reef biota: Is the Indonesian region a center of origin? *J. Biogeography* 29: 189–205. [13]

Santini, F. and Tyler, J.C. 2002. Phylogeny and biogeography of the extant species of triplespine fishes (Triacanthidae, Tetraodontiformes). *Zoologica Scripta* 31: 321–330. [13]

Sarkar, S. and C. Margules. 2002. Operationalizing biodiversity for conservation planning. *J. Biosciences* 27, S2: 299–308. [16]

Sarkar, S., A. Aggarwal, J. Garson, C. Margules and J. Zeidler. 2002. Place prioritization for biodiversity content. *J. Biosciences* 27, S2: 339–346 [16]

Sax, D. F. 2001. Latitudinal gradients and geographic ranges of exotic species: Implications for biogeography. *J. Biogeography* 28: 139–150. [9, 11; Parts 3, 5]

Sax, D. F. 2002. Native and naturalized plant diversity are positively correlated in scrub communities of California and Chile. *Diversity and Distributions* 8: 193–210. [17]

Sax, D. F. and S. D. Gaines. 2004. Species diversity: From global decreases to local increases. *Trends in Ecology and Evolution* 18: 561–566. [15; Part 3]

Sax, D. F., J. H. Brown and S. D. Gaines. 2002. Species invasions exceed extinctions on islands worldwide: A comparative study of plants and birds. *American Naturalist* 160: 766–783. [11, 12, 15, 17; Part 3]

Schall, J. J. and E. Pianka. 1978. Geographical trends in number of species. *Science* 201: 679–686. [10]

Schandelmeier, H. and P.O. Reynolds. 1997. Palaeogeographic-Palaeotectonic Atlas of North-Eastern Africa, Arabia, and Adjacent Areas: Late Proterozoic to Holocene. A. A. Balkema, Rotterdam. [1]

Scheiner, S. M. 2003. Six types of species-area curves. *Global Ecology and Biogeography* 12(6): 441. [11]

Scheiner, S. M. and J. M. Rey-Benayas. 1994. Global patterns of plant diversity. *Evolutionary Ecology* 8: 331–347. [9]

Schettino, A. and C. R. Scotese. 2004. Apparent polar wander paths for the major continents since the early Jurassic and global plate tectonic reconstructions. *J. Geophysical Research*, in press. [1]

Schlacher, T. A., P. Newell, J. Clavier, M. A. Schlacher-Hoenlinger, C. Chevillon and J. Britton. 1998. Soft-sediment benthic community structure in a coral reef lagoon: The prominence of spatial heterogeneity and "spot endemism." *Marine Ecology Progress Series* 174: 159–174. [14]

Schmarda, L. K. 1885. *Die Geographische Verbreitung der Thiere.* C. Gerold und Sohn, Wien. [Part 4]

Schoener, T. W. and A. Schoener. 1983. The time to extinction of a colonizing propagule of lizards increases with island area. *Nature* 302: 332–334. [12]

Schoener, T. W. and D. H. Janzen. 1968. Notes on environmental determinants of tropical versus temperate insect size patterns. *American Naturalist* 102: 207–224. [8]

Schuh, R. T. 2000. *Biological Systematics: Principles and Applications.* Comstock Publishing Associates, London, and Cornell University Press, Ithaca, NY. [4]

Schumm, S. A. 1991. *To Interpret the Earth: Ten Ways to be Wrong*. Cambridge University Press, Cambridge. [11]

Sclater, P. L. 1897. The distribution of marine mammals. *Proc. Zoological Society of London* 41: 347–359. [Part 4]

Scotese, C. R. 2001. *Atlas of Earth History*. PALEOMAP Project. 1. Department of Geology, University of Texas at Arlington. [1]

Scotese, C. R. and D. W. Baker. 1975. Continental drift reconstructions and animations. *J. Geological Education* 23: 167–171. [1]

Scotese, C. R. and S. F. Barrett. 1990. Gondwana's movement over the South Pole during the Palaeozoic: evidence from lithological indicators of climate. In W. S. McKerrow and C. R. Scotese (eds.), *Palaeozoic Biogeography and Palaeogeography*. Geological Society of London Memoir 12, pp. 75–86. [1]

Scotese, C. R. and J. Golonka. 1992. *Paleogeographic Atlas*. PALEOMAP Progress Report 20-0692. Department of Geology, University of Texas at Arlington. [1]

Scotese, C. R. and W. W. Sager (eds.). 1988. Mesozoic and Cenozoic plate tectonic reconstructions. *Tectonophysics* 155: 27–48. [1]

Scotese, C. R., R. K. Bambach, C. Barton, R. Van de Voo and A. M. Ziegler. 1979. Paleozoic base maps. *J. Geology* 87: 217–277. [1]

Scotese, C. R., L. M. Gahagan and R. L. Larson. 1988. Plate tectonic reconstructions of the Cretaceous and Cenozoic ocean basins. In C. R. Scotese and W. W. Sager (eds.), Mesozoic and Cenozoic plate reconstructions. *Tectonophysics* 155: 27–48. [1]

Scotese, C. R., S. Snelson, W. C. Ross and L. Dodge. 1981. A computer animation of continental drift. *J. Geomagnetism and Geoelectricity* 32 (suppl. III): 61–70. [1]

Scott, J. M., P. J. Heglund, M. L. Morrison, J. B. Haufler, M. G. Raphael, W. A. Wall and F. B. Samson (eds.). 2002. *Predicting Species Occurrences: Issues of Accuracy and Scale*. Island Press, Washington, D.C. [2, 16, 18]

Scott, J. M., T. H. Tear and F. W. Frank (eds.). 1996. *GAP Analysis: A Landscape Approach to Biodiversity Planning*. American Society of Photogrammetry and Remote Sensing, Bethesda, MD. [16]

Scott, M. C and G. S. Helfman. 2001. Native invasions, homogenization, and the mismeasure of integrity of fish assemblages. *Fisheries* 26(11): 6–15. [15]

Selley, R. C., ed. 1997. *African Basins. Sedimentary Basins of the World Vol.* 3. K. J. Hsu (series ed.), Elsevier, Amsterdam. [1]

Sengor, A. M. C. and B. A. Natalin. 1996. Paleotectonics of Asia: Fragments of a synthesis. In A. Yin and M. Harrison (eds.), *The Tectonic Evolution of Asia*, Cambridge University Press, Cambridge, pp. 486–641. [1]

Sepkoski, J. J., Jr. 1988. Alpha, beta, or gamma: Where does all the diversity go? *Paleobiology* 14: 221–234. [14]

Sepkoski, J. J., Jr. 1999. Rates of speciation in the fossil record. In A. E. Magurran and R. M. May (eds.), *Evolution of Biological Diversity*, pp. 260–282. Oxford University Press, Oxford. [13]

Serb, J. M., C. A. Phillips and J. B. Iverson. 2001. Molecular phylogeny and biogeography of *Kinosternon flavescens* based on complete mitochondrial control region sequences. *Molecular Phylogenetics and Evolution* 18: 149–162. [5]

Sharpe, S. E. 2002. Constructing seasonal climograph overlap envelopes from Holocene packrat midden contents, Dinosaur National Monument, Colorado. *Quaternary Research* 57: 306–313. [2]

Shepherd, U. L. 1998. A comparison of species diversity and morphological diversity across the North American latitudinal gradient. *J. Biogeo-graphy* 25: 19–29. [8]

Sheppard P. R., A. C. Comrie, G. D. Packin, K. Angersbach, M. K. Hughes. 2002. The climate of the U. S. Southwest. *Climate Research* 21: 219–238. [2]

Shields, O. 1979. Evidence for the initial opening of the Pacific Ocean in the Jurassic. *Palaeogeography, Palaeoclimatology, Palaeoecology* 26: 181–220. [4]

Shields, O. 1983. Trans-Pacific biotic links that suggest earth expansion. In S. W. Carey (ed.), *Expanding Earth Symposium*. University of Tasmania, Hobart, pp. 199–205. [4]

Shields, O. 1991. Pacific biogeography and rapid earth expansion. *J. Biogeography* 18: 583–585. [4]

Shields, O. 1996. Plate tectonics or an expanding earth? *J. Geological Society of India* 47: 399–408. [4]

Shilton, L. A., J. D. Altringham, S. G. Compton and R. J. Whittaker. 1999. Old World fruit bats can be long-distance seed dispersers through extended retention of viable seeds in the gut. *Proc. Royal Society of London Series B* 266: 219–223. [11]

Shine, R., R. N. Reed, S. Shetty, M. Lemaster and S. A. Karl. 2002. Reproductive isolating mechanisms between two sympatric sibling species of sea snakes. *Evolution* 56: 1655–1662. [13]

Shmida, A. and M. V. Wilson. 1985. Biological determinants of species diversity. *J. Biogeography* 12: 1–20. [15]

Shuman, B., P. Bartlein, N. Logar, P. Newby and T. Webb III. 2002a. Parallel climate and vegetation responses to the early Holocene collapse of the Laurentide Ice Sheet. *Quaternary Science Reviews* 21: 1793–1805. [3]

Shuman, B., T. Webb III, P. Bartlein and J. W. Williams. 2002b. The anatomy of a climatic oscillation: Vegetation changes in eastern North America during the Younger Dryas chronozone. *Quaternary Science Reviews* 21: 1777–1791. [3]

Signor, P. W. 1990. The geologic history of diversity. *Annual Review of Ecology and Systematics* 21: 509–539. [14]

Silvertown, J. 1985. History of a latitudinal diversity gradient: Woody plants in Europe 13,000–1000 years bp. *J. Biogeography* 12: 519–525. [9]

Simberloff, D. 1987. Calculating probabilities that cladograms match: A method of biogeographical inference. *Systematic Zoology* 36: 175–195. [7]

Simberloff, D. 1988. Effects of drift and selection on detecting similarities between large cladograms. *Systematic Zoology* 37: 56–59. [7]

Simberloff, D. 1996. Impacts of introduced species in the United States. *Consequences* 2: 13–24. [15]

Simberloff, D. S. and E. O. Wilson. 1970. Experimental zoogeography of islands: A two-year record of colonization. *Ecology* 51: 934–937. [11]

Simberloff, D. S. and L. G. Abele. 1976. Island biogeography theory and conservation practice. *Science* 191: 285–286. [17]

Simberloff, D. S. and L. G. Abele. 1982. Refuge design and island biogeographic theory: Effects of fragmentation. *American Naturalist* 120: 41–50. [17]

Simberloff, D., K. L. Heck, E. D. McCoy and E. F. Connor. 1980. There have been no statistical tests of cladistic biogeographic hypotheses. In G. Nelson and D. E. Rosen (eds.), *Vicariance Biogeography: A Critique*. Columbia University Press, New York, pp. 40–63. [7]

Simpson, G. G. 1940. Mammals and land bridges. *J. Washington Academy of Science* 30: 137–163. [Part 2]

Singleton, G. R., L. A. Hinds, C. J. Krebs and D. Sm Spratt (eds.). 2003. *Rats, Mice, and People: Rodent Biology and Management*. Australia Centre for International Agricultural Research, Canberra. [16]

Singleton, G. R., L. A. Hinds, H. Leirs and Z. Zhang (eds.). 1999. *Ecologically Based Rodent Management*. Australia Centre for International Agricultural Research, Canberra. [16]

Slobodkin, L. B. 2001. The good, the bad, and the reified. *Evolutionary Ecology Research* 3: 1–13. [17]

Smith, A. B. 1994. *Systematics and the Fossil Record: Documenting Evolutionary Patterns*. Blackwell Scientific Publications, Oxford. [4, 6]

Smith, A. G., D. G. Smith and B. M. Funnell. 1994. *Atlas of Mesozoic and Cenozoic Coastlines*. Cambridge University Press, Cambridge. [1]

Smith, A. R. 1972. Comparison of fern and flowering plant distributions with some evolutionary interpretations for ferns. *Biotropica* 4: 4–9. [12]

Smith, E. A. 1890. Report on the marine molluscan fauna of the island of St. Helena. *Proc. Zoological Society of London for 1890*, pp. 246–317. [12]

Smith, F. A. and J. L. Betancourt. 1998. Response of bushy-tailed woodrats *Neotoma cinerea*. to late Quaternary climate change in the Colorado Plateau. *Quaternary Research* 50: 1–11. [2]

Smith, F. A. and J. L. Betancourt. 2003. The effect of Holocene temperature fluctuations on the evolution and ecology of *Neotoma* woodrats in Idaho and northwestern Utah. *Quaternary Research* 59: 160–171. [2]

Smith, F. A., J. L. Betancourt and J. H. Brown. 1995. Effects of global warming on woodrat *Neotoma cinerea*. body size during the last deglaciation. *Science* 270: 2012–2014. [2]

Smith, H. M. and R. B. Smith. 1966. *Herpetology of Mexico: An Annotated Checklist and Key to the Amphibians and Reptiles*. A reprint of Bulletins 187, 194, and 199 of the U.S. National Museum, with a list of subsequent taxonomic innovations. Eric Lundberg, Ashton, MD. [16]

Smith, J. F. 2001. High species diversity in fleshy-fruited tropical understory plants. *American Naturalist* 157: 646–653. [10]

Smith, J. T. 2000. *Extinction Dynamics of the Late Neogene Pectinidae in California*. Unpublished Master's Thesis, University of California, San Diego. [8]

Smith, W. H. F. and D. T. Sandwell. 1997. Global seafloor topography from satellite altimetry and ship depth soundings. *Science* 277: 1957–1962. [1]

Soberón, J. 1999. Linking biodiversity information sources. *Trends in Ecology and Evolution* 14: 291. [16]

Sokal R. R. and F. J. Rohlf 1969. *Biometry*. W.H. Freeman, San Francisco. [9]

Solem, A. 1973. Island size and species diversity in Pacific island land snails. *Malacologia* 14: 397–400. [12]

Sorokin, Y. I. 1995. *Coral Reef Ecology: Ecological Studies*, Vol. 102. Springer-Verlag, Berlin. [14]

Spaulding, W. G., J. L. Betancourt, L. K. Croft and K. L. Cole. 1990. Packrat middens: Their composition and methods of analysis. In J. L.

Betancourt, T. R. Van Devender and P. S. Martin (eds.), *Packrat Middens: The Last 40,000 Years of Biotic Change*. University of Arizona Press, Tucson, pp. 59–84. [2]

Spear, R. W., M. B. Davis and L. C. K. Shane. 1994. Late Quaternary history of low- and mid-elevation vegetation in the White Mountains of New Hampshire. *Ecological Monographs* 64: 85–109. [3]

Specht, R. L. 1981. Biogeography of halophytic angiosperms (salt-marsh, mangrove, and sea grass). In A. Keast (ed.), *Ecological Biogeography of Australia*, pp. 577–589. Dr. W. Junk, The Hague. [13]

Spironello, M. and D. R. Brooks. 2003. Dispersal and diversification in the evolution of *Simulium (Inseliellium)*, an archipelagic dipteran group. *J. Biogeography* 30: 1–11. [7]

Springer, V. G. 1962. A review of the Blenniid fishes of the genus *Ophioblennius* Gill. *Copeia* 1962: 426–433. [Part 4]

Srivastava, D. S. 1999. Using local-regional richness plots to test for species saturation: Pitfalls and potentials. *J. Animal Ecology* 68: 1–16. [14]

Stafford, T. W., Jr., H. A. Semken, Jr., R. W. Graham, W. F. Klippel, A. Markova, N. G. Smirnov, and J. Southon. 1999. First accelerator mass spectrometry ^{14}C dates documenting contemporaneity of nonanalog species in late Pleistocene mammal communities. *Geology* 27: 903–906. [Part 1]

Stanley, S. M. 1986. Population size, extinction, and speciation: The fission effect in Neogene Bivalvia. *Paleobiology* 12: 89–110. [8, 14]

Stanley, S. M. 1988. Adaptive morphology of the shell in bivalves and gastropods. In E. R. Trueman and M. R. Clarke (eds.), *The Mollusca. II. Form and Function*. Academic Press, San Diego, pp. 105–141. [8]

Stanley, S. M. 1990. Adaptive radiation and macroevolution. In P. D. Taylor and G. P. Larwood (eds.), *Major Evolutionary Radiations*. Systematics Association Special Volume 42, pp. 1–15. [14]

Steadman, D. W. 1995. Prehistoric extinctions of Pacific island birds: Biodiversity meets zooarcheology. *Science* 267: 1123–1131. [Part 1]

Stehli, F. G. 1968. Taxonomic diversity gradients in pole location: The Recent model. In E. T. Drake (ed.), *Evolution and Environment*. Yale University Press, New Haven, pp. 163–227. [14]

Stehli, F. G. and S. D. Webb. 1985. *The Great American Biotic Interchange*. Plenum Press, New York. [Part 3]

Stehli, F. G. and J. W. Wells. 1971. Diversity and age patterns in hermatypic corals. *Systematic Zoology* 20: 115–126. [13, 14]

Stehli, F. G., R. Douglas and I. Kafescegliou. 1972. Models for the evolution of planktonic foraminifera. In T. J. M. Schopf (ed.), *Models in Paleobiology*. W.H. Freeman, San Francisco, pp. 116–128. [14]

Stehli, F. G., R. G. Douglas and N. D. Newell. 1969. Generation and maintenance of gradients in taxonomic diversity. *Science* 164: 947–949. [14]

Stehli, F. G., A. L. McAlester and C. E. Helsley. 1967. Taxonomic diversity of Recent bivalves and some implications for geology. *Geological Society of America Bulletin* 78: 455–466. [14]

Stein, G. 1937. *Everybody's Autobiography*. Random House, New York. [3]

Stephenson, N. L. 1990. Climatic control of vegetation distribution: The role of the water balance. *American Naturalist* 135: 649–670. [9]

Steppan, S. J. 1998. Phylogenetic relationships and species limits within *Phyllotis* (Rodentia: Sigmodontinae): concordance between mtDNA sequence and morphology. *J. Mammalogy* 79: 573–593. [2]

Stockwell, D. R. B. and D. Peters. 1999. The GARP modeling system: Problems and solutions to automated spatial prediction. *International J. Geographical Information Science* 13: 143–158. [16]

Streelman, J. T., M. Alfaro, M. W. Wesneat, D. R. Bellwood and S. A. Karl. 2002. Evolutionary history of the parrotfishes: Biogeography, ecomorphology, and comparative diversity. *Evolution* 56: 961–971. [13]

Stuart, A. J. 1991. Mammalian extinctions in the Late Pleistocene of Northern Eurasia and North America. *Biological Reviews* 66: 453–562. [3]

Stuart, C. T., M. A. Rex and R. J. Etter. 2003. Large-scale spatial and temporal patterns of deep-sea benthic species diversity. In P. A. Tyler (ed.), *Ecosystems of the World: Ecosystems of the Deep Oceans*. Elsevier, Amsterdam, pp. 295–311. [Part 4]

Stuiver, M. and P. J. Reimer. 1993. Extended ^{14}C data base and revised CALIB 3.0 ^{14}C age calibration program. *Radiocarbon* 35: 215–230. [3]

Sugita, S. 1994. Pollen representation of vegetation in Quaternary sediments: Theory and method in patchy vegetation. *J. Ecology* 82: 881–897. [3]

Swenson, U., A. Backlund, S. McLoughlin and R. S. Hill. 2001. *Nothofagus*: Biogeography Revisited with Special Emphasis on the Enigmatic

Distribution of Subgenus *Brassospora* in New Caledonia. *Cladistics* 17: 28–47. [4]

Swetnam, T. W., C. D. Allen and J. L. Betancourt. 1999. Applied historical ecology: Using the past to manage for the future. *Ecological Applications* 64: 1189–1206. [2]

Swofford, D. L. 1998. *PAUP* [Phylogenetic Analysis Using Parsimony *and Other Methods]*, Version 4.0. Sinauer Associates, Sunderland, MA. [6]

Taberlet, P., L. Fumagalli, A. G. Wust-Saucy and J. F. Cosson. 1998. Comparative phylogeography and postglacial colonization routes in Europe. *Molecular Ecology* 7: 453–464. [5]

Taylor, E. H. 1934. Philippine land mammals. *Monographs, Bureau of Science, Manila* 30: 1–548. [18]

Taylor, F. B. 1910. Bearing of the Tertiary Mountain Belt on the origin of the Earth's plan. *Bulletin of the Geological Society of America* 21: 179–226. [4]

Taylor, J. D. and C. N. Taylor. 1997. Latitudinal distribution of predatory gastropods on the eastern Atlantic shelf. *J. Biogeography* 4: 73–81. [8]

Templeton, A. R., E. Routman and C. A. Phillips. 1995. Separating population structure from population history: A cladistic analysis of the geographical distribution of mitochondrial DNA haplotypes in the tiger salamander, *Ambystoma tigrinum*. *Genetics* 140: 767–782. [5]

Terborgh, J. 1973. On the notion of favorableness in plant ecology. *American Naturalist* 107: 481–501. [9]

Terman, M. R. 1997. Natural links: Naturalistic golf courses as wildlife habitat. *Landscape and Urban Planning* 38: 183–197. [17]

Terwilliger, V. J., J. L. Betancourt, S. W. Leavitt and P. K.Van de Water. 2002. Leaf cellulose ?D and ?^{18}O trends with elevation and climate in semi-arid species. *Geochimica et Coscomochima Acta* 66: 3887–3900. [2]

Thomas, C. D. and J. J. Lennon. 1999. Birds extend their ranges northwards. *Nature* 399: 213. [9]

Thompson, L. G., E. Mosley-Thompson, M. E. Davis, K. A. Henderson, H. H. Brecher, V. S. Zagorodnov, T. A. Mashiotta, P.-N. Lin, V. N. Mikhalenko, D. R. Hardy and J. Beer. 2002. Kilimanjaro ice core records: Evidence of Holocene climate change in tropical Africa. *Science* 298: 589–593. [3]

Thompson, R. S. 1988. Western North America: Vegetation dynamics in the western United States: Modes of response to climatic fluctuations. In B. Huntley and T. Webb III (eds.), *Vegetation History*. Kluwer Academic Publishers, Dordrecht, NL, pp. 415–458 [3]

Thompson, R. S. and K. H. Anderson. 2000. Biomes of western North America at 18,000, 6,000, and 0 ^{14}C yr BP reconstructed from pollen and packrat midden data. *J. Biogeography* 27: 555–584. [2]

Thompson, R. S., K. H. Anderson and P. J. Bartlein. 1999. *Quantitative Paleoclimatic Reconstructions from Late Pleistocene Plant Macrofossils of the Yucca Mountain Region*. U.S.G.S. Open-File Report 99-338, Reston, VA. [2]

Thompson, R. S., K. H. Anderson and P. J. Bartlein. 2000a. *Atlas of Relations Between Climatic Parameters and Distributions of Important Trees and Shrubs in North America: Introduction and Conifers*. U.S.G.S. Professional Paper 1650-A, Reston, VA. [2]

Thompson, R. S., K. H. Anderson and P. J. Bartlein. 2000b. *Atlas of Relations Between Climatic Parameters and Distributions of Important Trees and Shrubs in North America: Hardwoods*. U.S.G.S. Professional Paper 1650-B, Reston, VA. [2]

Thompson, R. S., K. H. Anderson, P. J. Bartlein and S. A. Smith. 2000. *Atlas of Relations Between Climatic Parameters and Distributions of Important Trees and Shrubs in North America: Additional Conifers, Hardwoods, and Monocots*. U.S.G.S. Professional Paper 1650-C, Reston, VA. [2]

Thompson, R. S., S. W. Hostetler, P. J. Bartlein and K. H. Anderson. 1998. *A Strategy for Assessing Potential Future Changes in Climate, Hydrology, and Vegetation in the Western United States*. U.S.G.S. Circular 1153, Reston, VA. [2]

Thompson, R. S., C. Whitlock, P. J. Bartlein, S. P. Harrison and W. G. Spaulding. 1993. Climatic changes in the Western United States since 18,000 yr BP. In H. E. Wright, Jr., J. E. Kutzbach, T. Webb, III, W. F. Ruddiman, F. A. Street-Perrott and P. J. Bartlein (eds.), *Global Climates Since the Last Glacial Maximum*. University of Minnesota Press, p. 468–513. [2]

Thornton, I. W. B., R. A. Zann and S. van Balen. 1993. Colonization of Rakata (Krakatau Island) by non-migrant land birds from 1883 to 1992 and implications for the value of island equilibrium theory. *J. Biogeography* 20: 441–452. [11]

Thorson, G. 1936. The larval development, growth and metabolism of Artic marine bottom invertebrates compared with those of other seas. *Meddelelser om Grønland* 100: 1–155. [10]

Thorson, G. 1946. Reproduction and larval development of Danish marine bottom invertebrates, with special reference to the planktonic larvae in the Sound (Øresund). *Meddelelser fra Kommissionen for Danmarks Fiskeri- og Havundersögelser, Serie Plankton* 4: 1–523. [10]

Thorson, G. 1950. Reproductive and larval ecology of marine bottom invertebrates. *Biological Reviews* 25: 1–45. [10]

Thorson, G. 1957. Bottom communities (sublittoral or shallow shelf). In J. W. Hedgpeth (ed.), *Treatise on Marine Ecology and Paleoecology*. Geological Society of America, New York, pp. 461–534. [14]

Thorson, G. 1965. The distribution of benthic marine Mollusca along the NE Atlantic shelf from Gibraltar to Murmansk. *Proc. First European Malacological Congress* 5–23. [10]

Tidemann, C. D., D. J. Kitchener, R. A. Zann and I. W. B. Thornton. 1990. Recolonization of the Krakatau Islands and adjacent areas of West Java, Indonesia, by bats Chiroptera. 1883–1986. *Philosophical Transactions of the Royal Society of London Series B* 328: 121–130. [11]

Tjørve, E. 2003. Shapes and functions of species-area curves: A review of possible models. *J. Biogeography* 30: 827–835. [11]

Todd, J. A., J. B. C. Jackson, K. G. Johnson, H. M. Fortunato, A. Heitz, M. Alvarez and P. Jung. 2002. The ecology of extinction: Molluscan feeding and faunal turnover in the Caribbean Neogene. *Proc. Royal Society of London Series B* 269: 571–577. [8]

Toft, C. A. and T. W. Schoener. 1983. Abundance and diversity of orb spiders on 106 Bahamian islands: Biogeography at an intermediate trophic level. *Oikos* 41: 411–426. [11]

Torchin, M. E., K. D. Lafferty, A. P. Dobson, V. J. McKenzie and A. M. Kuris. 2003. Introduced species and their missing parasites. *Nature* 421: 628–630. [12]

Transeau, E. N. 1935. The prairie peninsula. *Ecology* 16: 423–437. [3]

Tsuda, R. T., F. R. Fosberg and M.-H. Sachet. 1977. Distribution of sea grasses in Micronesia. *Micronesica* 13: 191–198. [12]

Turner, J. R. G. 1986. Why are there so few butterflies in Liverpool? Homage to Alfred Russel Wallace. *Antenna* 10: 18–24. [9]

Turner, J. R. G. 1992. Stochastic processes in populations: The horse behind the cart? In J. R. Berry, T. J. Crawford and G. M. Hewitt (eds.), *Genes in Ecology*. Blackwell Scientific, Oxford, pp. 29–53. [9]

Turner, J. R. G. 2004. Explaining the global biodiversity gradient: Energy, area, history, and natural selection. *Basic and Applied Ecology* (in press). [9]

Turner, J. R. G. and J. J. Lennon. 1989. Species richness and the energy theory. *Nature* 340: 351. [8]

Turner, J. R. G., C. M. Gatehouse and C. A. Corey. 1987. Does solar energy control organic diversity? Butterflies, moths, and the British climate. *Oikos* 48: 195–205. [9]

Turner, J. R. G., J. J. Lennon and J. A. Lawrenson. 1988. British bird species distributions and the energy theory. *Nature* 335: 539–541. [9, 11]

Turner, J. R. G., J. J. Lennon and J. J. D. Greenwood. 1996. Does climate cause the global biodiversity gradient? In M. E. Hochberg, J. Clobert and R. Barbault (eds.), *Aspects of the Genesis and Maintenance of Biological Diversity*. Oxford University Press, Oxford, pp. 199–220. [9]

Turner, W., W. A. Leitner and M. L. Rosenzweig. 2000. *Ws2m.exe*. http://eebweb.arizona.edu/diversity. [17]

Udvard, M. D. F. 1969. *Dynamic Zoogeography*. VanNostrand Reinhold, New York. [Part 5]

Ulmishek, G. F. and H. D. Klemme. 1990. Depositional Controls, Distribution, and Effectiveness of World's Petroleum Source Rocks. U.S.G.S. Bulletin 1931,, Denver. [1]

Upchurch, P. and C. A. Hunn. 2002. Time: The neglected dimension in cladistic biogeography? *Geobios Mémoire Special* 24: 277–286. [4]

Upchurch, P., C. A. Hunn and D. B. Norman. 2002. An analysis of dinosaurian biogeography: Evidence for the existence of vicariance and dispersal patterns caused by geological events. *Proc. Royal Society of London Series B* 269: 613–621. [4]

Upton, D. E. and R. W. Murphy. 1997. Phylogeny of the side-blotched lizards (Phrynosomatidae: *Uta*) based on mtDNA sequences: Support for a mid-peninsular seaway in Baja California. *Molecular Phylogenetics and Evolution* 8: 104–113. [5]

Valdovinos, C. R. 1999. Biodiversidad de moluscos chilenos: Base de datos taxonómica y distribucional. *Gayana Zoología* 63: 59–112. [10]

Valdovinos, C., S. A. Navarrete and P. A. Marquet. 2003. Mollusk species diversity in the southeastern Pacific: Why are there more species towards the pole? *Ecography* 26: 139–144. [8, 10]

Valentine, J. W. 1966. Numerical analysis of marine molluscan ranges on the extra-tropical northeastern Pacific shelf. *Limnology and Oceanography* 11: 198–211. [8]

Valentine, J. W. 1983. Seasonality: Effects in marine benthic communities. In M. J. S. Tevesz and P. L. McColl (eds.), *Biotic Interactions in Recent and Fossil Benthic Communities*. Plenum, New York, pp. 121–156. [8]

Valentine, J. W. 1984. Neogene marine climate trends: Implications for biogeography and evolution of the shallow-sea biota. *Geology* 12: 647–650. [10]

Valentine, J. W. 1995. Why no new phyla after the Cambrian? Genome and ecospace hypotheses revisited. *Palaios* 10: 190–194. [8]

Valentine, J. W. and D. Jablonski. 1983. Larval adaptations and patterns of Brachiopod diversity in space and time. *Evolution* 37: 1052–1061. [10]

Valentine, J. W., D. Jablonski and D. H. Erwin. 1999. Fossils, molecules and embryos: New perspectives on the Cambrian explosion. *Development* 126: 851–859. [8]

Valentine, J. W., K. Roy and D. Jablonski. 2002. Carnivore–non-carnivore ratios in northeastern Pacific marine gastropods. *Marine Ecology–Progress Series* 228: 153–163. [8]

Van de Water, P. D., S. W. Leavitt and J. L. Betancourt. 1994. Trends in stomatal density and $^{13}C/^{12}C$ ratios of *Pinus flexilis* needles during last glacial/interglacial cycle. *Science* 264: 239–243. [2]

Van de Water, P. D., S. W. Leavitt and J. L. Betancourt. 2002. Leaf cellulose $\delta^{13}C$ variability with elevation, slope aspect and precipitation in the U.S. Southwest. *Oecologia* 132: 332–343. [2]

van der Hammen, T. and H. Hooghiemstra. 2000. Neogene and Quaternary history of vegetation, climate, and plant diversity in Amazonia. *Quaternary Science Reviews* 19: 725–742. [Part 1]

van der Spoel, S. and A. C. Pierrot-Bults (eds.) 1979. *Zoogeography and Diversity in Plankton*. Halsted Press, New York. [Part 4]

Van der Voo, R. 1993. *Paleomagnetism of the Atlantic, Tethys, and Iapetus Oceans*. Cambridge University Press, Cambridge. [1]

Van Devender, T. R. 1977. Comments on "Macrofossil analysis of woodrat *Neotoma*. middens as a key to the Quaternary vegetation of arid North America," by P. V. Wells. *Quaternary Research* 8: 236–237. [2]

Van Devender, T. R. 1990. Late Quaternary vegetation and climate of the Sonoran Desert, United States and Mexico. In J. L. Betancourt, T. R. Van Devender and P. S. Martin (eds.), *Packrat Middens: The Last 40,000 Years of Biotic Change*. University of Arizona Press, Tucson, p. 134–165. [2]

Van Devender, T. R. and W. E. Hall. 1994. Holocene arthropods from the Sierra Bacha, Mexico, with emphasis on beetles Coleoptera. *Coleopterists Bulletin* 48: 30–50. [2]

Van Devender, T. R., G. L. Bradley and A. H. Harris. 1987. Late Quaternary mammals from the Hueco Mountains, El Paso and Hudspeth counties, Texas. *Southwestern Naturalist* 32: 179–195. [2]

Van Devender, T. R., P. S. Martin, R. S. Thompson, K. L. Cole, A. J. T. Jull, A. Long, L. J. Toolin and D. J. Donahue. 1985. Fossil packrat middens and the tandem accelerator mass spectrometer. *Nature* 317: 610–613. [2]

van Rensburg, B. J., S. L. Chown and K. J. Gaston. 2002. Species richness, environmental correlates, and spatial scale: A test using South African birds. *American Naturalist* 159: 566–577. [11]

van Soest, R. W. M. and E. Hajdu. 1997. Marine area relationships from twenty sponge phylogenies: A comparison of methods and coding strategies. *Cladistics* 13: 1–20. [6, 7]

Van Valkenburgh, B. and C. M. Janis. 1993. Historical diversity patterns in North American large herbivores and carnivores. In R. E. Ricklefs and D. Schluter (eds.), *Species Diversity in Ecological Communities*. University of Chicago Press, Chicago, pp. 330–340. [8]

van Veller, M. G. P. and D. R. Brooks. 2001. When simplicity is not parsimonious: *A priori* and *a posteriori* methods in historical biogeography. *J. Biogeography* 28: 1–11. [7]

van Veller, M. G. P., D. J. Kornet and M. Zandee. 2000. Methods in vicariance biogeography: Assessment of the implementations of assumptions zero, 1 and 2. *Cladistics* 16: 319–345. [7]

van Veller, M. G. P., D. J. Kornet and M. Zandee. 2002. *A posteriori* and *a priori* methodologies for testing hypotheses of causal processes in vicariance biogeography. *Cladistics* 18: 207 217. [5, 7]

van Veller, M. G. P., D. R. Brooks and M. Zandee. 2003. Cladistic and phylogenetic biogeography: The art and the science of discovery. *J. Biogeography* 30: 319–329. [5, 7; Part 2]

van Veller, M. G. P., M. Zandee and D. J. Kornet. 1999. Two requirements for obtaining valid common patterns under different assumptions in vicariance biogeography. *Cladistics* 15: 393–406. [7]

van Veller, M. G. P., M. Zandee and D. J. Kornet. 2001. Measures for obtaining inclusive solution sets under assumptions zero, 1 and 2 with different methods for vicariance biogeography. *Cladistics* 17: 248–259. [7]

Various Authors. 1995. *Paleogeographic Atlas of Australia*. Vols. 1–10. Cambrian to Cenozoic. Australia Geological Survey Organization, Canberra. [1]

Vasan, S., X. Zhang, A. Kapurniotu, J. Bernhagen, S. Teichberg, J. Basgen, D. Wagle, D. Shih, I. Terlecky, R. Bucala, A. Cerami, J. Egan and P. Ulrich. 1996. An agent cleaving glucose-derived protein cross links in vitro and in vivo. *Nature* 382: 275–278. [2]

Veevers, J. J. (ed.). 1984. *Phanerozoic Earth History of Australia.* Oxford Monographs on Geology and Geophysics No. 2. Oxford University Press, New York. [1]

Veevers, J. J. and C. McA. Powell. 1994. Permian-Triassic Pangean Basins and Foldbelts along the Panthalassan Margin of Gondwanaland. Geological Society of America Memoir 184. [1]

Venable, D. L. and C. E. Pake. 1999. Population ecology of Sonoran Desert annual plants. In R. Robichaux (ed.), *The Ecology of Sonoran Desert Plants and Plant Communities.* University of Arizona Press, Tucson, pp. 115–142. [2]

Vermeij, G. J. 1972. Endemism and environment: Some shore molluscs of the tropical Atlantic. *American Naturalist* 106: 89–101. [12]

Vermeij, G. J. 1978. *Marine Biogeography and Adaptation: Patterns of Marine Life.* Harvard University Press, Cambridge, MA. [8, 12; Part 4]

Vermeij, G. J. 1982. Gastropod shell form, repair, and breakage in relation to predation by the crab *Calappa. Malacologia* 23: 1–12. [12]

Vermeij, G. J. 1987a. *Evolution and Escalation: An Ecological History of Life.* Princeton University Press, Princeton, NJ. [8, 13]

Vermeij, G. J. 1987b. The dispersal barrier in the tropical Pacific: Implications for molluscan speciation and extinction. *Evolution* 41: 1046–1058. [12]

Vermeij, G. J. 1990. Tropical Pacific pelecypods and productivity. *Bulletin of Marine Science* 47: 62–67. [12]

Vermeij, G. J. 1992. Repaired breakage and shell thickness in gastropods of the genera *Littorina* and *Nucella* in the Aleutian Islands, Alaska. In J. Grahame, P. J. Mill and D. G. Reid (eds.), *Proc. Third International Symposium on Littorinid Biology. Malacological Society of London*, pp. 135–139. [12]

Vermeij, G. J. 1993. *A Natural History of Shells.* Princeton University Press, Princeton, NJ. [12]

Vermeij, G. J. 2001a. Distribution, history, and taxonomy of the *Thais* clade (Gastropoda: Muricidae) in the Neogene of tropical America. *J. Paleontology* 75: 797–705. [12]

Vermeij, G. J. 2001b. Innovation and evolution at the edge: Origins and fates of gastropods with a labral tooth. *Biological J. Linnean Society* 72: 461–508. [12]

Vermeij, G. J. 2002a. Characters in context: Molluscan shells and the forces that mold them. *Paleobiology* 28: 41–54. [12]

Vermeij, G. J. 2002b. Evolution in the consumer age: Predators and the history of life. In M. Kowalewski and P. H. Kelley (eds.), *The fossil record of predation*, pp. 375–393. Paleontological Society Papers 7. [12]

Vermeij, G. J. and P. W. Signor. 1992. The geographic, taxonomic and temporal distribution of determinate growth in marine gastropods. *Biological J. Linnaean Society* 47: 233–247. [8]

Vermeij, G. J. and M. A. Snyder. 2002. *Leucozonia* and related genera of fasciolariid gastropods: Shell-based taxonomy and relationships. *Proc. Academy of Natural Sciences of Philadelphia* 152: 23–44. [12]

Vermeij, G. J., E. A. Kay, L. G. Eldredge. 1984. Molluscs of the northern Mariana Islands, with special reference to the selectivity of oceanic dispersal barriers. *Micronesica* 19: 27–55. [12]

Vermeij, G. J., A. R. Palmer and D. R. Lindberg. 1990. Range limits and dispersal of mollusks in the Aleutian Islands, Alaska. *Veliger* 33: 346–354. [12]

Veron, J. E. N. 1995. *Corals in Space and Time: The Biogeography and Evolution of the Scleractinia.* University of New South Wales, Sydney, and Comstock/Cornell, London. [13, 14]

Vinogradov, A. P. and others. 1968a. *Atlas of Lithological-Paleogeographical Maps of the USSR, Volume 3: Triassic, Jurassic and Cretaceous.* Academy of Sciences U.S.S.R., Leningrad. [1]

Vinogradov, A. P. and others. 1968b. *Atlas of Lithological-Paleogeographical Maps of the USSR, Volume 4: Paleogene, Neogene, & Quaternary.* Academy of Sciences U.S.S.R., Leningrad. [1]

Visser, K., R. Thunnell and L. Stott. 2003. Magnitude and timing of temperature change in the Indo-Pacific warm pool during deglaciation. *Nature* 421: 121–122. [2]

Vitousek, P. M., C. M. D'Antonio, L. L. Loope and M. Rejmanek. 1997. Introduced species: A significant component of human-caused global change. *New Zealand J. Ecology* 21: 1–16. [17]

Vitug, M. D. 1993. *The Politics of Logging: Power from the Forest.* Philippine Center for Investigative Journalism, Manila. [18]

von Euler, F. 2001. Selective extinction and rapid loss of evolutionary history in the bird fauna. *Proc. Royal Society of London Series B* 268: 127–130. [8]

von Humboldt, A. 1808. *Ansichten der Natur mit Wissenschaftlichen Erlauterungen.* J. G. Cotta, Tübingen, Germany. [9]

Vrba, E. S. 1980. Evolution, species, and fossils: How does life evolve? *South African J. Science* 76: 61–84. [10]
Vrba, E. S. 1985. Environment and evolution: Alternative causes of the temporal distribution of evolutionary events. *South African J. Science* 81: 229–236. [6]
Vrba, E. S. 1987. Ecology in relation to speciation rates: Some case histories of Miocene-Recent mammal clades. *Evolutionary Ecology* 1: 283–300. [10]
Vrba, E. S. 1989. Levels of selection and sorting with special reference to the species level. *Oxford Surveys in Evolutionary Biology* 6: 11–168. [10]
Vrba, E. S. and S. J. Gould. 1986. The hierarchical expansion of sorting and selection: Sorting and selection cannot be equated. *Paleobiology* 12: 217–228. [10]
Vrielynck, B. and P. Bouyesse. 2001. Le Visage Changeant de la Terre, L'eclatement de la Pangee et la mobilite des continents au cours des derniers 250 millions d'annees en 10 cartes. Commission de la Carte Geologique du Monde, Paris. [1]

Wagner, W. L. and V. A. Funk (eds.). 1995. *Hawaiian Biogeography: Evolution on a Hot Spot Archipelago*. Smithsonian Institution Press, Washington, D.C. [6, 7, 11, 18; Part 2]
Waide, R. B., M. R. Willig, C. F. Steiner, G. Mittelbach, L. Gough, S. I. Dodson, G. P. Juday and R. Parmenter. 1999. The relationship between productivity and species richness. *Annual Review of Ecology and Systematics* 30: 257–300. [10]
Wallace, A. R. 1855. On the law which has regulated the introduction of new species. *Annals of the Magazine of Natural History*, Second Series 16: 184–196. [6]
Wallace, A. R. 1876. *The Geographical Distribution of Animals*. 2 volumes. Macmillan, London. [4, 6; Parts 4, 5]
Wallace, A. R. 1892. *Island Life, or the Phenomena and Causes of Insular Faunas and Floras Including a Revision and Attempted Solution of the Problem of Geological Climates*, 2nd and Revised Ed. MacMillan and Co., London. [3]
Walsh, D. B. 1996. Late Jurassic through Holocene Paleogeographic Evolution of the South Atlantic Borderlands. Master's Thesis, University of Texas at Arlington. [1]
Walter, H. 1998. Driving forces of island biodiversity: An appraisal of two theories. *Physical Geography* 19: 351–377. [11]
Walther, G.-R., E. Post, P. Convey, A. Menzel, C. Parmesan, T. J. C. Beebee, J.-M. Fromentin, O. Hoegh-Guldberg and F. Bairlein. 2002. Ecological responses to recent climate change. *Nature* 416: 389–395. [16]
Wang, H. 1985. *Atlas of the Paleogeography of China*. Institute of Geology, Chinese Academy of Sciences, Wuhan College of Geology. Cartographic Publishing House, Beijing. [1]
Wares, J. P. 2002. Community genetics in the Northwestern Atlantic intertidal. *Molecular Ecology* 11: 1131–1144. [5]
Watts, W. A. 1988. Europe. In B. Huntley and T. Webb III (eds.), *Vegetation History*. Kluwer Academic Publishers, Dordrecht, NL, pp. 155–192 [3]
Weaver, M. and M. Kellman. 1981. The effects of forest fragmentation on woodlot tree biotas in Southern Ontario. *J. Biogeography* 8: 199–210. [11]
Webb, T. III. 1986. Is vegetation in equilibrium with climate? How to interpret late-Quaternary pollen data. *Vegetatio* 67: 75–91. [3]
Webb, T. III. 1987. The appearance and disappearance of major vegetational assemblages: Long-term vegetational dynamics in eastern North America. *Vegetatio* 69: 177–187. [3]
Webb, T. III. 1988. Eastern North America. In B. Huntley and T. Webb III (eds.), *Vegetation History*. Kluwer Academic Publishers, Dordrecht, NL, pp. 385–414. [3]
Webb, T. III, K. H. Anderson, P. J. Bartlein and R. S. Webb. 1998. Late Quaternary climate change in eastern North America: A comparison of pollen-derived estimates with climate model results. *Quaternary Science Reviews* 17: 587–606. [3]
Webb, T. III, P. J. Bartlein and J. E. Kutzbach. 1987. Climatic change in eastern North America during the past 18,000 years: Comparisons of pollen data with model results. In W. F. Ruddiman and H. E. Wright, Jr. (eds.), *North America and Adjacent Oceans During the Last Deglaciation*. Geological Society of America, Boulder, CO, pp. 447–462. [3]
Webb, T. III, P. J. Bartlein, S. P. Harrison and K. H. Anderson. 1993. Vegetation, lake levels and climate in eastern North America for the past 18,000 years. In H. E. Wright, Jr., J. E. Kutzbach, T. Webb III, W. F. Ruddiman, F. A. Street-Perrott and P. J. Bartlein (eds.), *Global Climates Since the Last Glacial Maximum*. University of Minnesota Press, Minneapolis, pp. 415–467. [3]
Weber, L. I. and S. J. Hawkins. 2002. Evolution of the limpet *Patella candei* d'Orbigny (Mollusca, Patellidae) in Atlantic archipelagoes: Human intervention and natural processes. *Biological J. Linnean Society* 77: 341–353. [12]

Wegener, A. 1912. Die Entstehung der Kontinente. *Geologische Rundschau* 3: 276–292. [1]
Wegener, A. 1915. *Die Enstehung der Kontinente und Ozeane*. Sammlung Vieweg, nr. 23. F. Vieweg und Sohn, Braunschweig. [4]
Wegener, A. L. 1929. *The Origin of Continents and Oceans*. Dover Publications, New York. [Part 4]
Wells, P. V. 1976. Macrofossil analysis of woodrat, *Neotoma*, middens as a key to the Quaternary vegetational history of arid North America. *Quaternary Research* 12: 223–248. [2]
Wells, P. V. and C. D. Jorgensen. 1964. Pleistocene wood rat middens and climatic change in Mojave Desert: A record of juniper woodlands. *Science* 143: 1171–1174. [2]
Werner, R., K. Hoernle, P. van den Bogaard, C. Ranero, R. von Huene and D. Korich. 1999. Drowned 14-my-old Galápagos archipelago off the coast of Costa Rica: Implications for tectonic and evolutionary models. *Geology* 27: 499–502. [12]
West, R. G. 1964. Inter-relations of ecology and Quaternary paleobotany. *J. Ecology* 52(Suppl.): 47–57. [3]
Whitehead, D. R. 1981. Late-Pleistocene vegetational changes in northeastern North Carolina. *Ecological Monographs* 51: 451–471. [3]
Whitlock, C., A. M. Sarna-Wojcicki, P. J. Bartlein and R. J. Nickmann. 2000. Environmental history and tephrostratigraphy at Carp Lake, southwestern Columbia Basin, Washington, USA. *Palaeogeography, Palaeoclimatology, and Palaeoecology* 155: 7–29. [Part 1]
Whittaker, R. H. 1970. *Communities and Ecosystems*. The MacMillan Company, London. [15]
Whittaker, R. H. 1977. Evolution of species diversity in land communities. In M. K. Hecht, W. C. Steere and B. Wallace (eds.), *Evolutionary Biology*, Vol. 10. Plenum Press, New York, pp. 250–268. [14]
Whittaker, R. J. 1995. Disturbed island ecology. *Trends in Ecology and Evolution* 10: 421–425. [11]
Whittaker, R. J. 1998. *Island Biogeography: Ecology, Evolution, and Conservation*. Oxford University Press, Oxford. [11, 18; Part 3]
Whittaker, R. J. 2000. Scale, succession and complexity in island biogeography: Are we asking the right questions? *Global Ecology and Biogeography* 9: 75–85. [7, 11]
Whittaker, R. J. and R. Field. 2000. Tree species richness modelling: An approach of global applicability? *Oikos* 89: 399–402. [9]
Whittaker, R. J. and E. Heegaard. 2003. What is the observed relationship between species richness and productivity? Comment. *Ecology* 84: 3384–3390. [9, 11]
Whittaker, R. J. and S. H. Jones. 1994. The role of frugivorous bats and birds in the rebuilding of a tropical forest ecosystem, Krakatau, Indonesia. *J. Biogeography* 21: 689–702. [11]
Whittaker, R. J., M. B. Bush and K. Richards. 1989. Plant recolonization and vegetation succession on the Krakatau Islands, Indonesia. *Ecological Monographs* 59: 59–123. [11; Part 3]
Whittaker, R. J., R. Field and T. Partomihardjo. 2000. How to go extinct: Lessons from the lost plants of Krakatau. *J. Biogeography* 27: 1049–1064. [11]
Whittaker, R. J., S. H. Jones and T. Partomihardjo. 1997. The re-building of an isolated rain forest assemblage: How disharmonic is the flora of Krakatau? *Biodiversity and Conservation* 6: 1671–1696. [11]
Whittaker, R. J., K. J. Willis and R. Field. 2001. Scale and species richness: Towards a general, hierarchical theory of species diversity. *J. Biogeography* 28: 453–470. [11]
Whittaker, R. J., K. J. Willis and R. Field. 2003. Climatic-energetic explanations of diversity: A macroscopic perspective. In T. M. Blackburn and K. J. Gaston (eds.), *Macroecology: Concepts and Consequences*. Blackwell Science, Oxford, pp. 107–129. [9, 11]
Wikramanayake, E., E. Dinerstein, C. J. Loucks, D. M. Olson, J. Morrision, J. Lamoureaux, M. McKnight and P. Hedao. 2002. *Terrestrial Ecoregions of the Indo-Pacific: A Conservation Assessment*. Island Press, Washington, DC. [18]
Wilcove, D. S., D. Rothstein, J. Dubow, A. Phillips and E. Losos. 1998. Quantifying threats to imperiled species in the United States. *BioScience* 48: 607–615. [15]
Wildlife Conservation Society of the Philippines. 1997. *Philippine Red Data Book*. Bookmark, Manila, Philippines. [18]
Wiley, E. O. 1981. *Phylogenetics: The Theory and Practice of Phylogenetic Systematics*. John Wiley and Sons, New York. [4, 5, 7; Part 2]
Wiley, E. O. 1986. Methods in vicariance biogeography. In P. Hovenkamp (ed.), *Systematics and Evolution: A Matter of Diversity*. University of Utrecht Press, NL, pp. 283–306. [4, 7]
Wiley, E. O. 1988a. Parsimony analysis and vicariance biogeography. *Systematic Zoology* 37: 271–290. [6, 7]
Wiley, E. O. 1988b. Vicariance biogeography. *Annual Review of Ecology and Systematics* 19: 513–542. [6, 7]
Wiley, E. O. and R. L. Mayden. 1985. Species and speciation in phylogenetic systematics, with examples from the North American fish fauna. *Annals of the Missouri Botanical Gardens* 72: 596–635. [6, 7]

Wiley, E. O., D. Siegel-Causey, D. R. Brooks and V. A. Funk. 1991. *The Compleat Cladist*. University of Kansas Museum of Natural History Special Publication 19. University of Kansas Press, Lawrence. [6]

Wilkins, J. S. 2003. How to be a chaste species pluralist-realist: The origins of species modes and the Synapomorphic Species Concept. *Biology and Philosophy* 18:621–638. [4]

Willerslev, E., A. J. Hansen, J. Binladen, T. B. Brand, M. T. P. Gilbert, B. Shapiro, M. Bunce, C. Wiuf, D. A. Gilichinsky and A. Cooper. 2003. Diverse plant and animal genetic records from Holocene and Pleistocene sediments. *Science* 300: 790–795. [3]

Williams, G. C. 1992. *Natural Selection: Domains, Levels, and Challenges*. Oxford University Press, Oxford. [7]

Williams, J. W., B. N. Shuman and T. Webb III. 2001. Dissimilarity analyses of late-Quaternary vegetation and climate in eastern North America. *Ecology* 82: 3346–3362. [3]

Williams, J. W., B. N. Shuman, T. Webb III, P. J. Bartlein and P. Leduc. 2003. Late Quaternary vegetation dynamics in North America: Scaling from taxa to biomes. *Ecological Monographs*, in press. [3]

Williams, P. H. 1998. Key sites for conservation: Area-selection methods for biodiversity. In G. M. Mace, A. Balmford and J. R. Ginsberg (eds.), *Conservation in a Changing World: Integrating Processes into Priorities for Actions*. Cambridge University Press, Cambridge, pp. 221–249. [16]

Williams, P. H. and C. J. Humphries. 1996. Comparing character diversity among biotas. In K. J. Gaston (ed.), *Biodiversity*. Blackwell Science, Oxford, pp. 54–76. [8]

Williams, P. H., D. Gibbons, C. Margules, A. Rebelo, C. Humphries and R. Pressey. 1996. A comparison of richness hotspots, rarity hotspots, and complementarity areas for conserving diversity of British birds. *Conservation Biology* 10: 155–174. [16]

Williamson, M. 1997. Marine biodiversity in its global context. In R. F. G. Ormond, J. D. Gage and M. V. Angel (eds.), *Marine Biodiversity: Pattern and Process*. Cambridge University Press, Cambridge, pp. 1–17. [14]

Williamson, M. H. 1988. Relationship of species number to area, distance and other variables. In A. A. Myers and P. S. Giller (eds.), *Analytical Biogeography, an Integrated Approach to the Study of Animal and Plant Distributions*, pp. 91–115. Chapman and Hall, London. [11]

Williamson, M., K. J. Gaston and W. M. Lonsdale. 2001. The species-area relationship does not have an asymptote! *J. Biogeography* 28: 827–830. [11]

Williamson, M., K. J. Gaston and W. M. Lonsdale. 2002. An asymptote is an asymptote and not found in species-area relationships. *J. Biogeography* 29: 1713. [11]

Willig, M. R., D. M. Kaufman and R. D. Stevens. 2003. Latitudinal gradients of biodiversity: Pattern, process, scale, and synthesis. *Annual Review of Ecology and Systematics* 34: 273–309. [9]

Willis, C. K., A. T. Lombard, R. M. Cowling, B. J. Heydenrych, C. J. Burgers. 1996. Reserve systems for limestone endemic flora of the Cape lowland fynbos: Iterative versus linear programming. *Biological Conservation* 77: 53–62. [16]

Willis, K. J. and R. J. Whittaker. 2002. Species diversity: Scale matters. *Science* 295: 1245–1248. [11]

Willis, K. J., A. Kleczkowski and S. J. Crowhurst. 1999. 124,000-year periodicity in terrestrial vegetation change during the late Pliocene epoch. *Nature* 397: 685–688. [Part 1]

Wills, M. A. 2001. Morphological disparity: A primer. In J. M. Adrain, G. D. Edegcombe and B. S. Lieberman (eds.), *Fossils, Phylogeny and Form*. Kluwer/Plenum, New York, pp. 55–144. [8]

Wilson, E. O. 1959. Adaptive shift and dispersal in a tropical ant fauna. *Evolution* 13: 122–144. [7]

Wilson, E. O. 1961. The nature of the taxon cycle in the Melanesian ant fauna. *American Naturalist* 95: 169–193. [7, 11, 12]

Wilson, E. O. 1999. Prologue to *Archipelago: The Islands of Indonesia*, G. Daws and M. Fujita (eds.). University of California Press, Berkeley. [Part 5]

Wilson, E. O. and E. O. Willis. 1975. Applied biogeography. In M. L. Cody and J. M. Diamond (eds.), *Ecology and Evolution of Communities*. Harvard University Press, Cambridge, MA, pp. 522–534. [11, 17]

Wilson, J. T. 1963. A possible origin of the Hawaiian Islands. *Canadian J. Physics* 41: 863–870. [1]

Wilson, M. E. J. and B. R. Rosen. 1998. Implications of paucity of corals in the Paleogene of Southeast Asia: Plate tectonics or center of origin? In R. Hall and J. D. Holloway (eds.), *Biogeography and Geological Evolution of Southeast Asia*. Backhuys, Leiden, NL, pp.165–195. [13, 14]

With, K. A. 2002. The landscape ecology of invasive spread. *Conservation Biology* 16: 1192–1203. [3]

Wojcicki, M. and D. R. Brooks. 2004. PACT: A simple and powerful algorithm for deriving general area cladograms. *J. Biogeography* 31. [6]

Wolfe, L. M. 2002. Why alien invaders succeed: Support for the escape-from-enemy hypothesis. *American Naturalist* 160: 705–711. [12]

Wood, R. 1999. *Reef Evolution*. Oxford University Press, Oxford. [13, 14]

Woodd-Walker, R. S., P. Ward and A. Clarke. 2002. Large-scale patterns in diversity and community structure of surface water copepods from the Atlantic Ocean. *Marine Ecology Progress Series* 236: 189–203. [14]

Woodland, D. J. 1983. Zoogeography of the Siganidae (Pisces): An interpretation of distribution and richness patterns. *Bulletin of Marine Science* 33: 713–717. [13]

Woodroffe, C. D. 1988. Pacific island mangroves: Distribution and environmental settings. *Pacific Science* 41: 166–185. [12]

Woodroffe, C. D. and J. Grindrod. 1991. Mangrove biogeography: The role of Quaternary environmental and sea-level change. *J. Biogeography* 18: 479–492. [12]

Woodward, F. I. and C. K. Kelly. 2003. Why are species not more widely distributed? Physiological and environmental limits. In T. M. Blackburn and K. J. Gaston (eds.), *Macroecology: Concepts and Consequences*. Blackwell Science, Oxford, pp. 239–255. [9]

Woodward, F. I., T. M. Smith and W. R. Emanuel. 1995. A global primary productivity and phytogeography model. *Global Biogeochemical Cycles* 9: 471–490. [2]

Worthy, T. H. and R. N. Holdaway. 2002. *The Lost World of the Moa: Prehistoric Life of New Zealand*. Indiana University Press, Bloomington, IN. [12]

Wright, D. H. 1983. Species-energy theory: An extension of species-area theory. *Oikos* 41: 496–506. [8, 9, 11]

Wright, D. H., D. J. Currie and B. A. Maurer. 1993. Energy supply and patterns of species richness on local and regional scales. In R. E. Ricklefs and D. Schluter (eds.), *Species Diversity in Ecological Communities*. University of Chicago Press, Chicago, pp. 66–74. [8]

Wright, H. E., Jr., J. E. Kutzbach, T. Webb III, W. F. Ruddiman, F. A. Street-Perrott and P. J. Bartlein (eds.). 1993. *Global Climates Since the Last Glacial Maximum*. University of Minnesota Press, Minneapolis. [3]

Wylie, J. L. and D. J. Currie. 1993. Species-energy theory and patterns of species richness. 1. Patterns of bird, angiosperm, and mammal richness on islands. *Biological Conservation* 63: 137–144. [9]

Yan, X., L. Zhenyu, W. P. Gregg and L. Dianmo. 2001. Invasive species in China: An overview. *Biodiversity and Conservation* 10: 1317–1341. [15]

Yin, A. and S. Nie. 1996. A Phanerozoic palinspastic reconstruction of China and its neighboring regions. In A. Yin and M. Harrison (eds.), *The Tectonic Evolution of Asia*, Cambridge University Press, Cambridge, pp. 442–486. [1]

Yoder, A. D., M. M. Burns, S. Zehr, T. Delefosse, G. Veron, S. M. Goodman and J. G. Flynn. 2003. Single origin of Malagasy Carnivora from an African ancestor. *Nature* 421: 734–737. [12]

Young, G. C. 1984. Comments on the phylogeny and biogeography of antiarchs (Devonian placoderm fishes) and the use of fossils in biogeography. *Proc. Linnaean Society of New South Wales* 107: 443–473. [4]

Young, G. C. 1995. Application of cladistics to terrane history: Parsimony analysis of qualitative geological data. *J. Southeast Asian Earth Sciences* 11: 167–176. [4]

Zachos, J., M. Pagani, L. Sloan, E. Thomas and K. Billups. 2001. Trends, rhythms, and aberrations in global climate 65 Ma to present. *Science* 292: 686–693. [14]

Zandee, M. and M. C. Roos. 1987. Component-compatibility in historical biogeography. *Cladistics* 3: 305–332. [4, 6, 7]

Zapata, F. A., K. J. Gaston and S. L. Chown. 2003. Mid-domain models of species richness gradients: Assumptions, methods and evidence. *J. Animal Ecology* 72: 677–690. [11]

Zbyszewski, G. and O. da Veiga Ferreira. 1962. *La Faune Miocène de l'Île de Santa Maria (Açores)*. Comunicações dos Serviços Geológicos, Portugal 46: 247–289. [12]

Zheng, X. G., B. S. Arbogast and G. J. Kenagy. 2003. Historical demography and genetic structure of sister species: Deermice (*Peromyscus*) in the North American temperate rain forest. *Molecular Ecology* 12: 711–724. [5]

Ziegler, A. M., C. R. Scotese and S. F. Barrett. 1983. Mesozoic and Cenozoic paleogeographic maps. In P. Broche and J. Sundermann (eds.), *Tidal Friction and the Earth's Rotation II*. Springer-Verlag, Berlin. [1]

Ziegler, P. A. 1989. *Evolution of Laurussia: A Study in Late Paleozoic Plate Tectonics*. Kluwer Academic Publishers, London. [1]

Ziegler, P. A. 1990. *Geological Atlas of Western and Central Europe*. 2nd and completely revised edition. Shell International Petroleum.

Maatschappij B.V. Geological Society Publishing House, Bath, England. [1]

Zink, R. M. 2002. Methods in comparative phylogeography and their application to studying evolution in the North American aridlands. *Integrative and Comparative Biology* 42: 953–959. [5]

Zink, R. M. and R. C. Blackwell. 1998a. Molecular systematics and biogeography of aridland gnatcatchers (genus *Polioptila*) and evidence supporting species status of the California Gnatcatcher (*Polioptila californica*). *Molecular Phylogenetics and Evolution* 9: 26–32. [5]

Zink, R. M. and R. C. Blackwell. 1998b. Molecular systematics of the scaled quail complex (genus *Callipepla*). *Auk* 115: 394–403. [5]

Zink, R. M. and R. C. Blackwell-Rago. 2000. Species limits and recent population history in the Curve-billed Thrasher. *Condor* 102: 881–886. [5]

Zink, R. M. and D. L. Dittmann. 1991. Evolution of brown towhees: Mitochondrial DNA evidence. *Condor* 93: 98–105. [5]

Zink, R. M., G. F. Barrowclough, J. L. Atwood and R. C. Blackwell-Rago. 2000a. Genetics, taxonomy and conservation of the threatened California Gnatcatcher. *Conservation Biology* 14: 1394–1405. [5]

Zink, R. M., R. C. Blackwell and O. Rojas-Soto. 1997. Species limits in the Le Conte's thrasher. *Condor* 99: 132–138. [5]

Zink, R. M., R. C. Blackwell-Rago and F. Ronquist. 2000b. The shifting roles of dispersal and vicariance in biogeography. *Proc. Royal Society of London Series B* 267: 497–503. [5]

Zink, R. M., D. L. Dittmann, J. Klicka and R. C. Blackwell-Rago. 1999. Evolutionary patterns of morphometrics, allozymes and mitochondrial DNA in thrashers (genus *Toxostoma*). *Auk* 116: 1021–1038. [5]

Zink, R. M., A. E. Kessen, T. V. Line and R. C. Blackwell-Rago. 2001. Comparative phylogeography of some aridland bird species. *Condor* 103: 1–10. [5]

Zink, R. M., S. J. Weller and R. C. Blackwell. 1998b. Molecular phylogenetics of the avian genus *Pipilo* and a biogeographic argument for taxonomic uncertainty. *Molecular Phylogenetics and Evolution* 10: 191–201. [5]

Zonenshain, L., M. I. Kuzmin and L. M. Natapov. 1990. *Geology of the U.S.S.R.: A Plate Tectonic Synthesis*. American Geophysical Union, Geodynamics Series 21, pp. 1–242. [1]

Index

Abies lasiocarpa var. *arizonica*, 40
Abrocoma, 28
Absolute Desert, 35, 36
Acacia, 37
Acanthaster planci, 252
Accelerator Mass Spectrometry (AMS), 50, 260
Accumulation curves, 336
Accumulation hypothesis, 257, 258, 265
ACEP (Atlantic-Caribbean-Eastern Pacific high-diversity focus), 282–283, 286
Acropora, 285, 286
Actual evapotranspiration (AET), 174, 225
Adirondacks, SLOSS example, 330
Adriatic promontory, 22
AET (actual evapotranspiration), 174
  Southern African woody plants, 225
Aetobatus narinari, 252
Age of taxa
  East Indies gradients, 258
  latitudinal gradients, 266–267
Albula neoguinaica, 252
Algae, diversity gradient, 274
The Alien Animals (Laycock), 298
Allocation rules, 216
Allopatric speciation
  Cenozoic diversification, 289
  center of origin hypothesis, 262
  high latitudes, 185
  relief, 182
  vs. sympatric speciation, 363
Allopatry, defined, 70
Alpha (within-habitat) diversity
  following biological invasions, 301–306
  human transport, 303–304
  latitudinal diversity gradient, 275, 279
Alps, formation, 24
Ambient energy hypothesis, 177–178
Amino acid racemization, 50
Ammonoid cephalopods, extinction and recovery, 167–168
Amphibians, environmental factors and diversity, 176
AMS (Accelerator Mass Spectrometry), 50
Amurian seaway, 21
Ancient DNA
  phylogenetic analysis, 45
  Quaternary population studies, 62–63
Animal diversity, energy effects, 189–190
Animals, species richness and climate gradients, 227
Anolis, 137
Antarctica, isolation of, 23, 284, 285
Appalachian Mountains, formation, 17
Arabia, rifting of, 23
Araucaria, 38
Archipelagic SPARs, 326, 328, 330–332
Area
  biological attributes and species richness, 193–194
  control in sampling, 222
  and diversity theory, 179–180
  species richness in differing taxa, 195

species richness on islands, 216–217
See also SPARs (Species-area relationships)
Area cladistics, 67–68
component analysis and consensus, 73–76
defined, 70
definition of area, 79
of Hawaiian Islands, 80–84
inferred geographical barriers, 80
method, 77–78
and time-slicing, 84–85
Area cladograms, 112, 114, 119, 129, 130, 133, 142
from BPA, 138–140
geo-dispersal, 118–119
Area Duplication Convention, 131–134
Areagram, 79
Artifacts, species richness models, 228
Asia
collision with India, 23–24
paleogeography, 21–22
Assumption 0, BPA, 131
Assumption 2, component analysis, 76
Assumptions, of cladistic biogeography, 130
Atacama Desert, Quaternary biogeography, 34–36
Atlantic-Caribbean-Eastern Pacific high-diversity focus (ACEP), 282–283, 286
Atlantic Islands, gastropod shell armor, 248–250
Atlantic Ocean
formation, 20–22
latitudinal diversity gradient, 275–277
Auirparus flaviceps, 107
Australia
collision with Indonesia, 24
collision with Southeast Asia, 284, 287
isolation of, 23
Austrocedrus, 38
Autocorrelation, 228
Banks, Joseph, 294
Barriers
comparing marine and terrestrial, 253
to marine dispersal, 244–246
Beetles, environmental factors and diversity, 176
Bergmann's Rule, 42–43
Bering Sea land bridge, 22, 118
Beta (between-habitat) diver-sity
following biological invasions, 304–305, 306–308
latitudinal diversity gradient, 277–278, 281
Between-region (delta) diversification, Cenozoic era, 285
BIOCLIM, modeling species distributions, 313–314
Bioclimatic modeling, 39–41, 44–45
Biodistributional factors, extinction selectivity, 169
Biodistributional units, other criteria, 161
Biodiversity
application of phylogenetic biogeography, 123
conservation in Philippines, 349–359
documenting, 311–312
hotspots, 268
recent origin of, 286
See also Diversity; Species richness
Biogeographic congruence. See Congruence
Biogeographic modeling
challenges of, 367
climate, 39–41
digital elevation model (DEM), 40
limitations, 44–45
USGS-OSU model, 39–40
Biogeographical islands, 327, 340
Biogeography
changing concepts, 363–364
definition, 211
future challenges, 365–368
and genetic variation, 62–63
history of, 361–362
mechanisms, 70
methodological advances, 272–273
new synthesis, 141–143
relationship to evolution and tectonics, 85
technological advances, 362–363
theoretical foundations, 367–368
twentieth century accomplishments, 125
twenty-first century goals, 86
See also Historical biogeography; Paleobiogeography; Panbiogeography; Phylogenetic biogeography; Phylogeography;
Biological diversity. See Biodiversity; Diversity; Species richness
Biological invasions
beta diversity, 304–305
broader scale impacts, 298–299
distribution of diversity, 309
as ecological experiments, 299
exotic species and SPARs, 332–333
island biogeography theory, 305–306
measuring diversity changes, 300–301
process of, 299–300
research on, 297–299
unified neutral theory, 305–308
Biome
limitations in diversity theory, 180
modeling, 40
BIOMES, 39
Biotic divergence, and area cladistics, 79

Birds
  diversity after biological invasions, 302–303
  diversity and energy, 178–179
  environmental factors and diversity, 177
  latitudinal diversity gradient, 172
  Lesser Antilles, 214
  species richness and AET, 175
  species richness and range in elevation, 183
Bivalves
  body size and provinciality, 161–163
  diversity gradients, 272
  latitudinal diversity gradient, 156
  species richness and temperature, 165
  See also Molluscs
Body size
  island adaptation, 242
  latitudinal gradient, 153
  marine islands, 251–252
  and species richness, 153–155
Boundaries, determinants of, 159–161
BPA (Brooks Parsimony Analysis)
  automated approaches, 138–140
  and biogeographic incongruence, 115
  in desert vertebrates, 102–107
  and extinction, 113
  four principles, 131–134
  and geo-dispersal, 117–118
  inclusive ORing, 132
  ontology of complexity, 134
  phylogeographic studies, 138
  primary and secondary, 132, 139, 140
  and reticulated area relationships, 128
Brachiopods, paleozoic extinction, 166
Bradypus, 43–44
Brazil, endemism in, 347, 349
Brooks Parsimony Analysis. *See* BPA
Bryozoans
  latitudinal diversity gradient, 275, 278
  paleozoic extinction, 166
Buccinum cinis, 250
Buffon's Law, 111
Bufo calamita, 341
Bulbometopon muricatum, 252
Bushy-tailed rat, 28
Butterflies, temperature and species diversity, 186, 187
Butterflies and moths, environmental factors and diversity, 176

C/NC (carnivorous to noncarnivorous ratio), gastropods, 158–159, 164
Cactus, columnar, 108
CAI. *See* Central American Isthmus
Calappa hepatica, 252
California gnatcatcher, 108
California watersheds, diversity after biological invasions, 302–303, 305
Campanian-Maastrichtian dinosaur fauna, 21
Campylorhynchus brunneicappilus, 107
Canadian Cordillera, formation, 22
Canarium spp., 246
Capacity rules, 215, 225
Caribbean islands, intra-island speciation, 137
Carnegiea gigantea, 33, 40
Carpilius maculatus, 252
Catastrophists, vs. gradualists, 69
Cellana, 244
Celtis, 37
Cenozoic era
  global cooling, 283–284
  marine diversity, 279, 283, 289, 290
  paleogeography, 23–25
  tectonics and climate, 23–25, 285–288
Centers of endemism, Philippines, 353, 356–357
Centers of origin
  centrifugal mechanism, 264–265
  conservation biology, 267–269
  and dispersal, 71, 258, 262, 266
  geography of speciation, 261–264
  latitudinal age gradient, 266–267
  question of existence, 255–260, 265–267, 289
  view of historical biogeography, 87
Central American Isthmus (CAI; Panamanian Isthmus), 23
  biotic changes, 168
  uplift of, 284, 288
Centrifugal speciation, centers of origin, 264–265
Cercidium, 37
Cereus, 37
Cheilinus undulatus, 252
Chinchilla rat, 28
Choeloepus, 43–44
Choisya dumosa, 29
Chozchori, 28
Cimmeria, 18–20
Cirrhitidae, centers of origin, 257
Cladistic biogeography
  compared to phylogenetic biogeography, 127–128
  definition, 93–94
  longitudinal patterns, 273
  ontology of simplicity, 134
  summarized, 88
  vicariant speciation, 126
  vs. phylogenetic biogeography, 72–73
  See also Area cladistics
Cladistics
  and circular reasoning, 72
  historical background, 69
  from phylogenetics to areas, 71
  See also Area cladistics

Cladograms, vs. phylogenetic trees, 72
Climate
  Atacama Desert, 34–36
  biogeographic modeling, 39–41
  and diversity gradients, 175
  Monte Desert, 38–39
  North American deserts, 30–32
  and provincial boundaries, 161
  See also Paleoclimate
Climate change, consequences for species distributions, 322
Climate gradients
  animal species richness, 227
  species richness models, 224–226
  See also Latitudinal diversity gradient
Climatic variability
  niche-assembly theory, 185
  palynological studies, 52–55
CO_2 levels, 56
  rodent midden analysis and, 41
Colonization
  recolonization after glaciation, 186–187
  and reticulated relationships, 127
  See also Dispersal
Commission for the Study and Use of Biodiversity (CONABIO), Mexico, 312
Community, in neutral theory, 172
Community composition, environmental shifts, 59–60
Comparative phylogeography, 96
  relationship to historical biogeography, 100
  vicariance and population genetics, 99–101
Competition, marine islands, 250–251
Competitive speciation, center of origin hypothesis, 263
COMPONENT, 88
Component analysis
  and biogeographic incongruence, 115
  use in area cladistics, 70, 73–76
Component compatibility, 76
Condalia, 37
Congruence
  definition, 113
  dispersal and migration, 74
  interpretations of area cladistics, 73–76
  need for, 85
  role of vicariance, 112, 113
  using areagrams, 79
Consensus trees, 74, 75
Conservation
  hotspots, 269
  importance of marine centers of origin, 267–269
  threats in Philippines, 349–350
Conservation biogeography
  environmental indicators, 334, 342
  identifying priority areas, 317–318, 321, 356–359
  models, 353–356, 358–359
  overview, 293–296
  reconciliation ecology, 338–341, 342–343
Continental corridors, 21–23
Continental drift
  historical background, 69, 88, 234
  See also Plate tectonics
Continental shelf, area and species richness, 203
Continents
  collisions, 24–25
  latitude and diversity, 179
  mapping the past, 12–14
  spatial variation of species richness, 222–230
  sunken, 234
  uncertainties in mapping, 14–16
Conus spp., 246
Cook, Captain James, 294
Corals
  centers of origin, 257, 258
  environmental factors and diversity, 176
  Indo-Pacific disjunctions, 273
Coral reefs
  glacioeustatic sea level fluctuations, 282
  species richness, 278, 287
Coralliophila violacea, 252
Coris aygula, 252
Corkbark fir, 40
Creosote bush, 27, 34, 37, 41–42
Cretaceous
  mass extinction of molluscs, 166–168
  paleogeography, 22–23
Crustaceans, development strategies and diversity, 204–205
Cypraeidae, centers of origin, 257
Cyrtulus, 244

Dana, James Dwight, 233
Darwin, Charles, 145–146, 265, 294, 298
Data matrix, geo-dispersal and vicariance, 118–120
Dating methods, 50–52
Deconstruction
  definition, 206
  diversity theories, 192–193
  relational theories of diversity, 206–209
Deep-sea studies, 236
Deep time, 137–138
Deforestation, threat to biodiversity, 311, 321
Delta (between-region) diversification, Cenozoic era, 285
DEM (digital elevation model), 40
Desert granivory, 38
Desert vertebrates, BPA applied to, 102–107
Deserts
  North American Quaternary, 30–34

Pangean, 19
rodent midden analysis, 27
South American Quaternary, 34–39
Development strategy, and diversity, 204–205
Dietz, Robert S., 234
Digital elevation model (DEM), 40
Dinaride Mountains, formation, 24
Dinosaurs
Campanian-Maastrichtian fauna, 21
climate and, 22
Diodon hystrix, 252
Dipodomys merriami, 107
Disjunct species, sampling artifacts, 228
Disjunction events, 108–109
Dispersal
in biotic histories, 87–88
center of origin hypothesis, 71, 258, 262, 266
defined, 70
departures from congruence, 74
differing meanings, 98–99
and diversity, 204–205
extinction and speciation, 201
geography of speciation, 109–110, 262
to islands, 241–242
Krakatau system, 218
macroevolutionary dynamics, 201
marine barriers to, 244–246
molecular studies in marine organisms, 235, 237
planktonic stages, 247
and range expansion, 115–116
role in historical biogeography, 137
See also Colonization; Geo-dispersal; Vicariance
Dispersal-vicariance analysis (DIVA), 89, 130–131
Distribution
dispersal and vicariance, 68
history of spruce, 58
and migration routes, 72
modern biota, 72
redundant, 108
widespread taxa, 107–108
See also Congruence; Dispersal; Geo-dispersal; Latitudinal diversity gradient; Vicariance
Distribution patterns, and interrelationships of areas, 69
DIVA (Dispersal-vicariance analysis), 89, 130–131
Diversity
at high latitudes, 178–179
other measures, 151–152
phylogenetic and extinction selectivity, 167
question of latitudinal diversity gradient, 171
selective extinction and recovery, 165–168
species-energy and productivity, 174–179
See also
Alpha/Beta/Gamma/Delta diversity; Animal diversity; Biodiversity; Centers of origin; Latitudinal diversity gradient; Morphological diversity; Plant diversity; Species richness
Diversity dynamics, morphological, functional and historical factors, 168–170
Diversity gradients, historical legacy, 290
Diversity theories
determining area, 179–180
future prospects, 148
hierarchical framework, 223
historical overview, 145–147
need for deconstruction, 192
prospects for, 228–230
questions and issues, 148
relational, 206–209
types of, 191–192
See also Latitude diversity gradient; MacArthur and Wilson theory; Neutral theory; Species-energy theory
DNA
sequences from rodent middens, 43
See also Ancient DNA; mtDNA
DOLY, 39
DOMAIN, modeling species distributions, 314
Drake Passage, 24, 285
Drosophila, 80
Drosophilid flies, 241
Drupa spp., 249
Drupella cornus, 252
Ducula bicolor, 217
Dynamic Earth, 68
time scales, 7–8
Dynamic models, future of, 227–228
Dynamic theories
definition, 146–147, 212
equilibrium on islands, 217, 219
evaluation of MacArthur and Wilson model, 214–215
limitations, 222
species richness on islands, 211–222

East African Rift system, 23, 24
East Gondwana, 21
East Indies Triangle center of origin
conservation priority, 269
evidence for, 258–260, 266
extent, 268
geological history, 261
species diversity, 256–258
See also IWP region
Easter Island, endemic species, 244
Eastern Indian Ocean, formation, 21
Echinometra, 259
Ecological drift, 172

Ecological niche modeling, potential distributions, 315–317
Ecological niche theory, habitat tracking in the Quaternary, 57
Ecological roles, limitations on islands, 241–243
The Ecology of Invasions by Animals and Plants (Elton), 294, 298
Ekman, Sven, 234, 256
El Niño, 30
Elevation
  latitude and diversity, 181–183
  modeling, 40
  vs. latitude, 364
Elton, Charles, 294, 298
Emperor seamount chain, 13
Endeavor (ship), 294
Endemism
  area analyses, 85
  and evolutionary success, 268
  on islands, 240–242, 345–346, 359
  marine islands, 244
  on mountain ranges, 356
  in Philippines, 346–347, 349, 353, 355–359
  Spain and Brazil, 347, 349
Energy, and productivity, 174–177, 177–178
Environmental change
  and evolution, 61–62
  multivariate model, 59
  species distributions, 57–58
  through time, 55–57
Environmental indicators, National Research Council list, 334, 342
Environmental lumpiness, Quaternary, 60–61
Epinephelus lanceolatus, 252
Equilibrium theory
  MacArthur and Wilson island theory, 136–137, 146, 214–220, 239
  subcontinental scale, 228–230
  unified neutral theory, 173
Euryozius, 244
Event stratigraphy, 50
Evolution
  centers of origin, 267
  and Quaternary environmental change, 61–62
  time trumps space, 141
Evolutionary adjustment, to environmental change, 57
Evolutionary biogeography, 87
Evolutionary radiations
  Cenozoic era, 288–289, 290
  marine diversity, 279
Evolutionary success
  center of origin hypothesis, 261–262, 264
  and endemism, 268
Exotic species
  species-area relationships, 332–333
  See also Biological invasions
Extent, 273
Extinction
  center of origin hypothesis, 259
  and phylogenetic biogeography, 113–115
  and Quaternary environmental change, 61–62
  See also Mass extinction
Extinction rates
  and diversity gradients, 287, 289, 290
  and diversity of lineages, 364
  on islands, 241
  MacArthur and Wilson model, 214–215
Extinction selectivity, spatial patterns of biodiversity, 165–168, 169

Fagus grandifolia, 57
FAUNMAP project, 51
Fiji, macrobenthos diversity, 277–278
Fish (freshwater), environmental factors and diversity, 176
Fitch optimization, 138
Fitch parsimony, 118
Flightlessness, island endemics, 242
FloraMap, modeling species distributions, 314
Foraminifera, benthic
  Cenozoic radiation, 288–289
  latitudinal diversity gradient, 276–278
Foraminifera, environmental factors and diversity, 176
Forbes, Edward, 233
Forest arthropods, BPA, 136
Formula 3, estimating SPARs, 336
Formula 5, estimating SPARs, 336–339
Forster, Johann Reinhold, 145
Fossils, Quaternary assemblages, 51–52
  See also Rodent midden analysis
Fossil record, in phylogenetic biogeography, 114
Founder principle, 264
Fragmentation, SLOSS question, 328–332
Frost, adaptation to, 188
Fruit bat, 217
Fruit-pigeon, 217
Functional groups
  and latitudinal diversity gradient, 155–159
  testing diversity hypothesis, 164–165
Fundamental biodiversity number (θ), neutral theory, 173

Galápagos Islands, species diversity, 241
Gamma (regional) diversity
  geological history, 281–286
  latitudinal diversity gradient, 275, 278–281
GAP analysis, modeling species distributions, 313, 321
GARP (genetic algorithm for rule prediction)
  modeling species distributions, 314–315

NPA model in Mexico, 319
Gastropods
C/NC ratio, 158–159, 164
Cenozoic radiation, 288
competition on islands, 250
latitudinal diversity gradient, 156, 158
morphological diversity and species richness, 154, 155
nature of marine barriers, 246–247
predation and shell armor, 247–250
species richness and temperature, 165
See also Molluscs
Genetic algorithms. *See* GARP
Genetic diversity, centers of origin, 259
Genetic drift, 172
Genetic studies, rodent midden analysis, 43–44
Genetic variation, centers of origin, 267
Geo-dispersal, 112
data matrix, 118–120
definition, 115–116
desert vertebrates, 108
distinguishing from vicariance, 137–138
historical antecedents, 116
and range expansion, 115–116
and reticulated relationships, 127
tree, 117–118
widespread taxa, 108
within phylogenetic biogeography, 117–122
Geographic barriers, rise and fall, 116–118
Geographic gradients, spatial scale, 222–224
Geographical congruence. *See* Congruence
Geologic record, 14
Geology, model for biogeography, 6
Geometric constraints, 228
Geospiza, 241
Gibberulus gibberulus, 246
GIS (Geographic Information Systems), species distribution modeling with, 312, 321
Glacial maxima, Quaternary, 54
Glaciation
recovery and latitudinal gradients, 186–187
sea-level change, 17–18
Glacioeustatic sea level fluctuations, 282, 289
Global cooling, Cenozoic trends, 283–284
Global diversity gradient. *See* Latitudinal diversity gradient
Global ice volume, 56
Global warming, testing species-energy theory, 190
Glyptodonts, persistence in South America, 243
Gondwana, 20, 21; Plates 5–8
Gradualists, vs. catastrophists, 69
Grain, 273
Grand cline. *See* Latitudinal diversity gradient
Granivorous rodents, 38–39
Great American Interchange, 23
Greater Antilles island arc, in Cretaceous, 22
Greater Gondwana, 11
"Greater Somalia" hypothesis, 21–22
Greenland
in Cenozoic era, 23
temperature history, 55
Ground sloths
genetic analysis, 43–44
Monte Desert, 38
Groundfinch species, Galápagos Islands, 241
Gulf of Aden, formation, 24–25
Gulf of California, formation, 23, 25
Gulf of Mexico, formation, 20
Habitat diversity, latitudinal gradients, 181
Habitat loss, modeling species distributions, 315–316, 349–350
Habitat specialization, and endemism, 357
Habitat tracking, 57, 322
recovery from glaciation, 186
Hawaiian Islands, 13
area cladistics, 80–84
biotic connections, 246
divergences of biota, 84
historical positions, 82
species diversity, 214, 241
Hellenide Mountains, formation, 24
Hennig, Willi, 71, 72, 73
Herbivores, roles on islands, 243
Hesperomannia, 80
Hess, Harry H., 234
Heterandria, 73, 74, 128
Heteroconchs, Cenozoic radiation, 288
Hierarchical links, trophic levels on islands, 219
High latitudes
energy and diversity, 178–179
niche-assembly theory, 183–184
Himalayas, formation, 17, 24
Historical biogeography, 6
controversies, 93–96
methodological bases, 142
and molecular markers, 62–63
new synthesis, 89
philosophies, 135–137
and phylogeography, 109–110
proliferation or convergence of methods, 88–89
Historical theories, definition, 211–212
Holocene Epoch, climate variation, 54
Holocene migration, desert species, 33–34
Homogocene, 333

Homology, "transformed cladistics," 72
Homoplasy, and species history, 76
Hot spots, 13
Hubbell, Stephen P., 171
Human impacts
  challenge of, 368
  islands, 219–220
  mass extinctions, 148, 364
  See also Conservation
Human transport, alpha (within-habitat) diversity, 303–304
Hydrographic conditions, biodistributional boundaries, 161

IBP-SES (International Biological Program-Structure of Ecosystems Subprogram), 37–38
Immigration rates, MacArthur and Wilson model, 214–215
Incongruence, in area cladistics, 77
India, collision with Eurasia, 23
Indian Ocean, formation, 20
Indo-Australian region, map, 347
Indo-West Pacific region. *See* IWP
Indonesia, collision with Australia, 25
Infauna diversity gradient, 274
Infauna/epifauna ratio, marine bivalves, 157, 160
Information, challenge of, 365–366
Insolation data, 56
Institute for Biodiversity (INBio), Costa Rica, 312
Interdisciplinary biogeography, challenge of, 366–367
Interim general model (IGM), southern Africa, 228
International Biogeography Society, 361
International Biological Program-Structure of Ecosystems Subprogram (IBP-SES), 37–38
Interprovincial SPARs, 326, 328
Intraprovincial SPARs, 326, 328
Invasive species. *See* Biological invasions
Inventory, species turnover on islands, 213–214
Island biogeography theory
  biological invasions, 305–306
  MacArthur and Wilson model, 136–137, 146, 212–220, 239
  population size and area, 173
Island biotas
  biological and evolutionary meaning, 253–254
  dispersal speciation, 262
Islands
  area and diversity, 179
  area and species richness, 193, 194, 195, 212–222, 240–241
  dispersal to, 241–242
  dynamic species richness theories, 211–222
  endemism, 240–241, 345–346, 359
  extinction rates, 241
  marine perspective, 239–240
  meaning of, 326–327, 340
  and SPARs (Species-area relationships), 212–222
  terrestrial perspective, 240–243
  See also Marine islands; Philippine Islands
IWP (Indo-West Pacific) region
  center of origin, 255–258, 265–266
  gastropod shells, 247–250
  geological history of, 286
  longitudinal gradient, 282–283
  modes of speciation, 263
  size of predators, 252
  See also East Indies Triangle

Jasus, 249
Johnston, Alexander K., 233
Juglans nigra, 57
Juniperus osteosperma, 30, 33
Jurassic period
  diversity gradients, 158
  paleogeography, 20–21

Kerguelen Island, endemic species, 244
Kerguelen plateau, 21
Kokia, 80
Krakatau system, species-richness models, 217–219

Ladd's accumulation hypothesis, 257
Lag, 273
Lagidium, 28
Land bridges
  Bering Strait, 22, 118
  Central American Isthmus (CAI), 23, 268, 284, 288
  historical views, 87, 234
Land use, reconciliation ecology, 340
Landscape dynamics, and species traits, 199–200
Large igneous provinces (LIPs), 13
Larrea, 27, 34, 37, 41–42
Latitude belt, determining area in diversity theory, 180
Latitudinal diversity gradient
  age of taxa, 266–267
  alpha and beta diversity, 279
  beta (between-habitat) diversity, 277–278
  birds, 172
  body size, 153
  causal processes, 290
  climate explanation, 228
  deep sea patterns, 275–277
  discordant gradients, 202
  environmental influence in marine taxa, 201–203

and functional groups, 155–159
glaciation and recent climate change, 186–187
habitats, 181
migration and diversity, 187–188
molluscs, 156–159, 201–204, 276, 279
morphological diversity, 152–155
niche-assembly theory, 182–185
Northern Hemisphere, 281–283
productivity or energy?, 174–179
question of, 171
regional scale patterns, 273–274
smaller scale patterns, 274–275
Southern Hemisphere, 282–283
speciation rate, 185–186
species-energy hypothesis, 163–164
theories, 188–190
time of development, 284–286
Laupa, 80
Laurasia, 20, 21; Plates 5–8
Laycock, George, 298
Leaf-eared mice, 28, 44
Lepidochelys spp., 259
Lesser Antilles, land birds, 214
Leucozonia nassa, 248
Linking mechanisms, macroscale models, 226–227
Links
in island models, 219, 221
plant diversity to animals, 181
Linnaean shortfall, 296
Linnaeus, Carolus, 145
Liomys spectabilis, potential distribution model, 316, 319–320
LIPs (large igneous provinces), 13
Lithology, indicators of paleolatitude, 14
Littorina spp., 249, 250
Longitudinal diversity gradient, 273, 282–283
morphological diversity, 153
Lophocereus, 108
Lyell, Charles, 69
Lystrosaurus, 19

MacArthur and Wilson theory
island biogeography, 146, 212–220, 239
phylogenetic biogeography, 136–137
MacArthur, Robert H., 239
Macrobenthos
beta (between-habitat) diversity, 277–278
taxonomic diversity gradient, 274, 275
Macroecology
role of functional biology, 169
unifying perspective, 209
Macroscale models
and climate, 224–226
future of, 227–228
linking mechanisms, 226–227
Madagascar
in Cretaceous period, 21
species diversity, 241
Magnetic anomalies, mid-ocean ridges, 12–13
Malay Archipelago, 262
Mammals
diversity in Philippines, 350–357
environmental factors and diversity, 177
NPA model in Transvolcanic Belt (Mexico), 319–321
Mapping
continental positions, 12–13
paleogeography, 16–18, 26
uncertainties, 13–14
MAPPS, 39
Maps
global biodiversity, 271–272
PALEOMAP project, 11–12, 20 ff!
sources, 18, 19
Marine biogeography
current issues, 290
historical overview, 233–235
molecular markers, 235, 237
Marine centers of origin. *See* Centers of origin
Marine invertebrates, species richness in Southeastern Pacific, 201–202
Marine islands
biotic characteristics, 240, 243, 246–252, 359
impoverished fauna, 243–244
predators, 240, 247–250, 252
Marine molluscs
environmental influence on diversity, 201–203
latitudinal diversity gradient, 156–159, 201–204
provincial boundaries, 159–163
testing the species-energy hypothesis, 163–164
See also Molluscs
Marine predators, size distribution, 251–252
Mass extinctions
anthropogenic, 148, 364
biotic factors, 167
reconciliation biology perspective, 343
Mean sea surface temperature (SST). *See* SST
Mesozoic era, marine diversity, 279
Metabolic model, species-energy hypothesis, 163–164, 178–179
Metacommunity
defined, 172–173
neutral theory, 179
Metapopulation model, landscape dynamics, 199–200

Michaelis-Menton equation, estimating SPARs, 336, 338, 339
Mid-ocean ridges, magnetic anomalies, 12–13
Middens, *see* Rodent midden analysis
MIGRATE, 96
Migration
  departures from congruence, 74
  desert species, 33–34
Migration routes
  continental corridors, 21–23
  and distribution, 72
Missing Data Coding Protocol, BPA, 131
Mitochondrial DNA. *See* mtDNA
Modeling
  mammalian diversity in Philippines, 355–356, 358
  species distributions, 313–317, 319, 321
  using museum bio-collections, 312–315
  water and energy in trees, 224–226
  See also Bioclimatic modeling; Biogeographic modeling
Molecular biogeography, 102
Molecular markers
  in historical biogeography, 62–63
  in marine studies, 235, 237, 257
  mtDNA, 257, 259
Molecular phylogeographic studies, woodrat, 45–46
Molluscs
  Atlantic latitudinal diversity gradient, 276
  environmental factors and diversity, 176
  latitudinal diversity gradient, 276, 279
  See also Bivalves; Gastropods; Marine molluscs
Monophyly, 72, 77
Monotaxis grandoculis, 252
Monte Desert, Quaternary biogeography, 37–39
Morphological diversity
  evolutionary timescales, 153
  geographic gradients, 152–153
  spatial patterns, 152–155
  and species richness in gastropods, 154, 155
Mountains
  endemism in, 356
  formation, 17
mtDNA (mitochondrial DNA)
  centers of origin hypothesis, 257, 259
  See also Ancient DNA; DNA; Molecular markers
Multi-collinearity, 228
Multivariate environmental change, conceptual model, 59
Murid rodents, diversification in Philippines, 350
Museum bio-collections
  baseline information, 318
  modeling species distributions, 312–315
Mutualists, absence on islands, 241
Mylodon, 44

Natterjack toads, 341
Natural protected areas (NPAs)
  criteria and implementation, 318
  mammals in Transvolcanic Belt (Mexico), 319–321
Neogastropods, Cenozoic radiation, 288
Neogene climate cycles, 282, 289
Neotoma. See Woodrat
Nested clade analysis, 96
Neurophyllodes, 80
Neutral theory
  of biodiversity, 171–174, 179
  biological invasions, 305–308
  deconstructive perspective, 208
  testing variables, 189
New Caledonia
  coral reef diversity, 278
  marine molluscs, 243
New Guinea
  collision with Southeast Asia, 284, 287
  mountain formation, 24
New Pangaea, 333
New synthesis
  biogeography, 141–143
  historical biogeography, 89
  evolutionary theory, 146
Niche-assembly theory
  latitudinal diversity gradient, 182–185
  limitations, 189
"No-analog" pollen assemblages, 60
Nodochila, 244
Non-equilibrium dynamic models, 221
NONA, 80
North American deserts
  phylogenetic biogeography of vertebrates, 102–107
  Quaternary biogeography, 30–34
  rodent middens, 28
North Atlantic Ocean, formation, 23
Northeastern Pacific, molluscan diversity dimensions, 161
Northern Hemisphere, latitudinal diversity gradient, 281–283
Norwegian Sea, biodiversity, 276, 278
Nothofagus, 38, 77, 78
Nothrotheriops shastensis, 43–44
NPAs. *See* Natural protected areas
NPP (Net primary productivity), Southern African woody plants, 225
Nucella, 249, 250

Ocean basins
  formation of, 20–23
  paleomapping, 13

sea-level change, 17–18
Ocean currents, marine dispersal, 245
Ocean floor, tectonic fabric, 13–14
Oceanic birds, diversity after biological invasions, 302–303
Oceanic circulation, Cenozoic developments, 284–288
Oceanic islands, *see* Marine islands
Oceanic plants, diversity after biological invasions, 302–303
Oceans, diversity gradients, 274
Octodontomys, 28
Offshore flows, biodistributional boundaries, 161
Onshore-to-offshore replacement process, 266
Ontology of complexity, BPA, 134
Ontology of simplicity, cladistic biogeography, 134
Optically stimulated luminescence, 50
Optimality criteria, of BPA and area cladistics, 130–134
Opuntia, 37
Ordovician mass extinction, Paleozoic bryozoans, 166
Ordovician period, marine diversity, 279
The Origin of Continents and Oceans (Wegener), 234
The Origin of Species (Darwin), 146

Pacific Decadal Oscillation (PDO), 30
Paleobiogeography
arid lands, 27–46
definition, 5
and geo-dispersal, 137
interdisciplinary nature, 5–6
phylogenetic, 123
and rodent midden analysis, 44–46
sources, 6
Paleobotany, rodent midden analysis, 27–28
See also Palynology
Paleoclimate, 36, 52–55
Cenozoic era, 24
Cretaceous period, 22
See also Climate
Paleoecology
fossil evidence vs. biogeography, 49
and genetic variation, 62–63
problems in Quaternary studies, 49–52
radiocarbon dating, 51–52
Paleogeography
Cenozoic, 23–25
Cretaceous, 21–23
Jurassic, 20–21
mapping, 16–18, 25
methods, 12–18
sources, 19
Triassic, 18–20
Paleomagnetism, continental mapping, 13
PALEOMAP project, 11–12, Plates 1–18
Paleomapping
ocean basins, 13
uncertainties, 13–14
websites, 25
Paleontology, and phylogeny, 112–113
Paleophylogeography, genetic studies, 45–46
Paleozoic extinctions, 166
Palinurus, 252
Palynology, 50–51
rodent midden analysis and, 27–28
Quaternary climate studies and, 48
PAN (annual rainfall), Southern African woody plants, 225
Panamanian Isthmus. *See* Central American Isthmus (CAI)
Panbiogeography, 71, 73, 88
Pangea, Plates 1–3
history of, 18–24
tectonic tree diagram, 10–11
Pannotia, 11, 15, 16
Paralogy, in area cladistics, 77–78
Parameter-Elevation Regressions on Independent Slopes Model (PRISM), 44
Parapatric speciation, center of origin hypothesis, 262–263
Parasites, absence on islands, 241–242
Parsimony criteria, phylogenetic vs. cladistic methods, 134
Pascahinnites, 244
Patella, 244
Pathogens, absence on islands, 241
PDO (Pacific Decadal Oscillation), 30
PEMIN (minimum monthly potential evapotranspiration), Southern African woody plants, 225
Penaeus, 259
Peripatric speciation, centers of origin, 264–265
PET (potential evapotranspiration)
Southern African woody plants, 226
species-energy theory, 174–175, 178–179
and species richness, 195–197
Philippine Islands
biodiversity threatened, 349–350
endemism in, 346–347, 349, 353, 355–359
geography, 346–348
mammalian diversity, 350–357
priorities for conservation, 356–358
tectonic history, 353–355
Phloeomys clade, 352
Phyllotis, 28, 44
Phylogenetic analysis
with ancient DNA, 45

marine dispersal, 245
Phylogenetic biogeography, 87
compared to cladistic biogeography, 67–68
definition, 93–94, 126–127
summarized, 88
use of fossil record, 114
use of phylogeographic data, 102–107, 107–109
vs. cladistic biogeography, 72–73
vs. component analysis, 76
Phylogenetic diversity, extinction selectivity, 167
Phylogenetic gradients, in IWP, 259
Phylogenetic information, in latitudinal diversity analyses, 190
Phylogenetic paleobiogeography, 123
Phylogenetic trees, taxon-area cladograms, 128, 129, 133
Phylogeny, and paleontology, 112–113
Phylogeography
ancient DNA analysis, 45
applying BPA, 138
as bridge discipline, 94, 96–98
compared to phylogenetic biogeography, 95–96
definition, 7
departures from congruence, 74
depth in time, 101–102
in marine studies, 234–235, 237
microevolution and macroevolution, 96–98
overview, 94–96
sequential approach, 109–110
vicariance and dispersal, 98
vicariance and population genetics, 99–101
See also Paleophylogeography
The Physical Atlas of Natural Phenomena (Johnston), 233
Picea
extinction in Quaternary, 61
historical distribution, 58
Pinus (pinyon species)
geographic distribution, 58
rodent midden studies, 32–33
Pinyon-juniper woodlands, Quaternary history, 32–33
Pinyon pine, and rodent midden analysis, 29
Planktonic development, and diversity, 204–205
Planktonic stages, dispersal, 247
Plant communities, shifting composition, 60
Plant diversity
after biological invasions, 302–303
environmental factors and, 176
link to animals, 181
Plant extinctions, Quaternary, 61–62
Plant macrofossils, Quaternary history, 51
Plants
macroscale models, 224–226
tree models vs. other forms, 225–226
Plate tectonics
biogeographical methodology, 272–273
the geologic record, 5–25
in historical biogeography, 71, 73, 88, 234
ocean floor, 13–14
tree diagram, 10–11
Platydesma, 80
Pleistocene sea levels, South East Asia, 347
Point scale SPARs, 326, 336
Polioptila californica, 108
Pollen analysis, *see* Palynology
Pollen assemblages, "no-analog," 60
Population genetics
in comparative phylogeography, 99–101
evaluating redundant distributions, 108
widespread taxa, 107–108
Potential evapotranspiration. *See* PET
Precipitation variation, Quaternary, 52–54
Predation
gastropod shells, 247–250
marine islands, 240, 243, 247–250, 252
Predators
roles on islands, 243, 252
size of marine, 251–252
Predictive biogeography, goals, applications, and future research, 321–322
Primary BPA, 132
PRISM (Parameter-Elevation Regressions on Independent Slopes Model), 44
Productivity
on islands, 242
marine island invertebrates, 250–251
Productivity gradients, 174
Productivity hypothesis, 177–178
difficulties, 189–190
Productivity theory. *See* Species-energy theory
Prognathogryllus, 80
Prosobranchs, environment and diversity, 176
Prosopis, 37
Provincial boundaries
and body size, 161–163
latitudinal correlation, 159–163
Pseudocongruence, 109
Pteropus vampyrus, 217
Punctuated equilibria, 264–265
Pyrenees Mountains, formation, 24

Quaternary communities, shifting compositions, 59–60
Quaternary environmental change, extinction and evolution, 61–62

Quaternary period
biogeography and paleoecology, 40
climate variation, 52–55
environmental lumpiness, 60–61
types and limitations of evidence, 49–52
Queensland, BPA of forest arthropods, 136
Quoyula madreporarum, 252

Radiocarbon dating
paleoecology, 51–52
Quaternary sediments, 50
Radiometric dating, 50
Rainfall, annual, 225
Range expansion
cyclic phenomena, 117
dispersal and geo-dispersal, 115–116
See also Species ranges
Rarefaction methodology, 276
Recolonization, after glaciation, 186–187
Reconciliation ecology, 338–341, 342–343
Recovery from extinction, 167–168
Red Sea, formation,23, 25
Redundant distributions, evaluating, 108
Reef fishes, centers of origin, 257, 260, 289
Regional (gamma) diversity
geological history, 281–286
latitudinal diversity gradient, 275, 278–281
Relief, and diversity, 181–182
Reptiles, environment and diversity, 177
Rescue effect, 215, 252
Reserves, size of (SLOSS controversy), 328–332, 340
Reticulated area relationships, models needed, 142–143
Reticulated relationships
common, 137
mechanisms of, 127
Rhynchomys clade, 352
Rhyncogonus, 214
Rodent midden analysis
Atacama Desert, 34–36
bioclimatic modeling, 39–41
calibration of, 30
discovery of, 27–28
future opportunities, 46
genetic studies, 43–46
North American deserts, 30–34
and paleobiogeography, 44–46
South American deserts, 34–39
strengths and weaknesses, 28–30
website, 29
Rodinia, 11, 15, 16

Saguaro, 33, 40
San Andreas Fault, 25
Satellite altimetry, ocean floor, 13–14
Scale
of biodiversity phenomena, 273, 326–328
and equilibrium in theories, 212
island ecology, 213
reticulated area relationships, 141
species richness, 274–275
temporal and spatial, 141
See also Macroscale; Spatial scale; Time scales
Scale effects, and island theory, 221
Scaling
of biogeographical processes, 5, 273–275, 326–328
Quaternary challenges, 63–64
spatial, 125–126
Schmarda, Ludwig K., 234
Science Citation Index
historical biogeography, 95, 99
phylogeography, 97, 101
SCOPE (Scientific Committee on Problems of the Environment), study of biological invasions, 298
Sea level change, 17–18
Cenozoic, 24
Cretaceous, 22
Pleistocene, Southeast Asia, 347, 354
Sea of Japan, formation, 23
Sea surface temperature. *See* SST
Sea urchins, centers of origin, 257
Seafloor spreading, 12–13, 234
Secondary BPA, 132
Semaphoronts, Hennig's, 72
Shallow time, phylogeography, 138
Shasta ground sloth, 43
Shelf (continental) area, and species richness, 203
Shell armor, distribution of marine gastropods, 247–249
Shorelines, history, 17
Shu-Yuan du Trade Route, 297
Siganidae, centers of origin, 257
Similarity. *See* Beta diversity
Single taxon studies, in phylogeography, 98
SLOSS (Single Large Or Several Small) controversy, 328–332, 342
Small island effect, 221
Solar radiation, species-energy relationship, 163–165, 174
South America, rodent middens, 28
South American deserts, Quaternary biogeography, 34–39
South Atlantic Ocean, formation, 21
Southeastern Pacific, marine mollusc diversity gradients, 201–204
Southern Hemisphere, latitudinal diversity gradient, 282–283

Southwestern U.S., vegetation history, 31
Spain, endemism in, 347, 349
SPARs (Species-area relationships)
bias of samples, 335–338, 341
biological meaning of, 220–221
environmental indicators, 334–335
estimation methods, 336–339, 341–342
human impacts on islands, 219–220
impact of exotic species, 332–333
island models, 212–214
reconciliation ecology, 338–341
and the SLOSS controversy, 328–332
types, 326–328
Spatial distribution. *See* Distribution
Spatial patterns, selective extinction and recovery, 165–168
Spatial scale
diversity and environmental variables, 223–224
effect on phylogenetic patterns, 125–126
geographic gradients, 222–224
and temporal scale, 141
Speciation
geography of, 109–110, 261–264
in marine environments, 235, 244
See also Allopatric speciation
Speciation modes, center of origin hypothesis, 262–265
Speciation rate
and diversity of lineages, 364
latitudinal diversity gradient, 185–186
Species, equivalence of, 195–196
Species-abundance distribution, MacArthur and Wilson theory, 215–216
Species accumulation curves, 336
Species-area relationships. *See* SPARs
Species density, evaluating environmental quality, 334–335
Species distributions, quantitative modeling, 313–315
Species diversity. *See* Biodiversity; Diversity; Species richness
Species-energy theory
functional biology, 169
latitudinal diversity gradient, 163–164
metabolic model, 163–164, 178–179
species richness and body size, 153
testing, 163–164, 190
Species equivalence, and diversity, 206
Species ranges
and climate, 224
and functional biodiversity, 161, 364
time scales, 7–8
See also Range expansion
Species richness
and body size, 153–155
capacity for (table), 216
centers of origin hypothesis, 258
climate gradients, 224–226
components, 273
deconstruction of, 192–205
and elevation, 356
and environmental factors, 176–177, 196
evaluating environmental quality, 334–335
geographic gradients, 222–230
on islands, 193–195, 212–222, 216–218, 240–241, 363
local and regional relationship, 279–281
morphological diversity in gastropods, 154, 155
in neutral theory, 173
and PET, 197
three dimensional model, 193–198
See also Biodiversity; Diversity; Latitudinal diversity gradient
Species richness-productivity relationships (SRPR), shape of, 227
Species saturation point, factors on islands, 215–217
Species traits
extinction and speciation rates, 201
and landscape dynamics, 199–200
Species turnover, islands, 213–214, 218
Spruce. *See Picea*
SRPR (Species richness-productivity relationships), shape of, 227
SST (mean sea surface temperature)
and crustacean diversity, 205
latitudinal gradient, 163–165
Starleaf Mexican-orange, 29
Static equilibrium, on islands, 217
Stehli, F.G., 271–272
Stomatal density, from rodent midden analysis, 41
Stramontia spp., 249
Structural theory, latitudinal diversity gradient, 182–185
Sub-tree analysis, in area cladistics, 77
Supertrees, 89
Sylvatic plague, 295
Symmetric theory, neutral theory, 208

Sympatric speciation, center of origin hypothesis, 263
Sympatry, 70
Synapomorphy, 72, 76

Tasmania, in Cretaceous, 23
TASS, 80
Taxon-area cladogram, 128
Taxon cycles, 135–136
Taxon pulse radiations, 135–136, 137
Temperate overspill, 188
Temperature
  species diversity and, 165, 186, 187
  species-energy relationship, 163–165, 174, 178–179
  See also SST (mean sea surface temperature)
Temperature change, as cause of provinciality, 161
  See also Climate change; Climate variability
Temperature history, Greenland, 55
Temporal scale, reticulated area relationships, 141
Tennessee watersheds, diversity after invasions, 302–303, 305
Tethys Ocean
  closing, 24, 284, 286
  in Jurassic period, 20
  tropical diversity center, 261
Tetragnatha, 80
Text recognition software, 139
Thais spp., 248–249
Thorson's rule, 204
3-Item analysis, 76–77
3item, 80–81
"Three Rule" Rule, BPA, 134
Threes rule, 98
Tibetan Plateau, 23
Time scales
  diversity and environmental variables, 223–224
  dynamic Earth, 7–8
  Quaternary challenges, 63–64
  Quaternary environmental state, 54–55
Time-slicing, 79, 84–85
Tithonian stage, diversity gradients, 158
Toleration, of environmental change, 57
Transformed cladistics, 72–73, 73
Transvolcanic Belt (Mexico), ecological niche modeling, 316
Tree rings, precipitation estimates, 52–53
Tree species, island gene flow, 217–218
Trees
  compared to cladograms, 68
  water-energy models, 224–226
  See also Phylogenetic trees
Triassic paleogeography, 18–21
Tridacna, 259
Tridacnidae, centers of origin, 257
Trilobites, vicariance and geo-dispersal, 121–122
Tropical overspill, 187–188
Tropics
  diversity centers, 272
  latitudinal dispersal, 266
  niche-assembly theory, 183–184
  species richness gradient, 187–188
  taxonomic diversity, 274
Turbo setosus, 246
Turnover, island species, 213–214, 218
Type I and II relationships, 279–289

Unified neutral theory. *See* Neutral theory
The Unified Neutral Theory of Biodiversity and Biogeography (Hubbell), 171
USGS-OSU model, 39–40
Utah juniper, 30

Vegetation, North American deserts, 31–34
Vegetation composition, plant macrofossils, 51
Vertebrate extinctions, Quaternary, 61
Vicariance
  and area cladograms, 71
  in biotic histories, 88
  and congruence, 112
  as default explanation, 142
  definition, 70, 79
  geography of speciation, 109–110
  molecular divergence calculations, 109
  molecular studies in marine organisms, 235
Vicariance biogeography, 73
  background, 69, 234–235
Vicariance matrix, 118–120
Vicariance tree, 117–118
Vicariant speciation, centers of origin hypothesis, 262, 263–264
Vizcachas, 28
Volcanoes, hot spots, 13
von Humboldt, Alexander, 145
Vortex model, East Indies diversity, 257

Wallace, Alfred Russel, 145, 294
Wallacean shortfall, 296
Water and energy, macroscale patterns and climate, 224–226
Water availability, and diversity, 174–175
Water-energy balance, species-energy theory, 174
Websites
  paleogeography, 26
  rodent midden analysis, 29
  scientific collections, 321
  taxon-archipelago data, 330
Wegener, Alfred Lothar, 234
West Gondwana, 21
White-winged dove, 33–34

Widespread taxa, phylogeographic analysis, 107–108
Wilson, Edward O., 239, 294–295
  See also MacArthur and Wilson theory
Within habitat diversity. *See* Alpha diversity
Woodrat
  body size and climate change, 42–43
  fossil middens, 28, 34
  molecular phylogenetic studies, 45–46
Wright's species-energy theory. *See* Species-energy theory

Xiphophorus, 73, 74, 128

Yersinia pestis, 295
Younger Dryas interval, 54

z-value, definition, 327
Zagros Mountains, 24
Zenaida asiatica, 33–34
Ziziphus, 37
Zooplankton, diversity, 176